FORMULAS FROM ALGEBRA

Exponents

$$a^m a^n = a^{m+n}$$

$$(a^m)^n = a^{mn}$$

$$\frac{a^m}{a^n} = a^{m-n}$$

$$(ab)^n = a^n b^n$$

$$\left(\frac{a}{b}\right)^n = \frac{a^n}{b^n}$$

Radicals

$$(\sqrt[n]{a})^n = a$$

$$\sqrt[n]{a^n} = a, \text{ if } a \geq 0$$

$$\sqrt[n]{ab} = \sqrt[n]{a}\,\sqrt[n]{b}$$

$$\sqrt[n]{\frac{a}{b}} = \frac{\sqrt[n]{a}}{\sqrt[n]{b}}$$

Logarithms

$$\log_a MN = \log_a M + \log_a N$$

$$\log_a (M/N) = \log_a M - \log_a N$$

$$\log_a (N^p) = p \log_a N$$

Factoring Formulas

$$x^2 - y^2 = (x - y)(x + y)$$

$$x^3 - y^3 = (x - y)(x^2 + xy + y^2)$$

$$x^3 + y^3 = (x + y)(x^2 - xy + y^2)$$

$$x^2 + 2xy + y^2 = (x + y)^2$$

$$x^2 - 2xy + y^2 = (x - y)^2$$

$$x^3 + 3x^2 y + 3xy^2 + y^3 = (x + y)^3$$

Binomial Formula

$$(x + y)^n = {}_n C_0 x^n y^0 + {}_n C_1 x^{n-1} y^1 + \cdots + {}_n C_{n-1} x^1 y^{n-1} + {}_n C_n x^0 y^n$$

Quadratic Formula

The solutions to $ax^2 + bx + c = 0$ are $x = \dfrac{-b \pm \sqrt{b^2 - 4ac}}{2a}$

Complex Numbers

Multiplication: $(a + bi)(c + di) = (ac - bd) + (ad + bc)i$

Polar form: $a + bi = r(\cos \theta + i \sin \theta)$ where $r = \sqrt{a^2 + b^2}$

Powers: $[r(\cos \theta + i \sin \theta)]^n = r^n(\cos n\theta + i \sin n\theta)$

Roots: $\sqrt[n]{r}\left[\cos\left(\dfrac{\theta + k \cdot 360°}{n}\right) + i \sin\left(\dfrac{\theta + k \cdot 360°}{n}\right)\right]$ $k = 0, 1, 2, \cdots, n - 1$

Precalculus Mathematics

SECOND EDITION

Precalculus Mathematics

A PROBLEM-SOLVING APPROACH

Walter Fleming
Hamline University

Dale Varberg
Hamline University

Prentice Hall
Englewood Cliffs, New Jersey 07632

Library of Congress Cataloging-in-Publication Data

Fleming, Walter
   Precalculus mathematics: a problem-solving approach / Walter
Fleming, Dale Varberg.—2nd ed.
     p. cm.
  Includes index.
  ISBN 0-13-695008-6: $27.50 (est.)
   1. Mathematics—1961– I. Varberg, Dale. II. Title.
QA39.2.F58 1989
512'.1—dc19

87-31015
CIP

Editorial/production supervision: Zita de Schauensee
Interior design: Chistine Gehring-Wolf
Cover design: Maureen Eide
Manufacturing buyer: Paula Massenaro
Computer graphics cover art created by Genigraphics Corporation

© 1989, 1984 by Prentice-Hall, Inc.
A Division of Simon & Schuster
Englewood Cliffs, New Jersey, 07632

Printed in the United States of America
10 9 8 7 6 5 4 3 2 1

Credits for quotations in text: Pages x and 234: George
Polya, *Mathematical Discovery* (New York: John Wi-
ley & Sons, Inc., 1962), vol. I, pp. 66 and 6–7,
respectively. Page 89: Morris Kline, ''Geometry,'' in
Scientific American (Sept. 1964), p. 69. Page 443:
Richard Courant and Herbert Robbins, *What Is Mathematics?* (New York: Oxford University Press, 1978),
p. 198.

ISBN   0-13-695008-6

Prentice-Hall International (UK) Limited, *London*
Prentice-Hall of Australia Pty. Limited, *Sydney*
Prentice-Hall Canada Inc., *Toronto*
Prentice-Hall Hispanoamericana, S.A., *Mexico*
Prentice-Hall of India Private Limited, *New Delhi*
Prentice-Hall of Japan, Inc., *Tokyo*
Prentice-Hall of Southeast Asia Pte. Ltd., *Singapore*
Editora Prentice-Hall do Brasil, Ltda., *Rio de Janeiro*

Contents

7 The Complex Number System

8 Theory of Polynomials

9 Systems of Equations and Inequalities

10 Sequences and Mathematical Induction

11 Analytic Geometry 443

Appendix 503

Answers to Odd-Numbered Problems 517

Index of Teaser Problems 557

Index of Names and Subjects 559

Preface

As its title indicates, we designed this book to prepare students for calculus. After a chapter on basic algebra (which many instructors will either omit or review quickly), we plunge into what must be the heart of any precalculus course—the study of functions and their graphs. We look in turn at linear functions, quadratic functions, polynomial functions, rational functions, exponential functions, logarithmic functions, and trigonometric functions with occasional asides to consider more special functions such as the absolute value function and the greatest integer function. Our concern is to develop in students both a geometric feeling for these functions and a solid understanding of their properties.

With our eye on the calculus course that students will take later, we attend to such topics as difference quotients (pages 30 and 110), composition and decomposition of functions (page 107), inverse functions (page 113), the natural logarithm function (page 154), and partial fractions (page 341). Because complex numbers play almost no role in a first calculus course, some instructors will prefer to omit discussion of these numbers in a precalculus course. We make this possible by postponing treatment of the complex numbers until Chapter 7 and by labeling thereafter all problems that use these numbers with the symbol Ⓘ.

We have made two significant content changes for this edition. First, we have added a chapter titled "Systems of Equations and Inequalities." This chapter includes an introduction to matrices and determinants. Second, the last chapter, retitled "Analytic Geometry," has been rewritten and enlarged. It features a complete treatment of the conic sections (now defined in the traditional way) and considerable material on polar coordinates and parametric equations.

Our organization and format is similar to that in our other books: *Algebra and Trigonometry* (third edition), *College Algebra* (third edition), and *Plane Trigonometry* (second edition). Here are some of the special features.

LIVELY OPENING DISPLAYS

Each section begins with a challenging problem, a historical anecdote, a famous quotation, or an appropriate cartoon. These displays are designed to spark the readers' curiosity, to draw them into the section.

INFORMAL WRITING STYLE

We avoid the ponderous theorem-proof style found in many mathematics books. We are more interested in explaining than in proving and we shun unneeded technical jargon like the plague. Yet we are careful to state definitions and theorems correctly and we do proceed in a reasoned logical manner. Our sections are broken into smaller subsections with descriptive titles but we do not chop the text into pieces with the constant use of headings like theorem, proof, definition, example, remark, and so on. We want students to read our book, so our goal was to make it readable.

EXAMPLES

We believe that students learn mathematics by studying examples and doing problems. Our book is full of examples. First there are plenty of examples within the textual discussion itself though they are not explicitly labeled as such. Then in the problem sets we usually offer several more examples, always accompanied by a group of related problems. We think that placing examples and problems together is good pedagogy. In any case, *the examples in the problem sets are an integral part of the text and should be studied carefully.*

PROBLEM-SOLVING EMPHASIS

We think the activity that most characterizes mathematics is problem solving. Problem solving is more than mere answer finding; it is much more than simply substituting in a formula or following a recipe. As we use the term, it involves formulating a problem clearly, organizing the given information (perhaps in a diagram), collecting the tools that will be needed, using one's wits to develop a strategy, following a reasoned process to a solution, and writing the results in a clear organized manner. It is the activity that George Polya (1887–1985) wrote about so wisely in his many books and from whom we offer this quotation.

"Solving a problem is similar to building a house. We must collect the right material, but collecting the material is not enough; a heap of stones is not yet a house. To construct the house or the solution, we must put together the parts and organize them into a purposeful whole." From *Mathematical Discovery* (vol. 1, p. 66).

If emphasis on problem solving characterized the first edition of this book, this emphasis is even more evident in the present edition. Every section of the book ends with an extensive problem set, and each of these is in two parts. First there is a set of basic problems, problems intended to develop the skills and reinforce the main ideas of the section. Following the basic problems, there is a set of miscellaneous problems. *These have been completely reworked for this edition.* Our aim for the miscellaneous problems is to give the student an exciting and challenging tour through the applications of the ideas of the section. We begin with a few easy review type problems, move in a carefully graded manner to harder and more interesting application, and conclude with a teaser problem.

THE TEASER PROBLEMS

In our combined seventy years of teaching, we have collected a large number of intriguing problems. Many of them are our own creation; some are part of current mathematical lore; others come from mathematical history. As a special attraction for this edition, we have inserted one of these problems at the end of each section. These teasers may appear difficult at first glance but in most cases become surprisingly easy when looked at the right way. In each case, the teaser relates to the ideas of the section in which it appears. As a group, the teasers form a collection of problems that we think would please even George Polya.

How should the teasers be used in class? We suggest that teachers might select some of their favorite teasers and offer a prize to the student offering the best set of solutions at the end of the term. Or teachers might use these problems as the basis for a weekly problem-solving session (perhaps as an addition to the regular class sessions). Or a teacher might select from them a problem of the week, offering a small prize for the best solution. Or they can simply be treated as extra stimulation for the very best students.

CALCULATORS

Calculators are by now standard equipment for most college students. They greatly aid in the solution of problems that involve heavy calculations. In trigonometry, use of these devices allows us to largely dispense with the traditional emphasis on use of tables. However, we warn students that calculators can never substitute for clear thinking, and we hope our problem sets will make this abundantly clear. Problems whose solution is substantially eased by use of a calculator are labeled with the symbol ⒞.

FLEXIBILITY

This book can be used in either a one or a two term course. To aid in the designing of a syllabus, we include the following dependence chart.

DEPENDENCE CHART

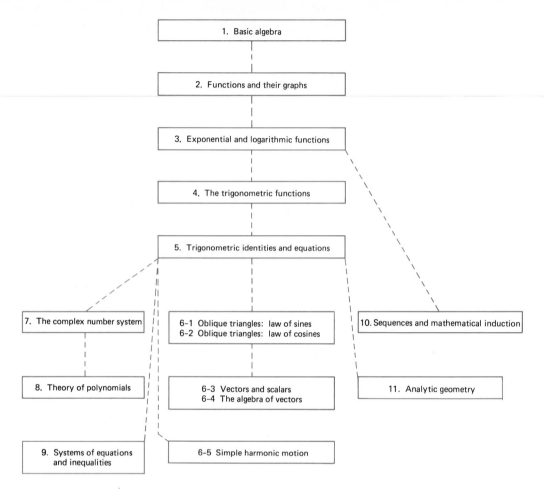

1. Basic algebra

2. Functions and their graphs

3. Exponential and logarithmic functions

4. The trigonometric functions

5. Trigonometric identities and equations

7. The complex number system

6-1 Oblique triangles: law of sines
6-2 Oblique triangles: law of cosines

10. Sequences and mathematical induction

8. Theory of polynomials

6-3 Vectors and scalars
6-4 The algebra of vectors

11. Analytic geometry

9. Systems of equations and inequalities

6-5 Simple harmonic motion

SUPPLEMENTARY MATERIALS

An extensive variety of instructional aids is available from Prentice Hall.

Instructor's Manual The instructor's manual was prepared by the authors of the textbook. It contains the following items.
   (a) Answers to all the even-numbered problems (answers to the odd problems appear at the end of the textbook).
   (b) Complete solutions to the last four problems in each problem set. This includes the teaser problem.
   (c) Six versions of a chapter test for each chapter together with an answer key for these tests.
   (d) A test bank of more than 1300 problems with answers designed to aid an instructor in constructing examinations.
   (e) A set of transparencies that illustrate key ideas.

Prentice Hall Test Generator   The test bank of more than 1300 problems is available on floppy disk for the IBM PC. This allows the instructor to generate examinations by choosing individual problems, editing them, and if desired by creating completely new problems.

Videotapes   Approximately five hours of videotaped lectures covering selected topics in college algebra are available with a qualified adoption. Contact your local Prentice Hall representative for details.

Student Solutions Manual   This manual has worked-out solutions to every third problem (not including teaser problems).

Function Plotter Software   A one-variable function plotter for the IBM PC is available with a qualified adoption. Contact your local Prentice Hall representative for details.

"How to Study Math"   Designed to help your students overcome math anxiety and to offer helpful hints regarding study habits, this useful booklet is available free with each copy sold. To request copies for your students in quantity, contact your local Prentice Hall representative.

ACKNOWLEDGMENTS

This and previous editions have profited from the warm praise and constructive criticism of many reviewers. We offer our thanks to the following people who gave helpful suggestions.

Karen E. Barker, *Indiana University, South Bend*
Chris Boldt, *Eastfield College*
Donald S. Coram, *Oklahoma State University, Stillwater*
Thomas Cupillari, *Keystone Junior College*
Phillip M. Eastman, *Boise State University*
Dianne S. Ellis, *Boise State University*
Laurene Fausette, *Florida Institute of Technology*
William J. Gordon, *State University of New York, Buffalo*
Judith J. Grasso, *University of New Mexico, Albuquerque*
B. C. Horne, Jr., *Virginia Polytechnic Institute and State University*
Eli Maor, *University of Wisconsin, Eau Claire*
Helen I. Medley, *Kent State University*
Ann B. Megaw, *University of Texas, Austin*
Maurice L. Monahan, *South Dakota State University, Brookings*
Stephen Peterson, *University of Notre Dame*
Donald Potts, *California State University at Northridge*
William W. Smith, *University of North Carolina, Chapel Hill*
Paul D. Trembeth, *Delaware Valley College*

The staff at Prentice Hall is to be congratulated on another fine production job. The authors wish to express appreciation especially to Priscilla McGeehon (mathematics editor), Zita de Schauensee (project editor), and Christine Gehring-Wolf (designer) for their exceptional contributions.

Walter Fleming
Dale Varberg

*As the sun eclipses the stars
by its brilliancy, so the man
of knowledge will eclipse the
fame of others in the assem-
blies of the people if he pro-
poses algebraic problems, and
still more if he solves them.*

Brahmagupta

CHAPTER 1

Basic Algebra

Let us make no mistake about it: mathematics is and always has been the numbers game par excellence.

Philip J. Davis

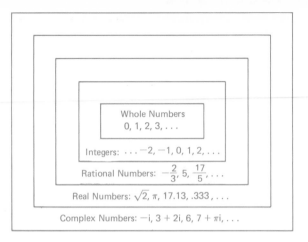

1-1   The Real Number System

Mathematics students are number crunchers. That popular image is both profoundly right and terribly wrong. Most mathematicians disdain adding up long columns of numbers, finding square roots, or doing long division. They have relegated such tasks to electronic calculators and computers. But it is still true that numbers play a fundamental role in most of mathematics. Certainly this is true in precalculus.

It would take too long to describe the long, tortuous road traveled by mankind in going from the whole numbers to the complex numbers. It is enough to call your attention to the five classes of numbers in our opening display, to suggest that they were developed roughly in the order listed, and to recall a few facts about the real numbers. The real numbers will, in fact, be the principal characters in this book until Chapter 7, where we give a thorough discussion of the complex numbers.

THE REAL NUMBERS

Given a prescribed unit of length, we can (at least theoretically) measure the length of any line segment. The set of all numbers that can measure lengths, together with their negatives and zero, constitute the **real numbers.** The number $\sqrt{2}$ is a real number since it measures the length of the hypotenuse of a right triangle with legs of unit length (Figure 1). So is π; it measures the circumference of a circle of unit diameter ($C = \pi d$). Every **rational number** (a number which can be expressed as a ratio of two integers p/q) is a real number.

The best way to visualize the system of real numbers is as a set of labels for points on a line. Consider a horizontal line and select an arbitrary point to

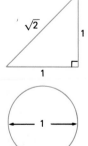

Figure 1

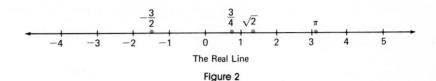

The Real Line

Figure 2

be labeled with 0. Then label a point one unit to the right with 1, a point one unit to the left with −1, and so on. The process is so familiar that we omit further details and draw a picture (Figure 2).

Of course, we cannot show all the labels, but we want you to imagine that each point has a number label, or **coordinate**, that measures its distance to the right or left of 0. We refer to the resulting coordinate line as the **real line.**

If $b - a$ is positive, we say that a is less than b. We write $a < b$, which is called an **inequality**. On the real line, $a < b$ simply means that a is to the left of b (Figure 3). Similarly, if $b - a$ is positive or zero, we say a is less than or equal to b and write $a \leq b$. It is correct to say $5 < 6$, $5 \leq 6$, and $5 \leq 5$.

The symbol $|a|$, read the **absolute value** of a, is defined by

$$|a| = \begin{cases} a & \text{if } 0 \leq a \\ -a & \text{if } a < 0 \end{cases}$$

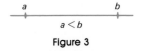

Figure 3

Geometrically, $|a|$ is the (undirected) distance from 0 to a. Some would say that $|a|$ is the magnitude of a without regard to its sign. Be careful with this: It is correct to say $|+5| = 5$ and $|-5| = 5$, but it is not necessarily true that $|-x| = x$ (try $x = -2$). Finally, note that the distance between b and a on the real line is $|b - a|$; this is correct whether a is to the left or to the right of b (Figure 4).

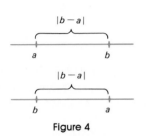

Figure 4

DECIMALS

There is another important way to describe the real numbers. We must first review a basic idea. Recall that

$$.4 = \frac{4}{10}$$

$$.42 = \frac{4}{10} + \frac{2}{100} = \frac{40}{100} + \frac{2}{100} = \frac{42}{100}$$

$$.731 = \frac{7}{10} + \frac{3}{100} + \frac{1}{1000} = \frac{700}{1000} + \frac{30}{1000} + \frac{1}{1000} = \frac{731}{1000}$$

Clearly, each of these decimals represents a rational number.

Conversely, if we are given a rational number, we can find its decimal expansion by long division. For example, the division in Figure 5 shows that $\frac{7}{8} = .875$. When we try the same procedure on $\frac{2}{11}$, something different happens (Figure 6). The decimal just keeps on going; it is a **nonterminating decimal.**

```
   .875
8) 7.000
   6 4
   ──
    60
    56
    ──
     40
     40
     ──
```

Figure 5

```
    .18181
11) 2.00000
    1 1
    ──
     90
     88
     ──
      20
      11
      ──
       90
       88
       ──
        20
        11
```

Figure 6

Actually, the decimal .875 can be thought of as nonterminating if we adjoin zeros. Thus

$$\frac{7}{8} = .8750000\ldots \qquad = .875\overline{0}$$

$$\frac{2}{11} = .181818\ldots \qquad = .\overline{18}$$

$$\frac{2}{7} = .285714285714\ldots = .\overline{285714}$$

Note that in each case, the decimal has a repeating pattern. This is indicated by putting a bar over the group of digits that repeat. Now we state a remarkable fact about the rational numbers (ratios of integers).

The rational numbers are precisely those numbers that can be represented as repeating nonterminating decimals.

What about nonrepeating decimals like

.12112111211112...

They represent the **irrational numbers,** of which $\sqrt{2} = 1.4142135\ldots$ and $\pi = 3.1415926\ldots$ are the best-known examples. Together, the rational numbers and the irrational numbers make up the real numbers. Thus we may say that:

The real numbers are those numbers that can be represented as non-terminating decimals.

THE REAL NUMBERS

Rational numbers (the repeating decimals)	Irrational numbers (the nonrepeating decimals)

While it is true that the real numbers are the fundamental numbers of precalculus (and calculus), in practical situations we work with a very small subset of them. Who can calculate with nonterminating decimals? Neither humans nor electronic calculators. Calculators are, in fact, restricted to decimals of a certain length (perhaps 8 or 10 digits). Thus in practical calculations, most real numbers must be **rounded**. For example, π is often rounded to 3.141593, or perhaps to 3.14159. Our rule for rounding is to round down if the first discarded digit is 4 or less and round up if it is 5 or more. Thus π is 3.1416 rounded to four decimal places, 3.142 rounded to three decimal places, and 3.14 rounded to two decimal places.

OPERATIONS AND THEIR PROPERTIES

Addition and multiplication are the fundamental operations; subtraction and division are offshoots of them. These operations are **binary** operations; that is, they work on two numbers at a time. Thus $3 + 4 + 5$ is technically meaningless. We ought to write either $3 + (4 + 5)$ or $(3 + 4) + 5$. Luckily, it really does not matter; we get the same answer either way. Addition is **associative**.

$$a + (b + c) = (a + b) + c$$

Thus we can write $3 + 4 + 5$ or even $3 + 4 + 5 + 6 + 7$ without ambiguity. The answer will be the same regardless of which addition is done first.

Addition and multiplication are like Siamese twins: What is true for one is quite likely to be true for the other. Thus multiplication, too, is associative.

$$a \cdot (b \cdot c) = (a \cdot b) \cdot c$$

If we wish, we can write $a \cdot b \cdot c$ with no parentheses at all.

It makes some difference whether you first put on your slippers and then take a bath or vice versa—the difference between wet and dry slippers. But for addition or multiplication, order does not matter. Both operations are **commutative**.

$$a + b = b + a$$

$$a \cdot b = b \cdot a$$

Thus

$$3 + 4 = 4 + 3$$

$$3 + 4 + 6 = 4 + 3 + 6 = 6 + 4 + 3$$

and

$$7 \cdot 8 \cdot 9 = 8 \cdot 7 \cdot 9 = 9 \cdot 8 \cdot 7$$

When we indicate an addition and a multiplication in the same expression, we face a problem. For example, what do we mean by $3 + 4 \cdot 2$? If we mean $(3 + 4) \cdot 2$, the answer is 14; if we mean $3 + (4 \cdot 2)$, the answer is 11. Most of us would not give the first answer. We are so familiar with a convention that we use it without thinking, but now is a good time to emphasize it. In any expression involving additions and multiplications which has no parentheses, we agree to do all the multiplications first. Thus

$$4 \cdot 5 + 3 = 20 + 3 = 23$$

and

$$4 \cdot 5 + 6 \cdot 2 = 20 + 12 = 32$$

We can always overrule this agreement by inserting parentheses. For example,

$$4 \cdot (5 + 3) = 4 \cdot 8 = 32$$

The agreement about the order in which operations are carried out is a matter of convenience; no law forces it upon us. However, the **distributive property,**

$$a \cdot (b + c) = a \cdot b + a \cdot c$$
$$(b + c) \cdot a = b \cdot a + c \cdot a$$

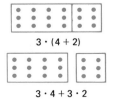

$$3 \cdot (4 + 2)$$

$$3 \cdot 4 + 3 \cdot 2$$

Figure 7

is not a matter of choice or convenience. Rather, it is another of the fundamental properties of numbers. That it must hold for the positive integers is almost obvious, as may be seen by examining the diagram in Figure 7. But we assert that it is true for all real numbers—for example, that

$$\frac{1}{2} \cdot (\sqrt{2} + \pi) = \frac{1}{2} \cdot \sqrt{2} + \frac{1}{2} \cdot \pi$$

There are two very special numbers, 0 and 1, called the **additive identity** and the **multiplicative identity,** respectively. They satisfy

$$a + 0 = 0 + a = a$$
$$a \cdot 1 = 1 \cdot a = a$$

For every real number a, there is a real number $-a$, called its **additive inverse** (or negative); similarly, for every real $a \neq 0$, there is a real number $1/a$, called its **multiplicative inverse** (or reciprocal). They satisfy

$$a + (-a) = (-a) + a = 0$$
$$a \cdot \frac{1}{a} = \frac{1}{a} \cdot a = 1$$

Now we can define subtraction and division by

$$a - b = a + (-b)$$

and

$$a \div b = a \cdot \frac{1}{b}$$

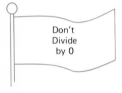

Don't Divide by 0

Figure 8

All elementary algebra rests on the boxed rules. Two consequences worth mentioning deal with zero. First, $0 \cdot a = a \cdot 0 = 0$, and second, division by 0 is forbidden (Figure 8). To see how these follow from the boxed rules, see Problems 55 and 56.

ROOTS

We give a complete treatment of general nth roots in Section 3-1. In the meantime, we feel free to use the two symbols $\sqrt{a}$ and $\sqrt[3]{a}$. The first makes sense if $a \geq 0$, in which case it is used to denote the nonnegative, or **principal**, square root of a. Thus $\sqrt{4} = 2$ and $\sqrt{9} = 3$, but $\sqrt{9} \neq -3$. Of course, the number 9 has two square roots—namely, 3 and -3—but $\sqrt{9}$ denotes just the positive one. In general,

$$\sqrt{a^2} = |a|$$

The symbol $\sqrt[3]{a}$ makes good sense whether a is positive or negative, and it is unambiguous since a has only one real cube root. For example, $\sqrt[3]{27} = 3$ and $\sqrt[3]{-8} = -2$.

The symbols $\sqrt{}$, $\sqrt[3]{}$, and $|\ |$ denote **unary** operations because they operate on just one number. They are in contrast to $+$ and $\cdot$, which are **binary** operations since they operate on two numbers.

Problem Set 1-1

In Problems 1–12, replace the symbol # by the appropriate symbol $<$, $>$, *or* $=$.

1. $1.5 \mathbin{\#} -1.6$
2. $-2 \mathbin{\#} -3$
3. $\sqrt{2} \mathbin{\#} 1.4$
4. $\pi \mathbin{\#} 3.15$
5. $\frac{1}{5} \mathbin{\#} \frac{1}{6}$
6. $-\frac{1}{5} \mathbin{\#} -\frac{1}{6}$
7. $5 - \sqrt{2} \mathbin{\#} 5 - \sqrt{3}$
8. $\sqrt{2} - 5 \mathbin{\#} \sqrt{3} - 5$
9. $-\frac{3}{16}\pi \mathbin{\#} -\frac{3}{17}\pi$
10. $\frac{1}{3} \mathbin{\#} .3$
11. $|-\pi + (-2)| \mathbin{\#} |-\pi| + |-2|$
12. $|\pi - 2| \mathbin{\#} \pi - 2$

13. Order the following numbers from least to greatest. (*Hint:* One way to do this is to change all of them to decimals.)

$$\frac{3}{4}, \ -2, \ \sqrt{2}, \ -\frac{\pi}{2}, \ -\frac{3}{2}\sqrt{2}, \ \frac{43}{24}$$

14. Order the following numbers from least to greatest.

$$5 - 5, \ .37, \ -\sqrt{3}, \ \frac{14}{33}, \ -\frac{7}{4}, \ -\frac{49}{35}, \ \frac{3}{8}$$

EXAMPLE A (Reducing Fractions) Reduce each of the following, leaving your answer with a positive denominator.

(a) $\dfrac{24}{36}$ (b) $\dfrac{3 + 9\sqrt{2}}{3 + 6}$ (c) $\dfrac{72 + 9\sqrt{3}}{-45}$

Solution. You can cancel common factors (which are multiplied) from numerator and denominator. Do not make the common mistake of trying to cancel common terms (which are added).

(a) $\dfrac{24}{36} = \dfrac{\cancel{12}\cdot 2}{\cancel{12}\cdot 3} = \dfrac{2}{3}$

(b) $\dfrac{3 + 9\sqrt{2}}{3 + 6} = \dfrac{\cancel{3} + 9\sqrt{2}}{\cancel{3} + 6} = \dfrac{9\sqrt{2}}{6}$ Wrong

$\dfrac{3 + 9\sqrt{2}}{3 + 6} = \dfrac{\cancel{3}(1 + 3\sqrt{2})}{\cancel{3}\cdot 3} = \dfrac{1 + 3\sqrt{2}}{3}$ Right

(c) $\dfrac{72 + 9\sqrt{3}}{-45} = \dfrac{\cancel{9}(8 + \sqrt{3})}{\cancel{9}(-5)} = \dfrac{-1(8 + \sqrt{3})}{-1(-5)} = \dfrac{-8 - \sqrt{3}}{5}$

Reduce each of the following, giving your answer with a positive denominator.

15. $\dfrac{24}{27}$ 16. $\dfrac{16}{36}$ 17. $\dfrac{45}{-60}$ 18. $\dfrac{63}{-81}$

19. $\dfrac{8 + 24\sqrt{5}}{12}$ 20. $\dfrac{6 - 27\sqrt{2}}{-9}$ 21. $\dfrac{8 - \sqrt{2}}{-4}$ 22. $\dfrac{4 + 28\sqrt{3}}{14}$

EXAMPLE B (Operating on Fractions)   Perform the indicated operations and simplify.

(a) $\frac{8}{9} + \frac{5}{12}$ (b) $\frac{3}{4}\cdot\frac{16}{27}$ (c) $\dfrac{\frac{3}{4}}{\frac{9}{16}}$

Solution.   (a) Fractions with the same denominator are added by adding the numerators and using the common denominator. In the case under consideration, we must first write the two fractions with a common denominator.

$$\frac{8}{9} + \frac{5}{12} = \frac{32}{36} + \frac{15}{36} = \frac{47}{36}$$

(b) We multiply fractions by multiplying numerators and denominators.

$$\frac{3}{4}\cdot\frac{16}{27} = \frac{3\cdot 16}{4\cdot 27} = \frac{\cancel{3}\cdot\cancel{4}\cdot 4}{\cancel{4}\cdot\cancel{3}\cdot 9} = \frac{4}{9}$$

(c) To divide fractions, invert the divisor and multiply.

$$\dfrac{\frac{3}{4}}{\frac{9}{16}} = \frac{3}{4}\cdot\frac{16}{9} \qquad \frac{\cancel{3}\cdot\cancel{4}\cdot 4}{\cancel{4}\cdot\cancel{3}\cdot 3} = \frac{4}{3}$$

CAUTION

$3\dfrac{x}{y} = \dfrac{3x}{3y}$

$3\dfrac{x}{y} = \dfrac{3x}{y}$

Perform the indicated operations and simplify.

23. $\frac{5}{12} + \frac{7}{18} - \frac{1}{6}$ 24. $\frac{11}{15} + \frac{3}{4} - \frac{5}{6}$ 25. $\frac{-5}{27} + \frac{5}{12} + \frac{3}{4}$ 26. $\frac{23}{30} + \frac{2}{25} - \frac{3}{5}$

27. $\frac{9}{11}\cdot\frac{33}{5}\cdot\frac{15}{18}$ 28. $\frac{3}{4}\cdot\frac{6}{15}\cdot\frac{5}{2}$ 29. $\dfrac{\frac{5}{6}}{\frac{8}{12}}$ 30. $\dfrac{\frac{9}{24}}{\frac{15}{12}}$

31. $\dfrac{\frac{3}{4}}{2}$ 32. $\dfrac{6}{\frac{7}{9}}$ 33. $\dfrac{\frac{2}{3} + \frac{3}{4}}{\frac{7}{12}}$ 34. $\dfrac{\frac{3}{5} - \frac{3}{4}}{\frac{9}{20}}$

EXAMPLE C (Removing Grouping Symbols)   Simplify

$$\frac{1}{2}\left[\frac{2}{3} - 3\left(\frac{1}{4} - \sqrt{2}\right)\right]$$

Solution.   The basic tool in removing parentheses is the distributive law, but be sure to begin with the innermost parentheses.

$$\frac{1}{2}\left[\frac{2}{3} - 3\left(\frac{1}{4} - \sqrt{2}\right)\right] = \frac{1}{2}\left[\frac{2}{3} - \frac{3}{4} + 3\sqrt{2}\right] = \frac{1}{3} - \frac{3}{8} + \frac{3\sqrt{2}}{2}$$

$$= \frac{8}{24} - \frac{9}{24} + \frac{36\sqrt{2}}{24} = \frac{-1 + 36\sqrt{2}}{24}$$

Remove all grouping symbols and simplify.

35. $-4[3(-6 + \sqrt{2}) - 2\sqrt{2}]$

36. $-4[\sqrt{3} - 2(3 - \sqrt{3}) + 4]$

37. $-\frac{2}{3}(\frac{5}{4} - \frac{1}{12})$

38. $\frac{3}{2}(\frac{8}{9} - \frac{5}{6})$

39. $\frac{1}{3}[\frac{1}{2}(\frac{1}{4} - \frac{1}{3}) + \frac{1}{6}]$

40. $-\frac{1}{3}[\frac{2}{5} - \frac{1}{2}(\frac{1}{3} - \frac{1}{5})]$

41. $\frac{14}{33}(\frac{2}{3} - \frac{1}{7})^2$

42. $(\frac{5}{7} + \frac{7}{9}) \div \frac{3}{2}$

43. $1 \div (2 + \frac{3}{4})$

44. $15(2 - \frac{3}{5})$

EXAMPLE D (Hierarchy of Operations in a Calculator)   Most scientific calculators (with algebraic logic) perform operations in the following order:

1. Unary operations, such as taking square roots.
2. Multiplications and divisions from left to right.
3. Additions and subtractions from left to right.

   Parentheses are used, just as in ordinary algebra. Pressing the $\boxed{=}$ key will cause all pending operations to be performed. Calculate

$$\frac{3.12 + (4.15)(5.79)}{5.13 - 3.76}$$

Solution.   This can be done in more than one way, and calculators may vary. On a model such as ours, either of the following sequences of keys will give the correct result.

$$3.12\boxed{+}4.15\boxed{\times}5.79\boxed{=}\boxed{\div}\boxed{(}\ 5.13\boxed{-}3.76\boxed{)}\boxed{=}$$
$$5.13\boxed{-}3.76\boxed{=}\boxed{1/x}\boxed{\times}\boxed{(}\ 3.12\boxed{+}4.15\boxed{\times}5.79\boxed{)}\boxed{=}$$

The result is 19.816423, or 19.82 rounded to two decimal places.

ⓒ *Calculate each of the following rounding your answer to two decimal places.*

45. $\dfrac{43.25 - 21.32}{14.34 - 18.39}$

46. $4.32(10.57 - 3.21) + 15.97$

47. $\dfrac{5.67(1.51 - .98)}{3.57 + 1.92}$

48. $11.91 + \dfrac{15.32 - 9.62}{31.34}$

49. $\dfrac{12\sqrt{3.49} + 15.21}{14.32 - \sqrt{21.31}}$

50. $\dfrac{\sqrt{14.21} - 3.19}{1.91 + \sqrt{3.79}}$

51. $\sqrt{31.27} + \dfrac{14.91}{5.32}$

52. $\dfrac{\sqrt{51.79} - \sqrt{44.37}}{\sqrt{21.32} - 4.79}$

MISCELLANEOUS PROBLEMS

53. Perform the indicated operations and simplify.

(a) $\frac{5}{6} + \frac{1}{9} - \frac{2}{3} + \frac{5}{18}$

(b) $2(\frac{5}{6} + \frac{2}{9}) - \frac{1}{3}(2 - \frac{5}{6})$

(c) $-\frac{4}{3}[\frac{1}{2} - \frac{5}{4}(\frac{2}{5} - \frac{3}{11})]$

(d) $[4 - (2 \div \frac{6}{7})]\frac{2}{11}$

(e) $\dfrac{\frac{5}{6} - \frac{2}{7}}{\frac{5}{6} + \frac{2}{7}} + \frac{24}{47}$

(f) $\frac{1}{18} + \dfrac{1}{2 - (3 - \frac{1}{4})}$

(g) $(1 - \frac{1}{2})(1 - \frac{1}{3})(1 - \frac{1}{4}) \cdots (1 - \frac{1}{19})$

(h) $(1 + \frac{3}{4})(1 + \frac{3}{5})(1 + \frac{3}{6}) \cdots (1 + \frac{3}{19})$

[c] **54.** Calculate.

(a) $\dfrac{5\sqrt{2} + 4}{(3 - \sqrt{2})^2}$

(b) $\sqrt{3 + \sqrt{3 - \sqrt{5}}}$

55. We claimed that we could show that $a \cdot 0 = 0$ for all a using only the properties displayed in boxes. Justify each of the following equalities by one of those properties.

$$0 = -(a \cdot 0) + a \cdot 0$$
$$= -(a \cdot 0) + a \cdot (0 + 0)$$
$$= -(a \cdot 0) + (a \cdot 0 + a \cdot 0)$$
$$= [-(a \cdot 0) + a \cdot 0] + a \cdot 0$$
$$= 0 + a \cdot 0$$
$$= a \cdot 0$$

56. Let p be a nonzero real number. Show that $p/0$ is meaningless. (*Hint:* If $p/0 = q$, then $p = 0 \cdot q$).

57. Demonstrate that $(-a) \cdot b = -(a \cdot b)$ and $(-a) \cdot (-b) = a \cdot b$ using only the properties displayed in boxes and the result of Problem 55.

58. Let # denote the exponentiation operation, that is, $a \# b = a^b$.

(a) Calculate 4 # 3, 2(3 # 2), and 2 # (3 # 2)

(b) Is # commutative?

(c) Is # associative?

59. Which of the following operations commute with each other, that is, which give the same answer when performed on a number in either order?

(a) Doubling and tripling.

(b) Doubling and cubing.

(c) Squaring and cubing.

(d) Squaring and taking the negative.

(e) Squaring and taking the reciprocal.

(f) Adding 3 and taking the reciprocal.

(g) Doubling and taking the absolute value.

(h) Multiplying by 3 and dividing by 2.

60. Here are the main properties of absolute value.

$$\text{(i) } |a \cdot b| = |a| \cdot |b|; \qquad \text{(ii) } \left| \frac{a}{b} \right| = \frac{|a|}{|b|}; \qquad \text{(iii) } |a + b| \le |a| + |b|$$

Which of the following are always true?

(a) $|-x| = |x|$ (b) $|1 - x| = x - 1$

(c) $|x^2| = |x|^2$ (d) $|x + 1| > |x|$

(e) $|x - y| \le |x| - |y|$ (f) $-|x| \le x \le |x|$

61. Note that $|x| < c$ is equivalent to $-c < x < c$. Write each of the following as a double inequality without absolute values.

(a) $|x| < 4$ (b) $|x - 2| < 4$ (c) $|3x - 2| < 7$

62. Write the repeating decimal $.\overline{45}$ as the ratio of two integers. *Hint:* Let $x = .\overline{45}$. Then $100x = 45.\overline{45}$. Subtract the first equation from the second and solve for x.

63. Write $.\overline{216}$ as the ratio of two integers. See Problem 62.

64. Show that $\sqrt{2} + \frac{2}{3}$ is irrational. *Hint:* Take it as a known fact that $\sqrt{2}$ is irrational. Let $r = \sqrt{2} + \frac{2}{3}$ and suppose that r is rational. By adding $-\frac{2}{3}$ to both sides, arrive at a contradiction. Conclude that r must be irrational. This is called a *proof by contradiction*.

65. Show that the sum of a rational number and an irrational number is irrational. See Problem 64.

66. Show that $\frac{2}{3}\sqrt{2}$ is irrational.

67. Show that the product of a nonzero rational number and an irrational number is irrational.

68. Show by example that the sum and the product of two irrational numbers can be rational.

69. Which of the following numbers are rational?

(a) $\sqrt{2} + \sqrt{12}$ (b) $\sqrt{2}\sqrt{8}$ (c) $\sqrt{2}(\sqrt{2} + 2)$

(d) $(.12)(.\overline{12})$ (e) $.12\sqrt{2}$ (f) $(3 + \sqrt{2})(3 - \sqrt{2})$

70. **TEASER** In Figure 9, $ABCD$ is a square with sides of unit length. Also $\overline{AH} = \overline{BE} = \overline{CF} = \overline{DG} = p < 1$. Show that $RSTU$ is a square and express its area in terms of p.

Figure 9

Algebra has been called the art of using symbols. If that be true, the French lawyer, Francois Vieta, can be called the father of algebra. He was the first to use letters systematically and purposefully both to represent unknowns and as symbols for arbitrary numbers. Arithmetic dealt with numbers, he said, but algebra was a method for operating on the forms of things.

Francois Vieta (1540–1603)

1-2   Integral Exponents

Of the many advances in algebraic symbolism, certainly the introduction of exponents was one of the most useful. Historically, the full development of this fruitful idea took several centuries. Here we present the simple case of integral exponents. Rational and real exponents will be discussed in Section 3-2.

Rather than the cumbersome expression

$$2 \cdot 2 \cdot 2 \cdot 2 \cdot 2 \cdot 2 \cdot 2 \cdot 2 \cdot 2 \cdot 2 \cdot 2 \cdot 2 \cdot 2 \cdot 2 \cdot 2$$

most ordinary people and all mathematicians write 2^{15}. The number 15 is called an **exponent**; it tells you how many 2's to multiply together. The number 2^{15} is called a **power** of 2, and we read "2 to the 15th power." In the general case where b is any number and n is a positive integer,

$$b^n = \underbrace{b \cdot b \cdot b \cdots b}_{n \text{ factors}}$$

RULES FOR EXPONENTS

The behavior of exponents is excellent, being governed by a few simple rules that are easy to remember. Consider multiplication first. If we multiply 2^5 by 2^8, we have

$$2^5 \cdot 2^8 = \underbrace{(2 \cdot 2 \cdot 2 \cdot 2 \cdot 2)}_{5}\underbrace{(2 \cdot 2 \cdot 2 \cdot 2 \cdot 2 \cdot 2 \cdot 2 \cdot 2)}_{8}$$

$$= 2 \cdot 2 \cdot 2 \cdot 2 \cdot 2 \cdot 2 \cdot 2 \cdot 2 \cdot 2 \cdot 2 \cdot 2 \cdot 2 \cdot 2$$

$$\underbrace{}_{13}$$

$$= 2^{13}$$

$$= 2^{5+8}$$

It suggests that to find the product of powers of 2, you should add the exponents. There is nothing special about 2; it could just as well be 5, $\frac{2}{3}$, or π. We can put the general rule in a nutshell by using the symbols of algebra.

$$b^m \cdot b^n = b^{m+n}$$

Here b can stand for any number, but (for now) think of m and n as positive integers. Be careful with this rule. If you write

$$3^4 \cdot 3^5 = 3^9$$

or

$$\pi^9 \cdot \pi^{12} \cdot \pi^2 = \pi^{23}$$

that is fine. But do not try to use the rule on $2^4 \cdot 3^5$ or $a^2 \cdot b^3$; it just does not apply.

Next consider the problem of raising a power to a power. By definition, $(2^{10})^3 = 2^{10} \cdot 2^{10} \cdot 2^{10}$, which allows us to apply the rule above. Thus

$$(2^{10})^3 = 2^{10} \cdot 2^{10} \cdot 2^{10} = 2^{10+10+10} = 2^{10 \cdot 3}$$

It appears that to raise a power to a power we should multiply the exponents; in symbols,

$$(b^m)^n = b^{m \cdot n}$$

Try to convince yourself that this rule is true for any number b and for any positive integer exponents m and n.

Sometimes we need to simplify quotients like

$$\frac{8^6}{8^6} \qquad \frac{2^9}{2^5} \qquad \frac{10^4}{10^6}$$

The first one is easy enough; it equals 1. Furthermore,

$$\frac{2^9}{2^5} = \frac{2^5 \cdot 2^4}{2^5} = 2^4 = 2^{9-5}$$

and

$$\frac{10^4}{10^6} = \frac{10^4}{10^4 \cdot 10^2} = \frac{1}{10^2} = \frac{1}{10^{6-4}}$$

These illustrate the general rules.

$$\frac{b^m}{b^n} = 1 \qquad \text{if } m = n$$

$$\frac{b^m}{b^n} = b^{m-n} \qquad \text{if } m > n$$

$$\frac{b^m}{b^n} = \frac{1}{b^{n-m}} \qquad \text{if } n > m$$

In each case, we assume $b \neq 0$.

We did not put a box around these rules simply because we are not happy with them. It took three lines to describe what happens when you divide powers of the same number. Surely we can do better than that, but first we shall have to extend the notion of exponents to numbers other than positive integers.

ZERO AND NEGATIVE EXPONENTS

So far, symbols like 4^0 and 10^{-3} have not been used. We want to give them meaning and do it in a way that is consistent with what we have already learned. For example, 4^0 must behave so that

$$4^0 \cdot 4^7 = 4^{0+7} = 4^7$$

This can happen only if $4^0 = 1$. More generally, we require that

$$\boxed{b^0 = 1}$$

Here b can be any number except 0 (0^0 will be left undefined).

What about 10^{-3}? If it is to be admitted to the family of powers, it too must abide by the rules. Thus we insist that

$$10^{-3} \cdot 10^3 = 10^{-3+3} = 10^0 = 1$$

This means that 10^{-3} has to be the reciprocal of 10^3. Consequently, we are led to make the definition

$$\boxed{b^{-n} = \frac{1}{b^n} \qquad b \neq 0}$$

For powers of 10, these definitions provide the nice pattern you see in Figure 10. Also, they allow us to restate the complicated law for quotients of powers in a simple way, as we first illustrate with several examples.

$$\frac{b^4}{b^9} = \frac{b^4}{b^4 \cdot b^5} = \frac{1}{b^5} = b^{-5} = b^{4-9}$$

Powers of 10

$10^3 = 10 \cdot 10 \cdot 10 = 1000$

$10^2 = 10 \cdot 10 = 100$

$10^1 = 10$

$10^0 = 1$

$10^{-1} = \dfrac{1}{10^1} = .1$

$10^{-2} = \dfrac{1}{10^2} = .01$

$10^{-3} = \dfrac{1}{10^3} = .001$

Figure 10

$$\frac{b^5}{b^5} = 1 = b^0 = b^{5-5}$$

$$\frac{b^{-3}}{b^{-9}} = \frac{1/b^3}{1/b^9} = \frac{b^9}{b^3} = b^6 = b^{-3-(-9)}$$

In fact, for any choice of integers m and n, we find that

$$\frac{b^m}{b^n} = b^{m-n} \qquad b \neq 0$$

What about the two rules we learned earlier? Are they still valid when m and n are arbitrary (possibly negative) integers? The answer is yes. A few illustrations may help convince you.

$$b^{-3} \cdot b^7 = \frac{1}{b^3} \cdot b^7 = \frac{b^7}{b^3} = b^4 = b^{-3+7}$$

$$(b^{-5})^2 = \left(\frac{1}{b^5}\right)^2 = \frac{1}{b^5} \cdot \frac{1}{b^5} = \frac{1}{b^{10}} = b^{-10} = b^{(-5)(2)}$$

A summary of the main rules of exponents follows. We include the three rules already discussed, plus two new rules that are just as easy to establish.

RULES FOR EXPONENTS

Given that a and b are any real numbers and m and n are any integers

1. $b^m b^n = b^{m+n}$
2. $(b^m)^n = b^{mn}$
3. $\dfrac{b^m}{b^n} = b^{m-n}$
4. $(ab)^n = a^n b^n$
5. $\left(\dfrac{a}{b}\right)^n = \dfrac{a^n}{b^n}$

Of course, in all these rules, we must avoid division by 0.

SCIENTIFIC NOTATION

The difficulty in working with very large or very small numbers is that there are so many zeros, zeros which serve only to place the decimal point. It is easy enough to calculate

$$\frac{(32)(284)}{128} = 71$$

But to calculate

$$P = \frac{(3,200,000,000)(.0000000284)}{.00000000128}$$

seems much harder, even though the answers in both cases have the same nonzero digits. The only difference is in the placement of the decimal point. To simplify calculations like this, we introduce a method of writing numbers called scientific notation. It is extensively used in all of modern science.

A positive number is in **scientific notation** when it is written in the form

$$c \times 10^n$$

where n is an integer and c is a real number satisfying the inequality $1 \leq c < 10$. For example, scientists would normally write the following.

Speed of light: 2.9979×10^{10} centimeters per second

Mass of proton: 1.67×10^{-24} grams

The digits of c are called **significant digits.**

For any decimal number, call the position right after the first nonzero digit the **standard position** for the decimal point. To put a decimal number in scientific notation, put the decimal point in standard position and then count the number of places to the right or left to the original position of the decimal point. This number, taken as positive if counted to the right and negative if counted to the left, is the exponent on the power of 10 which is used as multiplier. Thus to calculate the number P above, we write

$$3,200,000,000 = 3.2 \times 10^9$$

$$.0000000284 = 2.84 \times 10^{-8}$$

$$.00000000128 = 1.28 \times 10^{-9}$$

Then

$$P = \frac{(3.2 \times 10^9)(2.84 \times 10^{-8})}{1.28 \times 10^{-9}}$$

$$= \frac{(3.2)(2.84)}{1.28} \times 10^{9-8-(-9)}$$

$$= 7.1 \times 10^{10}$$

Scientific calculators handle numbers written in either ordinary decimal notation or scientific notation. Calculated results that are too large or too small for display in ordinary notation are automatically converted to scientific notation. Check the instruction book for your model to learn how to enter numbers in scientific notation. On our model, 3.12×10^{-11} would be entered as

3.12 $\boxed{\text{EE}}$ 11 $\boxed{+/-}$

Problem Set 1-2

Use the rules for exponents to simplify each of the following. Then calculate the result.

1. $\dfrac{3^2 3^5}{3^4}$ 2. $\dfrac{2^6 2^7}{2^{10}}$ 3. $\dfrac{(2^2)^4}{2^6}$ 4. $\dfrac{(5^4)^3}{5^9}$ 5. $\dfrac{(3^2 2^3)^3}{6^6}$ 6. $\dfrac{(5^3 2^4)^2}{10^6}$

EXAMPLE A (Removing Negative Exponents)   Rewrite without negative exponents and simplify.

(a) -4^{-2} (b) $(-4)^{-2}$ (c) $((\tfrac{3}{4})^{-1})^2$ (d) $2^5 3^{-2} 2^{-3}$

Solution.

(a) The exponent -2 applies just to 4.

$$-4^{-2} = -\frac{1}{4^2} = -\frac{1}{4 \cdot 4} = -\frac{1}{16}$$

(b) The exponent -2 now applies to -4.

$$(-4)^{-2} = \frac{1}{(-4)^2} = \frac{1}{16}$$

(c) First apply the rule for a power of a power.

$$\left[\left(\frac{3}{4}\right)^{-1}\right]^2 = \left(\frac{3}{4}\right)^{-2} = \frac{1}{(\frac{3}{4})^2} = \frac{1}{\frac{9}{16}} = \frac{16}{9}$$

(d) Note that the two powers of 2 can be combined.

$$2^5 3^{-2} 2^{-3} = 2^5 2^{-3} 3^{-2} = 2^2 \cdot \frac{1}{3^2} = \frac{4}{9}$$

Write without negative exponents and simplify.

7. 5^{-2} 8. $(-5)^{-2}$ 9. -5^{-2}

10. 2^{-5} 11. $(-2)^{-5}$ 12. $\left(\dfrac{1}{5}\right)^{-2}$

13. $\left(\dfrac{-2}{3}\right)^{-3}$ 14. $-\dfrac{2^{-3}}{3}$ 15. $\dfrac{2^{-2}}{3^{-3}}$

16. $\left[\left(\dfrac{2}{3}\right)^{-2}\right]^2$ 17. $\left[\left(\dfrac{3}{2}\right)^{-2}\right]^{-2}$ 18. $\dfrac{4^0 + 0^4}{4^{-1}}$

19. $\dfrac{2^{-2} - 4^{-3}}{(-2)^2 + (-4)^0}$ 20. $\dfrac{3^{-1} + 2^{-3}}{(-1)^3 + (-3)^2}$ 21. $3^3 \cdot 2^{-3} \cdot 3^{-5}$

22. $4^2 \cdot 4^{-4} \cdot 3^0$

EXAMPLE B (Rules for Products and Quotients)   Use Rules 4 and 5 to simplify

(a) $(2x)^6$;   (b) $\left(\dfrac{2x}{3}\right)^4$;   (c) $(x^{-1}y^2)^{-3}$.

Solution.

(a) $(2x)^6 = 2^6x^6 = 64x^6$

(b) $\left(\dfrac{2x}{3}\right)^4 = \dfrac{(2x)^4}{3^4} = \dfrac{2^4x^4}{3^4} = \dfrac{16x^4}{81}$

(c) $(x^{-1}y^2)^{-3} = (x^{-1})^{-3}(y^2)^{-3} = x^3y^{-6} = x^3 \cdot \dfrac{1}{y^6} = \dfrac{x^3}{y^6}$

Simplify, writing your answer without negative exponents.

23. $(3x)^4$

24. $\left(\dfrac{2}{y}\right)^5$

25. $(xy^2)^6$

26. $\left(\dfrac{y^2}{3z}\right)^4$

27. $\left(\dfrac{2x^2y}{w^3}\right)^4$

28. $\left(\dfrac{\sqrt{2}x}{3}\right)^4$

29. $\left(\dfrac{3x^{-1}y^2}{z^2}\right)^3$

30. $\left(\dfrac{2x^{-2}y}{z^{-1}}\right)^2$

31. $\left(\dfrac{\sqrt{5}}{x^{-2}}\right)^4$

32. $(\sqrt{3}x^{-2})^6$

33. $(4y^3)^{-2}$

34. $(x^3z^{-2})^{-1}$

35. $\left(\dfrac{5x^2}{ab^{-2}}\right)^{-1}$

36. $\left(\dfrac{3x^2y^{-2}}{2x^{-1}y^4}\right)^{-3}$

EXAMPLE C (Simplifying Complicated Expressions) Simplify

(a) $\dfrac{4ab^{-2}c^3}{a^{-3}b^3c^{-1}}$; (b) $\left[\dfrac{(2xz^{-2})^3(x^{-2}z)}{2xz^2}\right]^4$; (c) $(a^{-1}+b^{-2})^{-1}$.

Solution.

(a) $\dfrac{4ab^{-2}c^3}{a^{-3}b^3c^{-1}} = \dfrac{4a(1/b^2)\cdot c^3}{(1/a^3)b^3(1/c)} = \dfrac{4ac^3/b^2}{b^3/(a^3c)} = \dfrac{4ac^3}{b^2}\cdot\dfrac{a^3c}{b^3} = \dfrac{4a^4c^4}{b^5}$

In simplifying expressions like the one above, *a factor can be moved from numerator to denominator or vice versa by changing the sign of its exponent.* That is important enough to remember. Let us do part (a) again using this fact.

$$\dfrac{4ab^{-2}c^3}{a^{-3}b^3c^{-1}} = \dfrac{4aa^3c^3c}{b^2b^3} = \dfrac{4a^4c^4}{b^5}$$

(b) $\left[\dfrac{(2xz^{-2})^3(x^{-2}z)}{2xz^2}\right]^4 = \left[\dfrac{8x^3z^{-6}x^{-2}z}{2xz^2}\right]^4$

$= \left[\dfrac{8xz^{-5}}{2xz^2}\right]^4$

$= \left[\dfrac{4}{z^2z^5}\right]^4 = \dfrac{256}{z^{28}}$

(c) $(a^{-1}+b^{-2})^{-1} = \left(\dfrac{1}{a}+\dfrac{1}{b^2}\right)^{-1} = \left(\dfrac{b^2+a}{ab^2}\right)^{-1} = \dfrac{ab^2}{b^2+a}$

Note the difference between a product and a sum.

$$(a^{-1} \cdot b^{-2})^{-1} = a \cdot b^2$$

but

$$(a^{-1} + b^{-2})^{-1} \ne a + b^2$$

Simplify, leaving your answer free of negative exponents.

37. $\dfrac{2x^{-3}y^2z}{x^3y^4z^{-2}}$

38. $\dfrac{3x^{-5}y^{-3}z^4}{9x^2yz^{-1}}$

39. $\left(\dfrac{-2xy}{z^2}\right)^{-1}(x^2y^{-3})^2$

40. $(4ab^2)^3\left(\dfrac{-a^3}{2b}\right)^2$

41. $\dfrac{ab^{-1}}{(ab)^{-1}} \cdot \dfrac{a^2b}{b^{-2}}$

42. $\dfrac{3(b^{-2}d)^4(2bd^3)^2}{(2b^2d^3)(b^{-1}d^2)^5}$

43. $\left[\dfrac{(3b^{-2}d)(2)(bd^3)^2}{12b^3d^{-1}}\right]^5$

44. $\left[\dfrac{(ab^2)^{-1}}{(ba^2)^{-2}}\right]^{-1}$

45. $(a^{-2} + a^{-3})^{-1}$

46. $a^{-2} + a^{-3}$

47. $\dfrac{x^{-1}}{y^{-1}} - \left(\dfrac{x}{y}\right)^{-1}$

48. $(x^{-1} - y^{-1})^{-1}$

[c] *Evaluate each of the following using the* $\boxed{y^x}$ *key. Round your answer to two decimal places.*

49. $(1.214)^3$

50. $(3.617)^{-2}$

51. $(1.34)^{12}(2.345)^3$

52. $(14.72)^{12}(59.31)^{-11}$

53. $\dfrac{2.51(11.63)^4}{(51.24)^3}$

54. $\dfrac{(3.72)^{-2}(2.41)^4}{(4.63)^3}$

Write each of the following numbers in scientific notation.

55. 341,000,000

56. 25 billion

57. .0000000513

58. .00000000012

59. .0000000001245

60. .0000000000012578

Perform the following calculations leaving your answers in scientific notation.

61. $\dfrac{(1.2 \times 10^9)(2.4 \times 10^5)}{3.2 \times 10^6}$

62. $\dfrac{(3.3 \times 10^{11})(1.4 \times 10^{-4})}{2.1 \times 10^3}$

[c] 63. $\dfrac{(1.325 \times 10^{-5})(42,000)}{6.6 \times 10^{15}}$

[c] 64. $\dfrac{(1.433 \times 10^6)(2.3 \times 10^{14})}{4,100,000}$

[c] 65. $\dfrac{(1.342 \times 10^{-4})^2}{2.578 \times 10^5}$

[c] 66. $\dfrac{(3.791 \times 10^7)^3}{4.325 \times 10^5}$

MISCELLANEOUS PROBLEMS

Simplify the expressions in Problems 67–76, leaving your answer free of any negative exponents.

67. $(\tfrac{2}{3}x^2y^{-3})^{-2}$

68. $(x + 2x^{-1})^2$

69. $\dfrac{3^{-2}}{1 + \dfrac{2^{-1}}{1 + 2^{-2}}}$

70. $[(\tfrac{1}{3} + \tfrac{1}{4})^{-1} + (\tfrac{1}{3} + \tfrac{3}{5})^{-1}]^{-1}$

71. $[(\tfrac{1}{3}a^{-3})^2(9ab^{-2})^2]^{-2}$

72. $\left(\dfrac{\sqrt{3}u^3v^{-2}}{uvw^{-1}}\right)^4$

73. $\dfrac{(2x^{-1}y^2)^2}{2xy} \cdot \left(\dfrac{x^{-3}}{y^3}\right)^{-2}$

74. $\left[\dfrac{4y^2z^{-3}}{x^3(2x^{-1}z^2)^3}\right]^{-2}$

75. $(x^{-1} + y^{-1})(x + y)^{-1}$

76. $[1 - (1 + x^{-1})^{-1}]^{-1}$

77. Express each of the following as a single power of 2.

    (a) $\frac{1}{2} \cdot \frac{1}{4} \cdot \frac{1}{8} \cdot \frac{1}{16} \cdot \frac{1}{32}$ (b) $\left(\frac{1}{2} + \frac{1}{4} + \frac{1}{8} + \frac{1}{16}\right)\left(\frac{32}{15}\right)$

78. Without using a calculator, decide which is larger: 2^{1000} or 100^{150}.

79. Take a huge sheet of paper .01 inch thick and fold it in half, then in half again, and so on, until you have made 40 folds. Assuming this can be done, about how high would the final stack of paper be? Use the approximation $2^{10} \approx 1000$ in making your calculation and give the answer in miles.

80. If the base of the final pile in Problem 79 has area 1 square inch, approximately what was the area of the sheet of paper with which you started?

81. G. P. Jetty has agreed to pay his new secretary according to the following plan: 1¢ the first day, 2¢ the second day, 4¢ the third day, and so on, doubling each day.

    (a) How much will the secretary make during the first 4 days? 5 days? 6 days?

    (b) From part (a), you should see a pattern. How much will the secretary make during the first n days?

    (c) Assume Jetty is worth \$2 billion and that his secretary started work on January 1. About when will Jetty go broke?

82. **TEASER** By a^{b^c}, mathematicians mean $a^{(b^c)}$; that is, in a tower of exponents we start at the top and work down. For example, $2^{2^{2^2}} = 2^{2^4} = 2^{16} = 65{,}536$. Arrange the following numbers (all made with four 2s) from smallest to largest. You should be able to do this without making use of a calculator.

$$2222, \quad 222^2, \quad 22^{22}, \quad 2^{222}, \quad 22^{2^2}, \quad 2^{2^{2^2}}$$

Examples of Polynomials

$3x^2 + 11x + 9$

$\dfrac{3}{2}x - 5$

$\sqrt{3}x^5 + \pi x^3 + 2x^2 + \dfrac{5}{4}$

$-3x^4 + 2x^3 - 3x$

$5x^{12}$

39

A real polynomial in x is any expression of the form

$$a_nx^n + a_{n-1}x^{n-1} + a_{n-2}x^{n-2} + \cdots + a_1x + a_0$$

where the a's are real numbers and n is a non-negative integer.

    A real rational expression in x is a quotient (ratio) of two polynomials in x.

Examples of Rational Expressions

$\dfrac{3x + 1}{x^2 + x + 3}$

$\dfrac{x^3 - \pi x + \sqrt{2}}{x - 1}$

$\dfrac{3}{x^2 - 4}$

1-3   Polynomials and Rational Expressions

This book is designed to prepare you for calculus. In calculus, almost every concept is first studied in the context of polynomials. This is because polynomials are the simplest of all mathematical expressions. Fundamentally, a **real polynomial in x** is any expression that can be obtained from the real numbers and x using only the operations of addition, subtraction, and multiplication. For example, we can get $3 \cdot x \cdot x \cdot x \cdot x$ or $3x^4$ by multiplication. We can get $2x^3$ by the same process and then have

$$3x^4 + 2x^3$$

by addition. We could never get $2x^{-2} = 2/x^2$ or $\sqrt{x}$; the first involves a division and the second involves taking a root. Thus $2/x^2$ and $\sqrt{x}$ are not polynomials. Try to convince yourself that all the expressions in the left box above are polynomials.

There is another way to define a polynomial in fewer words. A **real polynomial in x** is any expression of the form

$$a_n x^n + a_{n-1} x^{n-1} + a_{n-2} x^{n-2} + \cdots + a_1 x + a_0,$$

where the a's, called **coefficients**, are real numbers and n is a nonnegative integer. The **degree** of a polynomial is the largest exponent that occurs in the polynomial in a term with a nonzero coefficient. Here are some more examples.

1. $\frac{3}{2}x - 5$ is a first degree (or **linear**) polynomial in x.
2. $3y^2 - 2y + 16$ is a second degree (or **quadratic**) polynomial in y.
3. $\sqrt{3}\,t^5 - \pi t^3 + 2t^2 - 17$ is a fifth degree polynomial in t.
4. $5x^3$ is a third degree polynomial in x. It is also called a **monomial** since it has only one term.
5. 13 is a polynomial of degree zero. Does that seem strange to you? If it helps, think of it as $13x^0$. In general, any nonzero constant polynomial has degree zero. We do not define the degree of the zero polynomial.

OPERATIONS ON POLYNOMIALS

The whole subject of polynomials would be pretty dull and almost useless if it stopped with a definition. Fortunately, polynomials—like numbers—can be manipulated. In fact, they behave something like the integers. Just as the sum, difference, and product of two integers are integers, so the sum, difference, and product of two polynomials are polynomials.

Adding two polynomials is a snap. Treat x like a number and use the commutative, associative, and distributive properties freely. When you are all done, you will discover that you have just grouped like terms (that is, terms of the same degree) and added their coefficients. Here, for example, is how we add $x^3 + 2x^2 + 7x + 5$ and $x^2 - 3x - 4$.

$(x^3 + 2x^2 + 7x + 5) + (x^2 - 3x - 4)$

$\quad = x^3 + (2x^2 + x^2) + (7x - 3x) + (5 - 4)$ (associative and commutative properties)

$\quad = x^3 + (2 + 1)x^2 + (7 - 3)x + 1$ (distributive property)

$\quad = x^3 + 3x^2 + 4x + 1$

To subtract two polynomials, simply replace the subtracted polynomial by its negative and add. Thus

$$(3x^2 - 5x + 2) - (5x^2 + 4x - 4) = (3x^2 - 5x + 2) + (-5x^2 - 4x + 4)$$
$$= -2x^2 - 9x + 6$$

The distributive property is the basic tool in multiplication. Here is a simple example.

$$(3x^2)(2x^3 + 7) = (3x^2)(2x^3) + (3x^2)(7)$$
$$= 6x^5 + 21x^2$$

But things can get more complicated.

$$(3x - 4)(2x^3 - 7x + 8) = (3x)(2x^3 - 7x + 8) + (-4)(2x^3 - 7x + 8)$$
$$= (3x)(2x^3) + (3x)(-7x) + (3x)(8) + (-4)(2x^3)$$
$$+ (-4)(-7x) + (-4)(8)$$
$$= 6x^4 - 21x^2 + 24x - 8x^3 + 28x - 32$$
$$= 6x^4 - 8x^3 - 21x^2 + 52x - 32$$

Notice that each term of $3x - 4$ multiplies each term of $2x^3 - 7x + 8$.

If the process just illustrated seems unwieldy, you may find the following format helpful.

$$
\begin{array}{l}
2x^3 - 7x \; + 8 \\
\underline{3x \; - 4} \\
6x^4 \qquad\;\; - 21x^2 + 24x \\
\underline{\;\; - 8x^3 \qquad\qquad + 28x - 32} \\
6x^4 - 8x^3 - 21x^2 + 52x - 32
\end{array}
$$

When both polynomials are linear (that is, of the form $ax + b$), there is a handy shortcut. For example, just one look at $(x + 4)(x + 5)$ convinces us that the product has the form $x^2 + (\;\;) + 20$. It is the middle term that may cause a little trouble. Think of it this way.

$$(x + 4)(x + 5) = x^2 + (\;\;) + 20$$
$$= x^2 + 9x + 20$$

Similarly,

$$(2x - 3)(x + 5) = 2x^2 + (\;\;) - 15$$
$$= 2x^2 + 7x - 15$$

You should be able to find such simple products mentally. Some people find the FOIL method, illustrated in Figure 11, to be helpful.

The FOIL Method

$(x + 4)(x + 5) =$

F   O   I   L

$x^2 + 5x + 4x + 20$

The four terms are the products of the Firsts, Outers, Inners, and Lasts.

Figure 11

POLYNOMIALS IN SEVERAL VARIABLES

So far, we have considered only polynomials in a single variable. That is too restrictive for later work. Expressions such as

$$x^2y + 3xy + y \qquad u^3 + 3u^2v + 3uv^2 + v^3$$

are called **polynomials in two variables**. The rules for operating on them are so similar to those we have just given for polynomials in a single variable that we feel free to use them without further discussion.

FACTORING POLYNOMIALS

To factor 90 means to write it as a product of smaller numbers; to factor it completely means to write it as a product of primes—that is, numbers that cannot be factored further. Thus we have factored 90 when we write $90 = 9 \cdot 10$, but it is not factored completely until we write

$$90 = 2 \cdot 3 \cdot 3 \cdot 5$$

Similarly to **factor** a polynomial means to write it as a product of simpler polynomials; to **factor** a polynomial **completely** is to write it as a product of polynomials that cannot be factored further. Thus when we write

$$x^3 - 9x = x(x^2 - 9)$$

we have factored $x^3 - 9x$, but not until we write

$$x^3 - 9x = x(x + 3)(x - 3)$$

have we factored $x^3 - 9x$ completely.

Factoring and multiplying are reverse processes; they should be studied together. Thus we urge you to learn the formulas below forwards and backwards. One way, they are product formulas; the other way, they are factoring formulas.

> PRODUCT FORMULAS $\longrightarrow$
> $\longleftarrow$ FACTORING FORMULAS
> 1. $a(x + y + z) = ax + ay + az$
> 2. $(x + a)(x + b) = x^2 + (a + b)x + ab$
> 3. $(x + y)^2 = x^2 + 2xy + y^2$
> 4. $(x + y)(x - y) = x^2 - y^2$
> 5. $(x + y)(x^2 - xy + y^2) = x^3 + y^3$
> 6. $(x - y)(x^2 + xy + y^2) = x^3 - y^3$
> 7. $(x + y)^3 = x^3 + 3x^2y + 3xy^2 + y^3$

The simplest factoring procedure is based on Formula 1—*taking out a common factor*. It should be tried first. Always factor out as much as you can. Taking 2 out of $4xy^2 - 6x^3y^4 + 8x^4y^2$ is not nearly enough, though 2, is a common factor; taking out $2xy$ is not enough either. You should take out $2xy^2$.

Then

$$4xy^2 - 6x^3y^4 + 8x^4y^2 = 2xy^2(2 - 3x^2y^2 + 4x^3)$$

Certain second degree (quadratic) polynomials are a breeze to factor. They are *perfect squares*.

$$x^2 + 10x + 25 = (x + 5)(x + 5) = (x + 5)^2$$
$$x^2 - 12x + 36 = (x - 6)^2$$
$$4x^2 + 12x + \ 9 = (2x + 3)^2$$

Each is modeled after Formula 3. We look for first and last terms that are squares. Then we ask if the middle term is two times (or minus two times) their product. Here are two more examples.

$$x^4 + 2x^2y^3 + y^6 = (x^2 + y^3)^2$$
$$y^2z^2 - 6ayz + 9a^2 = (yz - 3a)^2$$

Do you see a common feature in the following polynomials?

$$x^2 - 16 \qquad y^2 - 100 \qquad 4y^2 - 9b^2$$

Each is the *difference of two squares*. From Formula 4, we see that

$$x^2 - 16 = (x + 4)(x - 4)$$
$$y^2 - 100 = (y + 10)(y - 10)$$
$$4y^2 - 9b^2 = (2y + 3b)(2y - 3b)$$

Sometimes, factoring a quadratic polynomial depends on the trial-and-error method. What does not work is systematically eliminated; eventually, effort is rewarded. Let us see how this process works on $x^2 - 5x - 14$. We need to find numbers a and b such that

$$x^2 - 5x - 14 = (x + a)(x + b)$$

Since ab must equal -14, two possibilities immediately occur to us: $a = 7$ and $b = -2$ or $a = -7$ and $b = 2$. Try them both to see if one works.

$$(x + 7)(x - 2) = x^2 + 5x - 14$$

$$(x - 7)(x + 2) = x^2 - 5x - 14 \qquad \text{Success!}$$

The brackets help us calculate the middle term, the crucial step in this kind of factoring.

Here is a tougher factoring problem: Factor $2x^2 + 13x - 15$. It is a safe bet that if $2x^2 + 13x - 15$ factors at all, then

$$2x^2 + 13x - 15 = (2x + a)(x + b).$$

Since $ab = -15$, we are likely to try combinations of 3 and 5 first.

$$(2x + 5)(x - 3) = 2x^2 - x - 15$$

$$(2x - 5)(x + 3) = 2x^2 + x - 15$$

$$(2x + 3)(x - 5) = 2x^2 - 7x - 15$$

$$(2x - 3)(x + 5) = 2x^2 + 7x - 15$$

Discouraging, isn't it? But that is a poor reason to give up. Maybe we have missed some possibilities. We have, since combinations of 15 and 1 might work.

$$(2x - 15)(x + 1) = 2x^2 - 13x - 15$$

$$(2x + 15)(x - 1) = 2x^2 + 13x - 15 \qquad \text{Success!}$$

When you have had a lot of practice, you will be able to speed up the process. You will simply write

$$2x^2 + 13x - 15 = (2x + ?)(x + ?)$$

and mentally try the various possibilities until you find the right one. Of course, it may happen, as in the case of $2x^2 - 4x + 5$, that you cannot find a factorization.

Finally, we can factor the *sum and difference of cubes*, based on Formulas 5 and 6. Here is an example of each.

$$8x^3 + 27 = (2x)^3 + 3^3 = (2x + 3)[(2x)^2 - (2x)(3) + 3^2]$$

$$= (2x + 3)(4x^2 - 6x + 9)$$

$$x^3z^3 - 1000 = (xz)^3 - 10^3 = (xz - 10)(x^2z^2 + 10xz + 100)$$

CAUTION

$x^3 - 8 = (x - 2)(x^2 + 4)$
$x^3 - 8 = (x - 2)(x^2 - 4x + 4)$
$x^3 - 8 = (x - 2)(x^2 + 2x + 4)$

Someone is sure to make a terrible mistake and write

$$x^3 + y^3 = (x + y)^3 \qquad \text{Wrong!}$$

Remember that

$$(x + y)^3 = x^3 + 3x^2y + 3xy^2 + y^3$$

RATIONAL EXPRESSIONS

A **rational expression** is a quotient of two polynomials. It is in **reduced form** if its numerator and denominator have no (nontrivial) common factor. For example, $x/(x + 2)$ is in reduced form, but $x^2/(x^2 + 2x)$ is not, since

$$\frac{x^2}{x^2 + 2x} = \frac{\cancel{x} \cdot x}{\cancel{x}(x + 2)} = \frac{x}{x + 2}$$

To reduce a rational expression, we factor numerator and denominator and divide out, or cancel, common factors. Here are two examples.

$$\frac{x^2 + 7x + 10}{x^2 - 25} = \frac{\cancel{(x + 5)}(x + 2)}{\cancel{(x + 5)}(x - 5)} = \frac{x + 2}{x - 5}$$

$$\frac{2x^2 + 5xy - 3y^2}{2x^2 + xy - y^2} = \frac{\cancel{(2x - y)}(x + 3y)}{\cancel{(2x - y)}(x + y)} = \frac{x + 3y}{x + y}$$

We add (or subtract) rational expressions by rewriting them so that they have the same denominator and then adding (or subtracting) the new numerators. Suppose we want to add

$$\frac{3}{x} + \frac{2}{x + 1}$$

The appropriate common denominator is $x(x + 1)$. Remember that we can multiply numerator and denominator of a fraction by the same thing. Accordingly,

$$\frac{3}{x} + \frac{2}{x + 1} = \frac{3(x + 1)}{x(x + 1)} + \frac{x \cdot 2}{x(x + 1)} = \frac{3x + 3 + 2x}{x(x + 1)} = \frac{5x + 3}{x(x + 1)}$$

The same procedure is used in subtraction. Thus

$$\frac{2}{x + 1} - \frac{3}{x} = \frac{x \cdot 2}{x(x + 1)} - \frac{3(x + 1)}{x(x + 1)} = \frac{2x - (3x + 3)}{x(x + 1)}$$

$$= \frac{2x - 3x - 3}{x(x + 1)} = \frac{-x - 3}{x(x + 1)}$$

Here is a more complicated example. Study each step carefully.

Begin by factoring

$$\frac{3x}{x^2 - 1} - \frac{2x + 1}{x^2 - 2x + 1} = \frac{3x}{(x - 1)(x + 1)} - \frac{2x + 1}{(x - 1)^2}$$

$(x - 1)^2 (x + 1)$ is the lowest common denominator

$$= \frac{3x(x - 1)}{(x - 1)^2(x + 1)} - \frac{(2x + 1)(x + 1)}{(x - 1)^2(x + 1)}$$

Forgetting the parentheses around $2x^2 + 3x + 1$ would be a serious blunder

$$= \frac{3x^2 - 3x - (2x^2 + 3x + 1)}{(x - 1)^2(x + 1)}$$

$$= \frac{3x^2 - 3x - 2x^2 - 3x - 1}{(x - 1)^2(x + 1)}$$

No cancellation is possible since $x^2 - 6x - 1$ doesn't factor over the integers

$$= \frac{x^2 - 6x - 1}{(x - 1)^2(x + 1)}$$

We multiply rational expressions in the same manner as we do rational numbers; that is, we multiply numerators and multiply denominators. For example,

$$\frac{3}{x+5} \cdot \frac{x+2}{x-4} = \frac{3(x+2)}{(x+5)(x-4)} = \frac{3x+6}{x^2+x-20}$$

Sometimes we need to reduce the product, if we want the simplest possible answer. Here is an illustration.

$$\frac{2x-3}{x+5} \cdot \frac{x^2-25}{6xy-9y} = \frac{(2x-3)(x^2-25)}{(x+5)(6xy-9y)}$$

$$= \frac{(2x-3)(x-5)(x+5)}{(x+5)(3y)(2x-3)}$$

$$= \frac{x-5}{3y}$$

This example shows that it is a good idea to do as much factoring as possible at the outset. That is what set up the cancellation.

There are no real surprises with division, as we simply invert the divisor and multiply. Here is a nontrivial example.

$$\frac{x^2-5x+4}{2x+6} \div \frac{2x^2-x-1}{x^2+5x+6} = \frac{x^2-5x+4}{2x+6} \cdot \frac{x^2+5x+6}{2x^2-x-1}$$

$$= \frac{(x^2-5x+4)(x^2+5x+6)}{(2x+6)(2x^2-x-1)}$$

$$= \frac{(x-4)(x-1)(x+2)(x+3)}{2(x+3)(x-1)(2x+1)}$$

$$= \frac{(x-4)(x+2)}{2(2x+1)}$$

Problem Set 1-3

Decide whether the given expression is a polynomial. If it is, give its degree.

1. $3x^2 - x - 2$ 2. $4x^5 - x$ 3. $\pi s^2 - \sqrt{2}$

4. $\dfrac{\pi}{2} - \dfrac{2}{3}t^3$ 5. $x^3 - 2\sqrt{x} + 5$ 6. $2y^2 - 5 + \dfrac{4}{y}$

Perform the indicated operations and simplify. Write your answer as a polynomial in descending powers of the variable.

7. $(2x - 7) + (8 - 4x)$
8. $(\frac{3}{2}x - \frac{1}{4}) + \frac{5}{6}x$
9. $(3y^2 - 6y + 5) - (-2y^2 + 3y - 4)$

10. $(s^3 - 2s) - (3s^2 + 5s - 4)$

11. $5x(7x - 11) - 2x^2 + 19$

12. $-x^2(7x^3 - 5x) + 1 - 5x^3$

13. $(x + 9)(x - 10)$

14. $(x - 13)(x - 7)$

15. $(2y + 3)(3y - 1)$

16. $(a + 4)(3a - 5)$

17. $(2z - 1)(z^2 - 3z + 4)$

18. $(u^2 + 2u)(3u - 5)$

19. $(v + 5)^2$

20. $(m - 3)^2$

21. $(2x - 3)^2$

22. $(4y + 7)^2$

23. $(5t + 3)(5t - 3)$

24. $(s^2 + 7)(s^2 - 7)$

25. $(h + 3)^3$

26. $(2x - 5)^3$

27. $(4y - 1)(16y^2 + 4y + 1)$

28. $(2x + 3)(4x^2 - 6x + 9)$

29. $(2x^5 - 3)(2x^5 + 3)$

30. $4x^2 + (1 + 2x)(1 - 2x)$

Each expression in Problems 31–44 involves more than one variable. Perform the indicated operations and simplify.

31. $(x + y)(2x - 3y)$

32. $(2u - 3v)(u + 4v)$

33. $(2a + 3b)^2$

34. $(3mn - 5)^2$

35. $(3x + \sqrt{2}y)(3x - \sqrt{2}y)$

36. $(5x + 2h^2)(5x - 2h^2)$

37. $(y + 2z)(y^2 - 2yz + 4z^2)$

38. $(2a - b)(a^2 + b) + b(b + 2a)$

39. $(x + h)^3 + 2(x + h) - (x^3 + 2x)$

40. $(2x + 2h - 3)^2 - (2x - 3)^2$

41. $[(a + b) + c][(a + b) - c]$

42. $[(2c + d) - 3]^2$

43. $2(x - 2uw)^2 + (x + 4uw)^2$

44. $(2x^2y^3 - 3z)^2$

*Factor completely **over the integers** (that is, allow only integer coefficients in your answers).*

45. $x^2 + 5x$

46. $y^3 + 4y^2$

47. $x^2 + 5x - 6$

48. $x^2 + 5x + 4$

49. $y^4 - 6y^3$

50. $t^4 + t^2$

51. $y^2 + 4y - 12$

52. $z^2 - 3z - 40$

53. $y^2 + 8y + 16$

54. $9x^2 + 24x + 16$

55. $4x^2 - 12xy + 9y^2$

56. $9x^2 - 6x + 1$

57. $4a^2 - 9$

58. $x^2 - 4y^2$

59. $4z^2 - 4z - 3$

60. $7x^2 - 19x - 6$

61. $6a^2 - ax - 2x^2$

62. $x^4 - 16a^2x^2$

63. $9x^3 - x^3y^2$

64. $a^3 - 27b^3$

65. $8x^3 + 27y^3$

66. $bcx^2 - 4bc$

67. $ax^2 + 2abx + ab^2$

68. $ab^2 - a^2b$

Factor each of the following over the integers. In each case it will help to make a (mental) substitution. For example, to factor $(2x^2y + 1)^2 - a^2$, think of $u^2 - a^2 = (u + a)(u - a)$ with $u = 2x^2y + 1$. The result is $(2x^2y + 1 + a)(2x^2y + 1 - a)$.

69. $x^4 - x^2 - 6$

70. $x^6 + 9x^3 + 14$

71. $y^4 - 16$

72. $(2a + 1)^2 - b^2$

73. $x^6 - 64$

74. $y^{10} + 4y^5 + 3$

   Hint: First let $u = x^3$.

75. $(x + 4y)^2 - (x + 4y) - 2$

76. $(m - n)^2 + 5(m - n) + 4$

Factor each of the following over the integers by first doing some judicious grouping of terms. For example,

$$x^3 + 6x^2 - 2x - 12 = (x^3 + 6x^2) - (2x + 12)$$
$$= x^2(x + 6) - 2(x + 6)$$
$$= (x^2 - 2)(x + 6)$$

77. $x^3 + x^2 + 3x + 3$ 78. $7x^3 - x^2 + 7x - 1$

79. $2xa + 10x - 3a - 15$ 80. $ax^2 + x^2 - 4a - 4$

81. $x^2 + 6x + 9 + ax + 3a$ 82. $x^2 + 2bx + b^2 - 16$

Reduce each of the following.

83. $\dfrac{x + 6}{x^2 - 36}$ 84. $\dfrac{x^2 - 9}{4x - 12}$ 85. $\dfrac{y^3 + 1}{y^2 + y}$

86. $\dfrac{x^2 - 7x + 6}{x^2 - 4x - 12}$ 87. $\dfrac{x^2z + 4xyz + 4y^2z}{x^2 + 3xy + 2y^2}$ 88. $\dfrac{x^3 - 27}{3x^2 + 9x + 27}$

Perform the indicated operations and simplify.

89. $\dfrac{5}{x - 2} + \dfrac{4}{x + 2}$ 90. $\dfrac{x}{x - 2} - \dfrac{x + 1}{x + 2}$

91. $\dfrac{1}{(x - 2)^2} - \dfrac{3}{x^2 - 4}$ 92. $\dfrac{x^2}{x^2 - x + 1} - \dfrac{x + 1}{x}$

EXAMPLE A (The Three Signs of a Fraction)   Simplify $\dfrac{x}{3x - 6} - \dfrac{2}{2 - x}$.

Solution.

$$\frac{x}{3x - 6} - \frac{2}{2 - x} = \frac{x}{3(x - 2)} - \frac{2}{2 - x}$$

Now we make a crucial observation. Notice that

$$-(2 - x) = -2 + x = x - 2$$

That is, $2 - x$ and $x - 2$ are negatives of each other. Thus the expression above may be rewritten as

$$\frac{x}{3(x - 2)} - \frac{2}{2 - x} = \frac{x}{3(x - 2)} - \frac{2}{-(x - 2)}$$

$$= \frac{x}{3(x - 2)} + \frac{2}{x - 2}$$

$$= \frac{x}{3(x - 2)} + \frac{6}{3(x - 2)}$$

$$= \frac{x + 6}{3(x - 2)}$$

CAUTION

$$\frac{5}{x - 2} - \frac{x + 2}{x - 2} = \frac{5 - x + 2}{x - 2}$$
$$= \frac{7 - x}{x - 2}$$

$$\frac{5}{x - 2} - \frac{x + 2}{x - 2} = \frac{5 - x - 2}{x - 2}$$
$$= \frac{3 - x}{x - 2}$$

When we replaced $-\dfrac{2}{-(x-2)}$ by $\dfrac{2}{x-2}$, we used the fact that

$$-\frac{a}{-b} = \frac{a}{b}$$

Keep in mind that a fraction has three sign positions: numerator, denominator, and total fraction. You may change any two of them without changing the value of the fraction. Thus

$$\frac{a}{b} = -\frac{a}{-b} = -\frac{-a}{b} = \frac{-a}{-b}$$

Simplify.

93. $\dfrac{4}{2x-1} + \dfrac{x}{1-2x}$

94. $\dfrac{x}{6x-2} - \dfrac{3}{1-3x}$

95. $\dfrac{2}{6y-2} + \dfrac{y}{9y^2-1} - \dfrac{2y+1}{1-3y}$

96. $\dfrac{x}{4x^2-1} + \dfrac{2}{4x-2} - \dfrac{3x+1}{1-2x}$

Perform the following multiplications and divisions and simplify.

97. $\dfrac{5}{2x-1} \cdot \dfrac{x}{x+1}$

98. $\dfrac{3}{x^2-2x} \cdot \dfrac{x-2}{x}$

99. $\dfrac{x+2}{x^2-9} \cdot \dfrac{x+3}{x^2-4}$

100. $\left(1 + \dfrac{1}{x+2}\right)\left(\dfrac{4}{3x+9}\right)$

101. $\dfrac{\dfrac{5}{2x-1}}{\dfrac{x}{x+1}}$

102. $\dfrac{\dfrac{5}{2x-1}}{\dfrac{x}{4x^2-1}}$

103. $\dfrac{\dfrac{x+2}{x^2-4}}{x}$

104. $\dfrac{\dfrac{x+2}{x^2-3x}}{\dfrac{x^2-4}{x}}$

EXAMPLE B (A Quotient Arising in Calculus) Simplify

$$\frac{\dfrac{2}{x+h} - \dfrac{2}{x}}{h}$$

Solution. This expression may look artificial, but it is one you are apt to find in calculus. It represents the average rate of change in $2/x$ as x changes to $x + h$. We begin by simplifying the complicated numerator.

$$\frac{\dfrac{2}{x+h} - \dfrac{2}{x}}{h} = \frac{\dfrac{2x - 2(x+h)}{(x+h)x}}{h} = \frac{\dfrac{2x - 2x - 2h}{(x+h)x}}{\dfrac{h}{1}}$$

$$= \frac{-2h}{(x+h)x} \cdot \frac{1}{h} = \frac{-2}{(x+h)x}$$

Simplify each of the following.

105. $\dfrac{\dfrac{4}{x+h} - \dfrac{4}{x}}{h}$

106. $\dfrac{\dfrac{1}{2x+2h+3} - \dfrac{1}{2x+3}}{h}$

107. $\dfrac{\dfrac{x+h}{x+h+4} - \dfrac{x}{x+4}}{h}$

108. $\dfrac{\dfrac{1}{(x+h)^2} - \dfrac{1}{x^2}}{h}$

MISCELLANEOUS PROBLEMS

In Problems 109–116, perform the indicated operations and simplify.

109. $(2x - y)(x + 2y) - 2(x^2 - 2y^2)$

110. $(2c + d)^2 + (c - 2d)^2$

111. $2x(3x^2 - 6x + 4) - 3x[2x^2 - 4(x - 1)]$

112. $(y + 3)(2y - 5) - 2(3y - 2)(y + 2)$

113. $(2x + 3y)(4x^2 - 6xy + 9y^2) - 3(9y^3 - 2x)$

114. $(2x - y)^3 + 12x^2y - 6xy^2$

115. $(2x + 3c)^2(2x - 3c)^2 + 72x^2c^2$

116. $(2x + y)^4$

In Problems 117–128, factor completely over the integers.

117. $4 - 9m^2$

118. $9m^2 + 6m + 1$

119. $4x^2 - 8x$

120. $3x^3 + 9x^2 - 30x$

121. $6x^2 - 11x - 10$

122. $21z^2 - 2z - 8$

123. $3x^4 - 81x$

124. $(x^2 - 2x)^2 - 2(x^2 - 2x) - 3$

125. $4b^4 - 65b^2 + 16$

126. $(x^3 + 2)^2 + 5(x^3 + 2) - 6$

127. $(x - 3y)(x + 5y)^4 - 4(x - 3y)(x + 5y)^2$

128. $2x^3 + ax^2 - 2b^2x - ab^2$

In Problems 129–136, perform the indicated operations and simplify.

129. $\dfrac{1 - 2x}{x - 4} + \dfrac{6x + 2}{3x - 4}$

130. $x + y + \dfrac{y^2}{x - y}$

131. $\dfrac{x}{x^2 - 5x + 6} + \dfrac{3}{x^2 - 7x + 12}$

132. $\dfrac{x}{x^2 + 11x + 30} - \dfrac{5}{x^2 + 9x + 20}$

133. $\dfrac{\dfrac{a - b}{a + b} - \dfrac{a + b}{a - b}}{\dfrac{ab}{a - b}}$

134. $\dfrac{a^2 + 4a + 3}{a} - \dfrac{2a + 2}{a - 1}$

135. $\dfrac{4(x + h) - \dfrac{1}{x + h} - 4x + \dfrac{1}{x}}{h}$

136. $\left(x + \dfrac{1}{2x}\right)\left(x^4 - \dfrac{x^2}{2} + \dfrac{1}{4} - \dfrac{1}{8x^2} + \dfrac{1}{16x^4}\right)$

137. If $r > 0$ and $(r + r^{-1})^2 = 5$, find $r^3 + r^{-3}$.

138. If $a + b = 1$ and $a^2 + b^2 = 2$, find $a^3 + b^3$.

139. Calculate each of the following the easy way.

(a) $(547)^2 - (453)^2$ (b) $\dfrac{2^{20} - 2^{17} + 7}{2^{17} + 1}$

(c) $\left(1 - \dfrac{1}{2^2}\right)\left(1 - \dfrac{1}{3^2}\right)\left(1 - \dfrac{1}{4^2}\right) \cdots \left(1 - \dfrac{1}{29^2}\right)$

140. **TEASER**   If $a + b + c = 1$, $a^2 + b^2 + c^2 = 2$, and $a^3 + b^3 + c^3 = 3$ find $a^4 + b^4 + c^4$. *Hint:* This generalizes Problem 138.

Common Notions

1. Things equal to the same thing are equal to each other.
2. If equals be added to equals the wholes are equal.
3. If equals be subtracted from equals the remainders are equal.
4. Things that coincide with one another are equal to one another.

Euclid's *Elements*, composed about 300 B.C., has been reprinted in several hundred editions. It is, next to the Bible, the most influential book ever written. The common notions at the left are four of the ten basic axioms with which Euclid begins his book.

1-4   Equations

It was part of Euclid's genius to recognize that our usage of the word *equals* is fundamental to all that we do in mathematics. But to describe that usage may not be as simple as Euclid thought. When we write

$$.25 + \tfrac{3}{4} + \tfrac{1}{3} - (.3333 \ldots) = 1$$

we certainly do not mean that the symbol on the left coincides with the one on the right. Instead, we mean that both symbols, the complicated one and the simple one, stand for (or name) the same number. That is the basic meaning of *equals* as used in this book.

But having said that, we must make another distinction. When we write

$$x^2 - 25 = (x - 5)(x + 5)$$

and

$$x^2 - 25 = 0$$

> **Equality?**
>
> "We hold these truths to be self-evident, that all men are created equal ... "

we have two quite different things in mind. In the first case, we are making an assertion. We claim that no matter what number x represents, the expressions on the left and right of the equality stand for the same number. This certainly cannot be our meaning in the second case. There we are asking a question: What numbers can x symbolize so that both sides of the equality $x^2 - 25 = 0$ stand for the same number?

An equality that is true for all values of the variable is called an **identity**. One that is true only for some values is called a conditional **equation**. And here are the corresponding jobs for us to do. We **prove** identities, but we **solve** (or find the solutions of) equations. Both tasks are important in mathematics and will play a significant role in calculus. We discuss solving equations now; proving identities will be a major topic in Chapter 5.

How do we solve equations? Our general strategy is to modify an equation one step at a time until it is in a form in which the solution is obvious. Of course, we must be careful that the modifications we make do not change the solutions. Here, Euclid pointed the way.

RULES FOR MODIFYING EQUATIONS

1. Adding the same quantity to (or subtracting the same quantity from) both sides of an equation does not change its solutions.
2. Multiplying (or dividing) both sides of an equation by the same non-zero quantity does not change its solutions.

LINEAR EQUATIONS

The simplest equation to solve is one in which the *variable* (also called the *unknown*) occurs only to the first power. Consider

$$12x - 9 = 5x + 5$$

Our procedure is to use the rules for modifying equations to bring all the terms in x to one side and the constant terms to the other and then to divide by the coefficient of x. The result is that we have x all alone on one side of the equation and a number (the solution) on the other.

Given equation:	$12x - 9 = 5x + 5$
Add 9:	$12x = 5x + 14$
Subtract $5x$:	$7x = 14$
Divide by 7:	$x = 2$

It is always a good idea to check your answer. In the original equation, replace x by the value that you found to see if a true statement results.

$$12(2) - 9 \overset{?}{=} 5(2) + 5$$

$$15 = 15$$

An equation of the form $ax + b = 0$ $(a \neq 0)$ is called a **linear equation**. It has one solution, $x = -b/a$. Many equations not initially in this form can be transformed to it using the rules we have learned (see Example A).

QUADRATIC EQUATIONS

The next case to consider is the second degree, or **quadratic**, equation—that is, an equation of the form

$$ax^2 + bx + c = 0 \qquad (a \neq 0)$$

Here are four examples.

$$\text{(i)} \qquad x^2 = 7$$

$$\text{(ii)} \qquad x^2 - x - 6 = 0$$

$$\text{(iii)} \qquad 8x^2 - 2x - 1 = 0$$

$$\text{(iv)} \qquad x^2 = 6x - 2$$

While only (ii) and (iii) are initially in the required form, we accept the others because they can easily be put in that form. For example, (iv) may be written as $x^2 - 6x + 2 = 0$.

To solve (i), simply take the square root of both sides recognizing that 7 has two square roots. We obtain $x = \pm\sqrt{7}$.

Equations (ii) and (iii) can be solved by the *method of factoring*. We write them in the following form.

$$\text{(ii)} \qquad (x - 3)(x + 2) = 0$$

$$\text{(iii)} \qquad (4x + 1)(2x - 1) = 0$$

CAUTION

$x^2 - 5x = 0$
$x^2 = 5x$
$x = 5$

$x^2 - 5x = 0$
$x(x - 5) = 0$
$x = 0$ or $x = 5$

Then we recall that a product of two numbers is 0 if and only if one (or both) of the factors is 0. Hence, the solutions to (ii) and (iii) are obtained by setting each factor equal to 0 and solving the resulting linear equations. The solutions to (ii) are $x = 3$ and $x = -2$; the solutions to (iii) are $x = -\frac{1}{4}$ and $x = \frac{1}{2}$.

Equation (iv) presents new difficulties, since we do not know how to factor $x^2 - 6x + 2$. We use a method called *completing the square*. First, write the equation as

$$x^2 - 6x = -2$$

Now add 9 to both sides, making the left side a perfect square, and factor.

$$x^2 - 6x + 9 = -2 + 9$$

$$(x - 3)^2 = 7$$

This means that $x - 3$ must be one of the two square roots of 7. That is,

$$x - 3 = \pm\sqrt{7}$$

Hence

$$x = 3 + \sqrt{7} \quad \text{or} \quad x = 3 - \sqrt{7}$$

You may ask how we knew that we should add 9. Any expression of the form $x^2 + px$ becomes a perfect square when $(p/2)^2$ is added, since

$$x^2 + px + \left(\frac{p}{2}\right)^2 = \left(x + \frac{p}{2}\right)^2$$

For example, $x^2 + 10x$ becomes a perfect square when we add $(10/2)^2$, or 25.

$$x^2 + 10x + 25 = (x + 5)^2$$

The method of completing the square works on any quadratic equation. But there is a way of doing this process once and for all. Consider the general quadratic equation

$$ax^2 + bx + c = 0$$

with real coefficients $a \neq 0$, b, and c. First add $-c$ to both sides and then divide by a to obtain

$$x^2 + \frac{b}{a}x = -\frac{c}{a}$$

Next complete the square by adding $(b/2a)^2$ to both sides and then simplify.

$$x^2 + \frac{b}{a}x + \left(\frac{b}{2a}\right)^2 = -\frac{c}{a} + \left(\frac{b}{2a}\right)^2$$

$$\left(x + \frac{b}{2a}\right)^2 = -\frac{c}{a} + \frac{b^2}{4a^2}$$

$$\left(x + \frac{b}{2a}\right)^2 = \frac{b^2 - 4ac}{4a^2}$$

Finally take the square root of both sides.

$$x + \frac{b}{2a} = \pm \frac{\sqrt{b^2 - 4ac}}{2a}$$

or

$$x = \frac{-b}{2a} \pm \frac{\sqrt{b^2 - 4ac}}{2a}$$

The result is the **quadratic formula,** shown below.

$$x = \frac{-b \pm \sqrt{b^2 - 4ac}}{2a}$$

Let us see how the quadratic formula works on example (iv).

$$x^2 - 6x + 2 = 0$$

Here $a = 1$, $b = -6$, and $c = 2$. Thus

$$x = \frac{-(-6) \pm \sqrt{36 - 4 \cdot 2}}{2}$$

$$= \frac{6 \pm \sqrt{28}}{2}$$

$$= \frac{6 \pm \sqrt{4 \cdot 7}}{2}$$

$$= \frac{6 \pm 2\sqrt{7}}{2}$$

$$= \frac{2(3 \pm \sqrt{7})}{2}$$

$$= 3 \pm \sqrt{7}$$

You will note that in simplifying $\sqrt{4 \cdot 7}$, we used the fact that

$$\sqrt{ab} = \sqrt{a}\sqrt{b}$$

which is valid for all nonnegative numbers a and b.

The expression $b^2 - 4ac$ that appears under the square root sign in the quadratic formula is called the **discriminant**. It determines the character of the solutions.

1. If $b^2 - 4ac > 0$, there are two real solutions.
2. If $b^2 - 4ac = 0$, there is one real solution.
3. If $b^2 - 4ac < 0$, there are no real solutions.

SYSTEMS OF EQUATIONS

Sometimes a practical or a theoretical problem involves more than one unknown. To determine values for these unknowns will generally require that we have as many equations as unknowns. A typical case is a system of two linear equations in two unknowns, for example,

$$3x + y = 3$$
$$2x + 4y = -8$$

To solve—that is, to find values for x and y that satisfy both equations simultaneously—we need another rule.

RULE 3

You may add one equation to another (or subtract one equation from another) without changing the simultaneous solutions of a system of equations.

Then we use the method of *elimination of one unknown*.

Multiply the first equation by -4: $\qquad -12x - 4y = -12$

Write the second equation: $\qquad\qquad\quad 2x + 4y = -8$

Add the two equations: $\qquad\qquad\qquad -10x \qquad\quad = -20$

Multiply by $-\frac{1}{10}$: $\qquad\qquad\qquad\quad \boxed{x \qquad\quad = 2}$

Substitute $x = 2$ into one of the original
equations, say the first one: $\qquad\qquad 3(2) + y = 3$

Add -6: $\qquad\qquad\qquad\qquad\qquad\quad \boxed{y = -3}$

An alternative method is the method of *substitution*. Solve one of the
equations for one of the unknowns in terms of the other unknown; then substitute in the other equation. Here is how it works on the system just discussed.

Solve the first equation for y: $\qquad\qquad\qquad\qquad y = 3 - 3x$

Substitute in the second equation: $\quad 2x + 4(3 - 3x) = -8$

Solve the resulting equation: $\qquad\quad 2x + 12 - 12x = -8$

$$-10x = -20$$

$$\boxed{x = 2}$$

Substitute in the first equation: $\qquad\qquad 3(2) + y = 3$

$$\boxed{y = -3}$$

Problem Set 1-4

Determine which of the following are identities and which are conditional equations.

1. $3(2x - \frac{2}{3}) = 6x - 2$ $\qquad\qquad$ 2. $2x - 4 - \frac{2}{3}x = \frac{4}{3}x - 4$
3. $(x + 2)^2 = x^2 + 4$ $\qquad\qquad\qquad$ 4. $x^2 - 5x + 6 = (x - 1)(x - 6)$

Solve each of the following equations.

5. $3(x - 2) = 5(x - 3)$ $\qquad\qquad$ 6. $4(x + 1) = 2(x - 3)$
7. $\sqrt{3}z + 4 = -\sqrt{3}z + 8$ $\qquad\quad$ 8. $\sqrt{2}x + 1 = x + \sqrt{2}$

© 9. $3.23x - 6.15 = 1.41x + 7.63$

$\qquad$ (First obtain $x = \dfrac{7.63 + 6.15}{3.23 - 1.41}$ and then use the calculator.)

© 10. $42.1x + 11.9 = 1.03x - 4.32$

Solve by first clearing fractions. For example, in Problem 11 multiply both sides by 40.

11. $\frac{9}{10}x + \frac{5}{8} = \frac{1}{5}x + \frac{9}{20}$ 12. $\frac{1}{3}x + \frac{1}{4} = \frac{1}{5}x + \frac{1}{6}$

13. $\frac{3}{4}(x - 2) = \frac{9}{5}$ 14. $\frac{x}{8} = \frac{2}{3}(2 - x)$

EXAMPLE A (Equations That Can Be Changed to Linear Form)   Solve the following equations.

$$\text{(a)} \quad \frac{2}{x + 1} = \frac{3}{2x - 2} \qquad \text{(b)} \quad \frac{3x}{x - 3} = 1 + \frac{9}{x - 3}$$

Solution.   Both equations become linear when cleared of fractions.

(a) Multiply by $(x + 1)(2x - 2)$.

$$2(2x - 2) = 3(x + 1)$$
$$4x - 4 = 3x + 3$$
$$x = 7$$

Check in original equation.

$$\frac{2}{7 + 1} \overset{?}{=} \frac{3}{14 - 2}$$

$$\frac{2}{8} = \frac{3}{12}$$

It checks; 7 is the solution.

(b) Multiply by $x - 3$.

$$3x = x - 3 + 9$$
$$2x = 6$$
$$x = 3$$

Check in original equation:

$$\frac{3 \cdot 3}{3 - 3} \overset{?}{=} 1 + \frac{9}{3 - 3}$$

Nonsense—there is no solution. What went wrong? It happened in the first step; if x = 3, then (without realizing it), we multiplied both sides by zero, an *illegal step*.

In each case, we multiplied both sides of the equation by an expression involving the unknown. This was good strategy; in fact, we had to do it. But it is absolutely essential in such cases that we check the apparent solution as it may be extraneous (that is, not a solution at all).

Solve each of the following equations by first clearing fractions. Be sure to check that your apparent solution is actually a solution.

15. $\frac{5}{x + 2} = \frac{2}{x - 1}$ 16. $\frac{10}{2x - 1} = \frac{14}{x + 4}$

17. $\frac{x}{x - 2} = 2 + \frac{2}{x - 2}$ 18. $\frac{x}{2x - 4} - \frac{2}{3} = \frac{7 - 2x}{3x - 6}$

EXAMPLE B (Solving for one variable in terms of others)   Solve $I = nE/(R + nr)$ for n.

Solution.   This is a typical problem in science in which an equality relates several variables and we want to solve for one of them in terms of the others. To do this, we proceed as if the other variables were simply numbers, which, after all, is what every variable represents.

$$I = \frac{nE}{R + nr} \qquad \text{(original equality)}$$

$$(R + nr)I = nE \qquad \text{(multiply by } R + nr)$$

$$RI + nrI = nE \qquad \text{(distributive property)}$$

$$nrI - nE = -RI \qquad \text{(subtract } nE \text{ and } RI)$$

$$n(rI - E) = -RI \qquad \text{(factor)}$$

$$n = \frac{-RI}{rI - E} \qquad \text{(divide by } rI - E)$$

Solve for the indicated variable in terms of the remaining variables.

19. $A = P + Prt$ for P

20. $R = \dfrac{E}{L - 5}$ for L

21. $I = \dfrac{nE}{R + nr}$ for r

22. $mv = Ft + mv_0$ for m

23. $A = 2\pi r^2 + 2\pi rh$ for h

24. $F = \frac{9}{5}C + 32$ for C

25. $R = \dfrac{R_1 R_2}{R_1 + R_2}$ for R_1

26. $\dfrac{1}{R} = \dfrac{1}{R_1} + \dfrac{1}{R_2} + \dfrac{1}{R_3}$ for R_2

Solve the following quadratic equations by the method of factoring.

27. $x^2 - x - 2 = 0$

28. $x^2 + 13x + 22 = 0$

29. $9y^2 - 25 = 0$

30. $3z^2 - 12 = 0$

31. $3x^2 + 5x - 2 = 0$

32. $3x^2 + 7x - 20 = 0$

Solve by taking square roots of both sides.

33. $(x - 3)^2 = 16$

34. $(x + 4)^2 = 49$

35. $(2x + 5)^2 = 100$

36. $(3y - \frac{1}{3})^2 = 25$

Solve by completing the square.

37. $x^2 + 8x = 9$

38. $x^2 - 12x = 45$

39. $z^2 - z = \frac{3}{4}$

40. $x^2 + 5x = 2\frac{3}{4}$

Solve by using the quadratic formula.

41. $x^2 + 8x + 12 = 0$

42. $x^2 - 2x - 15 = 0$

43. $x^2 + 5x + 3 = 0$

44. $z^2 - 3z - 8 = 0$

45. $3x^2 - 6x - 11 = 0$

46. $4t^2 - t - 3 = 0$

47. $y^2 - 2y - 5 = 0$

48. $2u^2 - 2u - 5 = 0$

49. $2x^2 - \pi x - 1 = 0$

50. $3x^2 - \sqrt{2}x - 3\pi = 0$

In each of the following, find the values for the two unknowns that satisfy both equations. Use whichever method you prefer.

51. $2x + 3y = 13$
    $y = 13$

52. $2x - 3y = 7$
    $x = -4$

53. $y = -2x + 11$
    $y = 3x - 9$

54. $x = 5y$
    $x = -3y - 24$

55. $x - y = 14$
    $x + y = -2$

56. $2x - 3y = 8$
    $4x + 3y = 16$

57. $2s - 3t = -10$
    $5s + 6t = 29$

58. $2w - 3z = -23$
    $8w + 2z = -22$

59. $5x - 4y = 19$
    $7x + 3y = 18$

60. $4x - 2y = 16$
    $6x + 5y = 24$

Hint: In problem 59, multiply top equation by 3 and the bottom one by 4; then add.

61. $\frac{2}{3}x + y = 4$
    $x + 2y = 5$

62. $\frac{3}{4}x - \frac{1}{2}y = 12$
    $x + y = -8$

63. $\dfrac{4}{x} + \dfrac{3}{y} = 17$

    $\dfrac{1}{x} - \dfrac{3}{y} = -7$

64. $\dfrac{4}{x} + \dfrac{2}{y} = 12$

    $\dfrac{5}{x} + \dfrac{1}{y} = 8$

Hint: Let $u = 1/x$ and $v = 1/y$. Solve for u and v and then find x and y.

EXAMPLE C (Distance-Rate Problems)   An airplane flew with the wind for 1 hour and returned the same distance against the wind in $1\frac{1}{4}$ hours. If the speed of the plane in still air is 300 miles per hour, find the speed of the wind.

Solution.   We are sure to need the formula D = RT, which stands for *distance equals rate multiplied by time*. But first we assign a letter to the unknown, the speed of the wind, and we do it very precisely. *Let x be the speed of the wind in miles per hour.* Then

$$300 + x = \text{speed of the plane with the wind}$$

$$300 - x = \text{speed of the plane against the wind}$$

Now we apply the formula $D = RT$ to the two trips.

$$(300 + x) \cdot 1 = \text{distance flown with the wind}$$

$$(300 - x) \cdot \tfrac{5}{4} = \text{distance flown against the wind}$$

These two distances are the same. We equate them and solve for x.

$$300 + x = (300 - x)\tfrac{5}{4}$$

$$1200 + 4x = 1500 - 5x$$

$$9x = 300$$

$$x = 33\tfrac{1}{3}$$

The speed of the wind is $33\frac{1}{3}$ miles per hour.

65. At noon, Slowpoke left Kansas City traveling due east at 50 miles per hour. Two hours later, Speedy started after him, going 70 miles per hour. When will Speedy catch up with Slowpoke?

66. Two long-distance runners start out from the same spot on an oval track, which is $\frac{1}{2}$ mile around. If one runs at 8 miles per hour and the other at $6\frac{1}{2}$ miles per hour, when will the faster runner be one lap ahead of the slower runner?

67. The city of Harmony is 430 miles from the city of Dissension. At 12:00 noon, Paul Haymaker leaves Harmony traveling at 60 miles per hour toward Dissension. One hour later, Nick Ploughman starts from Dissension heading toward Harmony, managing only 45 miles per hour. When will they meet? How far is the meeting point from Harmony?

68. Luella can row 5 miles upstream in the same amount of time that it takes her to row 9 miles downstream. If the rate of the current is $2\frac{1}{2}$ miles per hour, how fast can she row in still water?

MISCELLANEOUS PROBLEMS

Solve for x in Problems 69–82.

69. $\frac{3}{4}x - \frac{4}{3} = \frac{1}{3}x + \frac{5}{6}$

70. $4(3x - \frac{1}{2}) = 5x + \frac{1}{2}$

71. $(x - 2)(x + 5) = (x - 3)(x + 1)$

72. $\dfrac{2}{x - 1} = \dfrac{9}{2x - 3}$

73. $\dfrac{2x}{4x + 2} = \dfrac{x + 1}{2x - 1}$

74. $\dfrac{2}{(x + 2)(x - 1)} = \dfrac{1}{x + 2} + \dfrac{3}{x - 1}$

75. $(2x - 1)^2 = \frac{9}{4}$

76. $3x^2 = 1 + 2x$

77. $2x^2 = 4x - 2$

78. $2x^2 + 3x = x^2 + 2x + 12$

79. $x^2 + 2x - 4 = 0$

80. $9x^2 - 3x - 1 = 0$

81. $x^4 - 7x^2 + 6 = 0$

82. $x - 4\sqrt{x} + 3 = 0$

83. Solve for x in terms of a.

$$1 + \cfrac{1}{1 + \cfrac{1}{a + \cfrac{1}{x}}} = \frac{1}{a}$$

84. Solve for x in terms of a and b.

$$\frac{x - a}{x + b} = \frac{a - ab}{a + b^2}$$

85. Solve for x and y if $2x - 3y = -8$ and $3x + 4y = 5$.

86. Solve for r and s if $\frac{1}{3}r + \frac{2}{9}s = 24$ and $\frac{2}{9}r + \frac{1}{3}s = 26$.

87. The sidewalk around a square garden is 3 feet wide and has area 249 square feet. What is the width of the garden?

88. What number must be added to both numerator and denominator of $\frac{3}{37}$ to bring this fraction up to $\frac{7}{24}$?

89. In a certain math class, the average weight for the girls was 128 pounds, for the boys 160 pounds, and for the class as a whole 146 pounds. How many people are in the class, given that there are 14 girls?

90. Sam Slugger got only 50 hits in his first 200 times at bat of the baseball season. During the rest of the season, he claims he will hit .300 and finish with a season average of .280. How many times at bat does he expect to have during the whole season?

91. When asked to make a contribution to the building fund, Janet Goodenrich pledged to give $1000 more than the average of all givers. The other 88 givers contributed $44,000. How much did Janet give?

92. To harvest a rectangular field of wheat, which is 720 meters by 960 meters, a farmer cut swaths around the outside, thus forming a steadily growing border of cut wheat and leaving a steadily shrinking rectangle of uncut wheat in the middle. How wide was the border when the farmer was half through?

93. At noon, Tom left point A walking due north; an hour later, Mary left point A walking due east. Both walked at 4 miles per hour and both carried walkie-talkies with a range of 8 miles. At what time did they lose contact with each other?

94. Mary Larson rows 14 miles downstream on a river and then rows back. The rate of the current is 2.5 miles per hour. If it takes her 9 hours in all, how fast can she row in still water?

95. (The golden ratio). As Euclid suggested, divide a line segment into two unequal parts so that the ratio of the whole to the longer part is equal to the ratio of the longer part to the shorter. Find the ratio of the longer to the shorter.

96. **TEASER** When Karen inherited a small amount of money, she decided to divide it among her four children. To Alice, she gave $2 plus one-third of the remainder; then to Brent she gave $2 plus one-third of the remainder. Next to Curtis, she gave $2 plus one-third of the remainder, and finally to Debra, she gave $2 plus one-third of the remainder. What was left, she divided equally among the four children. If the girls together got $35 more than the boys together, what was the size of Karen's inheritance and how much did each child get?

	Equation	Inequality

Each problem that I solved became a rule which served afterwards to solve other problems.
　　　　Descartes

Equation

$$-3x + 7 = 2$$
$$-3x = -5$$
$$x = \frac{5}{3}$$

Inequality

$$-3x + 7 < 2$$
$$-3x < -5$$
$$x > \frac{5}{3}$$

1-5 Inequalities

Solving an inequality is very much like solving an equation, as the example above demonstrates. However, there are dangers in proceeding too mechanically. It will be important to think at every step.

Recall the distinction we made between identities and equations in Section 1-4. A similar distinction applies to inequalities. An inequality which is true for all values of the variables is called an **unconditional inequality**. Examples are

$$(x - 3)^2 + 1 > 0$$

and

$$|x| \le |x| + |y|$$

Most inequalities (for example, $-3x + 7 < 2$) are true only for some values of the variables; we call them **conditional inequalities**. Our primary task in this section is to solve conditional inequalities, that is, to find all those numbers which make a conditional inequality true. Developing this skill is more important than you can appreciate now. In calculus, you will regularly be asked to solve inequalities, and there it will be assumed that you know how.

LINEAR INEQUALITIES

To solve the linear inequality $Ax + B < C$, we try to rewrite it in successive steps until the variable x stands by itself on one side of the inequality (see opening display). This depends primarily on the properties of inequalities stated next. These properties are also valid with $<$ replaced by $\le$ and $>$ replaced by $\ge$.

PROPERTIES OF INEQUALITIES

1. (Transitivity).   If $a < b$ and $b < c$, then $a < c$.
2. (Addition).   If $a < b$, then $a + c < b + c$.
3. (Multiplication).   If $a < b$ and $c > 0$, then $a \cdot c < b \cdot c$.
   If $a < b$ and $c < 0$, then $a \cdot c > b \cdot c$.

We illustrate the use of these properties, applied to $\le$ rather than $<$, by solving the following inequality.

$$-2x + 6 \le 18 + 4x$$

Add $-4x$:	$-6x + 6 \le 18$
Add -6:	$-6x \le 12$
Multiply by $-\frac{1}{6}$:	$x \ge -2$

By rights, we should check this solution. All we know so far is that any value of x that satisfies the original inequality satisfies $x \ge -2$. Can we go in the opposite direction? Yes, because every step is reversible. For example, starting with $x \ge -2$, we can multiply by -6 to get $-6x \le 12$. In practice, we do not actually carry out this check as we recognize that Property 2 can be restated:

$$a < b \text{ is equivalent to } a + c < b + c$$

There are similar restatements of Property 3.

$x \geq -2$

Figure 12

The set of solutions to an inequality is, in general, an interval (or perhaps a union of intervals). The diagram in Figure 12 shows how we display this set for the example just completed. The heavy dot at -2 indicates that -2 is part of the solution set. A small circle, as in our opening panel, indicates that the corresponding point is not part of the solution set.

Here is a practical question requiring the solution of an inequality. When is the reading on a Celsius thermometer higher than the corresponding reading on a Fahrenheit thermometer? To answer, recall that the Celsius temperature C is related to the Fahrenheit temperature F by the formula $F = \frac{9}{5}C + 32$. Thus we need to solve the inequality $C > \frac{9}{5}C + 32$.

Multiply by 5:	$5C > 9C + 160$
Add $-9C$:	$-4C > 160$
Multiply by $-\frac{1}{4}$:	$C < -40$

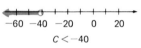

$C < -40$

Figure 13

We conclude that the Celsius temperature is higher than the Fahrenheit temperature whenever the Celsius temperature is less than $-40°$. The corresponding solution set is shown in Figure 13.

QUADRATIC INEQUALITIES

It is not difficult to solve a quadratic inequality

$$ax^2 + bx + c > 0$$

if we can factor the left side. To solve

$$x^2 - 2x - 3 > 0$$

we first factor, obtaining

$$(x + 1)(x - 3) > 0$$

Next we ask ourselves when the product of two numbers is positive. There are two cases; either both factors are negative or both factors are positive.

Case 1 (Both negative) We want to know when both factors are negative; that is, we seek to solve $x + 1 < 0$ and $x - 3 < 0$ simultaneously. The first gives $x < -1$ and the second gives $x < 3$. Together they give $x < -1$.

Case 2 (Both positive) Both factors are positive when $x + 1 > 0$ and $x - 3 > 0$—that is, when $x > -1$ and $x > 3$. These give $x > 3$.

The solution set for the original inequality is the union of the solution sets for the two cases. In set notation, it may be written either as

$$\{x : x < -1 \quad \text{or} \quad x > 3\}$$

or as

$$\{x : x < -1\} \cup \{x : x > 3\}$$

The chart in Figure 14 summarizes what we have learned.

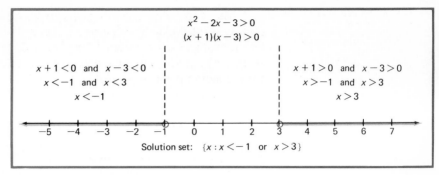

$$x^2 - 2x - 3 > 0$$
$$(x + 1)(x - 3) > 0$$

$x + 1 < 0$ and $x - 3 < 0$ $x + 1 > 0$ and $x - 3 > 0$

$x < -1$ and $x < 3$ $x > -1$ and $x > 3$

$x < -1$ $x > 3$

Solution set: $\{x : x < -1 \text{ or } x > 3\}$

Figure 14

SPLIT-POINT METHOD

The preceding example could have been approached in a slightly different way using the notion of split-points. The solutions of the equation $(x + 1)(x - 3) = 0$, which are -1 and 3, serve as split-points that divide the real line into the three intervals: $x < -1$, $-1 < x < 3$, and $3 < x$. Since $(x + 1)(x - 3)$ can change sign only at a split-point, it must be of one sign (that is, be either always positive or always negative) on each of these intervals. To determine which of them make up the solution set of the inequality $(x + 1)(x - 3) > 0$, all we need to do is pick a single (arbitrary) point from each interval and test it for inclusion in the solution set. If it passes the test, the entire interval from which it was drawn belongs to the solution set.

To show how this method works, let us consider the third degree inequality

$$(x + 2)(x - 1)(x - 4) < 0$$

The solutions of the corresponding equation

$$(x + 2)(x - 1)(x - 4) = 0$$

are -2, 1, and 4. They break the real line into the four intervals $x < -2$, $-2 < x < 1$, $1 < x < 4$, and $4 < x$. Suppose we pick -3 as the test point for the interval $x < -2$. We see that -3 makes each of the three factors $x + 2$, $x - 1$, and $x - 4$ negative, and so it makes their product $(x + 2)(x - 1) \times (x - 4)$ negative. You should pick test points from each of the other three intervals to verify the results shown in Figure 15 on the next page. The split-points are checked separately.

INEQUALITIES WITH ABSOLUTE VALUES

One basic fact to remember is that

$$|x| < a$$

means the same thing as the double inequality

$$-a < x < a$$

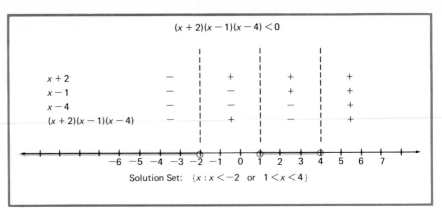

$$(x + 2)(x - 1)(x - 4) < 0$$

$x + 2$	$-$	$+$	$+$	$+$
$x - 1$	$-$	$-$	$+$	$+$
$x - 4$	$-$	$-$	$-$	$+$
$(x + 2)(x - 1)(x - 4)$	$-$	$+$	$-$	$+$

Solution Set: $\{x : x < -2 \text{ or } 1 < x < 4\}$

Figure 15

A second fact is that

$$|x| > a, \qquad a > 0$$

means

$$x < -a \quad \text{or} \quad x > a$$

To solve the inequality $|3x - 2| < 4$, proceed as follows.

Given inequality:	$|3x - 2| < 4$
Remove absolute value:	$-4 < 3x - 2 < 4$
Add 2:	$-2 < 3x \quad < 6$
Multiply by $\frac{1}{3}$:	$-\frac{2}{3} < x \quad < 2$

Here, as in the preceding examples, we could use the split-point method. The solutions of the equation $|3x - 2| = 4$, which are $-\frac{2}{3}$ and 2, would be the split-points. The solution would again be $-\frac{2}{3} < x < 2$.

AN APPLICATION

A student wishes to get a grade of B in Mathematics 16. On the first four tests, he got 82 percent, 63 percent, 78 percent, and 90 percent, respectively. A grade of B requires an average between 80 percent and 90 percent, inclusive. What grade on the fifth test would qualify this student for a B?

Let x represent the grade (in percent) on the fifth test. The inequality to be satisfied is

$$80 \leq \frac{82 + 63 + 78 + 90 + x}{5} \leq 90$$

This can be rewritten successively as

$$80 \leq \frac{313 + x}{5} \leq 90$$

$$400 \leq 313 + x \leq 450$$

$$87 \leq \qquad x \leq 137$$

A score greater than 100 is impossible, so the actual solution to this problem is $87 \le x \le 100$.

Problem Set 1-5

Which of the following inequalities are unconditional and which are conditional?

1. $x \ge 0$
2. $x^2 \ge 0$
3. $x^2 + 1 > 0$
4. $x^2 > 1$
5. $x - 2 < -5$
6. $2x + 3 > -1$
7. $x(x + 4) \le 0$
8. $(x - 1)(x + 2) > 0$
9. $(x + 1)^2 > x^2$
10. $(x - 2)^2 \le x^2$
11. $(x + 1)^2 > x^2 + 2x$
12. $(x - 2)^2 > x(x - 4)$

Solve each of the following inequalities and show the solution set on the real number line.

13. $3x + 7 < x - 5$
14. $-2x + 11 > x - 4$
15. $\frac{2}{3}x + 1 > \frac{1}{2}x - 3$
16. $3x - \frac{1}{2} < \frac{1}{2}x + 4$

Hint: First get rid of the fractions.

17. $\frac{3}{4}x - \frac{1}{2} < \frac{1}{6}x + 2$
18. $\frac{2}{7}x + \frac{1}{3} \le -\frac{2}{3}x + \frac{15}{14}$
19. $(x - 2)(x + 5) \le 0$
20. $(x + 1)(x + 4) \ge 0$
21. $(2x - 1)(x + 3) > 0$
22. $(3x + 2)(x - 2) < 0$
23. $x^2 - 5x + 4 \ge 0$
24. $x^2 + 4x + 3 \le 0$

Hint: Factor the left side.

25. $2x^2 - 7x + 3 < 0$
26. $3x^2 - 5x - 2 > 0$
27. $|2x + 3| < 2$
28. $|2x - 4| \le 3$
29. $|-2x - 1| \le 1$
30. $|-2x + 3| > 2$
31. $(x + 4)x(x - 3) \ge 0$
32. $(x + 3)x(x - 3) \ge 0$
33. $(x - 2)^2(x - 5) < 0$
34. $(x + 1)^2(x - 1) > 0$

EXAMPLE A (Inequalities Involving Quotients) Solve the following inequality.

$$\frac{3}{x - 2} > \frac{2}{x}$$

Solution. We rewrite the inequality as follows:

$$\frac{3}{x - 2} - \frac{2}{x} > 0 \qquad \text{(add } -2/x \text{ to both sides)}$$

$$\frac{3x - 2(x - 2)}{(x - 2)x} > 0 \qquad \text{(combine fractions)}$$

$$\frac{x + 4}{(x - 2)x} > 0 \qquad \text{(simplify numerator)}$$

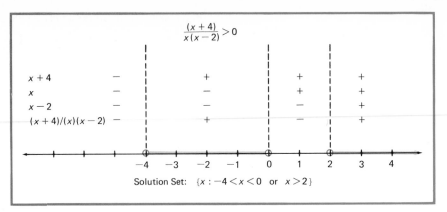

Figure 16

The factors $x + 4$, x, and $x - 2$ in the numerator and denominator determine the three split-points -4, 0, and 2. The chart in Figure 16 shows the signs of $x + 4$, x, $x - 2$, and of $(x + 4)/(x - 2)x$ on each of the intervals determined by the split-points, as well as the solution set.

Solve each of the following inequalities.

35. $\dfrac{x - 5}{x + 2} \leq 0$

36. $\dfrac{x + 3}{x - 2} > 0$

37. $\dfrac{x(x + 2)}{x - 5} > 0$

38. $\dfrac{x - 1}{(x - 3)(x + 3)} \geq 0$

39. $\dfrac{5}{x - 3} > \dfrac{4}{x - 2}$

40. $\dfrac{-3}{x + 1} < \dfrac{2}{x - 4}$

EXAMPLE B (Rewriting with Absolute Values)   Write $-2 < x < 8$ as an inequality involving absolute values.

Solution.   Look at this interval on the number line (Figure 17). It is 10 units long and its midpoint is at 3. A number x is in this interval provided that it is within a radius of 5 of this midpoint, that is, if

$$|x - 3| < 5$$

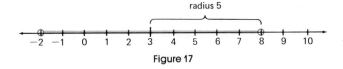

Figure 17

We can check that $|x - 3| < 5$ is equivalent to the original inequality by writing it as

$$-5 < x - 3 < 5$$

and then adding 3 to each member.

Write each of the following as an inequality involving absolute values.

41. $0 < x < 6$ 42. $0 < x < 12$ 43. $-1 \leq x \leq 7$

44. $-3 \leq x \leq 7$ 45. $2 < x < 11$ 46. $-10 < x < -3$

EXAMPLE C (Quadratic Inequalities That Cannot Be Factored by Inspection) Solve the inequality

$$x^2 - 5x + 3 \geq 0$$

Solution.   Even though $x^2 - 5x + 3$ cannot be factored by inspection, we can solve the quadratic equation $x^2 - 5x + 3 = 0$ by use of the quadratic formula. We obtain the two solutions $(5 - \sqrt{13})/2 \approx .7$ and $(5 + \sqrt{13})/2 \approx 4.3$. These two numbers split the real line into three intervals from which the numbers 0, 1, and 5 could be picked as test points. Notice that $x = 0$ makes $x^2 - 5x + 3$ positive, $x = 1$ makes it negative, and $x = 5$ makes it positive. This gives us the solution set

$$\{x : x \leq (5 - \sqrt{13})/2 \text{ or } x \geq (5 + \sqrt{13})/2\}$$

which can be pictured as shown in Figure 18. The split-points are included in the solution set because our inequality $x^2 - 5x + 3 \geq 0$ includes equality.

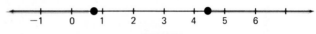

Figure 18

Solve each of the following inequalities and display the solution set on the number line.

47. $x^2 - 7 < 0$ 48. $x^2 - 12 > 0$

49. $x^2 - 4x + 2 \geq 0$ 50. $x^2 - 4x - 2 \leq 0$

ⓒ 51. $x^2 + 6.32x + 3.49 > 0$ ⓒ 52. $x^2 + 4.23x - 2.79 < 0$

EXAMPLE D (Finding Least Values)   Find the least value that $x^2 - 4x + 9$ can take on.

Solution.   We use the method of completing squares to write $x^2 - 4x + 9$ as the sum of a perfect square and a constant.

$$x^2 - 4x + 9 = (x^2 - 4x + 4) + 5 = (x - 2)^2 + 5$$

Since the smallest value $(x - 2)^2$ can take on is zero, the smallest value $(x - 2)^2 + 5$ can assume is 5.

Find the least value each of the following expressions can take on.

53. $x^2 + 8x + 20$ 54. $x^2 + 10x + 40$

55. $x^2 - 2x + 101$ 56. $x^2 - 4x + 104$

MISCELLANEOUS PROBLEMS

Solve the inequalities in Problems 57–72.

57. $\frac{1}{2}x + \frac{3}{4} > \frac{2}{3}x - \frac{4}{3}$ 58. $3(x - \frac{2}{5}) \leq 5x - \frac{1}{3}$

59. $2x^2 + 5x - 3 < 0$

60. $x^2 + x - 1 \le 0$

61. $(x + 1)^2(x - 1)(x - 4)(x - 8) < 0$

62. $(x - 3)(x^3 - x^2 - 6x) \ge 0$

63. $\dfrac{1}{x - 2} + 1 < \dfrac{2}{x + 2}$

64. $\dfrac{2}{x - 2} < \dfrac{3}{x - 3}$

65. $|4x - 3| \ge 2$

66. $|2x - 1| > 3$

67. $|3 - 4x| < 7$

68. $|x| < x + 3$

69. $|x - 2| < |x + 3|$

70. $||x + 2| - x| < 5$

71. $|x^2 - 2x - 4| > 4$

72. $|x^2 - 14x + 44| < 4$

73. For what values of k will the following equations have real solutions?
   (a) $x^2 + 4x + k = 0$
   (b) $x^2 - kx + 9 = 0$
   (c) $x^2 + kx + k = 0$
   (d) $x^2 + kx + k^2 = 0$

74. Amy scored 73, 82, 69, and 94 on four 100-point tests in Mathematics 16. Suppose that a grade of B requires an average between 75 percent and 85 percent.
   (a) What score on a 100-point final exam would qualify Amy for a B in the course?
   (b) What score on a 200-point final exam would qualify Amy for a B in the course?

75. The mathematics department at Podunk University has a staff of 6 people with an average salary of $32,000 per year. An additional full professor is to be hired. A salary S in what range can be offered if the department average must be between $31,000 and $35,000?

76. Company A will loan out a car for $35 per day plus 10 cents per mile, whereas company B charges $30 per day plus 12 cents per mile. I need a car for 5 days. For what range of mileage M will I be ahead to rent a car from Company B?

77. A ball thrown upward with a velocity of 64 feet per second from the top edge of a building 80 feet high will be at height $(-16t^2 + 64t + 80)$ feet above the ground after t seconds.
   (a) What is the greatest height attained by the ball (see Example D)?
   (b) During what time period is the ball higher than 96 feet?
   (c) At what time did the ball hit the ground, assuming it missed the building on the way down?

78. For what numbers is it true that the sum of the number and its reciprocal is greater than 2?

79. Let a, b, and c denote the lengths of the two legs and hypotenuse of a right triangle so that $a^2 + b^2 = c^2$. Show that if n is an integer greater than 2, then $a^n + b^n < c^n$.

80. **TEASER** Sophus Slybones, the proprieter of a specialty coffee shop, weighs coffee for customers by using the two-pan balance shown in Figure 19. Long ago, he dropped the balance on the concrete floor and now the left arm is slightly longer than the right one, that is, $a > b$. This poses no problem for Sophus. When a customer orders two kilograms of coffee, he first places a one-kilogram weight in the left pan balancing it with some coffee in the right pan, which he then pours in a sack. Next, he places the one-kilogram weight in the right pan balancing it with coffee in the left pan, which he also pours in the sack. That makes exactly 2 kilograms says Sophus, a claim you must analyze. Assume the balance arms are weightless but that the pans weigh 100 grams each.
   (a) Show that Sophus has been cheating himself all these years.

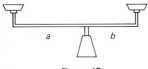

Figure 19

(b) If the amount Sophus gives a customer is actually 2.01 kilograms, determine a/b.

If you want to make the problem a little harder, show that the conclusion in (a) is valid even if you don't assume the balance arms are weightless.

Algebra   Geometry

"As long as algebra and geometry traveled separate paths, their advance was slow and their applications limited. But when the two sciences joined company, they drew from each other fresh vitality and thenceforward marched on at a rapid pace toward perfection."

Joseph-Louis Lagrange

1-6   Algebra and Geometry United

The Greeks were preeminent geometers but poor algebraists. Though they were able to solve a host of geometry problems, their limited algebraic skills kept others beyond their grasp. By 1600, geometry was a mature and eligible bachelor. Algebra was a young woman only recently come of age. Two Frenchmen, Pierre de Fermat (1601–1665) and René Descartes (1596–1650), were the matchmakers; they brought the two together. The resulting union is called **analytic geometry,** or coordinate geometry.

The brilliant idea which made this possible is that of a coordinate system. Fermat, a lawyer who made mathematics his hobby, used coordinates to describe points and curves as early as 1629. Descartes, the famed philosopher and scientist, published *La Géométrie* in 1637. By virtue of having the idea first and more explicitly, Fermat ought to get the major credit. History can be a fickle friend; coordinates are known as Cartesian coordinates, named after René Descartes.

No matter who gets the credit, it was an idea whose time had come. It made possible the invention of calculus, one of the greatest inventions of the human mind. That invention was to come in 1665 at the hands of a 23-year-old genius named Isaac Newton. You will probably study calculus later on. There you will use the ideas of this chapter over and over.

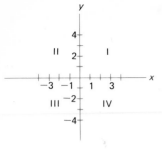

Figure 20

CARTESIAN COORDINATES

In the plane, draw two number lines, one horizontal and the other vertical, intersecting at their zero points. The two lines are called **coordinate axes;** their intersection is called the **origin** and labeled O. By convention, the horizontal line is called the **x-axis** and the vertical line is called the **y-axis.** The positive half of the x-axis is to the right; the positive half of the y-axis is upward. The coordinate axes divide the plane into four regions called **quadrants,** labeled I, II, III, and IV, as shown in Figure 20.

Each point in the plane can now be assigned a pair of numbers called its **Cartesian coordinates.** The point A in Figure 21, which is 4 units to the right of the y-axis and 1 unit above the x-axis, has coordinates $(4, 1)$. The point B, 5 units to the left of the y-axis and 2 units below the x-axis, has coordinates $(-5, -2)$. In general, the coordinates (a, b) of a point P are its directed distances from the y-axis and x-axis, respectively. We call (a, b) an *ordered* pair because it makes a difference which number is first. The first number a is the **x-coordinate** (or abscissa); the second number b is the **y-coordinate** (or ordinate).

Why was the introduction of coordinates so important? You will see.

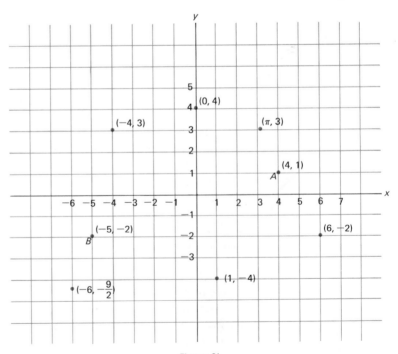

Figure 21

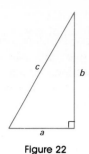

Figure 22

THE DISTANCE FORMULA

The first important consequence is a simple formula for the distance between any two points. It is based on the **Pythagorean theorem,** which says that if a and b measure the two legs of a right triangle and c measures its hypotenuse (Figure 22), then

$$a^2 + b^2 = c^2$$

This relationship between the three sides of a triangle holds only for a right triangle.

Now consider two points P and Q, with coordinates (x_1, y_1) and (x_2, y_2), respectively. Together with $R(x_2, y_1)$, they are vertices of a right triangle (see Figure 23). The lengths of PR and RQ are $|x_2 - x_1|$ and $|y_2 - y_1|$, respectively. When we apply the Pythagorean theorem and take the square root of both sides, we obtain the following expression for $d(P, Q)$, the distance between P and Q.

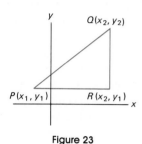

Figure 23

$$d(P, Q) = \sqrt{(x_2 - x_1)^2 + (y_2 - y_1)^2}$$

This is called the **distance formula.**

As an example, the distance between $(-2, 3)$ and $(4, -1)$ is

$$\sqrt{(4 - (-2))^2 + (-1 - 3)^2} = \sqrt{36 + 16} = \sqrt{52} \approx 7.21$$

The formula is correct even if the two points lie on the same horizontal or vertical line. For example, the distance between $(-2, 2)$ and $(6, 2)$ is

$$\sqrt{(-2 - 6)^2 + (2 - 2)^2} = \sqrt{64} = 8$$

THE EQUATION OF A CIRCLE

It is a small step from the distance formula to the equation of a circle. A **circle** is the set of points which lie at a fixed distance r (called the radius) from a fixed point (called the center). Consider, for example, the circle of radius 3 with center at $(0, 0)$ shown in Figure 24. Let (x, y) denote any point on this circle (the key step). By the distance formula,

$$\sqrt{(x - 0)^2 + (y - 0)^2} = 3$$

When we square both sides, we obtain

$$x^2 + y^2 = 9$$

which we call the *equation* of this circle.

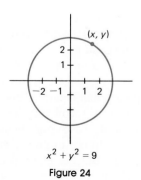

$x^2 + y^2 = 9$

Figure 24

What is significant here is that we have transformed a geometric object (a circle) into an algebraic object (an equation). That and the reverse process (graphing an equation) constitute the giant idea that John Stuart Mill called "the greatest single step ever made in the progress of the exact sciences."

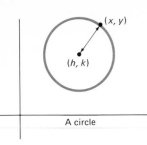

A circle

Figure 25

We can handle the general circle of radius r centered at (h, k) just as easily (see Figure 25). Its equation, called the **standard equation of the circle,** is

$$(x - h)^2 + (y - k)^2 = r^2$$

To illustrate, let us find the equation of a circle of radius 6 centered at $(2, -1)$. We use the boxed equation with $h = 2$, $k = -1$, and $r = 6$. This gives

$$(x - 2)^2 + (y + 1)^2 = 36$$

Conversely,

$$(x + 3)^2 + (y - 4)^2 = 49$$

is the equation of a circle with center at $(-3, 4)$ and radius 7.

THE GRAPH OF AN EQUATION

An equation is an algebraic object. By means of a coordinate system, it can be transformed into a curve, a geometric object. Here is how it is done.

Consider the equation $y = x^2 - 3$. Its set of solutions is the set of ordered pairs (x, y) that satisfy the equation. These ordered pairs are the coordinates of points in the plane. The set of all such points is called the **graph** of the equation. The graph of an equation in two variables x and y will usually be a curve.

To obtain this graph, we follow a definite procedure.

1. Obtain the coordinates of a few points.
2. Plot those points in the plane.
3. Connect the points with a smooth curve in the order of increasing x values.

The best way to do step 1 is to make a **table of values.** Assign values to one of the variables, say x, determine the corresponding values of the other, and then list the pairs of values in tabular form. The whole three-step procedure is illustrated in Figure 26 for the previously mentioned equation, $y = x^2 - 3$.

Of course, you need to use common sense and even a little faith. When you connect the points you have plotted with a smooth curve, you are assuming that the curve behaves nicely between consecutive points; that is faith. This is why you should plot enough points so the outline of the curve seems very clear; the more points you plot, the less faith you will need. Also you should recognize that you can seldom display the whole curve. In our example, the curve has infinitely long arms opening wider and wider. But our graph does show the essential features. This is what we always aim to do—show enough of the graph so the essential features are visible.

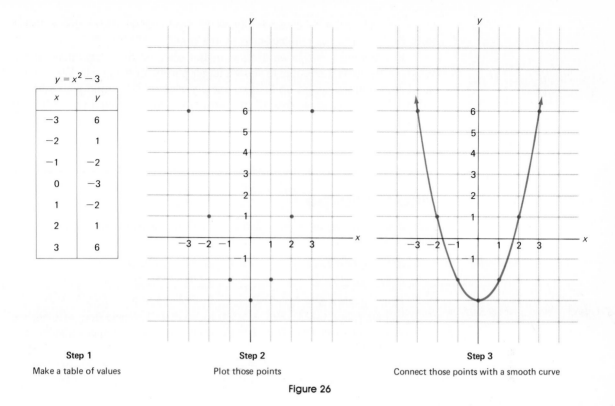

$y = x^2 - 3$	
x	y
-3	6
-2	1
-1	-2
0	-3
1	-2
2	1
3	6

Step 1
Make a table of values

Step 2
Plot those points

Step 3
Connect those points with a smooth curve

Figure 26

SYMMETRY OF A GRAPH

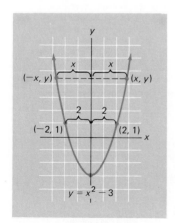

Figure 27

The graph of $y = x^2 - 3$, drawn earlier and again in Figure 27, has a nice property of symmetry. If the coordinate plane were folded along the y-axis, the two branches would coincide. For example, $(3, 6)$ would coincide with $(-3, 6)$; $(2, 1)$ would coincide with $(-2, 1)$; and, more generally, (x, y) would coincide with $(-x, y)$. Algebraically, this corresponds to the fact that we may replace x by $-x$ in the equation $y = x^2 - 3$ without changing it. More precisely, the equations $y = x^2 - 3$ and $y = (-x)^2 + 3$ are equivalent equations.

Whenever an equation is unchanged by replacing (x, y) with $(-x, y)$, the graph of the equation is **symmetric with respect to the y-axis.** Likewise, if the equation is unchanged when (x, y) is replaced by $(x, -y)$, its graph is **symmetric with respect to the x-axis.** The equation $x = 1 + y^2$ is of the latter type; its graph is shown in Figure 28 on the next page.

A third type of symmetry is **symmetry with respect to the origin.** It occurs whenever replacing (x, y) by $(-x, -y)$ produces no change in the equation. The equation $y = x^3$ is a good example as $-y = (-x)^3$ is equivalent

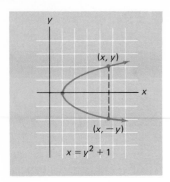

x = y² + 1

Figure 28

to $y = x^3$. The graph is shown in Figure 29. Note that the dotted line segment from $(-x, -y)$ to (x, y) is bisected by the origin.

In graphing $y = x^3$, we used a smaller scale on the y-axis than on the x-axis. This made it possible to show a larger portion of the graph. We suggest that before putting scales on the two axes, you should examine your table of values. Choose scales so that all of your points can be plotted and still keep your graph of reasonable size.

Problem Set 1-6

In Problems 1–6, plot the points P and Q in the Cartesian plane and find the distance between them.

1. $P(1, 5), Q(-2, 1)$
2. $P(-1, 12), Q(4, 0)$
3. $P(2, -2), Q(4, 3)$
4. $P(-1, 5), Q(2, 6)$
5. $P(\pi, -1.517), Q(-\sqrt{2}, \sqrt{3})$
6. $P(2.134, 2.612), Q(\sqrt{5}, \sqrt{7})$
7. Plot the points $A(1, 1), B(6, 3), C(4, 8),$ and $D(-1, 6)$ and show that they are vertices of a rectangle. *Hint:* It is not enough to show that opposite sides have equal lengths.
8. Show that $A(1, 3), B(2, 6), C(4, 7),$ and $D(3, 4)$ are vertices of a parallelogram.

$y = x^3$

x	y
-3	-27
-2	- 8
-1	- 1
0	0
1	1
2	8
3	27

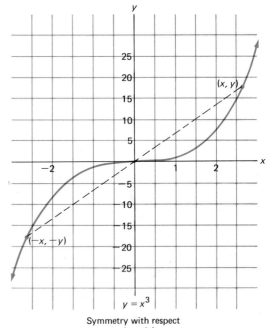

Symmetry with respect
to the origin

Figure 29

9. $A(2, -1)$ and $C(8, 7)$ are two of the vertices of a rectangle with sides parallel to the coordinate axes. Find the coordinates of the other two vertices.

10. Show that $A(2, -4)$, $B(8, -2)$, and $C(4, 0)$ are vertices of a right triangle.

11. Write the equation of the circle with center at $(1, -2)$ and radius 3.

12. Write the equation of the circle with center $(-\pi, 1)$ and radius $\sqrt{2}$.

13. Find the center and radius of the circle with equation $(x + 5)^2 + (y - 3)^2 = 16$.

14. Find the center and radius of the circle with equation $x^2 + (y + 4)^2 = \frac{9}{4}$.

EXAMPLE A (More on Circles)   When $(x - h)^2 + (y - k)^2 = r^2$ is expanded, it takes the form $x^2 + y^2 + Ax + By = C$. We therefore expect an equation of the latter form to represent a circle. Find the center and radius of the circle with equation

$$x^2 + y^2 - 2x + 6y = 6$$

Solution.   We rearrange the terms and then complete the squares by adding the same number to both sides of the equation.

$$(x^2 - 2x \quad) + (y^2 + 6y \quad) = 6$$
$$(x^2 - 2x + 1) + (y^2 + 6y + 9) = 6 + 1 + 9$$
$$(x - 1)^2 + (y + 3)^2 = 16$$

The last equation is in standard form; it is the equation of a circle centered at $(1, -3)$ with radius 4.

Use the method just described to find the centers and radii of circles with the following equations.

15. $x^2 + y^2 - 8x + 5y = 1$ 16. $x^2 + y^2 - 3x + 9y = 0$

17. $2x^2 + 2y^2 - 8x + 16y = 4$ 18. $3x^2 + 3y^2 - 6x + 9y = -1$

19. $x^2 + y^2 + \pi x - 2y = 0$ 20. $x^2 + y^2 - 6x + 4y = -13$

21. Find the distance between $A(1, 1)$ and $B(9, 1)$ and write the equation of the circle that has AB as a diameter.

22. Write the equation of the circle with center $(-4, 2)$ which is tangent to the y-axis.

Graph each of the following equations, showing enough of the graph to bring out its essential features. Note any of the three kinds of symmetry discussed in the text.

23. $y = 3x - 2$ 24. $y = 2x + 1$ 25. $y = -x^2 + 4$

26. $y = -x^2 - 2x$ 27. $y = x^2 - 4x$ 28. $y = x^3 + 2$

29. $y = -x^3$ 30. $y = \dfrac{12}{x^2 + 4}$ 31. $y = \dfrac{4}{x^2 + 1}$

32. $y = x^3 + x$ 33. $x^2 + y^2 = 25$ 34. $(x - 1)^2 + y^2 = 9$

EXAMPLE B (More Graphing)   Graph the equation $x = y^2 - 2y + 4$.

Solution.   Assign values to y and calculate the corresponding x. Then plot the corresponding points and draw the graph (Figure 30). Note that the graph is symmetric with respect to the line $y = 1$.

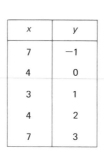

x	y
7	−1
4	0
3	1
4	2
7	3

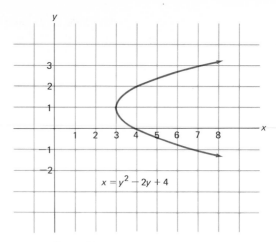

$x = y^2 - 2y + 4$

Figure 30

Graph each of the following equations.

35. $x = 2y - 1$ 36. $x = -3y + 1$ 37. $x = -2y^2$

38. $x = 2y - y^2$ 39. $x = y^3$ 40. $x = 8 - y^3$

MISCELLANEOUS PROBLEMS

41. Find x such that the distance between the points $(3, 0)$ and $(x, 3)$ is 5.

42. Find the perimeter of the triangle with vertices $(2, -4)$, $(3, 2)$, and $(-5, 1)$.

43. Write the equation of the circle with center at $(2, -3)$ that passes through $(5, 2)$.

44. The points $(2, 3)$, $(6, 3)$, $(6, -1)$, and $(2, -1)$ are the corners of a square. Find the equation of the inscribed circle (the circle tangent to all four sides) and the equation of the circumscribed circle (the circle that goes through all four corner points).

45. It is easy to show that the **midpoint** $P(x, y)$ of the line segment joining $A(x_1, y_1)$ and $B(x_2, y_2)$ has coordinates

$$x = \frac{x_1 + x_2}{2} \qquad y = \frac{y_1 + y_2}{2}$$

Given the points $A(2, 3)$ and $B(-4, 7)$, (a) find the midpoint of segment AB; (b) find C if B is the midpoint of AC.

46. Find the equation of the circle with diameter AB if A and B have coordinates $(3, -1)$ and $(-5, 7)$, respectively. *Hint:* See Problem 45.

47. Identify each of the following equations as corresponding to a circle (giving its radius and center), a point (giving its coordinates), or the empty set.
    (a) $x^2 + y^2 - 6x + 4y = -13$ (b) $2x^2 + 2y^2 - 2x + 6y = 3$
    (c) $4x^2 + 4y^2 - 8x - 4y = -7$ (d) $\sqrt{3}x^2 + \sqrt{3}y^2 - 6y = 2\sqrt{3}$

48. Find the equation of the circle of radius 5 with center in the first quadrant that passes through $(3, 0)$ and $(0, 1)$.

Sketch the graphs of the equations in Problems 49–52. Always begin by deciding if the graph has any of the three kinds of symmetry discussed in the text.

49. $y = \dfrac{12}{x}$

50. $y = 3 + \dfrac{4}{x^2}$

51. $y = 2(x - 1)^2$

52. $x = -y^2 + 8$

53. $x = 4y - y^2$

54. $x^2 + y^2 - 2y = 8$

ⓒ 55. Cities A, B, and C are located at $(0, 0)$, $(214, 17)$, and $(230, 179)$, respectively, with distances in miles. There are straight roads from A to B and from B to C. It costs \$3.71 per mile to ship a certain item by truck and \$4.81 per mile by air. Find the cheaper way to ship this item from A to C and calculate the savings made by choosing this method of shipment.

ⓒ 56. Point B is 8 miles downstream and on the opposite side of the river from point A. Starting at A, Ted will run 5 miles along the shore to an intermediate point C and then swim diagonally to B. If the river is $\frac{1}{8}$ mile wide, what is the total length of this path? How long will it take him if he runs at 8 miles per hour and swims at 3 miles per hour?

57. How far is it between the centers of the circles with equations $x^2 + y^2 + 4x - 6y = 2$ and $x^2 + y^2 - 8x + 10y = 0$?

58. A belt runs around two circular wheels with equations $(x - 1)^2 + (y - 1)^2 = 4$ and $(x - 8)^2 + (y - 7)^2 = 4$. How long is the belt?

59. A belt just fits around three wheels with equations $(x + 10)^2 + y^2 = 1$, $(x - 10)^2 + y^2 = 1$, and $x^2 + (y - 10\sqrt{3})^2 = 1$. Find the length of the belt.

60. **TEASER** Arnold Thinkhard has been ordered to attach four guy wires to the top of a tall pole. These four guy wires are anchored to four stakes A, B, C, and D, which form the vertices (in cyclic order) of a rectangle on the ground. After attaching wires of length 210, 60, and 180 feet from A, B, and C, Arnold discovers that the fourth wire from D is too short to reach. A new wire has to be cut, but how long should it be? Even though you do not know the dimensions of the rectangle or the height of the pole, it is your job to solve Arnold's problem. It is not enough to get the answer; you must demonstrate that your answer is correct before poor Arnold climbs back up the pole.

Chapter Summary

The **real numbers** can be visualized as labels for points along a horizontal line, called the **real line**. On this line, $a < b$ means a is to the left of b and $|a|$ denotes the distance between the points labeled a and 0 (the origin). Every real number can be expressed as an nonterminating **decimal**, which repeats if the number is rational and is nonrepeating if the number is irrational.

An **exponent** is a numerical superscript placed on a number to indicate a certain operation on that number. In particular,

$$b^4 = b \cdot b \cdot b \cdot b \qquad b^0 = 1 \qquad b^{-5} = \dfrac{1}{b^5}$$

Exponents satisfy five important laws called **rules for exponents**. We use exponents to write decimals in a special format called **scientific notation**.

A **real polynomial** in x is an expression of the form

$$a_n x^n + a_{n-1} x^{n-1} + \cdots + a_1 x + a_0$$

where n is a nonnegative integer and the a_i's are real numbers. Polynomials can be added, subtracted, and multiplied; and they can sometimes be **factored**— that is, written as a product of simpler polynomials. A quotient of two polynomials is called a **rational expression**.

To **solve** an **equation** in x means to find all values of x for which the equation is a true statement. Similarly, to solve a **system** of two equations in x and y means to find the pairs of values of x and y for which both equations are true statements. Of special interest are **linear** equations $ax + b = 0$ and **quadratic** equations $ax^2 + bx + c = 0$ $(a \neq 0)$. The latter can sometimes be solved by factoring and can always be solved by the **quadratic formula**

$$x = \frac{-b \pm \sqrt{b^2 - 4ac}}{2a}$$

If the **discriminant** $b^2 - 4ac$ is negative, the equation has no real solutions.

The procedure for solving a linear **inequality** such as $2x - 7 < 4x + 9$ is similar to that for a linear equation, except that we must remember to reverse the inequality sign, $<$, when multiplying both sides by a negative number. Higher degree inequalities are usually solved most easily by the **split-point** method.

By means of a coordinate system, we establish a one-to-one correspondence between points in the plane and **ordered pairs** (x, y) of real numbers, called the **Cartesian coordinates** of the points. This leads to the **distance formula**, which says that the distance between $P(x_1, y_1)$ and $Q(x_2, y_2)$ is

$$d(P, Q) = \sqrt{(x_2 - x_1)^2 + (y_2 - y_1)^2}$$

More importantly, this leads to the subject called **analytic geometry** with its two major activities: (1) graphing an equation and (2) finding the equation of a curve. An example of the latter is the equation of the **circle** with radius r and center (h, k), namely,

$$(x - h)^2 + (y - k)^2 = r^2$$

Chapter Review Problem Set

1. Order the following numbers from least to greatest.

$$\sqrt{3}, \ 1.7, \ \tfrac{170}{99}, \ \tfrac{27}{16}, \ 1.\overline{7}$$

2. Perform the indicated operations and simplify.

(a) $\frac{1}{3}[\frac{1}{2} - (\frac{5}{6} - \frac{2}{3})]$

(b) $3 + \dfrac{\frac{3}{4} - \frac{7}{8}}{\frac{5}{12}}$

(c) $\dfrac{3\sqrt{20} - 12}{6}$

□ 3. Calculate and round to two decimal places.

$$\left(\sqrt{31.62} - \frac{.14}{\sqrt{.62}}\right)^2$$

4. Which of the following are rational numbers?

(a) $\sqrt{\frac{18}{50}}$ (b) $(4\sqrt{3})\sqrt{3}$ (c) $\sqrt{2}(5 - \sqrt{2})$ (d) $.\overline{53}$

(e) $.01001000100001\ldots$

5. Simplify, leaving your answer free of negative exponents.

(a) $(\frac{5}{6})^2(\frac{5}{6})^{-4}$ (b) $(\frac{5}{6} + \frac{1}{3})^{-2}$ (c) $\dfrac{2x^{-2}y^2}{4xy^{-3}}$ (d) $(2a^{-1}b^3)^{-2}(3a^2b^{-2})^3$

6. Express $\dfrac{(144)10^4}{.00009}$ in scientific notation.

7. Perform the indicated operations and simplify.

(a) $5 - x^2 - 2x(3 - 5x)$ (b) $(y + 13)(y - 10) - y^2$

(c) $(2s - 1)(5s + 3)$ (d) $(3t^2 + 2)(3t^2 - 2)$

(e) $(x + 2y)(x - 2y + 5)$ (f) $(c - 2d)(c^2 + 2cd + 4d^2)$

(g) $(3xz - 5h^2)^2$ (h) $(x + 2m)^3$

8. Factor over the integers.

(a) $2x^4 - x^3 + 11x^2$ (b) $z^2 + 6z - 27$ (c) $(x^2 + 2)^2 - 9$

(d) $x^3y^6 - 27y^3$ (e) $(x^2 + x + 1)^2 + (x^2 + x + 1) - 12$

9. Perform the indicated operations and simplify.

(a) $\dfrac{2x + 12}{x^2 - 36}(3x - 18)$ (b) $\dfrac{3}{x} - \dfrac{2x + 1}{x - 1}$

(c) $\dfrac{\dfrac{4}{2x + 2h + 5} - \dfrac{4}{2x + 5}}{h}$ (d) $\left(x^2 + \dfrac{1}{x}\right)^{-1} \div \dfrac{3x^2}{x^2 - x + 1}$

10. Solve the following equations.

(a) $5(x - \frac{3}{4}) = x + \frac{1}{3}$ (b) $\dfrac{6}{x - 5} = \dfrac{21}{x}$

(c) $(3x + 2)^2 = 25$ (d) $x^2 + 9x + 20 = 0$

(e) $2x^2 + x - 2 = 0$ (f) $\dfrac{x - 2}{x} = \dfrac{3}{x + 2}$

11. Solve the equation $z = [2z/(m - 4)] + 3$ for

(a) z in terms of m;   (b) m in terms of z.

12. Solve for x and y:

$$2x + 3y = 6$$

$$y = 4x + 9$$

13. Solve the following inequalities.

(a) $4x - 7 \le x + 11$ (b) $|3x - 1| < 5$ (c) $2x^2 + 5x - 3 < 0$

14. Find the equation of the circle that has the line segment connecting $(0, 0)$ and $(6, 8)$ as a diameter.

15. Find the center and radius of the circle whose equation is

$$x^2 + y^2 - 6x + 14y = -54$$

16. Sketch the graph of each of the following.

(a) $3x + 5y = 0$ (b) $(x + 2)^2 + (y - 1)^2 = 16$

(c) $x = 2y^3$ (d) $y = \dfrac{2}{x^2 + 3}$

Mathematicians do not deal in objects, but in relations between objects; thus, they are free to replace some objects by others so long as the relations remain unchanged. Content to them is irrelevant: they are interested in form only.
Henri Poincaré

CHAPTER 2

Functions and Their Graphs

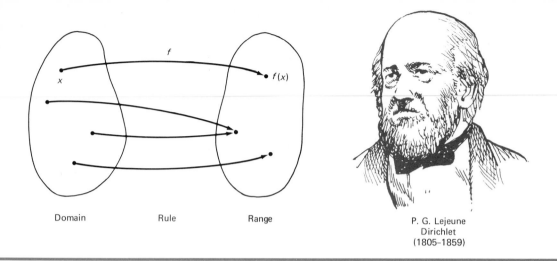

Domain   Rule   Range

P. G. Lejeune
Dirichlet
(1805–1859)

2-1   Functions and Functional Notation

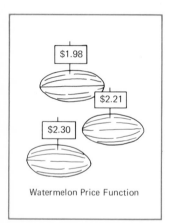

Watermelon Price Function

Figure 1

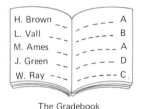

The Gradebook
Function

Figure 2

One of the most important ideas in mathematics is that of a function. For a long time, mathematicians and scientists wanted a precise way to describe the relationships that may exist between two variables. It is somewhat surprising that it took so long for the idea to crystallize into a clear, unambiguous concept. The French mathematician P. G. Lejeune Dirichlet (1805–1859) is credited with the modern definition of function.

DEFINITION

*A **function** is a rule which assigns to each element in one set (called the **domain** of the function) exactly one value from another set. The set of all assigned values is called the **range** of the function.*

Three examples will help clarify this idea. When a grocer puts a price tag on each of the watermelons for sale (Figure 1), a function is determined. Its domain is the set of watermelons, its range is the set of prices, and the rule is the procedure the grocer uses in assigning prices (perhaps a specified amount per pound.)

When a professor assigns a grade to each student in a class (Figure 2), he or she is determining a function. The domain is the set of students and the range is the set of grades, but who can say what the rule is? It varies from professor to professor; some may even prefer to keep it a secret.

A much more typical function from our point of view is the *squaring* function displayed in Figure 3. It takes a number from the domain {−2, −1, 0, 1, 2, 3} and squares it, producing a number in the range {0, 1, 4, 9}. This function is typical for two reasons: Both the domain and range are sets of

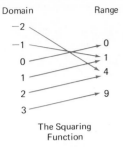

Domain   Range

-2
-1   0
0   1
1   4
2   9
3

The Squaring
Function

Figure 3

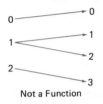

0   0
1   1
      2
2   3

Not a Function

Figure 4

numbers and the rule can be specified by giving an algebraic formula. Most functions in this book will be of this type.

The definition says that a function assigns exactly one value to each element of the domain. Thus Figure 4 is not the diagram for a function, since two values are assigned to the element 1.

FUNCTIONAL NOTATION

Long ago, mathematicians introduced a special notation for functions. A single letter like f (or g or h) is used to name the function. Then $f(x)$, read f of x or f at x, denotes the value that f assigns to x. Thus if f names the squaring function,

$$f(-2) = 4 \qquad f(2) = 4 \qquad f(-1) = 1$$

and, in general,

$$f(x) = x^2$$

We call this last result the *formula* for the function f. It tells us in a concise algebraic way what f does to any number. Notice that the given formula and

$$f(y) = y^2 \qquad f(z) = z^2$$

all say the same thing; the letter used for the domain variable is a matter of no significance, though it does happen that we shall usually use x. Many functions do not have simple formulas (see Problems 17 and 18), but in this book, most of them do.

For a further example, consider the function that cubes a number and then subtracts 1 from the result. If we name this function g, then

$$g(2) = 2^3 - 1 = 7$$
$$g(-1) = (-1)^3 - 1 = -2$$
$$g(.5) = (.5)^3 - 1 = -.875$$
$$g(\pi) = \pi^3 - 1 \approx 30$$

and, in general,

$$g(x) = x^3 - 1$$

Few students would have trouble using this formula when x is replaced by a specific number. However, it is important to be able to use it when x is replaced by anything whatever, even an algebraic expression. Be sure you understand the following calculations.

$$g(a) = a^3 - 1$$
$$g(y^2) = (y^2)^3 - 1 = y^6 - 1$$
$$g\left(\frac{1}{z}\right) = \left(\frac{1}{z}\right)^3 - 1 = \frac{1}{z^3} - 1$$

$$f(x) = x^2 + 4$$
$$f(3x) = 3x^2 + 4$$

$$f(x) = x^2 + 4$$
$$f(3x) = (3x)^2 + 4$$
$$= 9x^2 + 4$$

$$g(2 + h) = (2 + h)^3 - 1 = 8 + 12h + 6h^2 + h^3 - 1$$
$$= h^3 + 6h^2 + 12h + 7$$
$$g(x + h) = (x + h)^3 - 1 = x^3 + 3x^2h + 3xh^2 + h^3 - 1$$

DOMAIN AND RANGE

The rule of correspondence is the heart of a function, but a function is not completely determined until its domain is given. Recall that the **domain** is the set of elements to which the function assigns values. In the case of the squaring function f (reproduced in Figure 5), we gave the domain as the set $\{-2, -1, 0, 1, 2, 3\}$. We could just as well have specified the domain as the set of all real numbers; the formula $f(x) = x^2$ would still make perfectly good sense. In fact, there is a common agreement that if no domain is specified, it is understood to be the largest set of real numbers for which the rule for the function makes sense and gives real number values. We call it the **natural domain** of the function. Thus if no domain is specified for the function with formula $g(x) = x^3 - 1$, it is assumed to be the set of all real numbers. Similarly, for

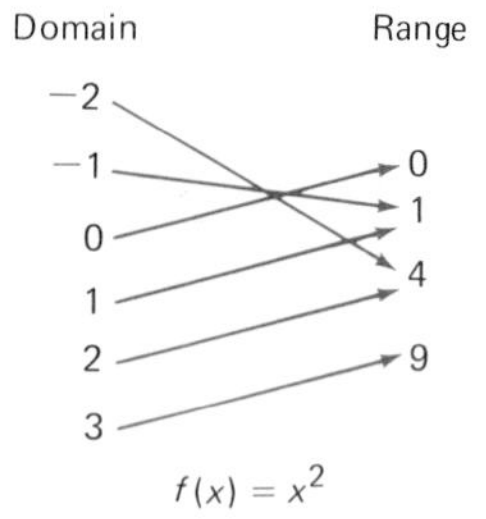

$$h(x) = \frac{1}{x - 1}$$

we would take the natural domain to consist of all real numbers except 1. Here, the number 1 is excluded to avoid division by zero.

Once the domain is understood and the rule of correspondence is given, the **range** of the function is determined. It is the set of values of the function. Here are several examples.

RULE	DOMAIN	RANGE
$F(x) = 4x$	All reals	All reals
$G(x) = \sqrt{x - 3}$	$\{x : x \geq 3\}$	Nonnegative reals
$H(x) = \dfrac{1}{(x - 2)^2}$	$\{x : x \neq 2\}$	Positive reals

INDEPENDENT AND DEPENDENT VARIABLES

Scientists like to use the language of variables in talking about functions. Let us illustrate by referring to an object falling under the influence of gravity near the earth's surface. If the object is very dense (so air resistance can be neglected), it will fall according to the formula

$$d = f(t) = 16t^2$$

Here d represents the distance in feet that the object falls during the first t seconds of falling. In this case, t is called the **independent variable** and d, which depends on t, is the **dependent variable**. Also, d is said to be a function of t.

Domain   Range

-2
-1   0
0   1
1   4
2   9
3

$$f(x) = x^2$$

Figure 5

FUNCTIONS OF TWO OR MORE VARIABLES

All functions illustrated so far have been functions of one variable. Suppose that a function f associates a value with each ordered pair (x, y). In this case, we say that f is a function of two variables. As an example, let

$$f(x, y) = x^2 + 3y^2$$

Then

$$f(3, -2) = 3^2 + 3(-2)^2 = 21$$

and

$$f(0, 6) = (0)^2 + 3(6)^2 = 108$$

The natural domain for this function is the set of all ordered pairs (x, y), that is, the whole Cartesian plane. Its range is the set of nonnegative numbers.

For a second example, let $g(x, y, z) = 2x^2 - 4y + z$. We say that g is a function of three variables. Note that

$$g(4, 2, -3) = 2(4)^2 - 4(2) + (-3) = 21$$

and

$$g\left(u^3, 2v, \frac{1}{w}\right) = 2(u^3)^2 - 4(2v) + \frac{1}{w}$$

$$= 2u^6 - 8v + \frac{1}{w}$$

VARIATION

Sometimes we describe a functional relationship between a dependent variable and one or more independent variables by using the language of variation (or proportion). To say that y **varies directly** as x (or that y is proportional to x) means that $y = kx$ for some constant k. To say that y **varies inversely** as x means that $y = k/x$ for some constant k. Finally, to say that z **varies jointly** as x and y means that $z = kxy$ for some constant k.

In variation problems, we are often given not only the form of the relationship but also a specific set of values satisfied by the variables involved. This allows us to evaluate the constant k and obtain an explicit formula for the dependent variable in terms of the independent variables. Here is an illustration.

It is known that y varies directly as x and that $y = 10$ when $x = -2$. Find an explicit formula for y in terms of x.

We substitute the given values in the equation $y = kx$ and get $10 = k(-2)$. Thus $k = -5$, and we can write the explicit formula $y = -5x$.

We offer a second example, this time using functional notation.

It is given that a function f varies jointly with x and y and that $f(4, 3) = 2$. Find the explicit formula for f and also evaluate $f(3, 12)$.

We know that $f(x, y) = kxy$. Also, $f(4, 3) = k(4)(3) = 12k = 2$, which implies that $k = \frac{1}{6}$. Thus the explicit formula for $f(x, y)$ is

$$f(x, y) = \tfrac{1}{6}xy$$

and therefore

$$f(3, 12) = \tfrac{1}{6}(3)(12) = 6$$

Problem Set 2-1

1. If $f(x) = x^2 - 4$, evaluate each expression.
   (a) $f(-2)$ (b) $f(0)$ (c) $f(\tfrac{1}{2})$ (d) $f(.1)$
   (e) $f(\sqrt{2})$ (f) $f(a)$ (g) $f(1/x)$ (h) $f(x + 1)$
2. If $f(x) = (x - 4)^2$, evaluate each expression in Problem 1.
3. If $f(x) = 1/(x - 4)$, evaluate each expression.
   (a) $f(8)$ (b) $f(2)$ (c) $f(\tfrac{9}{2})$ (d) $f(\tfrac{31}{8})$
   (e) $f(4)$ (f) $f(4.01)$ (g) $f(1/x)$ (h) $f(x^2)$
   (i) $f(2 + h)$ (j) $f(2 - h)$
4. If $f(x) = x^2$ and $g(x) = 2/x$, evaluate each expression.
   (a) $f(-7)$ (b) $g(-4)$ (c) $f(\tfrac{1}{4})$ (d) $1/f(4)$
   (e) $g(\tfrac{1}{4})$ (f) $1/g(4)$ (g) $g(0)$ (h) $g(1)f(1)$
   (i) $f(g(1))$ (j) $f(1)/g(1)$

EXAMPLE A (Finding Natural Domains) Find the natural domain of
   (a) $f(x) = 4x/[(x + 2)(x - 3)]$; (b) $g(x) = \sqrt{x^2 - 4}$.

Solution. We recall that the natural domain is the largest set of real numbers for which the formula makes sense and gives real number values. Thus in part (a), the domain consists of all real numbers except -2 and 3; in part (b) we must have $x^2 \geq 4$, which is equivalent to $|x| \geq 2$. Notice that if $|x| < 2$, we would be taking the square root of a negative number, so the result would not be a real number.

In Problems 5–16, find the natural domain of the given function.

5. $f(x) = x^2 - 4$

6. $f(x) = (x - 4)^2$

7. $g(x) = \dfrac{1}{x^2 - 4}$

8. $g(x) = \dfrac{1}{9 - x^2}$

9. $h(x) = \dfrac{2}{x^2 - x - 6}$

10. $h(x) = \dfrac{1}{2x^2 + 3x - 2}$

11. $F(x) = \dfrac{1}{x^2 + 4}$

12. $F(x) = \dfrac{1}{9 + x^2}$

13. $G(x) = \sqrt{x - 2}$

14. $G(x) = \sqrt{x + 2}$

15. $H(x) = \dfrac{1}{5 - \sqrt{x}}$

16. $H(x) = \dfrac{1}{\sqrt{x + 1} - 2}$

17. Not all functions arising in mathematics have rules given by simple algebraic formulas. Let $f(n)$ be the nth digit in the decimal expansion of

$$\pi = 3.14159265358979323846 \ldots$$

Thus $f(1) = 3$ and $f(3) = 4$. Find (a) $f(6)$; (b) $f(9)$; (c) $f(16)$. What is the natural domain for this function?

18. Let g be the function which assigns to each positive integer the number of factors in its prime factorization. Thus

$$g(2) = 1$$
$$g(4) = g(2 \cdot 2) = 2$$
$$g(36) = g(2 \cdot 2 \cdot 3 \cdot 3) = 4$$

Find (a) $g(24)$; (b) $g(37)$; (c) $g(64)$; (d) $g(162)$. Can you find a formula for this function?

EXAMPLE B (Functions Generated on a Calculator)   Show how the function $f(x) = 2\sqrt{x} + 5$ can be generated on a calculator. Then calculate $f(\pi)$.

Solution.   To avoid confusion with the times sign, we use # rather than x for an arbitrary number. Then on a typical algebraic logic calculator,

$$f(\#) = 2 \,\boxed{\times}\, \# \,\boxed{\sqrt{x}}\,\boxed{+}\, 5 \,\boxed{=}$$

In particular,

$$f(\pi) = 2 \,\boxed{\times}\, \pi \,\boxed{\sqrt{x}}\,\boxed{+}\, 5 \,\boxed{=}$$

which yields the value 8.544908.

[c] *In Problems 19–26, write the sequence of keys that will generate the given function, using # for an arbitrary number. Then use the sequence to calculate $f(2.9)$.*

19. $f(x) = (x + 2)^2$

20. $f(x) = \sqrt{x + 3}$

21. $f(x) = 3(x + 2)^2 - 4$

22. $f(x) = 4\sqrt{x + 3} - 11$

23. $f(x) = \left(3x + \dfrac{2}{\sqrt{x}}\right)^3$

24. $f(x) = \left(\dfrac{x}{3} + \dfrac{3}{x}\right)^5$

25. $f(x) = \dfrac{\sqrt{x^5 - 4}}{2 + 1/x}$

26. $f(x) = \dfrac{(\sqrt{x} - 1)^3}{x^2 + 4}$

[c] *In Problems 27–32, write the algebraic formula for the function that is generated by the given sequence of calculator keys.*

27. $\# \,\boxed{+}\, 5 \,\boxed{\sqrt{x}}\,\boxed{=}$

28. $\boxed{(}\, \# \,\boxed{+}\, 5 \,\boxed{)}\,\boxed{\sqrt{x}}\,\boxed{=}$

29. $2 \,\boxed{\times}\, \# \,\boxed{x^2}\,\boxed{+}\, 3 \,\boxed{\times}\, \# \,\boxed{=}$

30. $3 \,\boxed{+/-}\,\boxed{\times}\, \# \,\boxed{x^2}\,\boxed{-}\, 7 \,\boxed{=}$

31. $\boxed{(}\, 3 \,\boxed{\times}\,\boxed{(}\, \# \,\boxed{-}\, 2 \,\boxed{)}\,\boxed{x^2}\,\boxed{+}\, 9 \,\boxed{)}\,\boxed{\sqrt{x}}\,\boxed{=}$

32. $\boxed{(}\, \# \,\boxed{1/x}\,\boxed{\times}\, 2 \,\boxed{+}\, 4 \,\boxed{)}\,\boxed{\sqrt{x}}\,\boxed{=}$

Problems 33–44 deal with functions of two variables. Let $g(x, y) = 3xy - 5x$ and $G(x, y) = (5x + 3y)/(2x - y)$. Find each of the following.

33. $g(2, 5)$ 34. $g(-1, 3)$ 35. $g(5, 2)$ 36. $g(3, -1)$

37. $G(1, 1)$ 38. $G(3, 3)$ 39. $G(\frac{1}{2}, 1)$ 40. $G(5, 0)$

41. $g(2x, 3y)$ 42. $G(2x, 4y)$ 43. $g(x, 1/x)$ 44. $G(x - y, y)$

EXAMPLE C (More on Variation)   Suppose that w varies jointly as x and the square root of y and that $w = 14$ when $x = 2$ and $y = 4$. Find an explicit formula for w.

Solution.   We translate the first statement into mathematical symbols as

$$w = kx\sqrt{y}$$

To evaluate k, we substitute the given values for w, x, and y.

$$14 = k \cdot 2\sqrt{4} = 4k$$

or

$$k = \frac{14}{4} = \frac{7}{2}$$

Thus the explicit formula for w is

$$w = \frac{7}{2}x\sqrt{y}$$

In Problems 45–50, find an explicit formula for the dependent variable.

45. y varies directly as x, and $y = 12$ when $x = 3$.

46. y varies directly as x^2, and $y = 4$ when $x = 0.1$.

47. y varies inversely as x (that is, $y = k/x$), and $y = 5$ when $x = \frac{1}{5}$.

48. V varies jointly as r^2 and h, and $V = 75$ when $r = 5$ and $h = 9$. .

49. I varies directly as s and inversely as d^2, and $I = 9$ when $s = 4$ and $d = 12$.

50. W varies directly as x and inversely as the square root of yz, and $W = 5$ when $x = 7.5$, $y = 2$, and $z = 18$.

51. The maximum range of a projectile varies as the square of the initial velocity. If the range is 16,000 feet when the initial velocity is 600 feet per second,
    (a) write an explicit formula for R in terms of v, where R is the range in feet and v is the initial velocity in feet per second;
    (b) use this formula to find the range when the initial velocity is 800 feet per second.

52. Suppose that the amount of gasoline used by a car varies jointly as the distance traveled and the square root of the average speed. If a car used 8 gallons on a 100-mile trip going at an average speed of 64 miles per hour, how many gallons would that car use on a 160-mile trip at an average speed of 25 miles per hour?

MISCELLANEOUS PROBLEMS

53. If $f(x) = x^2 - (2/x)$, evaluate and simplify each expression.
    (a) $f(2)$ (b) $f(-1)$ (c) $f(\frac{1}{2})$

(d) $f(\sqrt{2})$ (e) $f(2 - \sqrt{2})$ (f) $f(.01)$

(g) $f(1/x)$ (h) $f(a^2)$ (i) $f(a + b)$

54. If $f(x, y) = (x^2 - xy)/(x + y)$, evaluate and simplify each expression.

(a) $f(2, 1)$ (b) $f(3, 0)$ (c) $f(-2, 1)$

(d) $f(3\sqrt{2}, \sqrt{2})$ (e) $f(1/x, 2y)$ (f) $f(a + b, b)$

55. Determine the natural domain of each function.

(a) $f(t) = \dfrac{t + 2}{t^2(t + 3)}$ (b) $g(t) = \sqrt{t^2 - 4}$

(c) $h(t) = \dfrac{3t + 1}{1 - \sqrt{2t}}$ (d) $k(s, t) = \dfrac{3st\sqrt{9 - s^2}}{t^2 - 1}$

56. Let $f(n)$ be the nth digit in the decimal expansion of $\frac{5}{13}$. Determine the domain and range of f.

57. A 2-mile race track has the shape of a rectangle with semicircular ends of radius x. If $A(x)$ is the area of the region inside the track, determine the domain and range of A.

58. Cut a yardstick in two pieces of length x and $3 - x$ and, together with a 1-foot stick, form a triangle of area $A(x)$. Determine the domain and range of A. *Note:* The triangle of maximum area will be isosceles.

59. Write a formula for $F(x)$ in each case.

(a) $F(x)$ is the area of an equilateral triangle of perimeter x.

(b) $F(x)$ is the area of a regular hexagon inscribed in a circle of radius x.

(c) $F(x)$ is the volume of water of depth x in a conical tank with vertex downward. The tank is 8 feet high and has diameter 6 feet at the top.

(d) $F(x)$ is the average cost per unit of producing x refrigerators in a day for a company that has daily overhead of $1300 and pays direct costs (labor and materials) of $240 to make each refrigerator.

(e) $F(x)$ is the dollar cost of renting a car for 10 days and driving x miles if the rental company charges $18 per day and 22 cents per mile for mileage beyond the first free 100 miles.

60. Some of the following equations determine a function f with formula of the form $y = f(x)$. For those that do, find $f(x)$. *Recall:* A function must associate just one value with each x.

(a) $x^2 + 3y^2 = 1$ (b) $xy + 2y = 5 - 2x$

(c) $x = \sqrt{2y - 1}$ (d) $2x = (y + 1)/y$

61. The safe load $S(x, y, z)$ of a horizontal beam supported at both ends varies directly as its breadth x and the square of its depth y and inversely as its length z. If a 2-inch by 6-inch white pine joist 10 feet long safely supports 1000 pounds when placed on edge (as shown in Figure 6), find an explicit formula for $S(x, y, z)$. Then determine its safe load when placed flatwise.

62. Which of the following functions satisfy $f(x + y) = f(x) + f(y)$ for all real numbers x and y?

(a) $f(t) = 2t$ (b) $f(t) = t^2$

(c) $f(t) = 2t + 3$ (d) $f(t) = -3t$

63. Determine the formula for $f(t)$ if for all real numbers x and y

$$f(x)f(y) - f(xy) = x + y$$

64. **TEASER** A function f satisfying $f(x + y) = f(x) + f(y)$ for all real numbers x and y is said to be an *additive function*. Prove that if f is additive, then there is a number m such that $f(t) = mt$ for all rational numbers t.

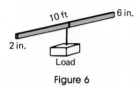

10 ft 6 in.

2 in.

Load

Figure 6

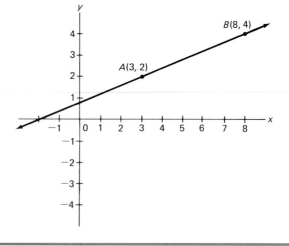

A point is that which has no part.

A line is breadthless length.

A straight line is a line which lies evenly with the points of itself.

Euclid

300 B.C.

2-2   Linear Functions

In this and the next few sections, we initiate the study of several important classes of functions. We are particularly interested in the graphs of these functions. By the graph of a function f, we mean the graph of the equation $y = f(x)$. For example, the graph of the function f with formula $f(x) = 3x - 2$ is the graph of the equation $y = 3x - 2$.

The first class of functions to be studied is the class of **linear functions.** A function f is linear if it has a formula of the form $f(x) = ax + b$, where a and b are constants. The name arises from the fact that the graph of a linear function is (as we shall see) a straight line. In fact, in order to understand linear functions, it will help to begin with a rather complete discussion of straight lines. The knowledge gained will be valuable for another reason. A thorough understanding of lines is indispensable in calculus.

What is a straight line? Euclid's definition in our opening panel is not very helpful, but neither are most of the alternatives we have heard. Fortunately, we all know what we mean by a straight line even if we cannot seem to describe it in terms of more primitive ideas. There is one thing on which we must agree: Given two points (for example, A and B above), there is one and only one straight line that passes through them. And contrary to Euclid, let us agree that the word *line* shall always mean straight line.

A line is a geometric object. When it is placed in a coordinate system, it ought to have an equation just as a circle does. How do we find the equation of a line? To answer this question we will need the notion of slope.

THE SLOPE OF A LINE

Consider the line in our opening diagram. From point A to point B, there is a **rise** (vertical change) of 2 units and a **run** (horizontal change) of 5 units. We say that the line has a slope of $\frac{2}{5}$. In general (see Figure 7), for a line through $A(x_1, y_1)$ and $B(x_2, y_2)$, where $x_1 \neq x_2$, we define the **slope** m of that line by

$$m = \frac{\text{rise}}{\text{run}} = \frac{y_2 - y_1}{x_2 - x_1}$$

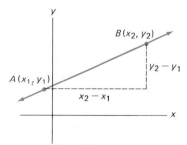

Figure 7

You should immediately raise a question. A line has many points. Does the value we get for the slope depend on which pair of points we use for A and B? The similar triangles in Figure 8 show us that

$$\frac{y_2' - y_1'}{x_2' - x_1'} = \frac{y_2 - y_1}{x_2 - x_1}$$

Thus points A' and B' would do just as well as A and B. It does not even matter whether A is to the left or right of B since

$$\frac{y_1 - y_2}{x_1 - x_2} = \frac{y_2 - y_1}{x_2 - x_1}$$

All that matters is that we subtract the coordinates in the same order in numerator and denominator.

The slope m is a measure of the steepness of a line, as Figure 9 illustrates.

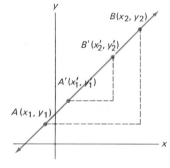

Figure 8

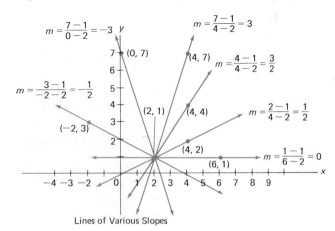

Lines of Various Slopes

Figure 9

CAUTION

A horizontal line has no slope.

A horizontal line has zero slope.
A vertical line has no slope.

Notice that a horizontal line has zero slope and a line that rises to the right has positive slope. The larger this positive slope is, the more steeply the line rises. A line that falls to the right has negative slope. The concept of slope for a vertical line makes no sense since it would involve division by zero. Therefore, the notion of slope for a vertical line is left undefined.

THE POINT-SLOPE FORM

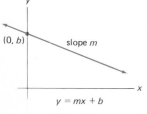

Figure 10

Consider again the line of our opening diagram; it is reproduced in Figure 10. We know

1. it passes through (3,2);
2. it has slope $\frac{2}{5}$.

Take any other point on that line, such as one with coordinates (x, y). If we use this point together with (3, 2) to measure slope, we must get $\frac{2}{5}$, that is,

$$\frac{y - 2}{x - 3} = \frac{2}{5}$$

or, after multiplying by $x - 3$,

$$y - 2 = \tfrac{2}{5}(x - 3)$$

Notice that this last equation is satisfied by all points on the line, even by (3, 2). Moreover, no points not on the line can satisfy this equation.

What we have just done in an example can be done in general. The line passing through the (fixed) point (x_1, y_1), with slope m has equation

$$y - y_1 = m(x - x_1)$$

We call this the **point-slope** form of the equation of a line.

Consider once more the line of our example. That line passes through (8, 4) as well as (3, 2). If we use (8, 4) as (x_1, y_1), we get the equation

$$y - 4 = \tfrac{2}{5}(x - 8)$$

which looks quite different from

$$y - 2 = \tfrac{2}{5}(x - 3)$$

However, both can be simplified to $5y - 2x = 4$; they are equivalent.

THE SLOPE-INTERCEPT FORM

The equation of a line can be expressed in various forms. Suppose we are given the slope m for a line and the y-intercept b (that is, the line intersects the y-axis at $(0, b)$ as in Figure 11). Choosing $(0, b)$ as (x_1, y_1) and applying the point-slope form, we get

$$y - b = m(x - 0)$$

Figure 11

which we can rewrite as

$$y = mx + b$$

The latter is called the **slope-intercept** form for the equation of a line.

Why get excited about that, you ask? Because any time we see an equation written this way, we recognize it as the equation of a line and can immediately read its slope and y-intercept. For example, consider the equation

$$3x - 2y + 4 = 0$$

If we solve for y, we get

$$y = \tfrac{3}{2}x + 2$$

It is the equation of a line with slope $\tfrac{3}{2}$ and y-intercept 2.

EQUATION OF A VERTICAL LINE

Vertical lines do not fit within the discussion above; they do not have slopes. But they do have equations, very simple ones. The line in Figure 12 has equation $x = \tfrac{5}{2}$, since every point on the line satisfies this equation. The equation of any vertical line can be put in the form

$$x = k$$

where k is a constant. It should be noted that the equation of a horizontal line can be written in the form $y = k$, since this is the same as $y = 0 \cdot x + k$.

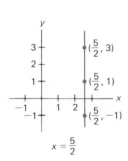

$x = \dfrac{5}{2}$

Figure 12

THE FORM $Ax + By + C = 0$

It would be nice to have a form that covered all lines including vertical lines. Consider for example,

(i) $y - 2 = -4(x + 2)$
(ii) $y = 5x - 3$
(iii) $x = 5$

These can be rewritten (by taking everything to the left side) as follows:

(i) $4x + y + 6 = 0$
(ii) $-5x + y + 3 = 0$
(iii) $x + 0y - 5 = 0$

All are of the form

$$Ax + By + C = 0$$

which we call the **general linear equation.** It takes only a moment's thought to see that the equation of any line can be put in this form. Conversely, the graph of $Ax + By + C = 0$ is always a line (if A and B are not both zero (see Problem 54)).

PARALLEL LINES

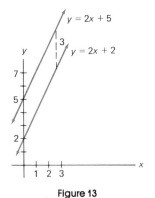

Figure 13

If two lines have the same slope, they are parallel. Thus $y = 2x + 2$ and $y = 2x + 5$ represent parallel lines; both have a slope of 2. The second line is 3 units above the first for every value of x (see Figure 13).

Similarly, the lines with equations $-2x + 3y + 12 = 0$ and $4x - 6y = 5$ are parallel. To see this, solve these equations for y (that is, find the slope-intercept form); you get $y = \frac{2}{3}x - 4$ and $y = \frac{2}{3}x - \frac{5}{6}$, respectively. Both have slope $\frac{2}{3}$; they are parallel.

We may summarize by stating that *two nonvertical lines are parallel if and only if they have the same slope*.

PERPENDICULAR LINES

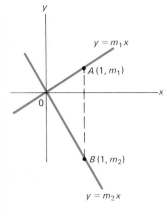

Figure 14

Is there a simple slope condition which characterizes perpendicular lines? Yes; *two nonvertical lines are perpendicular if and only if their slopes are negative reciprocals of each other.* To prove this, we show that the two lines $y = m_1 x$ and $y = m_2 x$ are perpendicular if and only if $m_1 m_2 = -1$. Choose the point $A(1, m_1)$ on the line $y = m_1 x$ and the point $B(1, m_2)$ on the line $y = m_2 x$, as shown in Figure 14. Clearly, the two lines are perpendicular if and only if

$$[d(A, B)]^2 = [d(0, A)]^2 + [d(0, B)]^2$$

that is,

$$(m_1 - m_2)^2 = (1 + m_1^2) + (1 + m_2^2)$$

or

$$m_1^2 - 2m_1 m_2 + m_2^2 = m_1^2 + 2 + m_2^2$$

The latter is equivalent to $m_1 m_2 = -1$, or $m_2 = -1/m_1$.

In our proof, choosing lines through the origin was a matter of convenience; the argument extends easily to arbitrary lines. Thus the lines $y = \frac{2}{3}x - 4$ and $y = -\frac{3}{2}x + 1$ are perpendicular. So are $5x - 3y = 5$ and $3x + 5y = -4$, since the first has slope $\frac{5}{3}$ and the second has slope $-\frac{3}{5}$.

LINES AND LINEAR FUNCTIONS

We claimed in our introduction that the graph of a linear function $f(x) = ax + b$ is always a line. Our discussion shows that it is the line with slope a and y-intercept b. For example, the graph of $f(x) = -2x + 3$ is a line with slope -2 and y-intercept 3. Conversely, every nonvertical line has a slope a and a y-intercept b and so is the graph of the linear function $f(x) = ax + b$.

Note the word *nonvertical* in the statement of the converse. A vertical line is not the graph of a function since a function f must associate just one value $f(x)$ with each x. However, any line, vertical or nonvertical, can be described by an equation. In fact, we have shown that every line has an equation of the form $Ax + By + C = 0$.

Problem Set 2-2

Find the slope of the line containing the given two points.

1. (2, 3) and (4, 8)
2. (4, 1) and (8, 2)
3. (−4, 2) and (3, 0)
4. (2, −4) and (0, −6)
5. (3, 0) and (0, 5)
6. (−6, 0) and (0, 6)
[c] 7. (−1.732, 5.014) and (4.315, 6.175)
[c] 8. (π, $\sqrt{3}$) and (1.642, $\sqrt{2}$)

Find an equation for each of the lines in problems 9–18. Then write your answer in the form $Ax + By + C = 0$.

9. Through (2, 3) with slope 4.
10. Through (4, 2) with slope 3.
11. Through (3, −4) with slope −2.
12. Through (−5, 2) with slope −1.
13. With y-intercept 4 and slope −2.
14. With y-intercept −3 and slope 1.
15. Through (2, 3) and (4, 8).
16. Through (4, 1) and (8, 2).
17. Through (1, 1) and (1, 6).
18. Through (1, 1) and (6, 1).

EXAMPLE A (Linear Functions)   Find a formula for the linear function f that satisfies $f(-2) = 1$ and $f(3) = 16$. Then calculate $f(5)$.

Solution.   The graph of f goes through the points (−2, 1) and (3, 16). The line through these points has slope $(16 - 1)/(3 + 2) = 3$. The equation of this line (in point-slope form) is

$$y - 1 = 3(x + 2)$$

Solved for y, this gives $y = 3x + 7$. Thus $f(x) = 3x + 7$ and $f(5) = 3(5) + 7 = 22$.

In Problems 19–24, find a formula for $f(x)$ if f is a linear function satisfying the given conditions. Then calculate $f(5)$.

19. $f(0) = -4; f(2) = 6$
20. $f(0) = 2; f(2) = -4$
21. $f(-2) = 4; f(3) = 2$
22. $f(1) = -2; f(6) = 7$
23. $f(-1) = 2; f(3) = 8$
24. $f(1) = 2; f(3) = -4$

In Problems 25–32, find the slope and y-intercept of each line.

25. $y = 3x + 5$ 26. $y = 6x + 2$

27. $3y = 2x - 4$ 28. $2y = 5x + 2$

29. $2x + 3y = 6$ 30. $4x + 5y = -20$

31. $y + 2 = -4(x - 1)$ 32. $y - 3 = 5(x + 2)$

33. Write the equation of the line through $(3, -3)$
   (a) parallel to the line $y = 2x + 5$;
   (b) perpendicular to the line $y = 2x + 5$;
   (c) parallel to the line $2x + 3y = 6$;
   (d) perpendicular to the line $2x + 3y = 6$;
   (e) parallel to the line through $(-1, 2)$ and $(3, -1)$;
   (f) parallel to the line $x = 8$;
   (g) perpendicular to the line $x = 8$.

34. Find the value of k for which the line $4x + ky = 5$
   (a) passes through the point $(2, 1)$;
   (b) is parallel to the y-axis;
   (c) is parallel to the line $6x - 9y = 10$;
   (d) has equal x- and y-intercepts;
   (e) is perpendicular to the line $y - 2 = 2(x + 1)$.

35. Write the equation of the line through $(0, -4)$ that is perpendicular to the line $y + 2 = -\frac{1}{2}(x - 1)$.

36. Find the value of k such that the line $kx - 3y = 10$
   (a) is parallel to the line $y = 2x + 4$;
   (b) is perpendicular to the line $y = 2x + 4$;
   (c) is perpendicular to the line $2x + 3y = 6$.

EXAMPLE B (Intersection of Two Lines)   Find the coordinates of the point of intersection of the lines $3x - 4y = 5$ and $x + 2y = 5$ (Figure 15).

Solution.   We simply solve the two equations simultaneously. Multiply the second equation by -3 and then add the two equations.

$$3x - 4y = 5$$
$$\underline{-3x - 6y = -15}$$
$$-10y = -10$$
$$y = 1$$
$$x = 3$$

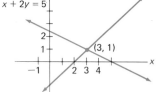

Figure 15

Find the coordinates of the point of intersection in each problem below. Then write the equation of the line through that point perpendicular to the line given first.

37. $2x + 3y = 4$
      $-3x + y = 5$

38. $4x - 5y = 8$
     $2x + y = -10$

39. $3x - 4y = 5$
     $2x + 3y = 9$

40. $5x - 2y = 5$
     $2x + 3y = 6$

EXAMPLE C (Distance from a Point to a Line)   It can be shown that the distance d from the point $(x_1\ y_1)$ to the line $Ax + By + C = 0$ is

$$d = \frac{|Ax_1 + By_1 + C|}{\sqrt{A^2 + B^2}}$$

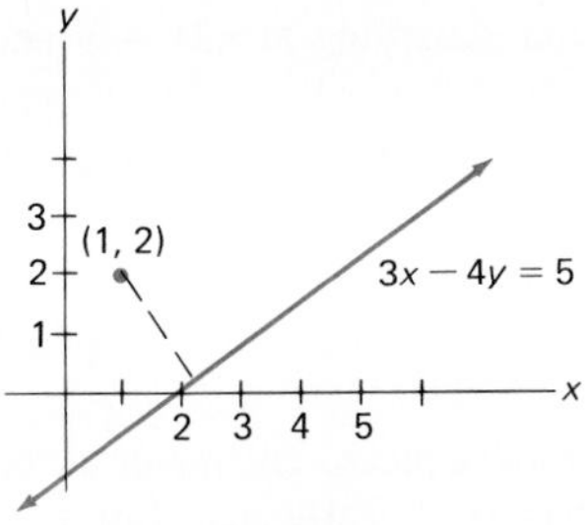

Find the distance from $(1, 2)$ to $3x - 4y = 5$ (see Figure 16).

Solution. First write the equation as $3x - 4y - 5 = 0$. The formula gives

$$d = \frac{|3 \cdot 1 - 4 \cdot 2 - 5|}{\sqrt{3^2 + (-4)^2}} = \frac{|-10|}{5} = 2$$

Figure 16

In each case, find the distance from the given point to the given line.

41. $(-3, 2)$, $3x + 4y = 6$

42. $(4, -1)$, $2x - 2y + 4 = 0$

43. $(-2, -1)$, $5y = 12x + 1$

44. $(3, -1)$, $y = 2x - 5$

Find the (perpendicular) distance between the given parallel lines. Hint: First find a point on one of the lines.

45. $3x + 4y = 6$, $3x + 4y = 12$

46. $5x + 12y = 2$, $5x + 12y = 7$

MISCELLANEOUS PROBLEMS

47. Which pairs of the lines below are parallel, which are perpendicular, and which are neither?
    (a) $y = 3x - 2$, $6x - 2y = 0$
    (b) $x = -2$, $y = 4$
    (c) $x = 2(y - 2)$, $y = -\frac{1}{2}(x - 1)$
    (d) $2x + 5y = 3$, $10x - 4y = 7$

48. Find the equation of the line through the point $(2, -1)$ that:
    (a) passes through $(-3, 5)$;
    (b) is parallel to $2x - 3y = 5$;
    (c) is perpendicular to $x + 2y = 3$;
    (d) is perpendicular to the y-axis.

49. Find the equation of the line passing through the intersection of the lines $x + 2y = 1$ and $3x + 2y = 5$ that is parallel to the line $3x - 2y = 4$.

50. A line L is perpendicular to the line $2x + 3y = 6$ and passes through $(-3, 1)$. Where does L cut the y-axis?

51. Find the equation of the perpendicular bisector of the line segment connecting $(3, -2)$ and $(7, 6)$.

52. The center of the circle that circumscribes a triangle is at the intersection of the perpendicular bisectors of the sides. Use this fact to find the center of the circle that goes through the three points $(0, 8)$, $(6, 2)$, and $(12, 14)$.

53. Show that the equation of the line with x-intercept a and y-intercept b (both a and b nonzero) can be written in the form

$$\frac{x}{a} + \frac{y}{b} = 1$$

54. Show that the graph of $Ax + By + C = 0$ is always a line (provided A and B are not both 0). *Hint:* Consider two cases: (i) $B = 0$ and (ii) $B \neq 0$.

55. A certain line passes through $(3, 2)$ and its nonzero y-intercept is twice its x-intercept. Find the equation of the line.

56. For each k, the equation $2x - y + 4 + k(x + 3y - 6) = 0$ represents a line (why?). One value of k determines a line with slope $\frac{3}{4}$. Where does this line cut the y-axis?

57. Find the formula for $f(x)$ if f is a linear function satisfying $f(-2) = 8$ and $f(1) = -1$. Then find r such that $f(r) = 0$.

58. A small company estimates that every additional \$2 it spends on advertising will produce an additional \$15 in sales. Last year, when it spent only \$800 on advertising, its sales were \$60,000. Express its expected sales $S(x)$ in terms of an advertising budget of x dollars and use it to predict the yearly sales if it spends \$2000 on advertising.

59. The ABC company makes zeebos which it sells for \$20 a piece. The material and labor to make a zeebo cost \$16 and the company has fixed yearly costs (utilities, real estate taxes, and so on) of \$8500. Write an expression for the company's profit $P(x)$ in a year in which it makes and sells x zeebos. What is its profit in a year in which it makes only 2000 zeebos?

60. A piece of equipment purchased today for \$80,000 will depreciate linearly to a scrap value of \$2000 after 20 years. Write a formula for its value V after t years.

61. Find the distance between the parallel lines $12x - 5y = 2$ and $12x - 5y = 7$.

62. Find a formula for the distance between the lines $y = mx + b$ and $y = mx + B$ in terms of m, b, and B.

63. Show that the line through the midpoints of two sides of a triangle is parallel to the third side.

64. Let the three vertices of a triangle lie on a circle with two of them being the ends of a diameter. Show that the triangle is a right triangle. *Hint:* Place the triangle in the coordinate system so two of its vertices are $(-a, 0)$ and $(a, 0)$.

65. Find the two points in the plane that are simultaneously equidistant from the lines $y = \pm x$ and the point $(5, 3)$.

66. Using the same axes, sketch the graphs of $x^2 + y^2 = 1$ and $|x| + |y| = 1$ and compute the area of the region between the two graphs.

67. A line through $(4, 4)$ is tangent to the circle $x^2 + y^2 = 4$ at a point P in the fourth quadrant. Find the coordinates of P.

68. **TEASER**   Suppose that 2 million points in the plane are given. Show that there is a line with exactly 1 million of these points on each side of it. Can you always find a circle with exactly 1 million of these points inside it?

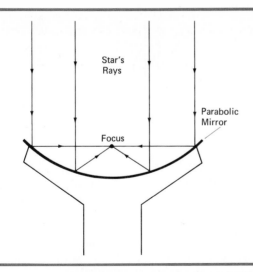

Star's
Rays

Parabolic
Mirror

Focus

Reflecting Telescopes

A reflecting telescope, such as the one at Mount Palomar in California, makes use of a parabolic mirror. The rays of light from a star (parallel lines) are focused by the mirror at a single point. This is a characteristic property of the mathematical curve that we call a parabola.

2-3   Quadratic Functions

We have seen that the graph of the linear function $f(x) = ax + b$ is always a line. Now we want to study the graph of the quadratic function $f(x) = ax^2 + bx + c$ $(a \neq 0)$. Its graph is, of course, the graph of the equation $y = ax^2 + bx + c$. As we will discover, it is always a smooth, cup-shaped curve something like the cross section of the mirror shown in the opening diagram. We call it a *parabola*.

The simplest case is $y = x^2$. The graph of this equation is shown in Figure 17. Two important features should be noted.

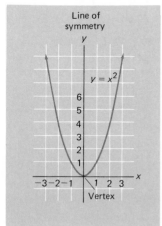

Line of
symmetry

$y = x^2$

Vertex

Figure 17

1. The curve is symmetric about the y-axis. This follows from the fact that the equation $y = x^2$ is not changed if we replace (x, y) by $(-x, y)$.
2. The curve reaches its lowest point at $(0, 0)$, the point where the curve intersects the line of symmetry. We call this point the **vertex** of the parabola.

Next we consider how the graph of $y = x^2$ is modified as we look successively at $y = ax^2$, $y = x^2 + k$, $y = (x - h)^2$, and $y = a(x - h)^2 + k$.

THE GRAPH OF $y = ax^2$

In Figure 18, we show the graphs of $y = x^2$, $y = 3x^2$, $y = -2x^2$, and $y = \frac{1}{2}x^2$. They suggest the following general facts. The graph of $y = ax^2$, $a \neq 0$, is a parabola with vertex at the origin, opening upward if $a > 0$ and downward if $a < 0$. Increasing $|a|$ makes the graph narrower.

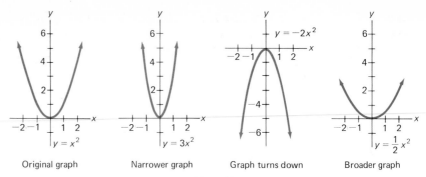

Original graph   Narrower graph   Graph turns down   Broader graph

Figure 18

THE GRAPHS OF $y = x^2 + k$ and $y = (x - h)^2$

The graphs of $y = x^2 + 4$, $y = x^2 - 6$, $y = (x - 2)^2$, and $y = (x + 1)^2$ can all be obtained by shifting (translating) the graph of $y = x^2$ while maintaining its shape. They are shown in Figure 19. The general situation is as follows.

The graph of $y = x^2 + k$ is obtained by shifting the graph of $y = x^2$ vertically $|k|$ units, upward if $k > 0$ and downward if $k < 0$. The vertex is at $(0, k)$.

The graph of $y = (x - h)^2$ is obtained by a horizontal shift of $|h|$ units, to the right if $h > 0$ and to the left if $h < 0$. The vertex is at $(h, 0)$.

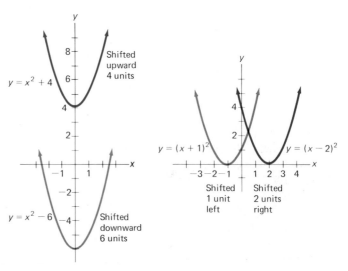

Figure 19

THE GRAPH OF $y = a(x - h)^2 + k$

We can get the graph of $y = 2(x - 3)^2 - 4$ by shifting the graph of $y = 2x^2$ 3 units to the right and 4 units down. This puts the vertex at $(3, -4)$. More generally, the graph of $y = a(x - h)^2 + k$ is the graph of $y = ax^2$ shifted horizontally $|h|$ units and vertically $|k|$ units, so that the vertex is at (h, k). The graphs in Figure 20 illustrate these facts.

What about the graph of $y = -2(x - 4)^2 + 6$? It has the same shape as the graph of $y = 2x^2$, but turns down with its vertex at $(4, 6)$.

Recall for the circle $x^2 + y^2 = r^2$ that replacing x by $x - h$ and y by $y - k$ to obtain $(x - h)^2 + (y - k)^2 = r^2$ had the effect of shifting the center from $(0, 0)$ to (h, k). A similar thing happens to the parabola $y = ax^2$. Replacing x by $x - h$ and y by $y - k$ changes $y = ax^2$ to $y - k = a(x - h)^2$ (which you will note is equivalent to $y = a(x - h)^2 + k$) and correspondingly shifts the vertex from $(0, 0)$ to (h, k).

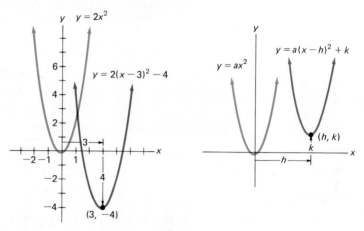

Figure 20

PARABOLAS AND QUADRATIC FUNCTIONS

The most general equation considered so far is $y = a(x - h)^2 + k$. The graph of this equation is a parabola with vertex (h, k) and line of symmetry $x = h$. If we expand $a(x - h)^2$ and collect terms on the right side, the equation takes the form $y = ax^2 + bx + c$. Conversely, $y = ax^2 + bx + c \, (a \neq 0)$ always represents a parabola with a vertical line of symmetry, as we shall now show. We use the method of completing the square to rewrite $y = ax^2 + bx + c$ as follows:

$$y = a\left(x^2 + \frac{b}{a}x + \quad\right) + c$$

$$= a\left[x^2 + \frac{b}{a}x + \left(\frac{b}{2a}\right)^2\right] + c - a\left(\frac{b}{2a}\right)^2$$

$$= a\left(x + \frac{b}{2a}\right)^2 + \left(c - \frac{b^2}{4a}\right)$$

This is the equation of a parabola with vertex at $(-b/2a, \, c - b^2/4a)$ and line of symmetry $x = -b/2a$.

What we have just said leads to an important conclusion. The graph of the quadratic function $f(x) = ax^2 + bx + c\,(a \neq 0)$ is always a parabola with vertical axis of symmetry. Moreover, we have the following facts.

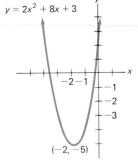
$y = 2x^2 + 8x + 3$

$(-2, -5)$

Figure 21

THE PARABOLA $y = f(x) = ax^2 + bx + c$

1. The x-coordinate of the vertex is $-b/2a$; the y-coordinate is easily obtained by substitution in the equation $y = f(x)$.
2. The parabola turns upward if $a > 0$ and downward if $a < 0$. It is a fat or thin parabola depending on whether $|a|$ is small or large.

As an example, consider $y = f(x) = 2x^2 + 8x + 3$. Its graph is a parabola with vertex at $x = -b/2a = -8/4 = -2$. The y-coordinate of the vertex, obtained by substituting $x = -2$ in the equation, is $y = -5$. The parabola turns upward, and it is rather thin. A sketch is shown in Figure 21.

APPLICATIONS

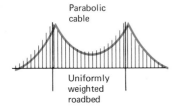
Parabolic cable

Uniformly weighted roadbed

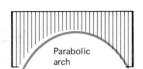
Parabolic arch

Figure 22

Our opening panel hinted at one important application of the parabola related to its optical properties. This application is treated in Chapter 11, where we again discuss parabolas but from a more geometric point of view.

If equal weights are equally spaced along a line and suspended from a thin cable, the cable will assume the approximate shape of a parabola. This fact is used in the design of suspension bridges. A related property is used in the design of arches that are to support a uniform load (see Figure 22).

We turn to a very different kind of application. From physics, we learn that the path of a projectile is a parabola. It is known, for example, that a projectile fired at an angle of 45° from the horizontal with an initial speed of $320\sqrt{2}$ feet per second follows a curve with equation

$$y = -\frac{1}{6400}x^2 + x$$

where the coordinate axes are placed as shown in Figure 23. Taking this for granted, we may ask two questions.

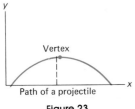
Vertex

Path of a projectile

Figure 23

1. What is the maximum height attained by the projectile?
2. What is the range (horizontal distance traveled) of the projectile?

To find the maximum height is simply to find the y-coordinate of the vertex. First we find the x-coordinate.

$$x = \frac{-b}{2a} = -\frac{1}{-2/6400} = 3200$$

When we substitute this value in the equation, we get

$$y = -\frac{1}{6400}(3200)^2 + 3200 = -1600 + 3200 = 1600$$

The greatest height is thus 1600 feet.

The range of the projectile is the x-coordinate of the point where it lands. By symmetry, this is simply twice the x-coordinate of the vertex; that is,

$$\text{range} = 2(3200) = 6400 \text{ feet}$$

This value could also be obtained by solving the quadratic equation

$$-\frac{1}{6400}x^2 + x = 0$$

since the x-coordinate of the landing point is the value of x when $y = 0$.

Problem Set 2-3

The equations in Problems 1–10 represent parabolas. Sketch the graph of each parabola, indicating the coordinates of the vertex.

1. $y = 3x^2$
2. $y = -2x^2$
3. $y = x^2 + 5$
4. $y = 2x^2 - 4$
5. $y = (x - 4)^2$
6. $y = -(x + 3)^2$
7. $y = 2(x - 1)^2 + 5$
8. $y = 3(x + 2)^2 - 4$
9. $y = -4(x - 2)^2 + 1$
10. $y = \frac{1}{2}(x + 3)^2 + 3$

Write each of the following in the form $y = ax^2 + bx + c$.

11. $y = 2(x - 1)^2 + 7$
12. $y = -3(x + 2)^2 + 5$
13. $-2y + 5 = (x - 5)^2$
14. $3y + 6 = (x + 3)^2$

Sketch the graph of each function. Begin by plotting the vertex and at least one point on each side of the vertex. Recall that the x-coordinate of the vertex for $y = f(x) = ax^2 + bx + c$ is $x = -b/2a$.

15. $f(x) = x^2 + 2x$
16. $f(x) = 3x^2 - 6x$
17. $f(x) = -2x^2 + 8x + 1$
18. $f(x) = -3x^2 + 6x + 4$

EXAMPLE A (Horizontal Parabolas) Sketch the graphs of (a) $x = 2y^2$; (b) $x = 2(y - 3)^2 - 4$.

Solution. Note that the roles of x and y are interchanged when compared to earlier examples. Now x is a function of y; that is, each equation has the form $x = g(y)$. The graphs are parabolas with horizontal lines of symmetry (Figure 24). The vertex in (a) is (0,0); in (b), it is (−4, 3).

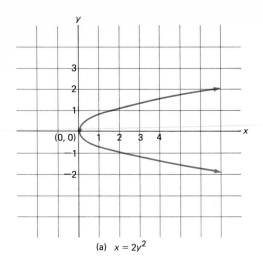

(a) $x = 2y^2$

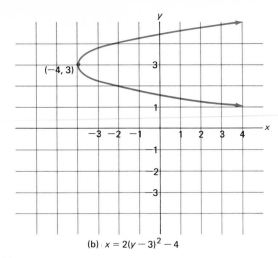

(b) $x = 2(y - 3)^2 - 4$

Figure 24

In Problems 19–24, sketch the graph of the function g and indicate the coordinates of the vertex.

19. $x = g(y) = -2y^2$ 20. $x = g(y) = -2y^2 + 8$
21. $x = g(y) = -2(y + 2)^2 + 8$ 22. $x = g(y) = 3(y - 1)^2 + 6$
23. $x = g(y) = y^2 + 4y + 2$.
Note: The *y*-coordinate of the vertex is at $-b/2a$.
24. $x = g(y) = 4y^2 - 8y + 10$

EXAMPLE B (Intersection of a Line and a Parabola)   Find the points of intersection of the line $y = -2x + 2$ and the parabola $y = 2x^2 - 4x - 2$ (see Figure 25).

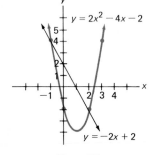

Figure 25

Solution.   We must solve the two equations simultaneously. This is easy to do by substituting the expression for *y* from the first equation into the second equation and then solving the resulting equation for *x*.

$$-2x + 2 = 2x^2 - 4x - 2$$
$$0 = 2x^2 - 2x - 4$$
$$0 = 2(x^2 - x - 2)$$
$$0 = 2(x - 2)(x + 1)$$
$$x = -1 \qquad x = 2$$

By substitution, we find the corresponding values of *y* to be 4 and -2; the intersection points are therefore $(-1, 4)$ and $(2, -2)$.

Find the points of intersection for each pair of equations.

25. $y = -x + 1$ 26. $y = -x + 4$
$\quad y = x^2 + 2x + 1$ $\quad y = -x^2 + 2x + 4$

27. $y = -2x + 1$
    $y = -x^2 - x + 3$

28. $y = -3x + 15$
    $y = 3x^2 - 3x + 12$

© 29. $y = 1.5x + 3.2$
    $y = x^2 - 2.9x$

© 30. $y = 2.1x - 6.4$
    $y = -1.2x^2 + 4.3$

EXAMPLE C (Using Side Conditions) Find the equation of the vertical parabola with vertex $(2, -3)$ and passing through the point $(9, -10)$.

Solution. The equation must have the form

$$y = a(x - 2)^2 - 3$$

To find a, we substitute $x = 9$ and $y = -10$.

$$-10 = a(7)^2 - 3$$

Thus $a = -\frac{1}{7}$ and the required equation is

$$y = -\frac{1}{7}(x - 2)^2 - 3.$$

In Problems 31–34, find the equation of the vertical parabola that satisfies the given conditions.

31. Vertex $(0, 0)$, passing through $(-6, 3)$.
32. Vertex $(0, 0)$, passing through $(3, -6)$.
33. Vertex $(-2, 0)$, passing through $(6, -8)$.
34. Vertex $(3, -1)$, passing through $(-2, 5)$.
35. Find the equation of the parabola passing through $(1, 1)$ and $(2, 7)$ and having the y-axis as the line of symmetry. *Hint:* The equation has the form $y = ax^2 + k$.
36. Find the equation of the parabola passing through $(1, 2)$ and $(-2, -7)$ and having the y-axis as the line of symmetry.

MISCELLANEOUS PROBLEMS

37. In each case, find the coordinates of the vertex and then sketch the graph. Use the same coordinate axes for all three graphs.
    (a) $y = 2x^2 - 8x + 4$ (b) $y = 2x^2 - 8x + 8$
    (c) $y = 2x^2 - 8x + 11$
38. Sketch the following parabolas using the same coordinate axes.
    (a) $y = -3x^2 + 6x + 9$ (b) $y = -3x^2 + 6x - 3$
    (c) $y = -3x^2 + 6x - 9$
39. Sketch the graphs of the following parabolas.
    (a) $y = \frac{1}{2}x^2 - 2x$ (b) $x = \frac{1}{2}y^2 - 2y$
40. Recall that the discriminant d of $ax^2 + bx + c$ is $d = b^2 - 4ac$ (see page 36). Calculate d for each of the parabolas of Problem 37. Then show that the graph of $y = ax^2 + bx + c$ will cross the x-axis, just touch the x-axis, or not meet the x-axis, according as $d > 0$, $d = 0$, or $d < 0$.
41. Find the points of the intesection (if any) of the following pairs of curves.
    (a) $y = x^2 - 2x + 6$ (b) $y = x^2 - 4x + 6$
    $y = -3x + 8$ $y = 2x - 3$

(c) $y = -x^2 + 2x + 4$ (d) $y = x^2 - 2x + 7$
$y = -2x + 9$ $y = 11 - x^2$

42. For what values of k does the parabola $y = x^2 - kx + 4$ have two x-intercepts?

43. The parabola $y = a(x - 2)(x - 8)$ passes through $(10, 40)$. Find a and the vertex of this parabola.

44. Find the equation of the vertical parabola that passes through $(-1, -2)$, $(0, 3)$, and $(2, 7)$.

45. If the curve shown in Figure 26 is part of a parabola, find the distance $\overline{PQ}$. *Hint:* Begin by finding the equation of the parabola, assuming its vertex is at the origin.

46. The parabolic cable for a suspension bridge is attached to the two towers at points 400 feet apart and 90 feet above the horizontal bridge deck. The cables drop to a point 10 feet above the deck. Find the xy-equation of the cable, assuming it is symmetric about the y-axis with vertex at $(0, 10)$.

47. A retailer has learned from experience that if she charges x dollars apiece for toy trucks, she can sell $300 - 100x$ of them. The trucks cost her \$2 each. Write a formula for her total profit P in terms of x. Then determine what she should charge for each truck to maximize her profit.

48. A company that makes fancy golf carts has fixed overhead costs of \$12,000 per year and direct costs (labor and materials) of \$80 per cart. It sells its carts to a certain retailer at a nominal price of \$120 each. However, the company offers a discount of 1 percent for 100 carts, 2 percent for 200 carts, and in general $x/100$ percent for x carts. Assume the retailer will buy as many carts as the company can produce but that its production facilities limit this to a maximum of 1800 units.
    (a) Write a formula for C, the cost of producing x carts.
    (b) Show that its total revenue R in dollars is $R = 120x - .012x^2$.
    (c) What is the smallest number of carts it can produce and still break even?
    [c] (d) **What number of carts will produce the maximum profit and what is this profit?**

49. Find the area (in terms of a and b) of the triangle shown in Figure 27. The point c is midway between a and b.

50. Starting at $(0, 0)$, a ball travels along the path $y = ax^2 + bx$. Find a and b if the ball reaches a height of 75 at $x = 50$ and its maximum height at $x = 100$.

51. In Figure 28, FG is parallel to the x-axis and RG is parallel to the y-axis. Show that the length $L = \overline{FR} + \overline{RG}$ is independent of where R is on the parabola by finding a formula for L in terms of p alone.

52. **TEASER** Let $\{a_1, a_2, \ldots, a_n\}$ and $\{b_1, b_2, \ldots, b_n\}$ be two sets of n numbers each. Show that the inequality

$$(a_1 b_1 + a_2 b_2 + \cdots + a_n b_n)^2$$
$$\leq (a_1^2 + a_2^2 + \cdots + a_n^2)(b_1^2 + b_2^2 + \cdots + b_n^2)$$

always holds. *Hint:* See if you can use the fact that if $Ax^2 + 2Bx + C \geq 0$ for all x, then $B^2 - AC \leq 0$ (see Problem 40).

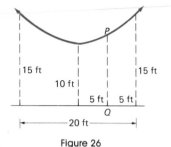

Figure 26

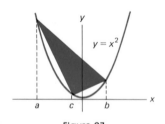

Figure 27

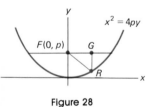

Figure 28

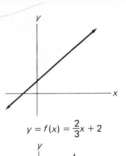

$$y = f(x) = \frac{2}{3}x + 2$$

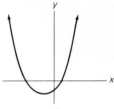

$$y = f(x) = x^2 + 2x - 3$$

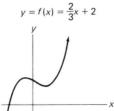

$$y = f(x) = x^3 - 3x + 4$$

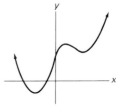

$$y = f(x) = x^4 - x^3 - 8x^2 + 16x + 20$$

Geometry, however, supplies sustenance and meaning to bare formulas . . . . One can still believe Plato's statement that "geometry draws the soul toward truth."

Morris Kline

2-4   Polynomial Functions

So far we have considered linear functions $f(x) = ax + b$ and quadratic functions $f(x) = ax^2 + bx + c$, $a \neq 0$. They are special cases of polynomial functions—that is, functions with formulas of the form

$$f(x) = a_n x^n + a_{n-1} x^{n-1} + \cdots + a_1 x + a_0$$

We are interested primarily in the graphs of these functions, for as Morris Kline has suggested, geometric pictures give meaning to bare formulas. We know from previous sections that the graph of a linear function is always a line and that the graph of a quadratic function is always a parabola. But what can we say about the graphs of cubic functions, quartic functions, and so on?

POLYNOMIAL FUNCTIONS OF DEGREE $n \geq 3$

The graphs of higher degree polynomial functions are harder to describe, but after we have studied two examples, we can offer some general guidelines. Consider first the cubic function

$$f(x) = x^3 - 3x + 4$$

With the help of a table of values, we sketch its graph (Figure 29).

Notice that for large positive values of x, the values of y are large and positive; similarly, for large negative values of x, y is large and negative. This is due to the dominance of the leading term x^3 for large $|x|$. This dominance is responsible for a drooping left arm and a right arm held high on the graph. Notice also that the graph has one hill and one valley. This is typical of the graph of a cubic function, though it is possible for it to have no hills or valleys. The graph of $y = x^3$ illustrates this latter behavior (see page 56).

x	y
−3	−14
−2	2
−1	6
0	4
1	2
2	6
3	22

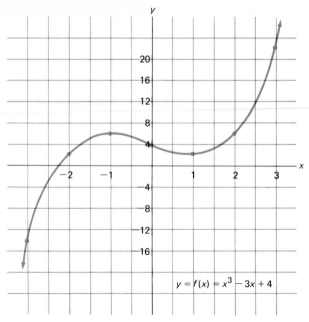

$y = f(x) = x^3 - 3x + 4$

Figure 29

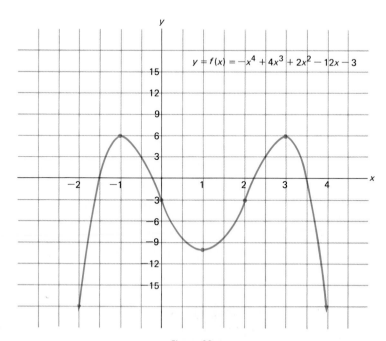

$y = f(x) = -x^4 + 4x^3 + 2x^2 - 12x - 3$

x	y
−2	−19
−1	6
0	−3
1	−10
2	−3
3	6
4	−19

Figure 30

Next consider a typical fourth degree polynomial function.

$$f(x) = -x^4 + 4x^3 + 2x^2 - 12x - 3$$

A table of values and the graph are shown in Figure 30.

The leading term $-x^4$, which is negative for all values of x, determines that the graph has two drooping arms. Note that there are two hills and one valley.

In general, we can make the following statements about the graph of

$$f(x) = a_n x^n + a_{n-1} x^{n-1} + \cdots + a_1 x + a_0, \qquad a_n \neq 0$$

1. If n is even and $a_n < 0$, the graph will have two drooping arms; if n is even and $a_n > 0$, it will have both arms raised. This is due to the dominance of $a_n x^n$ for large values of $|x|$.

2. If n is odd, one arm droops and the other points upward. Again, this is dictated by the dominance of $a_n x^n$.

3. The combined number of hills and valleys cannot exceed $n - 1$, although it can be less.

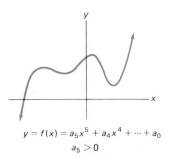

$y = f(x) = a_5 x^5 + a_4 x^4 + \cdots + a_0$

$a_5 > 0$

Figure 31

Based on these facts, we expect the graph of a fifth degree polynomial function with a positive leading coefficient to look something like the graph in Figure 31.

FACTORED POLYNOMIAL FUNCTIONS

The task of graphing can be simplified considerably if our polynomial is factored. The real solutions of $f(x) = 0$ correspond to the x-intercepts of the graph of $y = f(x)$—that is, to the x-coordinates of the points where the graph intersects the x-axis. If the polynomial is factored, these intercepts are easy to find.

Consider, as an example,

$$y = f(x) = x(x + 3)(x - 1)$$

The solutions of $f(x) = 0$ are 0, -3, and 1; these are the x-intercepts of the graph. Clearly, $f(x)$ cannot change signs between adjacent x-intercepts, since only at these points can any of the linear factors change sign. The signs of $f(x)$ on the four intervals determined by $x = -3$, $x = 0$, and $x = 1$ are shown in Figure 32 (to check this, try substituting an x-value from each of these intervals, as in the split-point method of Section 1-5).

With this information and just a few plotted points, we can easily sketch the graph. It is shown in Figure 33.

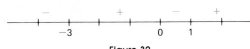

Figure 32

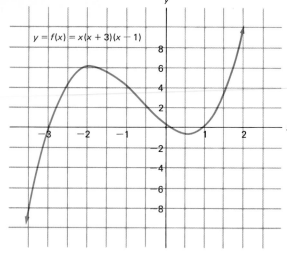

x	y
−2	6
−1	4
.5	−.88
2	10

Figure 33

FUNCTIONS WITH MULTI-PART RULES

Sometimes a function has polynomial components, even though it is not a polynomial function. Especially notable is the absolute value function $f(x) = |x|$, which has the two-part rule

$$f(x) = \begin{cases} -x & \text{if } x < 0 \\ x & \text{if } x \geq 0 \end{cases}$$

For $x < 0$, the graph of this function coincides with the line $y = -x$; for $x \geq 0$, it coincides with the line $y = x$ (Figure 34). Note the sharp corner at the origin.

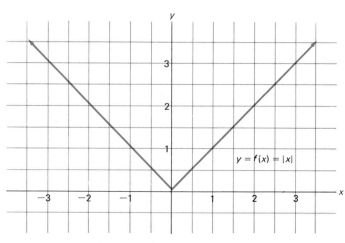

x	|x|
−3	3
−2	2
−1	1
0	0
1	1
2	2
3	3

Figure 34

Here is a more complicated example.

$$g(x) = \begin{cases} x + 2 & \text{if } x < 0 \\ x^2 & \text{if } 0 \le x \le 2 \\ 4 & \text{if } x > 2 \end{cases}$$

Though this way of describing a function may seem strange, it is not at all unusual in more advanced courses. The graph of g consists of three pieces (Figure 35).

1. A part of the line $y = x + 2$.
2. A part of the parabola $y = x^2$.
3. A part of the horizontal line $y = 4$.

Note the use of the open circle at $(0, 2)$ to indicate that this point is not part of the graph.

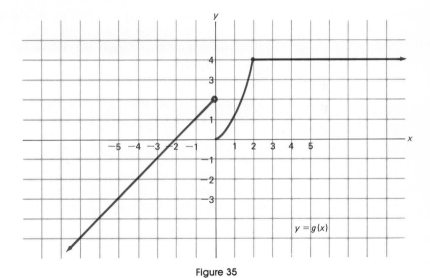

Figure 35

Problem Set 2-4

Graph each of the following polynomial functions. The first two are called constant functions.

1. $f(x) = 5$
2. $f(x) = -4$
3. $f(x) = -3x + 5$
4. $f(x) = 4x - 3$
5. $f(x) = x^2 - 5x + 4$
6. $f(x) = x^2 + 2x - 3$
7. $f(x) = x^3 - 9x$
8. $f(x) = x^3 - 16x$
9. $(x) = 2.12x^3 - 4.13x + 2$
10. $f(x) = -1.2x^3 + 2.3x^2 - 1.4x$

Graph each of the following functions.

11. $f(x) = 2|x|$

12. $f(x) = |x| - 2$

13. $f(x) = |x - 2|$

14. $f(x) = |x| + 2$

15. $f(x) = \begin{cases} x & \text{if } x < 0 \\ 2 & \text{if } x \geq 0 \end{cases}$

16. $f(x) = \begin{cases} -1 & \text{if } x \leq 0 \\ 2x & \text{if } x > 0 \end{cases}$

17. $f(x) = \begin{cases} -5 & \text{if } x \leq -3 \\ 4 - x^2 & \text{if } -3 < x \leq 3 \\ -5 & \text{if } x > 3 \end{cases}$

18. $f(x) = \begin{cases} 9 & \text{if } x < 0 \\ 9 - x^2 & \text{if } 0 \leq x \leq 3 \\ x^2 - 9 & \text{if } x > 3 \end{cases}$

EXAMPLE A (Symmetry Properties)   A function f is called an **even function** if $f(-x) = f(x)$ for all x in its domain. The graph of an even function is symmetric with respect to the y-axis. A function g is called an **odd function** if $g(-x) = -g(x)$ for all x in its domain; its graph is symmetric with respect to the origin. (See Section 1–6 for a full discussion of symmetry.) Graph the following two functions, observing their symmetries.

(a) $f(x) = x^4 + x^2 - 3$ (b) $g(x) = x^3 + 2x$

Solution.   Notice that

$$f(-x) = (-x)^4 + (-x)^2 - 3 = x^4 + x^2 - 3 = f(x)$$

$$g(-x) = (-x)^3 + 2(-x) = -x^3 - 2x = -g(x)$$

Thus f is even and g is odd. Their graphs are sketched in Figure 36. Note that a polynomial function involving only even powers of x is even, while one involving only odd powers of x is odd.

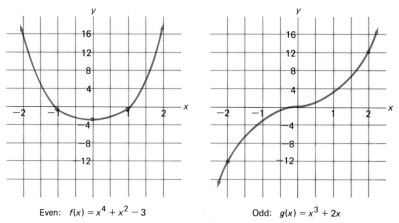

Even: $f(x) = x^4 + x^2 - 3$ Odd: $g(x) = x^3 + 2x$

Figure 36

Determine which of the following are even functions, which are odd functions, and which are neither. Then sketch the graphs of those that are even or odd, making use of the symmetry properties.

19. $f(x) = 2x^2 - 5$

20. $f(x) = -3x^2 + 2$

21. $f(x) = x^2 - x + 1$

22. $f(x) = -2x^3$

23. $f(x) = 4x^3 - x$

24. $f(x) = x^3 + x^2$

25. $f(x) = 2x^4 - 5x^2$

26. $f(x) = 3x^4 + x^2$

EXAMPLE B (More on Factored Polynomials)   Graph

$$f(x) = (x - 1)^2(x - 3)(x + 2)$$

Solution.   The x-intercepts are at $1, 3$, and -2. The new feature is that $x - 1$ occurs as a square. The factor $(x - 1)^2$ never changes sign, so the graph does not cross the x-axis at $x = 1$; it merely touches the axis there (Figure 37). Note the entries in the table of values corresponding to $x = 0.9$ and $x = 1.1$.

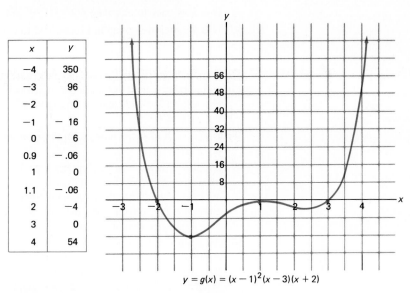

x	y
-4	350
-3	96
-2	0
-1	-16
0	-6
0.9	$-.06$
1	0
1.1	$-.06$
2	-4
3	0
4	54

$$y = g(x) = (x - 1)^2(x - 3)(x + 2)$$

Figure 37

Sketch the graph of each of the following.

27. $f(x) = (x + 1)(x - 1)(x - 3)$

28. $f(x) = x(x - 2)(x - 4)$

29. $f(x) = x^2(x - 4)$

30. $f(x) = x(x + 2)^2$

31. $f(x) = (x + 2)^2(x - 2)^2$

32. $f(x) = x(x - 1)^3$

MISCELLANEOUS PROBLEMS

Graph each of the functions in Problems 33–42.

33. $f(x) = -3$

34. $f(x) = 2x - 3$

35. $f(x) = x^2 - 1$

36. $f(x) = x^3 - 3x$

37. $f(x) = (x + 2)(x - 1)^2(x - 3)$ 38. $f(x) = x(x - 2)^3$

39. $f(x) = \begin{cases} 4 - x^2 & \text{if } -2 \le x < 2 \\ x - 1 & \text{if } \quad 2 \le x \le 4 \end{cases}$

40. $f(x) = \begin{cases} |x| & \text{if } -2 \le x \le 1 \\ x^2 & \text{if } \quad 1 < x \le 2 \\ 4 & \text{if } \quad 2 < x \le 4 \end{cases}$

41. $(x) = x^4 + x^2 + 2, -2 \le x \le 2$

42. $f(x) = x^5 + x^3 + x, -1 \le x \le 1$

43. The graphs of the three functions $f(x) = x^2$, $g(x) = x^4$, and $h(x) = x^6$ all pass through the points $(-1, 1)$, $(0, 0)$ and $(1, 1)$. Draw careful sketches of these three functions using the same axes. Be sure to show clearly how they differ for $-1 < x < 1$.

44. Sketch the graph of $f(x) = x^{50}$ for $-1 \le x \le 1$. Be sure to calculate $f(.5)$ and $f(.9)$. What simple figure does the graph resemble?

45. Notice that $f(x) = 3x^4 + 2x^3 - 3x^2 + x + 1$ can be written as

$$f(x) = [((3x + 2)x - 3)x + 1]x + 1$$

It is now easy to calculate $f(2), f(1.3), f(4.2)$ on a calculator. Do so.

46. Use the trick described in Problem 45 to evaluate $f(3), f(4.3)$, and $f(-1.6)$ for

$$f(x) = 4x^5 - 3x^4 + 2x^3 - x^2 + 7x - 3$$

47. The function $f(x) = [x]$ is called the **greatest integer function**. It assigns to each real number x the largest integer which is less than or equal to x. For example, $[\frac{5}{2}] = 2$, $[13] = 13$, and $[-14.25] = -15$. Graph this function on the interval $-2 \le x \le 6$.

48. Graph each of the following on the interval $-2 \le x \le 2$.
    (a) $f(x) = 3[x]$ (b) $g(x) = [3x]$

49. Suppose that the cost of shipping a package is 15¢ for anything weighing less than an ounce and 25¢ for anything weighing at least 1 ounce but less than 2 ounces. Beyond that the pattern continues with the cost increased 10¢ for each additional ounce or fraction thereof. Write a formula for the cost $C(x)$ of shipping a package weighing x ounces (using the symbol [ ]) and graph this function.

50. An open box is to be made from a piece of 12-inch by 18-inch cardboard by cutting a square of side x inches from each corner and turning up the sides. Express the volume $V(x)$ of the box in terms of x and graph the resulting function. What is the domain of V? Use your graph to help you find the value of x that makes $V(x)$ a maximum. What is this maximum value?

51. The function $f(x) = \langle x \rangle$ denotes the **distance to nearest integer function.** It assigns to each real number x the distance to the integer nearest to x. For example, $\langle 1.2 \rangle = .2$, $\langle 1.7 \rangle = .3$, and $\langle 2 \rangle = 0$. Graph this function on the interval $0 \le x \le 4$ and then find the area of the region between this graph and the x-axis.

52. **TEASER**   Consider the function $f(x) = \langle x \rangle / 10^{[x]}$ on the infinite interval $x \ge 0$. Sketch the graph of this function and calculate the *total* area of the region between the graph and the x-axis. Write this area first as an unending decimal and then as a ratio of two integers.

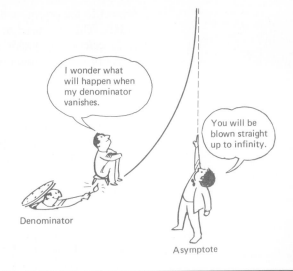

Rational Functions

The graph of the rational function

$$f(x) = \frac{p(x)}{q(x)}$$

exhibits spectacular behavior whenever the denominator $q(x)$ nears 0. It must either blow up to plus infinity or down to minus infinity.

Denominator

Asymptote

2-5 Rational Functions

If $f(x)$ is given by

$$f(x) = \frac{p(x)}{q(x)}$$

where $p(x)$ and $q(x)$ are polynomials, then f is called a **rational function.** For simplicity, we shall assume that $f(x)$ is in reduced form, that is, that $p(x)$ and $q(x)$ have no common nontrivial factors. Typical examples of rational functions are

$$f(x) = \frac{x + 1}{x^2 - x - 6} = \frac{x + 1}{(x - 3)(x + 2)}$$

$$g(x) = \frac{(x + 2)(x - 5)}{(x + 3)^3}$$

Graphing a rational function can be tricky, primarily because of the denominator $q(x)$. Whenever it is zero, something dramatic is sure to happen to the graph. That is the point of our opening cartoon.

THE GRAPHS OF $1/x$ AND $1/x^2$

Let us consider two simple cases.

$$f(x) = \frac{1}{x} \qquad g(x) = \frac{1}{x^2}$$

Notice that f is an odd function ($f(-x) = -f(x)$), while g is even

$(g(-x) = g(x))$. These facts imply that the graph of f is symmetric with respect to the origin, and that the graph of g is symmetric with respect to the y-axis. Thus we need to use only positive values of x to calculate y-values. Each calculation yields two points on the graph. Observe particularly the behavior of each graph near $x = 0$ (Figure 38).

In both cases, the x- and y-axes play special roles; we call them *asymptotes* for the graphs. If, as a point moves away from the origin along a curve, the distance between it and a line becomes closer and closer to zero, then that line is called an **asymptote** for the curve. Clearly the line $x = 0$ is a vertical asymptote for both of our curves and the line $y = 0$ is a horizontal asymptote for both of them.

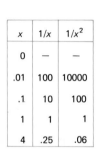

x	$1/x$	$1/x^2$
0	—	—
.01	100	10000
.1	10	100
1	1	1
4	.25	.06

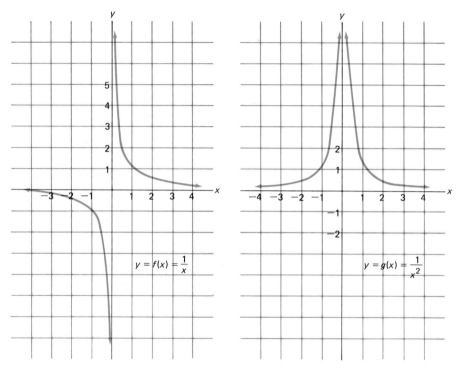

Figure 38

THE GRAPHS OF $1/(x - 2)$ AND $1/(x - 2)^2$

If we replace x by $x - 2$ in our two functions, we get two new functions.

$$h(x) = \frac{1}{x - 2} \qquad k(x) = \frac{1}{(x - 2)^2}$$

Their graphs are just like those of f and g except that they are moved two units to the right (Figure 39).

Two observations should be made. The vertical asymptote (dotted line) occurs where the denominator is zero; that is, it is the line $x = 2$. The horizontal asymptote is again the line $y = 0$.

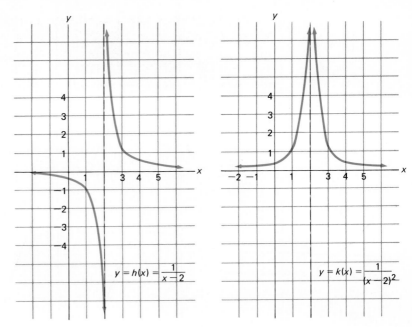

Figure 39

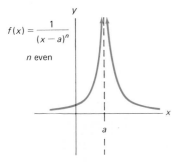

$$f(x) = \frac{1}{(x-a)^n}$$

n even

Figure 40

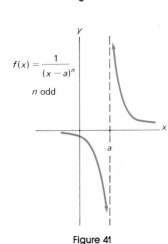

$$f(x) = \frac{1}{(x-a)^n}$$

n odd

Figure 41

In general, the graph of $f(x) = 1/(x-a)^n$ has the line $y = 0$ as a horizontal asymptote and the line $x = a$ as a vertical asymptote. The behavior of the graph near $x = a$ for n even and n odd is illustrated in Figures 40 and 41.

MORE COMPLICATED EXAMPLES

Consider next the rational function determined by

$$y = f(x) = \frac{x}{x^2 + x - 6} = \frac{x}{(x-2)(x+3)}$$

We expect its graph to have vertical asymptotes at $x = 2$ and $x = -3$. Again, the line $y = 0$ will be a horizontal asymptote since, as $|x|$ gets large, the term x^2 in the denominator will dominate, so that y will behave much like x/x^2 or $1/x$ and thus will approach zero. The graph crosses the x-axis where the numerator is zero, namely, at $x = 0$. Finally, with the help of a table of values, we sketch the graph, shown in Figure 42 at the top of the next page.

Lastly we consider

$$y = f(x) = \frac{2x^2 + 2x}{x^2 - 4x + 4} = \frac{2x(x+1)}{(x-2)^2}$$

The graph will have one vertical asymptote, at $x = 2$. To check on a horizontal asymptote, we note that for large $|x|$, the numerator behaves like $2x^2$ and the denominator behaves like x^2. It follows that $y = 2$ is a horizontal asymptote. The graph crosses the x-axis where the numerator $2x(x+1)$ is zero, namely, at $x = 0$ and $x = -1$. It is shown in Figure 43 on the next page.

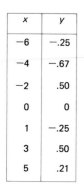

x	y
−6	−.25
−4	−.67
−2	.50
0	0
1	−.25
3	.50
5	.21

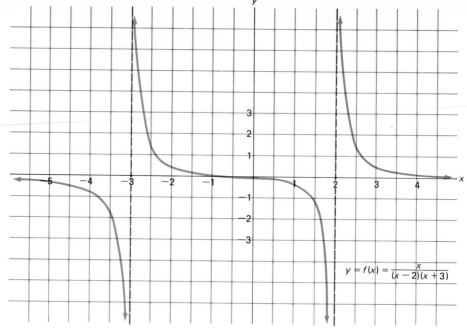

$$y = f(x) = \frac{x}{(x-2)(x+3)}$$

Figure 42

x	y
−30	1.70
−10	1.25
−6	.94
−4	.67
−1	0
0	0
1	4
3	24
6	4.5
10	3.44
20	2.59

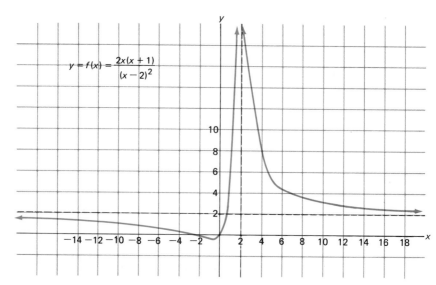

$$y = f(x) = \frac{2x(x+1)}{(x-2)^2}$$

Figure 43

A GENERAL PROCEDURE

Here is an outline of the procedure for graphing a rational function

$$y = f(x) = \frac{p(x)}{q(x)}$$

which is in reduced form.

1. Check for symmetry with respect to the y-axis and the origin.
2. Factor the numerator and denominator.
3. Determine the vertical asymptotes (if any) by checking where the denominator is zero. Draw a dotted line for each asymptote. Be sure to examine the behavior of the graph near a vertical asymptote.
4. Determine the horizontal asymptote (if any) by asking what y approaches as $|x|$ becomes large. This is accomplished by examining the quotient of the leading terms from numerator and denominator. Indicate any horizontal asymptote with a dotted line.
5. Determine the x-intercepts (if any). These occur where the numerator is zero.
6. Make a small table of values and plot corresponding points.
7. Sketch the graph.

Problem Set 2-5

Sketch the graph of each of the following functions.

1. $f(x) = \dfrac{2}{x + 2}$

2. $f(x) = \dfrac{-1}{x + 2}$

3. $f(x) = \dfrac{2}{(x + 2)^2}$

4. $f(x) = \dfrac{1}{(x - 3)^2}$

5. $f(x) = \dfrac{2x}{x + 2}$

6. $f(x) = \dfrac{x + 2}{x - 3}$

7. $f(x) = \dfrac{1}{(x + 2)(x - 1)}$

8. $f(x) = \dfrac{3}{x^2 - 9}$

9. $f(x) = \dfrac{x + 1}{(x + 2)(x - 1)}$

10. $f(x) = \dfrac{3x}{x^2 - 9}$

11. $f(x) = \dfrac{2x^2}{(x + 2)(x - 1)}$

12. $f(x) = \dfrac{x^2 - 4}{x^2 - 9}$

EXAMPLE A (No Vertical Asymptotes)   Sketch the graph of

$$f(x) = \frac{x^2 - 4}{x^2 + 1} = \frac{(x - 2)(x + 2)}{x^2 + 1}$$

Solution.   Note that f is an even function, so the graph will be symmetric with respect to the y-axis. The denominator is not zero for any real x, so there are no vertical asymptotes. The line $y = 1$ is a horizontal asymptote, since for large $|x|$, $f(x)$ behaves like x^2/x^2. The x-intercepts are $x = 2$ and $x = -2$. The graph is shown in Figure 44 on the next page.

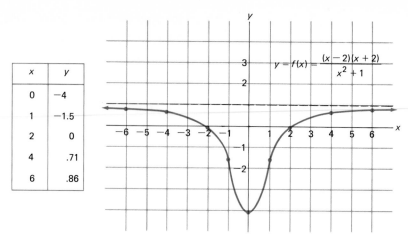

x	y
0	-4
1	-1.5
2	0
4	.71
6	.86

$$y = f(x) = \frac{(x - 2)(x + 2)}{x^2 + 1}$$

Figure 44

Now sketch the graph of each of the following.

13. $f(x) = \dfrac{1}{x^2 + 2}$

14. $f(x) = \dfrac{x^2 - 2}{x^2 + 2}$

15. $f(x) = \dfrac{x}{x^2 + 2}$

16. $f(x) = \dfrac{x^3}{x^2 + 2}$

EXAMPLE B (Oblique Asymptotes)   Sketch the graph of

$$f(x) = \frac{x^2}{x + 1}$$

Solution.   From our earlier discussion, we expect a vertical asymptote at $x = -1$. There is no horizontal asymptote. However, when we do a long division (as in Figure 45), we find that

$$\begin{array}{r} x - 1 \\ x + 1 \overline{)x^2} \\ \underline{x^2 + x} \\ -x \\ \underline{-x - 1} \\ 1 \end{array}$$

Figure 45

$$f(x) = x - 1 + \frac{1}{x + 1}$$

As $|x|$ gets larger and larger, the term $1/(x + 1)$ tends to zero, and so $f(x)$ gets closer and closer to $x - 1$. This means that the line $y = x - 1$ is an asymptote, called an **oblique asymptote.** Its significance is indicated on the graph in Figure 46. In general, we can expect an oblique asymptote for the graph of a rational function whenever the degree of the numerator is exactly one more than that of the denominator.

Sketch the graphs of the following rational functions.

17. $f(x) = \dfrac{2x^2 + 1}{2x}$

18. $f(x) = \dfrac{x^2 - 2}{x}$

19. $f(x) = \dfrac{x^2}{x - 1}$

20. $f(x) = \dfrac{x^3}{x^2 + 1}$

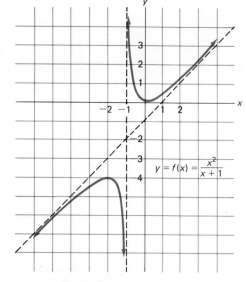

x	y
−4	−5.3
−2	−4
−1.5	−4.5
− .5	.5
0	0
1	.5
4	3.2

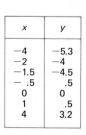

Figure 46

EXAMPLE C (Rational Functions That Are Not in Reduced Form)   Sketch the
graph of

$$f(x) = \frac{x^2 + x - 6}{x - 2}$$

Solution.   Notice that

$$f(x) = \frac{(x + 3)(x - 2)}{x - 2}$$

You have the right to expect that we will cancel the factor $x - 2$ from
numerator and denominator and graph

$$g(x) = x + 3$$

But note that 2 is in the domain of g but not in the domain of f. Thus f
and g and their graphs are exactly alike except at one point, namely, at
$x = 2$. Both graphs are shown in Figure 47 at the top of the next page.
You will notice the hole in the graph of $y = f(x)$ at $x = 2$. This technical
distinction is occasionally important.

*Sketch the graph of each of the following rational functions, which, you will note, are
not in reduced form.*

21. $f(x) = \dfrac{(x + 2)(x - 4)}{x + 2}$

22. $f(x) = \dfrac{x^2 - 4}{x - 2}$

23. $f(x) = \dfrac{x^3 - x^2 - 12x}{x + 3}$

24. $f(x) = \dfrac{x^3 - 4x}{x^2 - 2x}$

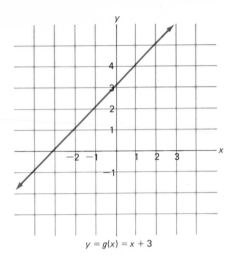

$y = g(x) = x + 3$

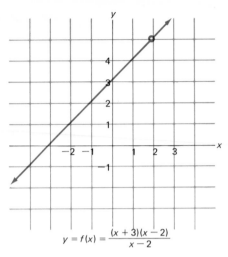

$y = f(x) = \dfrac{(x + 3)(x - 2)}{x - 2}$

Figure 47

MISCELLANEOUS PROBLEMS

Sketch the graphs of the rational functions in Problems 25–30.

25. $f(x) = \dfrac{x}{x + 5}$

26. $f(x) = \dfrac{x - 2}{x + 3}$

27. $f(x) = \dfrac{x^2 - 9}{x^2 - x - 2}$

28. $f(x) = \dfrac{x - 2}{(x + 3)^2}$

29. $f(x) = \dfrac{x^2 - 9}{x^2 - x - 6}$

30. $f(x) = \dfrac{x - 2}{x^2 + 3}$

31. Sketch the graphs of $f(x) = x^n/(x^2 + 1)$ for $n = 1, 2,$ and 3, being careful to show all asymptotes.

32. Where does the graph of $f(x) = (x^3 + x^2 - 2x + 1)/(x^3 + 2x^2 - 2)$ cross its horizontal asymptote?

33. Determine all asymptotes (vertical, horizontal, oblique) for the graph of $f(x) = x^4/(x^n - 1)$ in each case.
    (a) $n = 1$ (b) $n = 2$ (c) $n = 3$
    (d) $n = 4$ (e) $n = 5$ (f) $n = 6$

34. Consider the graph of the general rational function

$$f(x) = \frac{a_n x^n + a_{n-1} x^{n-1} + \cdots + a_1 x + a_0}{b_m x^m + b_{m-1} x^{m-1} + \cdots b_1 x + b_0}, \qquad a_n \neq 0, b_m \neq 0$$

Identify its horizontal asymptote in each case.
    (a) $m > n$ (b) $m = n$ (c) $m < n$

35. A manufacturer of gizmos has overhead of \$20,000 per year and direct costs (labor and material) of \$50 per gizmo. Write an expression for $U(x)$, the average cost per unit, if the company makes x gizmos per year. Graph the function U and then draw some conclusions from your graph.

36. A cylindrical can is to contain 10π cubic inches. Write a formula for $S(r)$, the total surface area, in terms of the radius r. Graph the function S and use it to estimate the radius of the can that will require the least material to make.

37. Find a formula for $f(x)$ if f is a rational function whose graph goes through $(2, 5)$ and has exactly two asymptotes, namely, $y = 2x + 3$ and $x = 3$.

38. TEASER Sketch the graphs of $f(x) = [1/x]$ and $g(x) = (1/x)$ for $0 < x \le 1$. The symbols [ ] and ( ) were defined in Problems 47 and 51 of Section 2-4.

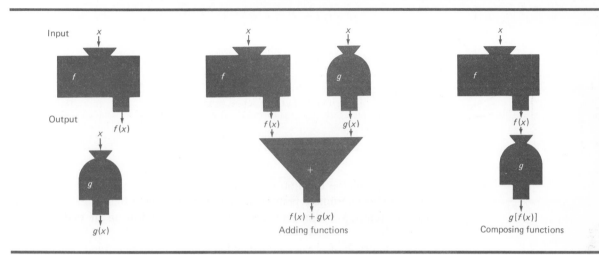

f(x) + g(x)
Adding functions

g[f(x)]
Composing functions

2-6   Putting Functions Together

There is still another way to visualize a function. Think of the function named f as a machine. It accepts a number x as input, operates on it, and then presents the number $f(x)$ as output. Machines can be hooked together to make more complicated machines; similarly, functions can be combined to produce more complicated functions. That is the subject of this section.

SUMS, DIFFERENCES, PRODUCTS, AND QUOTIENTS

The simplest way to make new functions from old ones is to use the four arithmetic operations on them. Suppose, for example, that the functions f and g have the formulas

$$f(x) = \frac{x - 3}{2} \qquad g(x) = \sqrt{x}$$

We can make a new function $f + g$ by having it assign to x the value $(x - 3)/2 + \sqrt{x}$; that is,

$$(f + g)(x) = f(x) + g(x) = \frac{x - 3}{2} + \sqrt{x}$$

Of course, we must be a little careful about domains. Clearly, x must be a number on which both f and g can operate. In other words, the domain of $f + g$ is the intersection (common part) of the domains of f and g.

The functions $f - g, f \cdot g,$ and f/g are defined in a completely analogous way. Assuming that f and g have their respective natural domains—namely, all reals and the nonnegative reals, respectively—we have the following.

FORMULA   DOMAIN

$$(f + g)(x) = f(x) + g(x) = \frac{x - 3}{2} + \sqrt{x} \qquad x \geq 0$$

$$(f - g)(x) = f(x) - g(x) = \frac{x - 3}{2} - \sqrt{x} \qquad x \geq 0$$

$$(f \cdot g)(x) = f(x) \cdot g(x) = \frac{x - 3}{2} \sqrt{x} \qquad x \geq 0$$

$$(f/g)(x) = f(x)/g(x) = \frac{x - 3}{2\sqrt{x}} \qquad x > 0$$

To graph the function $f + g$, it is often best to graph f and g separately in the same coordinate plane and then add the y-coordinates together along vertical lines. We illustrate this method (called **addition of ordinates**) in Figure 48.

The graph of $f - g$ can be handed similarly; simply graph f and g in the same coordinate plane and subtract ordinates. We can even graph $f \cdot g$ and f/g in the same manner, but that is harder.

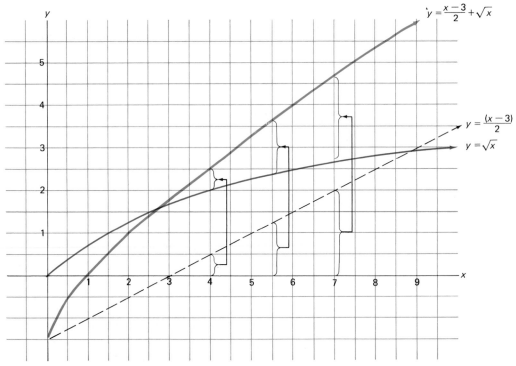

Figure 48

COMPOSITION OF FUNCTIONS

To compose functions is to string them together in tandem. Part of our opening display (reproduced in Figure 49) shows how this is done. If f operates on x to produce $f(x)$ and then g operates on $f(x)$ to produce $g(f(x))$, we say that we have composed g and f. The resulting function, called the **composite of g with f,** is denoted by $g \circ f$. Thus

$$(g \circ f)(x) = g(f(x))$$

Recall our earlier examples, $f(x) = (x - 3)/2$ and $g(x) = \sqrt{x}$. We may compose them in two ways.

$$(g \circ f)(x) = g(f(x)) = g\left(\frac{x-3}{2}\right) = \sqrt{\frac{x-3}{2}}$$

$$(f \circ g)(x) = f(g(x)) = f(\sqrt{x}) = \frac{\sqrt{x} - 3}{2}$$

We note one thing right away: Composition of functions is not commutative; $g \circ f$ and $f \circ g$ are not the same. We must also be careful in describing the domain of a composite function. The domain of $g \circ f$ is that part of the domain of f for which g can acccept $f(x)$ as input. In our example, the domain of $g \circ f$ is $x \geq 3$, not all x or $x \geq 0$ as we might have thought at first glance. Figure 50 offers another view of these matters. Note that the shaded portion of the domain of f is not in the domain of $g \circ f$; for x in this portion, $f(x)$ is outside the domain of g.

In calculus, we shall often need to take a given function and decompose it, that is, break it into composite pieces. Usually, this can be done in several ways. Take $p(x) = \sqrt{x^2 + 3}$ for example. We may think of it as

$$p(x) = g(f(x)) \quad \text{where} \quad g(x) = \sqrt{x} \quad \text{and} \quad f(x) = x^2 + 3$$

or as

$$p(x) = g(f(x)) \quad \text{where} \quad g(x) = \sqrt{x + 3} \quad \text{and} \quad f(x) = x^2$$

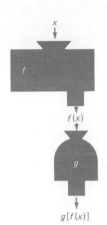

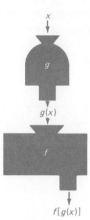

Figure 49

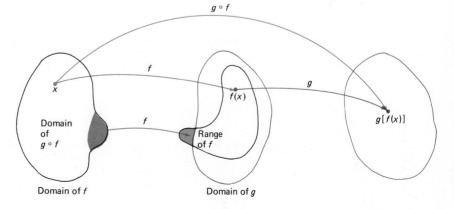

Figure 50

TRANSLATIONS

Observing how a function is built up from simpler ones using the operations of this section can be a big aid in graphing. This is especially true of *translatons* which result from the composition $f(x - h)$ and/or the simple addition of a constant k.

Consider, for example, the graphs of

$$y = f(x) \qquad y = f(x - 3) \qquad y = f(x) + 2 \qquad y = f(x - 3) + 2$$

for the case $f(x) = |x|$. The four graphs are shown in Figure 51. Notice that all four graphs have the same shape; the last three are just translations (rigid movements) of the first. Replacing x by $x - 3$ translates the graph 3 units to the right; adding 2 translates the graph 2 units upward.

What happened with $f(x) = |x|$ is typical. In Figure 52, we give another illustration, this time for the function $f(x) = x^3 + x^2$.

Exactly the same principles apply in the general situation. Figure 53 shows what happens when h and k are both positive. If $h < 0$, the translation is to the left; if $k < 0$, the translation is downward.

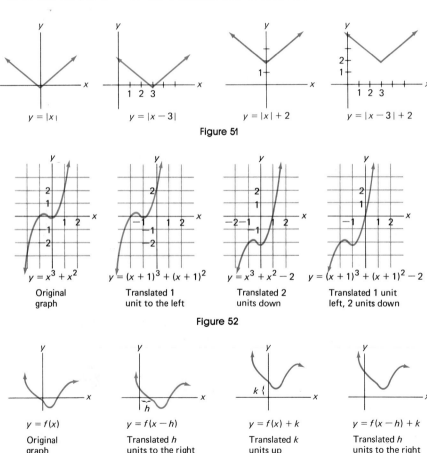

$y = |x|$ $y = |x - 3|$ $y = |x| + 2$ $y = |x - 3| + 2$

Figure 51

$y = x^3 + x^2$ $y = (x + 1)^3 + (x + 1)^2$ $y = x^3 + x^2 - 2$ $y = (x + 1)^3 + (x + 1)^2 - 2$

Original graph   Translated 1 unit to the left   Translated 2 units down   Translated 1 unit left, 2 units down

Figure 52

$y = f(x)$ $y = f(x - h)$ $y = f(x) + k$ $y = f(x - h) + k$

Original graph   Translated h units to the right   Translated k units up   Translated h units to the right and k units up

Figure 53

Problem Set 2-6

1. Let $f(x) = x^2 - 2x + 2$ and $g(x) = 2/x$. Calculate each of the following.
   (a) $(f + g)(2)$ (b) $(f + g)(0)$ (c) $(f - g)(1)$
   (d) $(f \cdot g)(-1)$ (e) $(f/g)(2)$ (f) $(g/f)(2)$
   (g) $(f \circ g)(-1)$ (h) $(g \circ f)(-1)$ (i) $(g \circ g)(3)$

2. Let $f(x) = 3x + 5$ and $g(x) = |x - 2|$. Perform the calculations in Problem 1 for these functions.

In each of the following, write the formulas for $(f + g)(x)$, $(f - g)(x)$, $(f \cdot g)(x)$, and $(f/g)(x)$ and give the domains of these four functions.

3. $f(x) = x^2$, $g(x) = x - 2$ 4. $f(x) = x^3 - 1$, $g(x) = x + 3$

5. $f(x) = x^2$, $g(x) = \sqrt{x}$ 6. $f(x) = 2x^2 + 5$, $g(x) = \dfrac{1}{x}$

7. $f(x) = \dfrac{1}{x - 2}$, $g(x) = \dfrac{x}{x - 3}$ 8. $f(x) = \dfrac{1}{x^2}$, $g(x) = \dfrac{1}{5 - x}$

For each of the following, write the formulas for $(g \circ f)(x)$ and $(f \circ g)(x)$ and give the domains of these composite functions.

9. $f(x) = x^2$, $g(x) = x - 2$ 10. $f(x) = x^3 - 1$, $g(x) = x + 3$

11. $f(x) = \dfrac{1}{x}$, $g(x) = x + 3$ 12. $f(x) = 2x^2 + 5$, $g(x) = \dfrac{1}{x}$

13. $f(x) = \sqrt{x - 2}$, $g(x) = x^2 - 2$ 14. $f(x) = \sqrt{2x}$, $g(x) = x^2 + 1$

15. $f(x) = 2x - 3$, $g(x) = \frac{1}{2}(x + 3)$ 16. $f(x) = x^3 + 1$, $g(x) = \sqrt[3]{x - 1}$

EXAMPLE A (Decomposing Functions)   In each of the following, H can be thought of as a composite function $g \circ f$. Write formulas for $f(x)$ and $g(x)$.

(a) $H(x) = (2 + 3x)^2$ (b) $H(x) = \dfrac{1}{(x^2 + 4)^3}$

Solution.
(a) Think of how you might calculate $H(x)$. You would first calculate $2 + 3x$ and then square the result. That suggests

$$f(x) = 2 + 3x \qquad g(x) = x^2$$

(b) Here there are two obvious ways to proceed. One way would be to let

$$f(x) = x^2 + 4 \qquad g(x) = \dfrac{1}{x^3}$$

Another selection, which is just as good, is

$$f(x) = \dfrac{1}{x^2 + 4} \qquad g(x) = x^3$$

We could actually think of H as the composite of four functions. Let

$$f(x) = x^2 \qquad g(x) = x + 4 \qquad h(x) = x^3 \qquad j(x) = \frac{1}{x}$$

Then

$$H = j \circ h \circ g \circ f$$

You should check this result.

In each of the following, write formulas for $g(x)$ and $f(x)$ so that $H = g \circ f$. The answer is not unique.

17. $H(x) = (x + 4)^3$

18. $H(x) = (2x + 1)^3$

19. $H(x) = \sqrt{x} + 2$

20. $H(x) = \sqrt[3]{2x + 1}$

21. $H(x) = \dfrac{1}{(2x + 5)^3}$

22. $H(x) = \dfrac{6}{(x + 4)^3}$

23. $H(x) = |x^3 - 4|$

24. $H(x) = |4 - x - x^2|$

Use the method of addition or subtraction of ordinates to graph each of the following. That is, graph $y = f(x)$ and $y = g(x)$ in the same coordinate plane and then obtain the graph of $f + g$ or $f - g$ by adding or subtracting ordinates.

25. $f + g$ where $f(x) = x^2$ and $g(x) = x - 2$.

26. $f + g$ where $f(x) = |x|$ and $g(x) = x$.

27. $f - g$ where $f(x) = 1/x$ and $g(x) = x$.

28. $f - g$ where $f(x) = x^3$ and $g(x) = -x + 1$.

EXAMPLE B (Graphing by Translation)   Graph the following functions by making use of translations.

$$\text{(a)} \ \ g(x) = x^2 + 2 \qquad \text{(b)} \ \ g(x) = \frac{4}{x - 3}$$

Solution.   (a) First we graph $f(x) = x^2$ and then translate 2 units up. (b) Here we graph $f(x) = 4/x$ and then translate 3 units to the right. The results are shown in Figure 54.

In each of the following, graph the function g by first graphing a simpler function f and then translating.

29. $g(x) = x^3 - 2$

30. $g(x) = 2x^2 - 5$

31. $g(x) = \dfrac{6}{x + 2}$

32. $g(x) = \sqrt{x + 4}$

33. $g(x) = (x - 1)^3 + 1$
   Hint: First graph $f(x) = x^3$.

34. $g(x) = 2(x + 1)^2 + 2$

35. $g(x) = \dfrac{6}{x - 1} + 2$

36. $g(x) = \sqrt{x - 3} + 3$

EXAMPLE C (Difference Quotients)   In calculus, the difference quotient

$$\frac{f(x + h) - f(x)}{h}$$

(a)

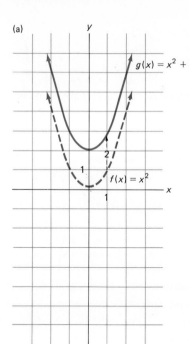

$g(x) = x^2 + 2$

$f(x) = x^2$

(b)

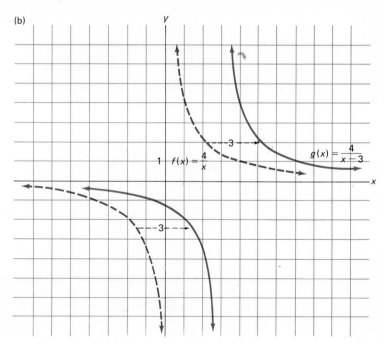

$f(x) = \dfrac{4}{x}$

$g(x) = \dfrac{4}{x-3}$

Figure 54

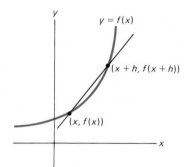

$y = f(x)$

$(x + h, f(x + h))$

$(x, f(x))$

Figure 55

CAUTION

$f(x) = \dfrac{1}{2x}$

$f(x + h) = \dfrac{1}{2x} + h$

$f(x) = \dfrac{1}{2x}$

$f(x + h) = \dfrac{1}{2(x + h)}$

$= \dfrac{1}{2x + 2h}$

arises repeatedly. It represents the slope of the line determined by the points $(x, f(x))$ and $(x + h, f(x + h))$ on the graph of f (see Figure 55). Calculate this expression for $f(x) = 3/x$, simplify it, and then evaluate it when $x = 1$ and $h = .5$.

Solution.

$$\frac{f(x + h) - f(x)}{h} = \frac{\dfrac{3}{x + h} - \dfrac{3}{x}}{h}$$

$$= \frac{3x - 3(x + h)}{(x + h)x} \cdot \frac{1}{h}$$

$$= \frac{-3}{(x + h)x}$$

When $x = 1$ and $h = .5$, this difference quotient has the value $-3/1.5 = -2$.

In each of the following, calculate the difference quotient and simplify it. Then find its value when $x = 3$ and $h = 1$.

37. $f(x) = x^2$

38. $f(x) = -3x^2$

39. $f(x) = 2x^2 - 5x$

40. $f(x) = -x^2 + 4x + 2$

41. $f(x) = 2x - 3$

42. $f(x) = -5x + 2$

43. $f(x) = \dfrac{2}{x + 1}$

44. $f(x) = \dfrac{3}{x^2}$

45. $f(x) = x^3 - 2$ 46. $f(x) = 2x^3$

MISCELLANEOUS PROBLEMS

47. Let $f(x) = 2x + 3$ and $g(x) = x^3$. Write formulas for each of the following.
  (a) $(f + g)(x)$ (b) $(g - f)(x)$ (c) $(f \cdot g)(x)$ (d) $(f/g)(x)$
  (e) $(f \circ g)(x)$ (f) $(g \circ f)(x)$ (g) $(f \circ f)(x)$ (h) $(g \circ g \circ g)(x)$

48. If $f(x) = 1/(x - 1)$ and $g(x) = \sqrt{x + 1}$, write formulas for $(f \circ g)(x)$ and $(g \circ f)(x)$ and give the domains of these composite functions.

49. If $f(x) = x^2 - 4$, $g(x) = |x|$, and $h(x) = 1/x$, write a formula for $(h \circ g \circ f)(x)$ and indicate its domain.

50. In general, how many different functions can be obtained by composing three different functions f, g, and h in different orders?

51. Evaluate the difference quotient of Problem 45 for $x = 2$ and (a) $h = 1$; (b) $h = .5$; (c) $h = .1$; (d) $h = .01$. What seems to be happening as h gets closer and closer to zero? Can you interpret this?

52. Follow the directions of Problem 51 for the difference quotient of Problem 44.

☐ 53. Let $f(x) = (1 + \sqrt{x})^3/(3x^2 + 1)$. Calculate each of the following.
  (a) $f(3.1)$ (b) $f(.03)$

☐ 54. Let $g(x) = (3 + 1/x)^2\sqrt{x^3 + 1}$. Calculate each of the following.
  (a) $g(4.2)$ (b) $g(-.91)$

55. Find $(f \circ g)(x)$ and $(g \circ f)(x)$ in each case.
  (a) $f(x) = x^2$, $g(x) = \sqrt{x}$ (b) $f(x) = x^3$, $g(x) = \sqrt[3]{x}$
  (c) $f(x) = x^2$, $g(x) = x^3$ (d) $f(x) = x^2$, $g(x) = 1/x^3$

56. Let $f(x) = \dfrac{x - 3}{x + 1}$. Show that if $x \neq \pm 1$, then $f(f(f(x))) = x$.

57. Let $f(x) = \left(\dfrac{1 - \sqrt{x}}{1 + \sqrt{x}}\right)^2$. Solve for x if $f(f(x)) = x^2 + \frac{1}{4}$ and $0 < x < 1$.

58. Let $f(x) = x^2 + 5x$. Solve for x if $f(f(x)) = f(x)$.

59. Sketch the graph of $f(x) = |x + 1| - |x| + |x - 1|$ on the interval $-2 \leq x \leq 2$. Then calculate the area of the region between this graph and the x-axis.

60. The *greatest integer function* [ ] was defined in Problem 47 of Section 2-4. Graph each of the following functions on the interval $-2 \leq x \leq 6$.
  (a) $f(x) = 2[x]$ (b) $g(x) = 2 + [x]$
  (c) $h(x) = [x - 2]$ (d) $k(x) = x - [x]$

61. Let f be an even function (meaning $f(-x) = f(x)$) and let g be an odd function (meaning $g(-x) = -g(x)$), both functions having the whole real line as their domains. Which of the following are even? Odd? Neither even nor odd?
  (a) $f(x)g(x)$ (b) $f(x)/g(x)$ (c) $[g(x)]^2$
  (d) $[g(x)]^3$ (e) $f(x) + g(x)$ (f) $g(g(x))$
  (g) $f(f(x))$ (h) $3f(x) + [g(x)]^2$ (i) $g(x) + g(-x)$

62. Show that any function f having the whole real line as its domain can be represented as the sum of an even function and an odd function. *Hint:* Consider $f(x) + f(-x)$ and $f(x) - f(-x)$.

63. The *distance to the nearest integer function* ⟨ ⟩ was defined in Problem 51 of Section 2-4.

(a) Sketch the graphs of $f(x) = \langle x \rangle$, $g(x) = \langle 2x \rangle/2$, $h(x) = \langle 4x \rangle/4$ and $F(x) = f(x) + g(x) + h(x)$ on the interval $0 \le x \le 4$.

(b) Find the area of the region between the graph of each of these functions and the x-axis on the interval $0 \le x \le 4$. Note that the sum of the first three areas is the fourth.

64. **TEASER** Generalize Problem 63 by considering the graph of

$$F_n(x) = \langle x \rangle + \frac{1}{2}\langle 2x \rangle + \frac{1}{4}\langle 4x \rangle + \cdots + \frac{1}{2^n}\langle 2^n x \rangle$$

on the interval $0 \le x \le 4$. Find a nice formula for the area A_n of the region between this graph and the x-axis. What happens to A_n as n grows without bound? *Note:* The limiting form of F_n plays an important role in advanced mathematics, giving an example of a function whose graph is continuous but does not have a tangent line at any point.

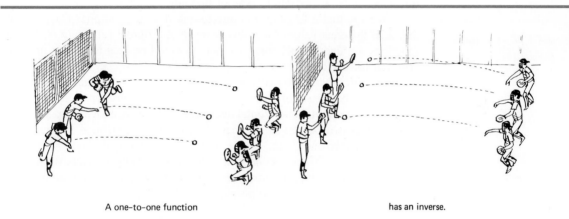

A one-to-one function   has an inverse.

2-7 Inverse Functions

Some processes are reversible; most are not. If I take off my shoes, I may put them back on again. The second operation undoes the first one and brings things back to the original state. But if I throw my shoes in the fire, I will have a hard time undoing the damage I have done.

A function f operates on a number x to produce a number $y = f(x)$. It may be that we can find a function g that will operate on y and give back x. For example, if

$$y = f(x) = 2x + 1$$

then

$$g(x) = \tfrac{1}{2}(x - 1)$$

is such a function, since

$$g(y) = g(f(x)) = \tfrac{1}{2}(2x + 1 - 1) = x$$

When we can find such a function g, we call it the *inverse* of f. Not all functions have inverses. Whether they do or not has to do with a concept called one-to-oneness.

ONE-TO-ONE FUNCTIONS

In Figure 56, we have reproduced an example we studied earlier, the squaring function with domain $\{-2, -1, 0, 1, 2, 3\}$. It is a perfectly fine function, but it does have one troublesome feature. It may assign the same value to two different x's. In particular, $f(-2) = 4$ and $f(2) = 4$. Such a function cannot possibly have an inverse g. For what would g do with 4? It would not know whether to give back -2 or 2 as the value.

In contrast, consider $f(x) = 2x + 1$, pictured in Figure 57. Notice that this function never assigns the same value to two different values of x. Therefore there is an unambiguous way of undoing it.

We say that a function f is **one-to-one** if $x_1 \neq x_2$ implies $f(x_1) \neq f(x_2)$; that is, if different values for x always result in different values for $f(x)$. Some functions are one-to-one; some are not. It would be nice to have a graphical criterion for deciding.

Consider the functions $f(x) = x^2$ and $f(x) = 2x + 1$ again, but now let the domains be the set of all real numbers. Their graphs appear in Figure 58. In the first case, certain horizontal lines (those which are above the x-axis) meet the graph in two points; in the second case, every horizontal line meets the graph in exactly one point. Notice on the first graph that $f(x_1) = f(x_2)$ even though $x_1 \neq x_2$. On the second graph, this cannot happen. Thus we have the important fact that *if every horizontal line meets the graph of a function f in at most one point, then f is one-to-one.*

INVERSE FUNCTIONS

Now we are ready to give a formal definition of the main idea of this section.

Domain   Range

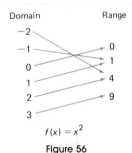

$f(x) = x^2$

Figure 56

Domain   Range

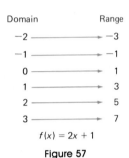

$f(x) = 2x + 1$

Figure 57

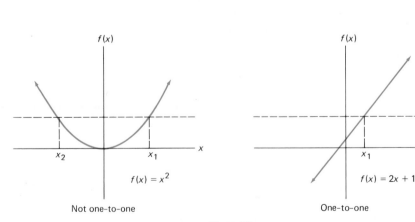

Not one-to-one   One-to-one

Figure 58

DEFINITION

Let f be a one-to-one function with domain X and range Y. Then the function g with domain Y and range X which satisfies

$$g(f(x)) = x$$

*for all x in X is called the **inverse of f**.*

We make several important observations. First, the boxed formula simply says that g undoes what f did. Second, if g undoes f, then f will undo g; that is,

$$f(g(y)) = y$$

CAUTION

$$f^{-1}(x) = \frac{1}{f(x)}$$

f^{-1} is the notation for an inverse function.

for all y in Y. Third, the function g is usually denoted by the symbol f^{-1}. You are cautioned to remember that f^{-1} does *not* mean $1/f$, as you have the right to expect. Mathematicians decided long ago that f^{-1} should stand for the inverse function (the undoing function). Thus

$$(f^{-1} \circ f)(x) = x \quad \text{and} \quad (f \circ f^{-1})(y) = y$$

For example, if $f(x) = 4x$, then $f^{-1}(y) = \frac{1}{4}y$ since

$$(f^{-1} \circ f)(x) = f^{-1}(f(x)) = f^{-1}(4x) = \frac{1}{4}(4x) = x$$

and

$$(f \circ f^{-1})(y) = f(f^{-1}(y)) = f(\tfrac{1}{4}y) = 4(\tfrac{1}{4}y) = y$$

The boxed results are illustrated pictorially in Figure 59.

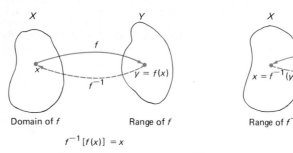

Figure 59

FINDING A FORMULA FOR f^{-1}

If f adds 2, then f^{-1} ought to subtract 2. To say it in symbols, if $f(x) = x + 2$, then we might expect $f^{-1}(y) = y - 2$. And we are right, for

$$f^{-1}[f(x)] = f^{-1}(x + 2) = x + 2 - 2 = x$$

If f divides by 3 and then subtracts 4, we expect f^{-1} to add 4 and multiply by 3. Symbolically, if $f(x) = x/3 - 4$, then we expect $f^{-1}(y) = 3(y + 4)$. Again we are right, for

$$f^{-1}[f(x)] = f^{-1}\left(\frac{x}{3} - 4\right) = 3\left(\frac{x}{3} - 4 + 4\right) = x$$

Note that we must undo things in the reverse order in which we did them (that is, we divided by 3 and then subtracted 4, so to undo this, we first add 4 and then multiply by 3).

When we get to more complicated functions, it is not always easy to find the formula for the inverse function. Here is an important way to look at it.

$$\boxed{x = f^{-1}(y) \quad \text{if and only if} \quad y = f(x)}$$

That means that we can get the formula for f^{-1} by solving the equation $y = f(x)$ for x. Here is an example. Let $y = f(x) = 3/(x - 2)$. Follow the steps below.

$$y = \frac{3}{x - 2}$$

$$(x - 2)y = 3$$

$$xy - 2y = 3$$

$$xy = 3 + 2y$$

$$x = \frac{3 + 2y}{y}$$

Thus

$$f^{-1}(y) = \frac{3 + 2y}{y}$$

In the formula for f^{-1} just derived, there is no need to use y as the variable. We might use u or t or even x. The formulas

$$f^{-1}(u) = \frac{3 + 2u}{u}$$

$$f^{-1}(t) = \frac{3 + 2t}{t}$$

$$f^{-1}(x) = \frac{3 + 2x}{x}$$

all say the same thing in the sense that they give the same rule. It is conventional to give formulas for functions using x as the variable, and so we would write $f^{-1}(x) = (3 + 2x)/x$ as our answer. Let us summarize. To find the formula for $f^{-1}(x)$, use the following steps.

> **THREE-STEP PROCEDURE FOR FINDING f⁻¹(X)**
>
> 1. Solve $y = f(x)$ for x in terms of y.
> 2. Use $f^{-1}(y)$ to name the resulting expression in y.
> 3. Replace y by x to get the formula for $f^{-1}(x)$.

THE GRAPHS OF f AND f⁻¹

Since $y = f(x)$ and $x = f^{-1}(y)$ are equivalent, the graphs of these two equations are the same. Suppose we want to compare the graphs of $y = f(x)$ and $y = f^{-1}(x)$ (where, you will note, we have used x as the domain variable in both cases). To get $y = f^{-1}(x)$ from $x = f^{-1}(y)$, we interchange the roles of x and y. Graphically, this corresponds to folding (reflecting) the graph across the 45° line—that is, across the line $y = x$ (Figure 60). This is the same as saying that if the point (a, b) is on one graph, then (b, a) is on the other.

Here is a simple example. Let $f(x) = x^3$; then $f^{-1}(x) = \sqrt[3]{x}$, the cube root of x. The graphs of $y = x^3$ and $y = \sqrt[3]{x}$ are shown in Figure 61, first separately and then on the same coordinate plane. Note that $f(2) = 8$ and $f^{-1}(8) = 2$.

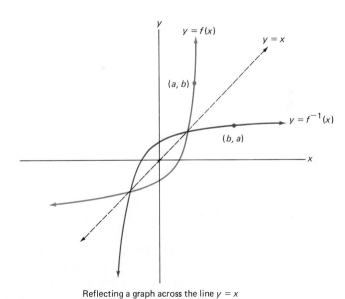

Reflecting a graph across the line $y = x$

Figure 60

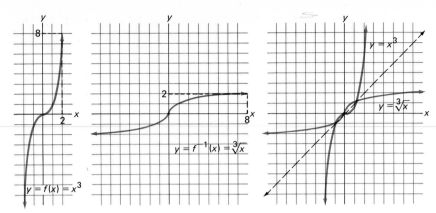

Figure 61

Problem Set 2-7

1. Examine the graphs in Figure 62.
   (a) Which of these are the graphs of functions with x as domain variable?
   (b) Which of these functions are one-to-one?
   (c) Which of them have inverses?

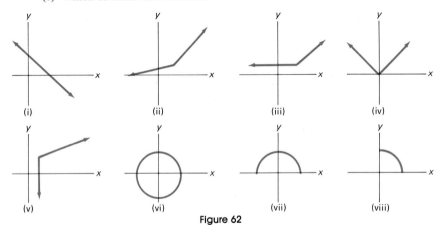

Figure 62

2. Let each of the following functions have their natural domains. Which of them are one-to-one? *Hint:* Consider their graphs.

   (a) $f(x) = x^4$ (b) $f(x) = x^3$ (c) $f(x) = \dfrac{1}{x}$

   (d) $f(x) = \dfrac{1}{x^2}$ (e) $f(x) = x^2 + 2x + 3$ (f) $f(x) = |x|$

   (g) $f(x) = \sqrt{x}$ (h) $f(x) = -3x + 2$

3. Let $f(x) = 3x - 2$. To find $f^{-1}(2)$, note that $f^{-1}(2) = a$ if $f(a) = 2$—that is, if $3a - 2 = 2$; and so $f^{-1}(2) = a = \frac{4}{3}$. Find each of the following.
   (a) $f^{-1}(1)$ (b) $f^{-1}(-3)$ (c) $f^{-1}(14)$

4. Let $g(x) = 1/(x - 1)$. Find each of the following.
   (a) $g^{-1}(1)$ (b) $g^{-1}(-1)$ (c) $g^{-1}(14)$

EXAMPLE A (Finding $f^{-1}(x)$) Use the three-step procedure to find $f^{-1}(x)$ if $f(x) = 2x^3 - 1$. Check your result by calculating $f(f^{-1}(x))$.

Solution.

Step 1 We solve $y = 2x^3 - 1$ for x in terms of y.

$$2x^3 = y + 1$$

$$x^3 = \frac{y + 1}{2}$$

$$x = \sqrt[3]{\frac{y + 1}{2}}$$

Step 2 Call the result $f^{-1}(y)$.

$$f^{-1}(y) = \sqrt[3]{\frac{y + 1}{2}}$$

Step 3 Replace y by x.

$$f^{-1}(x) = \sqrt[3]{\frac{x + 1}{2}}$$

Check: $\quad f(f^{-1}(x)) = 2\left(\sqrt[3]{\frac{x + 1}{2}}\right)^3 - 1 = 2\left(\frac{x + 1}{2}\right) - 1 = x$

Each of the functions in Problems 5–14 has an inverse (using its natural domain). Find the formula for $f^{-1}(x)$. Then check your result by calculating $f(f^{-1}(x))$.

5. $f(x) = 5x$
6. $f(x) = -4x$
7. $f(x) = 2x - 7$
8. $f(x) = -3x + 2$
9. $f(x) = \sqrt{x} + 2$
10. $f(x) = 2\sqrt{x} - 6$
11. $f(x) = \dfrac{x}{x - 3}$
12. $f(x) = \dfrac{x - 3}{x}$
13. $f(x) = (x - 2)^3 + 2$
14. $f(x) = \frac{1}{3}x^5 - 2$

15. In the same coordinate plane, sketch the graphs of $y = f(x)$ and $y = f^{-1}(x)$ for $f(x) = \sqrt{x} + 2$ (see Problem 9).

16. In the same coordinate plane, sketch the graphs of $y = f(x)$ and $y = f^{-1}(x)$ for $f(x) = x/(x - 3)$ (see Problem 11).

17. Sketch the graph of $y = f^{-1}(x)$ if the graph of $y = f(x)$ is as shown in Figure 63.

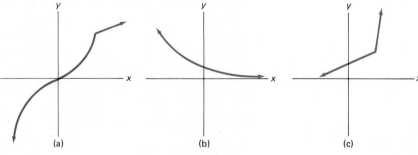

(a)   (b)   (c)

Figure 63

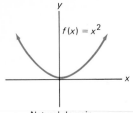

Natural domain

Figure 64

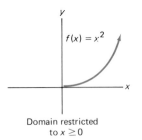

Domain restricted
to $x \geq 0$

Figure 65

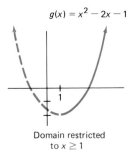

Domain restricted
to $x \geq 1$

Figure 66

18. Show that $f(x) = 2x/(x - 1)$ and $g(x) = x/(x - 2)$ are inverses of each other by calculating $f(g(x))$ and $g(f(x))$.

19. Show that $f(x) = 3x/(x + 2)$ and $g(x) = 2x/(3 - x)$ are inverses of each other.

20. Sketch the graph of $f(x) = x^3 + 1$ and note that f is one-to-one. Find a formula for $f^{-1}(x)$.

EXAMPLE B (Restricting the Domain)   The function $f(x) = x^2$ does not have an inverse if we use its natural domain (Figure 64). However, if we restrict its domain to $x \geq 0$ so that we are considering only its right branch (Figure 65), then it has an inverse, $f^{-1}(x) = \sqrt{x}$. Use the same idea to show that $g(x) = x^2 - 2x - 1$ has an inverse when its domain is appropriately restricted. Find $g^{-1}(x)$.

Solution.   The graph of $g(x)$ is shown in Figure 66; it is a parabola with vertex at $x = 1$. Accordingly, we can restrict the domain to $x \geq 1$. To find the formula for $g^{-1}(x)$, we first solve $y = x^2 - 2x - 1$ for x using an old trick, completing the square.

$$y + 1 = x^2 - 2x$$

$$y + 1 + 1 = x^2 - 2x + 1$$

$$y + 2 = (x - 1)^2$$

$$\pm\sqrt{y + 2} = x - 1$$

$$1 \pm \sqrt{y + 2} = x$$

Notice that there are two expressions for x; they correspond to the two halves of the parabola. We chose to make $x \geq 1$, so $x = 1 + \sqrt{y + 2}$ is the correct expression for $g^{-1}(y)$. Thus

$$g^{-1}(x) = 1 + \sqrt{x + 2}$$

If we had chosen to make $x \leq 1$, the correct answer would have been $g^{-1}(x) = 1 - \sqrt{x + 2}$.

In each of the following, restrict the domain so that f has an inverse. Describe the restricted domain and find a formula for $f^{-1}(x)$. Note: Different restrictions of the domain are possible.

21. $f(x) = (x - 1)^2$

22. $f(x) = (x + 3)^2$

23. $f(x) = (x + 1)^2 - 4$

24. $f(x) = (x - 2)^2 + 3$

25. $f(x) = x^2 + 6x + 7$

26. $f(x) = x^2 - 4x + 9$

27. $f(x) = |x + 2|$

28. $f(x) = 2|x - 3|$

29. $f(x) = \dfrac{(x - 1)^2}{1 + 2x - x^2}$

30. $f(x) = \dfrac{-1}{x^2 + 4x + 3}$

MISCELLANEOUS PROBLEMS

31. Sketch the graph of $f(x) = 1/(x - 1)$. Is f one-to-one? Calculate each of the following.

(a) $f(3)$ (b) $f^{-1}(\frac{1}{2})$ (c) $f(0)$

(d) $f^{-1}(-1)$ (e) $f^{-1}(3)$ (f) $f^{-1}(-2)$

32. If $f(x) = x/(x - 2)$, find the formula for $f^{-1}(x)$.

33. Find the formula for $f^{-1}(x)$ if $f(x) = 1/(x - 1)$ and sketch the graph of $y = f^{-1}(x)$. Compare this graph with the graph of $y = f(x)$ that you sketched in Problem 31.

34. Find the formula for $f^{-1}(x)$ if $f(x) = (x^3 + 2)/(x^3 + 3)$.

35. Sketch the graph of $f(x) = x^2 - 2x - 3$ and observe that it is not one-to-one. Restrict its domain so it is and then find a formula for $f^{-1}(x)$.

36. Let $f(x) = (x - 3)/(x + 1)$. Show that $f^{-1}(x) = f(f(x))$.

37. Let $f(x) = (2x^2 - 4x - 1)/(x - 1)^2$. If we restrict the domain so $x > 1$, then f has an inverse. Find the formula for $f^{-1}(x)$.

38. Suppose that f and g have inverses. Show that in this case $f \circ g$ has an inverse and that $(f \circ g)^{-1} = g^{-1} \circ f^{-1}$.

39. What must be true of the graph of f if f is *self-inverse* (meaning f is its own inverse)? What does this mean for the xy-equation determining f?

40. Let $y = f(x) = x/(x - 1)$. Show that the condition found in Problem 39 is satisfied. Now check that f is self-inverse by showing that $f(f(x)) = x$.

41. Let $f(x) = (ax + b)/(cx + d)$ where $bc - ad \neq 0$.
    (a) Find the formula for $f^{-1}(x)$.
    (b) Why did we impose the condition $bc - ad \neq 0$?
    (c) What relation connecting a and d will make f self-inverse?

42. **TEASER** Let $f_1(x) = x$, $f_2(x) = 1/x$, $f_3(x) = 1 - x$, $f_4(x) = 1/(1 - x)$, $f_5(x) = (x - 1)/x$, and $f_6(x) = x/(x - 1)$. Note that

$$f_4(f_3(x)) = \frac{1}{1 - f_3(x)} = \frac{1}{1 - (1 - x)} = \frac{1}{x} = f_2(x)$$

that is, $f_4 \circ f_3 = f_2$. In fact, if we compose any two of these six functions, we will get one of the six functions. Complete the indicated composition table and then use it to find each of the following (which will also be one of the six functions).

(a) $f_3 \circ f_3 \circ f_3 \circ f_3 \circ f_3$ (b) $f_1 \circ f_2 \circ f_3 \circ f_4 \circ f_5 \circ f_6$

(c) f_6^{-1}

(e) F if $f_2 \circ f_5 \circ F = f_5$ (d) $(f_3 \circ f_6)^{-1}$

$\circ$	f_1	f_2	f_3	f_4	f_5	f_6
f_1						
f_2						
f_3						
f_4		f_2				
f_5						
f_6						

Chapter Summary

A **function** f is a rule which assigns to each element x in one set (called the **domain**) a value $f(x)$ from another set. The set of all these values is called the **range** of the function. Numerical functions are usually specified by formulas (for example, $g(x) = (x^2 + 1)/(x + 1)$). The **natural domain** for such a function is the largest set of real numbers for which the formula makes sense and gives real values (thus, the natural domain for g consists of all real numbers except $x = -1$). Related to the notion of function is that of **variation.**

The **graph** of a function is simply the graph of the equation $y = f(x)$. Of special interest are the graphs of **polynomial functions,** and **rational functions.**

In graphing them, we should show the hills, the valleys, the **x-intercepts,** and, in the case of rational functions, the vertical and horizontal **asymptotes.**

Functions can be combined in many ways. Of these, composition is perhaps the most significant. The **composite** of f with g is defined by $(f \circ g)(x) = f(g(x))$.

Some functions are **one-to-one;** some are not. Those that are one-to-one have undoing functions called inverses. The **inverse** of f, denoted by f^{-1}, satisfies $f^{-1}(f(x)) = x$. Finding a formula for $f^{-1}(x)$ can be tricky; therefore, we described a definite procedure for doing it.

Chapter Review Problem Set

1. Let $f(x) = x^2 - 1$ and $g(x) = 2/x$. Calculate if possible.
   (a) $f(4)$ (b) $g(\frac{1}{2})$ (c) $g(0)$
   (d) $f(1)/g(3)$ (e) $f(g(4))$ (f) $g(f(4))$

2. Find the natural domain of f if $f(x) = \sqrt{x + 1}/(x - 1)$.

3. If y varies directly as the cube of x and $y = 1$ when $x = 2$, find an explicit formula for y in terms of x.

4. If z varies directly as x and inversely as the square of y, and if $z = 1$ when x and y are both 3, find z when $x = 16$ and $y = 2$.

5. Find the equation of the line satisfying the following conditions. Write your answer in the form $Ax + By + C = 0$.
   (a) Through $(-1, 2)$ and $(4, 7)$.
   (b) Through $(4, 7)$ and vertical.
   (c) Through $(-1, 2)$ and horizontal.
   (d) Through $(-1, 2)$ and parallel to $3x - 2y = 7$.
   (e) Through $(-1, 2)$ and perpendicular to $2x - 5y = 8$.

6. Find the points of intersection of the line $y = x - 3$ and the parabola $y = -x^2 + x - 2$.

7. Graph each of the following functions.
   (a) $f(x) = -2x + 5$ (b) $f(x) = -x^2 + 4x - 3$
   (c) $f(x) = (x - 2)^2$ (d) $f(x) = x^3 + 2x$
   (e) $f(x) = \dfrac{1}{x^2 - x - 2}$ (f) $f(x) = \begin{cases} 0 & \text{if } x \le 0 \\ x^2 & \text{if } 0 < x < 1 \\ 1 & \text{if } x \ge 1 \end{cases}$

8. Suppose that g is an even function satisfying $g(x) = \sqrt{x}$ for $x \ge 0$. Sketch its graph on $-4 \le x \le 4$.

9. Sketch the graph of $h(x) = x^2 + 1/x$ by first graphing $y = x^2$ and $y = 1/x$ and then adding ordinates.

10. If $f(x) = x^3 + 1$ and $g(x) = x + 2$, give formulas for each of the following.
    (a) $f(g(x))$ (b) $g(f(x))$ (c) $f(f(x))$
    (d) $g(g(x))$ (e) $f^{-1}(x)$ (f) $g^{-1}(x)$
    (g) $g(x + h)$ (h) $[g(x + h) - g(x)]/h$ (i) $f(3x)$

11. How does the graph of $y = f(x - 2) + 3$ relate to the graph of $y = f(x)$?

12. Which of the following functions are even? Odd? One-to-one?
    (a) $f(x) = 1/(x^2 - 1)$ (b) $f(x) = 1/x$
    (c) $f(x) = |x|$ (d) $f(x) = 3x + 4$

13. Let $f(x) = x/(x - 2)$. Find a formula for $f^{-1}(x)$. Graph $y = f(x)$ and $y = f^{-1}(x)$ using the same coordinate axes.

14. How could we restrict the domain of $f(x) = (x + 2)^2$ so that f has an inverse?

The method of logarithms, by reducing to a few days the labor of many months, doubles as it were, the life of the astronomer, besides freeing him from the errors and disgust inseparable from long calculation.

P. S. Laplace

CHAPTER 3

Exponential and Logarithmic Functions

Radicals on Top of Radicals

The discovery of the quadratic formula led to a long search for a corresponding formula for the cubic equation. Success had to wait until the sixteenth century, when the Italian school of mathematicians at Bologna found formulas for both the cubic and quartic equations. A typical result is Cardano's solution to $x^3 + ax = b$ which takes the form

$$x = \sqrt[3]{\sqrt{\frac{a^3}{27} + \frac{b^2}{4}} + \frac{b}{2}} - \sqrt[3]{\sqrt{\frac{a^3}{27} + \frac{b^2}{4}} - \frac{b}{2}}$$

Hieronimo Cardano
1501–1576

3-1 Radicals

Historically, interest in radicals has been associated with the desire to solve equations. Even the general cubic equation leads to very complicated radical expressions. Today, powerful iterative methods make results like Cardano's solution historical curiosities. Yet the need for radicals continues; it is important that we know something about them.

Raising a number to the 3rd power (or cubing it) is a process which can be undone. The inverse process—taking the 3rd root—is denoted by $\sqrt[3]{\ }$. We call $\sqrt[3]{a}$ a *radical* and read it "the cube root of a." Thus $\sqrt[3]{8} = 2$ and $\sqrt[3]{-125} = -5$ since $2^3 = 8$ and $(-5)^3 = -125$.

Our first goal is to give meaning to the symbol $\sqrt[n]{a}$ when n is any positive integer. Naturally, we require that $\sqrt[n]{a}$ be a number which yields a when raised to the nth power; that is

$$(\sqrt[n]{a})^n = a$$

When n is odd, that is all we need to say, since for any real number a, there is exactly one real number whose nth power is a.

When n is even, we face two serious problems, problems that are already apparent when $n = 2$. We have already discussed square roots, using the symbol $\sqrt{\ }$ rather than $\sqrt[2]{\ }$ (see Section 1-1). Recall that if $a < 0$, then $\sqrt{a}$ is not a real number. Even if $a > 0$, we are in trouble since there are always two real numbers with squares equal to a. For example, both -3 and 3 have squares equal to 9. We agree that in this ambiguous case, $\sqrt{a}$ shall always denote the positive square root of a. Thus $\sqrt{9}$ is equal to 3, not -3.

We make a similar agreement about $\sqrt[n]{a}$ for n an even number greater than 2. First, we shall avoid the case $a < 0$. Second, when $a \geq 0$, $\sqrt[n]{a}$ will always denote the nonnegative number whose nth power is a. Thus $\sqrt[4]{81} = 3$, $\sqrt[4]{16} = 2$, and $\sqrt[4]{0} = 0$; however, the symbol $\sqrt[4]{-16}$ will be assigned no meaning in this book. Let us summarize.

> If n is odd, $\sqrt[n]{a}$ is the unique real number satisfying $(\sqrt[n]{a})^n = a$.
> If n is even and $a \geq 0$, $\sqrt[n]{a}$ is the unique nonnegative real number satisfying $(\sqrt[n]{a})^n = a$.

The symbol $\sqrt[n]{a}$, as we have defined it, is called the **principal nth root of a**; for brevity, we often drop the adjective *principal*.

RULES FOR RADICALS

Radicals, like exponents, obey certain rules. The most important ones are listed below, where it is assumed that all radicals name real numbers.

> **RULES FOR RADICALS**
>
> 1. $(\sqrt[n]{a})^n = a$
> 2. $\sqrt[n]{a^n} = a$ $(a \geq 0)$
> 3. $\sqrt[n]{ab} = \sqrt[n]{a}\sqrt[n]{b}$
> 4. $\sqrt[n]{\dfrac{a}{b}} = \dfrac{\sqrt[n]{a}}{\sqrt[n]{b}}$

Rule 2 holds also for $a < 0$ if n is odd; for example, $\sqrt[5]{(-2)^5} = -2$.

These rules can all be proved, but we believe that the following illustrations will be more helpful to you than proofs.

$$(\sqrt[4]{7})^4 = 7$$
$$\sqrt[14]{3^{14}} = 3$$
$$\sqrt{2} \cdot \sqrt{18} = \sqrt{36} = 6$$
$$\frac{\sqrt[3]{750}}{\sqrt[3]{6}} = \sqrt[3]{\frac{750}{6}} = \sqrt[3]{125} = 5$$

SIMPLIFYING RADICALS

One use of the four rules given above is to simplify radicals. Here are two examples.

$$\sqrt[3]{54x^4y^6} \qquad \sqrt[4]{x^8 + x^4y^4}$$

We assume that x and y represent positive numbers.

In the first example, we start by factoring out the largest possible third power.

$$\sqrt[3]{54x^4y^6} = \sqrt[3]{(27x^3y^6)(2x)}$$
$$= \sqrt[3]{(3xy^2)^3(2x)}$$
$$= \sqrt[3]{(3xy^2)^3}\sqrt[3]{2x} \qquad \text{(Rule 3)}$$
$$= 3xy^2\sqrt[3]{2x} \qquad \text{(Rule 2)}$$

In the second example, it is tempting to write $\sqrt[4]{x^8 + x^4y^4} = x^2 + xy$, thereby pretending that $\sqrt[4]{a^4 + b^4} = a + b$. This is wrong, because $(a + b)^4 \neq a^4 + b^4$. Here is what we can do.

$$\sqrt[4]{x^8 + x^4y^4} = \sqrt[4]{x^4(x^4 + y^4)}$$
$$= \sqrt[4]{x^4}\sqrt[4]{x^4 + y^4} \qquad \text{(Rule 3)}$$
$$= x\sqrt[4]{x^4 + y^4} \qquad \text{(Rule 2)}$$

We were able to take x^4 out of the radical because it is a 4th power and a factor of $x^8 + x^4y^4$.

RATIONALIZING DENOMINATORS

For some purposes (including hand calculations), fractions with radicals in their denominators are considered to be needlessly complicated. Fortunately, we can usually rewrite a fraction so that its denominator is free of radicals. The process we go through is called **rationalizing the denominator.** Here are two examples.

$$\frac{1}{\sqrt[5]{x}} \qquad \frac{x}{\sqrt{x} + \sqrt{y}}$$

In the first case, we multiply numerator and denominator by $\sqrt[5]{x^4}$, which gives the 5th root of a 5th power in the denominator.

$$\frac{1}{\sqrt[5]{x}} = \frac{1 \cdot \sqrt[5]{x^4}}{\sqrt[5]{x} \cdot \sqrt[5]{x^4}} = \frac{\sqrt[5]{x^4}}{\sqrt[5]{x \cdot x^4}} = \frac{\sqrt[5]{x^4}}{\sqrt[5]{x^5}} = \frac{\sqrt[5]{x^4}}{x}$$

In the second case, we make use of the identity $(a + b)(a - b) = a^2 - b^2$. If we multiply numerator and denominator of the fraction by $\sqrt{x} - \sqrt{y}$, the radicals in the denominator disappear

$$\frac{x}{\sqrt{x} + \sqrt{y}} = \frac{x(\sqrt{x} - \sqrt{y})}{(\sqrt{x} + \sqrt{y})(\sqrt{x} - \sqrt{y})} = \frac{x\sqrt{x} - x\sqrt{y}}{x - y}$$

We should point out that this manipulation is valid provided $x \neq y$.

Problem Set 3-1

Simplify the following radical expressions. This will involve removing perfect powers from radicals and rationalizing denominators. Assume that all letters represent positive numbers.

1. $\sqrt{9}$
2. $\sqrt[3]{-8}$
3. $\sqrt[5]{32}$
4. $\sqrt[4]{16}$
5. $(\sqrt[3]{7})^3$
6. $(\sqrt{\pi})^2$
7. $\sqrt[3]{(\frac{3}{2})^3}$
8. $\sqrt[5]{(-2/7)^5}$
9. $(\sqrt{5})^4$
10. $(\sqrt[3]{5})^6$
11. $\sqrt{3}\sqrt{27}$
12. $\sqrt{2}\sqrt{32}$
13. $\sqrt[3]{16}/\sqrt[3]{2}$
14. $\sqrt[4]{48}/\sqrt[4]{3}$
15. $\sqrt[3]{10^{-6}}$
16. $\sqrt[4]{10^8}$
17. $1/\sqrt{2}$
18. $1/\sqrt{3}$
19. $\sqrt{10}/\sqrt{2}$
20. $\sqrt{6}/\sqrt{3}$
21. $\sqrt[3]{54x^4y^5}$
22. $\sqrt[3]{-16x^3y^8}$
23. $\sqrt[4]{(x+2)^4y^7}$
24. $\sqrt[4]{x^5(y-1)^8}$
25. $\sqrt{x^2+x^2y^2}$
26. $\sqrt{25+50y^4}$
27. $\sqrt[3]{x^6-9x^3y}$
28. $\sqrt[4]{16x^{12}+64x^8}$
29. $\sqrt[3]{x^4y^{-6}z^6}$
30. $\sqrt[4]{32x^{-4}y^9}$
31. $\dfrac{2}{\sqrt{x+3}}$
32. $\dfrac{4}{\sqrt{x-2}}$
33. $\dfrac{2}{\sqrt{x+3}}$
34. $\dfrac{4}{\sqrt{x-2}}$
35. $\dfrac{1}{\sqrt[4]{8x^3}}$
36. $\dfrac{1}{\sqrt[3]{5x^2y^4}}$
37. $\sqrt[3]{2x^{-2}y^4}\sqrt[3]{4xy^{-1}}$
38. $\sqrt[4]{125x^5y^3}\sqrt[4]{5x^{-9}y^5}$
39. $\sqrt{50}-2\sqrt{18}+\sqrt{8}$
40. $\sqrt[3]{24}+\sqrt[3]{375}$

CAUTION

$$\sqrt{a^4+a^4b^2} \;\cancel{=}\; a^2+a^2b$$

$$\sqrt{a^4+a^4b^2} = \sqrt{a^4(1+b^2)}$$
$$= a^2\sqrt{1+b^2}$$

EXAMPLE A (Equations Involving Radicals)   Solve the following equations.

(a) $\sqrt[3]{x-2}=3$ (b) $x=\sqrt{2-x}$

Solution.

(a) Raise both sides to the 3rd power and solve for x.

$$(\sqrt[3]{x-2})^3 = 3^3$$
$$x - 2 = 27$$
$$x = 29$$

(b) Square both sides and solve for x.

$$x^2 = 2 - x$$
$$x^2 + x - 2 = 0$$
$$(x-1)(x+2) = 0$$
$$x = 1 \qquad x = -2$$

Let us check our answers in part (b) by substituting them in the original equation. When we substitute these numbers for x in $x=\sqrt{2-x}$, we find that 1 works but -2 does not.

$$1 = \sqrt{2} - 1 \qquad -2 \neq \sqrt{2} - (-2)$$

In squaring both sides of $x = \sqrt{2} - x$, we introduced an extraneous solution. That happened because $a = b$ and $a^2 = b^2$ are not equivalent statements. Whenever you square both sides of an equation (or raise both sides of an equation to any even power), be sure to check your answers.

Solve each of the following equations.

41. $\sqrt{x - 1} = 5$ 　　42. $\sqrt{x + 2} = 3$ 　　43. $\sqrt[3]{2x - 1} = 2$

44. $\sqrt[3]{1 - 5x} = 6$ 　45. $\sqrt{\dfrac{x}{x + 2}} = 4$ 　46. $\sqrt[3]{\dfrac{x - 2}{x + 1}} = -2$

47. $\sqrt{x^2 + 4} = x + 2$ 　48. $\sqrt{x^2 + 9} = x - 3$

49. $\sqrt{2x + 1} = x - 1$ 　50. $\sqrt{x} = 12 - x$

EXAMPLE B (Combining Fractions Involving Radicals)　Sums and differences of fractions involving radicals occur often in calculus. It is usually desirable to combine these fractions. Do so in

(a) $\dfrac{1}{\sqrt[3]{x + h}} - \dfrac{1}{\sqrt[3]{x}}$; 　(b) $\dfrac{x}{\sqrt{x^2 + 4}} - \dfrac{\sqrt{x^2 + 4}}{x}$.

Solution.

(a) $\dfrac{1}{\sqrt[3]{x + h}} - \dfrac{1}{\sqrt[3]{x}} = \dfrac{\sqrt[3]{x}}{\sqrt[3]{x}\sqrt[3]{x + h}} - \dfrac{\sqrt[3]{x + h}}{\sqrt[3]{x}\sqrt[3]{x + h}} = \dfrac{\sqrt[3]{x} - \sqrt[3]{x + h}}{\sqrt[3]{x}\sqrt[3]{x + h}}$

(b) $\dfrac{x}{\sqrt{x^2 + 4}} - \dfrac{\sqrt{x^2 + 4}}{x} = \dfrac{x^2}{x\sqrt{x^2 + 4}} - \dfrac{\sqrt{x^2 + 4}\sqrt{x^2 + 4}}{x\sqrt{x^2 + 4}}$

$$= \dfrac{x^2 - (x^2 + 4)}{x\sqrt{x^2 + 4}} = \dfrac{-4}{x\sqrt{x^2 + 4}}$$

Combine the fractions in each of the following. Do not bother to rationalize denominators.

51. $\dfrac{2}{\sqrt{x + h}} - \dfrac{2}{\sqrt{x}}$ 　　52. $\dfrac{\sqrt{x}}{\sqrt{x + 2}} - \dfrac{1}{\sqrt{x}}$

53. $\dfrac{1}{\sqrt{x + 6}} + \sqrt{x + 6}$ 　54. $\dfrac{\sqrt{x + 1}}{\sqrt{x + 3}} - \dfrac{\sqrt{x + 3}}{x + 1}$

55. $\dfrac{\sqrt[3]{(x + 2)^2}}{2} - \dfrac{1}{\sqrt[3]{x + 2}}$ 　56. $\dfrac{\sqrt{x + 7}}{\sqrt{x - 2}} - \dfrac{\sqrt{x - 2}}{x + 7}$

57. $\dfrac{1}{\sqrt{x^2 + 9}} - \dfrac{\sqrt{x^2 + 9}}{x^2}$ 　58. $\dfrac{x}{\sqrt{x^2 + 3}} + \dfrac{\sqrt{x^2 + 3}}{x}$

MISCELLANEOUS PROBLEMS

59. Simplify each expression (including rationalizing denominators). Assume all letters represent positive numbers.

(a) $\sqrt[4]{16a^4b^8}$

(b) $\sqrt{27}\sqrt{3b^3}$

(c) $\sqrt{12} + \sqrt{48} - \sqrt{27}$

(d) $\sqrt{250a^4b^6}$

(e) $\sqrt[3]{\dfrac{-32x^2y^7}{4x^5y}}$

(f) $\left(\sqrt[3]{\dfrac{y}{2x}}\right)^6$

(g) $\sqrt{8a^5} + \sqrt{18a^3}$

(h) $\sqrt[4]{512} - \sqrt{50} + \sqrt[6]{128}$

(i) $\sqrt[4]{a^4 + a^4b^4}$

(j) $\dfrac{1}{\sqrt[3]{7bc^3}}$

(k) $\dfrac{2}{\sqrt{a} - b}$

(l) $\sqrt{a}\left(\sqrt{a} + \dfrac{1}{\sqrt{a^3}}\right)$

60. If a is *any* real number and n is even, then $\sqrt[n]{a^n} = |a|$. Use this to simplify each of the following.

(a) $\sqrt{a^4 + 4a^2}$ (b) $\sqrt[4]{a^4 + a^4b^4}$ (c) $\sqrt{(a - b)^2c^4}$

61. Solve each equation for x.

(a) $\sqrt[3]{1 - 5x} = -4$

(b) $\sqrt{4x + 1} = x + 1$

(c) $\sqrt{x + 3} = 2 + \sqrt{x - 5}$

(d) $\sqrt{12 + x} = 4 + \sqrt{4 + x}$

(e) $x - \sqrt{x} - 6 = 0$

(f) $\sqrt[3]{x^2} - 2\sqrt[3]{x} - 8 = 0$

C 62. Most scientific calculators have a key for roots (on some you must use the two keys $\boxed{\text{INV}}$ $\boxed{y^x}$). Calculate each of the following.

(a) $\sqrt[5]{31}$

(b) $\sqrt[3]{240}$

(c) $\sqrt[10]{78}$

(d) $\sqrt{282} - \sqrt{280}$

(e) $\sqrt[4]{.012}(\sqrt{30} - \sqrt{29})^2$

(f) $\dfrac{\sqrt[4]{29} + \sqrt[3]{6}}{\sqrt{14}}$

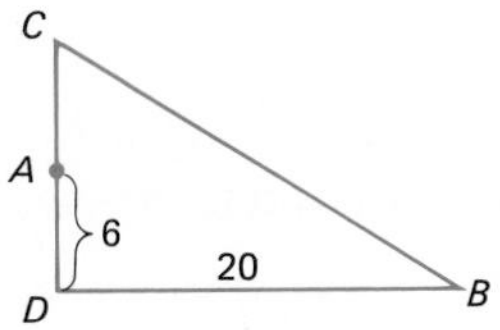

63. We know that $f(x) = x^5$ and $g(x) = \sqrt[5]{x}$ are inverse functions. Sketch their graphs using the same coordinate axes.

64. Rewrite $1/(\sqrt{2} + \sqrt{3} - \sqrt{5})$ with a rational denominator.

65. Figure 1 shows a right triangle. Determine $\overline{AC}$ so that the routes ACB and ADB from A to B have the same length.

66. In calculus, it is sometimes advantageous to rationalize the numerator. Rewrite each of the following with a rational numerator.

(a) $\dfrac{\sqrt{x} - \sqrt{y}}{\sqrt{x} + \sqrt{y}}$ (b) $\dfrac{\sqrt{x + h} - \sqrt{x}}{h}$ (c) $\dfrac{\sqrt[3]{x} - \sqrt[3]{y}}{x - y}$

67. Show each of the following to be true.

(a) $\dfrac{\sqrt{6} + \sqrt{2}}{2} = \sqrt{2 + \sqrt{3}}$ (b) $\sqrt{2 + \sqrt{3}} + \sqrt{2 - \sqrt{3}} = \sqrt{6}$

(c) $\sqrt[3]{9\sqrt{3} - 11\sqrt{2}} = \sqrt{3} - \sqrt{2}$

68. **TEASER** Find the exact value of x if $x = \sqrt[3]{9 + 4\sqrt{5}} + \sqrt[3]{9 - 4\sqrt{5}}$. *Hint*: You might guess at the answer by using your calculator but you will not be done until you have given an algebraic demonstration that your guess is correct.

Figure 1

One of the authors once asked a student to write the definition of $2^{4.6}$ on the blackboard. After thinking deeply for a minute, he wrote:

$$2^2 = 2 \cdot 2$$
$$2^3 = 2 \cdot 2 \cdot 2$$
$$2^4 = 2 \cdot 2 \cdot 2 \cdot 2$$
$$2^{4.6} = 2 \cdot 2 \cdot 2 \cdot 2 \cdot \raisebox{-2pt}{¿}$$

3-2 Exponents and Exponential Functions

After you have criticized the student mentioned above, ask yourself how you would define $2^{4.6}$. Of course, integral powers of 2 make perfectly good sense, although 2^{-3} and 2^0 became meaningful only after we had *defined* a^{-n} to be $1/a^n$ and a^0 to be 1 (see Section 1-2). Those were good definitions because they were consistent with the familiar rules of exponents. Now we ask what meaning we can give to powers like $2^{1/2}$, $2^{4.6}$, and even 2^{π} so that these familiar rules still hold.

RATIONAL EXPONENTS

We assume throughout this section that $a > 0$. If n is any positive integer, we want

$$(a^{1/n})^n = a^{(1/n) \cdot n} = a^1 = a$$

But we know that $(\sqrt[n]{a})^n = a$. Thus we define

$$\boxed{a^{1/n} = \sqrt[n]{a}}$$

For example, $2^{1/2} = \sqrt{2}$, $27^{1/3} = \sqrt[3]{27} = 3$, and $(16)^{1/4} = \sqrt[4]{16} = 2$.
Next, if m and n are positive integers, we want

$$(a^{1/n})^m = a^{m/n} \quad \text{and} \quad (a^m)^{1/n} = a^{m/n}$$

This forces us to define

$$\boxed{a^{m/n} = (\sqrt[n]{a})^m = \sqrt[n]{a^m}}$$

Accordingly,

$$2^{3/2} = (\sqrt{2})^3 = \sqrt{2}\,\sqrt{2}\,\sqrt{2} = 2\sqrt{2}$$

and

$$27^{2/3} = (\sqrt[3]{27})^2 = 3^2 = 9$$

Lastly, we define

$$a^{-m/n} = \frac{1}{a^{m/n}}$$

so that

$$2^{-1/2} = \frac{1}{2^{1/2}} = \frac{1}{\sqrt{2}}$$

and

$$4^{-3/2} = \frac{1}{4^{3/2}} = \frac{1}{(\sqrt{4})^3} = \frac{1}{8}$$

We have just succeeded in defining a^x for all rational numbers x (recall that a rational number is a ratio of two integers). What is more important is that we have done it in such a way that the rules of exponents still hold. Incidentally, we can now answer the question in our opening display.

$$2^{4.6} = 2^4 2^{.6} = 2^4 2^{6/10} = 16(\sqrt[10]{2})^6$$

CAUTION

$(-8)^{-1/3} = 8^{1/3} = 2$

$(-8)^{-1/3} = \dfrac{1}{(-8)^{1/3}} = \dfrac{1}{-2}$

For simplicity, we have assumed that a is positive in our discussion of $a^{m/n}$. But we should point out that the definition of $a^{m/n}$ given above is also appropriate for the case in which a is negative and n is odd. For example,

$$(-27)^{2/3} = (\sqrt[3]{-27})^2 = (-3)^2 = 9$$

REAL EXPONENTS

Irrational powers such as 2^π and $3^{\sqrt{2}}$ are intrinsically more difficult to define than are rational powers. Rather than attempt a technical definition, we ask you to consider what 2^π might mean. The decimal expansion of π is 3.14159 . . . . Thus we could look at the sequence of rational powers

$$2^3, \ 2^{3.1}, \ 2^{3.14}, \ 2^{3.141}, \ 2^{3.1415}, \ 2^{3.14159}, \ . . .$$

As you should suspect, when the exponents get closer and closer to π, the corresponding powers of 2 get closer and closer to a definite number. We shall call the number 2^π.

The process of starting with integral exponents and then extending to rational exponents and finally to real exponents can be clarified by means of three graphs (Figure 2 on the next page). Note the table of values in the margin.

The first graph suggests a curve rising from left to right. The second graph makes the suggestion stronger. The third graph leaves nothing to the imagination; it is a continuous curve and it shows 2^x for all values of x, rational and irrational. As x increases in the positive direction, the values of 2^x increase

x	2^x
-3	$\frac{1}{8}$
-2	$\frac{1}{4}$
-1	$\frac{1}{2}$
0	1
$\frac{1}{2}$	$\sqrt{2} \approx 1.4$
1	2
$\frac{3}{2}$	$2\sqrt{2} \approx 2.8$
2	4
3	8

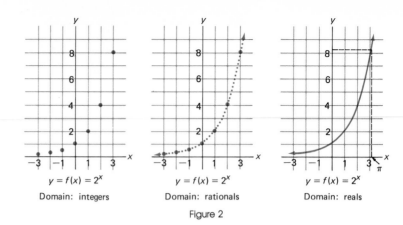

$y = f(x) = 2^x$ $y = f(x) = 2^x$ $y = f(x) = 2^x$

Domain: integers   Domain: rationals   Domain: reals

Figure 2

without bound; in the negative direction, the values of 2^x approach 0. Notice that 2^π is a little less than 9; its value correct to seven decimal places is

$$2^\pi = 8.8249778$$

See if your calculator gives this value.

EXPONENTIAL FUNCTIONS

The function $f(x) = 2^x$, graphed above, is one example of an exponential function. But what has been done with 2 can be done with any positive real number a. In general, the formula

$$f(x) = a^x$$

determines a function called an **exponential function with base a.** Its domain is the set of all real numbers and its range is the set of positive numbers.

Let us see what effect the size of a has on the graph of $f(x) = a^x$. We choose $a = 2$, $a = 3$, $a = 5$, and $a = \frac{1}{3}$, showing all four graphs in Figure 3.

x	3^x	$(\frac{1}{3})^x$
-2	$1/9$	9
-1	$1/3$	3
0	1	1
1	3	$1/3$
2	9	$1/9$

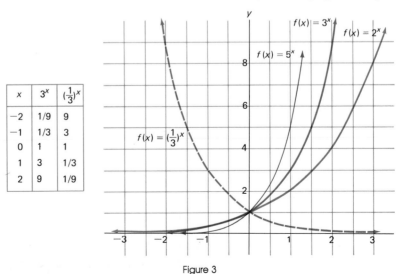

Figure 3

The graph of $f(x) = 3^x$ looks much like the graph of $f(x) = 2^x$, although it rises more rapidly. The graph of $f(x) = 5^x$ is even steeper. All three of these functions are *increasing functions,* meaning that the values of $f(x)$ increase as x increases; more formally, $x_2 > x_1$ implies $f(x_2) > f(x_1)$. The function $f(x) = (\frac{1}{3})^x$, on the other hand, is a *decreasing function*. In fact, you can get the graph of $f(x) = (\frac{1}{3})^x$ by reflecting the graph of $f(x) = 3^x$ about the y-axis. This is because $(\frac{1}{3})^x = 3^{-x}$.

We can summarize what is suggested by our discussion as follows.

If $a > 1$, $f(x) = a^x$ is an increasing function.

If $0 < a < 1$, $f(x) = a^x$ is a decreasing function.

In both of these cases, the graph of f has the x-axis as an asymptote. The case $a = 1$ is not very interesting since it yields the constant function $f(x) = 1$.

PROPERTIES OF EXPONENTIAL FUNCTIONS

It is easy to describe the main properties of exponential functions, since they obey the rules we learned in Section 1-2. Perhaps it is worth repeating them, since we do want to emphasize that they now hold for all *real* exponents x and y (at least for the case where a and b are both positive).

1. $a^x a^y = a^{x+y}$

2. $\dfrac{a^x}{a^y} = a^{x-y}$

3. $(a^x)^y = a^{xy}$

4. $(ab)^x = a^x b^x$

5. $\left(\dfrac{a}{b}\right)^x = \dfrac{a^x}{b^x}$

Here are a number of examples that are worth studying.

$$3^{1/2} 3^{3/4} = 3^{1/2+3/4} = 3^{5/4}$$

$$\frac{\pi^4}{\pi^{5/2}} = \pi^{4-5/2} = \pi^{3/2}$$

$$(2^{\sqrt{3}})^4 = 2^{4\sqrt{3}}$$

$$(5^{\sqrt{2}} 5^{1-\sqrt{2}})^{-3} = (5^1)^{-3} = 5^{-3}$$

Problem Set 3-2

Write each of the following as a power of 7.

1. $\sqrt[3]{7}$ 2. $\sqrt[5]{7}$ 3. $\sqrt[3]{7^2}$ 4. $\sqrt[5]{7^3}$ 5. $\dfrac{1}{\sqrt[3]{7}}$

6. $\dfrac{1}{\sqrt[5]{7}}$ 7. $\dfrac{1}{\sqrt[3]{7^2}}$ 8. $\dfrac{1}{\sqrt[5]{7^3}}$ 9. $7\sqrt[3]{7}$ 10. $7\sqrt[5]{7}$

Rewrite each of the following using exponents instead of radicals. For example,
$\sqrt[5]{x^3} = x^{3/5}$.

11. $\sqrt[3]{x^2}$ 12. $\sqrt[4]{x^3}$ 13. $x^2\sqrt{x}$ 14. $x\sqrt[3]{x}$

15. $\sqrt{(x+y)^3}$ 16. $\sqrt[3]{(x+y)^2}$ 17. $\sqrt{x^2+y^2}$ 18. $\sqrt[3]{x^3+8}$

Rewrite each of the following using radicals instead of fractional exponents. For example, $(xy^2)^{3/7} = \sqrt[7]{x^3y^6}$

19. $4^{2/3}$ 20. $10^{3/4}$ 21. $8^{-3/2}$

22. $12^{-5/6}$ 23. $(x^4+y^4)^{1/4}$ 24. $(x^2+xy)^{1/2}$

25. $(x^2y^3)^{2/5}$ 26. $(3ab^2)^{2/3}$ 27. $(x^{1/2}+y^{1/2})^{1/2}$

28. $(x^{1/3}+y^{2/3})^{1/3}$

Simplify each of the following. Give your answer without any exponents.

29. $25^{1/2}$ 30. $27^{1/3}$

31. $8^{2/3}$ 32. $16^{3/2}$

33. $9^{-3/2}$ 34. $64^{-2/3}$

35. $(-.008)^{2/3}$ 36. $(-.027)^{5/3}$

37. $(.0025)^{3/2}$ 38. $(1.44)^{3/2}$

39. $5^{2/3}5^{-5/3}$ 40. $4^{3/4}4^{-1/4}$

41. $16^{7/6}16^{-5/6}16^{-4/3}$ 42. $9^2 9^{2/3} 9^{-7/6}$

43. $(8^2)^{-2/3}$ 44. $(4^{-3})^{3/2}$

EXAMPLE A (Simplifying Expressions Involving Exponents)   Simplify and
write the answer without negative exponents.

(a) $\dfrac{x^{1/3}(8x)^{-2/3}}{x^{-3/4}}$ (b) $\left(\dfrac{2x^{-1/2}}{y}\right)^4\left(\dfrac{x}{y}\right)^{-1}(3x^{10/3})$

Solution.

(a) $\dfrac{x^{1/3}(8x)^{-2/3}}{x^{-3/4}} = x^{1/3}8^{-2/3}x^{-2/3}x^{3/4} = \dfrac{x^{1/3-2/3+3/4}}{8^{2/3}} = \dfrac{x^{5/12}}{4}$

(b) $\left(\dfrac{2x^{-1/2}}{y}\right)^4\left(\dfrac{x}{y}\right)^{-1}(3x^{10/3}) = \left(\dfrac{16x^{-2}}{y^4}\right)\left(\dfrac{y}{x}\right)(3x^{10/3})$

$= \dfrac{48x^{-2-1+10/3}}{y^{4-1}} = \dfrac{48x^{1/3}}{y^3}$

Simplify, writing your answer without negative exponents.

45. $(3a^{1/2})(-2a^{3/2})$

46. $(2x^{3/4})(5x^{-3/4})$

47. $(2^{1/2}x^{-2/3})^6$

48. $(\sqrt{3}x^{-1/4}y^{3/4})^4$

49. $(xy^{-2/3})^3(x^{1/2}y)^2$

50. $(a^2b^{-1/4})^2(a^{-1/3}b^{1/2})^3$

51. $\dfrac{(2x^{-1}y^{2/3})^2}{x^2y^{-2/3}}$

52. $\left(\dfrac{a^{1/2}b^{1/3}}{c^{5/6}}\right)^{12}$

53. $\left(\dfrac{x^{-2}y^{3/4}}{x^{1/2}}\right)^{12}$

54. $\dfrac{x^{1/3}y^{-3/4}}{x^{-2/3}y^{1/2}}$

55. $y^{2/3}(2y^{4/3} - y^{-5/3})$

56. $x^{-3/4}\left(-x^{7/4} + \dfrac{2}{\sqrt[4]{x}}\right)$

57. $(x^{1/2} + y^{1/2})^2$

58. $(a^{3/2} + \pi)^2$

EXAMPLE B (Combining Fractions)   Perform the following addition.

$$\frac{(x + 1)^{2/3}}{x} + \frac{1}{(x + 1)^{1/3}}$$

Solution.

$$\frac{(x + 1)^{2/3}}{x} + \frac{1}{(x + 1)^{1/3}} = \frac{(x + 1)^{2/3}(x + 1)^{1/3}}{x(x + 1)^{1/3}} + \frac{x}{x(x + 1)^{1/3}}$$

$$= \frac{x + 1 + x}{x(x + 1)^{1/3}} = \frac{2x + 1}{x(x + 1)^{1/3}}$$

Combine the fractions in each of the following.

59. $\dfrac{(x + 2)^{4/5}}{3} + \dfrac{2x}{(x + 2)^{1/5}}$

60. $\dfrac{(x - 3)^{1/3}}{4} - \dfrac{1}{(x - 3)^{2/3}}$

61. $(x^2 + 1)^{1/3} - \dfrac{2x^2}{(x^2 + 1)^{2/3}}$

62. $(x^2 + 2)^{1/4} + \dfrac{x^2}{(x^2 + 2)^{3/4}}$

EXAMPLE C (Mixing Radicals of Different Orders)   Express $\sqrt{2}\sqrt[3]{5}$ using just one radical.

Solution.   Square roots and cube roots mix about as well as oil and water, but exponents can serve as a blender. They allow us to write both $\sqrt{2}$ and $\sqrt[3]{5}$ as sixth roots.

$$\sqrt{2}\sqrt[3]{5} = 2^{1/2} \cdot 5^{1/3}$$

$$= 2^{3/6} \cdot 5^{2/6}$$

$$= (2^3 \cdot 5^2)^{1/6}$$

$$= \sqrt[6]{200}$$

Express each of the following in terms of at most one radical in simplest form.

63. $\sqrt{2}\sqrt[3]{2}$

64. $\sqrt[3]{2}\sqrt[4]{2}$

65. $\sqrt[4]{2}\sqrt[6]{x}$

66. $\sqrt[3]{5}\sqrt{x}$

67. $\sqrt[3]{x\sqrt{x}}$

68. $\sqrt{x\sqrt[3]{x}}$

    69. $2^{1.34}$ 70. $2^{-.79}$ 71. $\pi^{1.34}$ 72. π^{π}

    73. $(1.46)^{\sqrt{2}}$ 74. $\pi^{\sqrt{2}}$ 75. $(.9)^{50.2}$ 76. $(1.01)^{50.2}$

Sketch the graph of each of the following functions.

    77. $f(x) = 4^x$ 78. $f(x) = 4^{-x}$ 79. $f(x) = \left(\frac{2}{3}\right)^x$

    80. $f(x) = \left(\frac{2}{3}\right)^{-x}$ 81. $f(x) = \pi^x$ 82. $f(x) = (\sqrt{2})^x$

MISCELLANEOUS PROBLEMS

83. Rewrite using exponents in place of radicals and simplify.

    (a) $\sqrt[5]{b^3}$ (b) $\sqrt[8]{x^4}$ (c) $\sqrt[3]{a^2 + 2ab + b^2}$

84. Simplify.

    (a) $(32)^{-6/5}$ (b) $(-.008)^{2/3}$ (c) $(5^{-1/2}\, 5^{3/4}\, 5^{1/8})^{16}$

85. Simplify, writing your answer without either radicals or negative exponents.

    (a) $(27)^{2/3}(.0625)^{-3/4}$ (b) $\sqrt[3]{4}\sqrt{2} + \sqrt[6]{2}$

    (c) $\sqrt[3]{a^2}\sqrt[4]{a^3}$ (d) $\sqrt{a\sqrt[3]{a^2}}$

    (e) $[a^{3/2} + a^{-3/2}]^2$ (f) $[a^{1/4}(a^{-5/4} + a^{3/4})]^{-1}$

    (g) $\left(\dfrac{\sqrt[3]{a^3 b^2}}{\sqrt[4]{a^6 b^3}}\right)^{-1}$ (h) $\left(\dfrac{a^{-2} b^{2/3}}{b^{-1/2}}\right)^{-4}$

    (i) $\left[\dfrac{(27)^{4/3} - (27)^0}{(3^2 + 4^2)^{1/2}}\right]^{3/4}$ (j) $(16a^2 b^3)^{3/4} - 4ab^2(a^2 b)^{1/4}$

    (k) $(\sqrt{3})^{3\sqrt{3}} - (3\sqrt{3})^{\sqrt{3}} + (\sqrt{3}^{\sqrt{3}})^{\sqrt{3}}$

    (l) $(a^{1/3} - b^{1/3})(a^{2/3} + a^{1/3} b^{1/3} + b^{2/3})$

86. Combine and simplify, writing your answer without negative exponents.

    (a) $4x^2(x^2 + 2)^{-2/3} - 3(x^2 + 2)^{1/3}$

    (b) $x^3(x^3 - 1)^{-3/4} - (x^3 - 1)^{1/4}$

87. Solve for x.

    (a) $4^{x+1} = (1/2)^{2x}$ (b) $5^{x^2-x} = 25$

    (c) $2^{4x} 4^{x-3} = (64)^{x-1}$ (d) $(x^2 + x + 4)^{3/4} = 8$

    (e) $x^{2/3} - 3x^{1/3} = -2$ (f) $2^{2x} - 2^{x+1} - 8 = 0$

88. Using the same axes, sketch the graph of each of the following.

    (a) $f(x) = 2^x$ (b) $g(x) = -2^x$ (c) $h(x) = 2^{-x}$

    (d) $k(x) = 2^x + 2^{-x}$ (e) $m(x) = 2^{x-4}$

89. Sketch the graph of $f(x) = 2^{-|x|}$.

□ 90. Using the same axes, sketch the graphs of $f(x) = x^{\pi}$ and $g(x) = \pi^x$ on the interval $2 \le x \le 3.5$. One solution of $x^{\pi} = \pi^x$ is π. Use your graphs to help you find another one (approximately).

91. Give a simple argument to show that an exponential function $f(x) = a^x (a > 0, a \ne 1)$ is not equivalent to any polynomial function.

92. **TEASER** If a and b are irrational, does it follow that a^b is irrational? *Hint:* Consider $\sqrt{2}^{\sqrt{2}}$ and $(\sqrt{2}^{\sqrt{2}})^{\sqrt{2}}$.

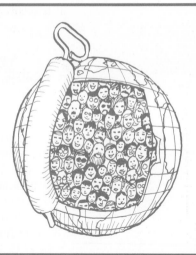

A Packing Problem

World population is growing at about 2 percent per year. If this continues indefinitely, how long will it be until we are all packed together like sardines? In answering, assume that ''sardine packing'' for humans is one person per square foot of land area.

3-3 Exponential Growth and Decay

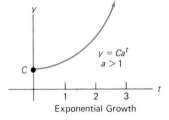

$y = Ca^t$
$a > 1$

Exponential Growth

Figure 4

The phrase *exponential growth* is used repeatedly by professors, politicians, and pessimists. Population, energy use, mining of ores, pollution, and the number of books about these things are all said to be growing exponentially. Most people probably do not know what exponential growth means, except that they have heard it guarantees alarming consequences. For students of this book, it is easy to explain its meaning. For y to grow exponentially with time t means that it satisfies the relationship

$$y = Ca^t$$

for constants C and a, with $C > 0$ and $a > 1$ (see Figure 4). Why should so many ingredients of modern society behave this way? The basic cause is population growth.

POPULATION GROWTH

Simple organisms reproduce by cell division. If, for example, there is one cell today, that cell may split so that there are two cells tomorrow. Then each of those cells may divide giving four cells the following day (Figure 5). As this process continues, the numbers of cells on successive days form the sequence

$$1, 2, 4, 8, 16, 32, \ldots$$

If we start with 100 cells and let $f(t)$ denote the number present t days from now, we have the results indicated in the table of Figure 6.

Figure 5

t	0	1	2	3	4	5
$f(t)$	100	200	400	800	1600	3200

Figure 6

It seems that

$$f(t) = (100)2^t$$

A perceptive reader will ask if this formula is really valid. Does it give the right answer when $t = 5.7$? Is not population growth a discrete process, occurring in unit amounts at distinct times, rather than a continuous process as the formula implies? The answer is that the exponential growth model provides a very good approximation to the growth of simple organisms, provided the initial population is large.

The mechanism of reproduction is different (and more interesting) for people, but the pattern of population growth is similar. World population is presently growing at about 2 percent per year. In 1975, there were about 4 billion people. Accordingly, the population in 1976 in billions was $4 + 4(.02) = 4(1.02)$, in 1977 it was $4(1.02)^2$, in 1978 it was $4(1.02)^3$, and so on. If this trend continues, there will be $4(1.02)^{20}$ billion people in the world in 1995, that is, 20 years after 1975. It appears that world population obeys the formula.

$$p(t) = 4(1.02)^t$$

where $p(t)$ represents the number of people (in billions) t years after 1975.

In general, if $A(t)$ is the amount at time t of a quantity growing exponentially at the rate of r (written as a decimal), then

$$A(t) = A(0)(1 + r)^t$$

Here $A(0)$ is the initial amount—that is, the amount present at $t = 0$.

DOUBLING TIMES

One way to get a feeling for the spectacular nature of exponential growth is via the concept of **doubling time**; this is the length of time required for an exponentially growing quantity to double in size. It is easy to show that if a quantity doubles in an initial time interval of length T, it will double in size in *any* time interval of length T. Consider the world population problem as an example. By the table of Figure 7, $(1.02)^{35} \approx 2$, so world population doubles in 35 years. Since it was 4 billion in 1975, it should be 8 billion in 2010, 16 billion in 2045, and so on. This alarming information is displayed graphically in Figure 8.

Now we can answer the question about sardine packing in our opening display. There are slightly more than 1,000,000 billion square feet of land area on the surface of the earth. Sardine packing for humans is about 1 square foot per person. Thus we are asking when $4(1.02)^t$ billion will equal 1,000,000 billion. This leads to the equation

$$(1.02)^t = 250,000$$

To solve this exponential equation, we use the following approximations

t	$(1.02)^t$
5	1.104
10	1.219
15	1.346
20	1.486
25	1.641
30	1.811
35	2.000
40	2.208
45	2.438
50	2.692
55	2.972
60	3.281
65	3.623
70	4.000
75	4.416
80	4.875
85	5.383
90	5.943

Figure 7

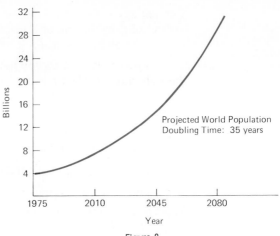

Figure 8

$$(1.02)^{35} \approx 2 \qquad 250{,}000 \approx 2^{18}$$

Our equation can then be rewritten as

$$[(1.02)^{35}]^{t/35} = 2^{18}$$

or

$$2^{t/35} = 2^{18}$$

We conclude that

$$\frac{t}{35} = 18$$

$$t = (18)(35) = 630$$

Thus, after about 630 years, we will be packed together like sardines. If it is any comfort, war, famine, or birth control will change population growth patterns before then.

COMPOUND INTEREST

One of the best practical illustrations of exponential growth is money earning compound interest. Suppose that Amy puts $1000 in a bank today at 8 percent interest compounded annually. Then at the end of one year the bank adds the interest of $(.08)(1000) = \$80$ to her $1000, giving her a total of $1080. But note that $1080 = 1000(1.08)$. During the second year, $1080 draws interest. At the end of that year, the bank adds $(.08)(1080)$ to the account, bringing the total to

$$1080 + (.08)(1080) = (1080)(1.08)$$

$$= 1000(1.08)(1.08)$$

$$= 1000(1.08)^2$$

Continuing in this way, we see that Amy's account will have grown to $1000(1.08)^3$ by the end of 3 years, $1000(1.08)^4$ by the end of 4 years, and so on. By the end of 15 years, it will have grown to

$$1000(1.08)^{15} \approx 1000(3.172169)$$

$$= \$3172.17$$

To calculate $(1.08)^{15}$, we used Table 2 at the end of the problem set. We could have used a calculator.

How long would it take for Amy's money to double—that is, when will

$$1000(1.08)^t = 2000$$

This will occur when $(1.08)^t = 2$. According to the table just mentioned, this happens at $t \approx 9$, or in about 9 years.

EXPONENTIAL DECAY

Fortunately, not all things grow; some decline or decay. In fact, some things—notably the radioactive elements—decay exponentially. This means that the amount y present at time t satisfies

$$y = Ca^t$$

for some constants C and a with $C > 0$ and $0 < a < 1$ (see Figure 9).

Here an important idea is that of **half-life,** the time required for half of a substance to disappear. For example, radium decays with a half-life of 1620 years. Thus if 1000 grams of radium are present now, 1620 years from now 500 grams will be present, $2(1620) = 3240$ years from now only 250 grams will be present, and so on.

The precise nature of radioactive decay is used to date old objects. If an object contains radium and lead (the product to which radium decays) in the ratio 1 to 3, then it is believed that an original amount of pure radium has decayed to $\frac{1}{4}$ its original size. The object must be two half-lives, or 3240 years, old. Two important assumptions have been made: (1) decay of radium is exactly exponential over long periods of time; and (2) no lead was originally present. Recent research raises some question about the correctness of such assumptions.

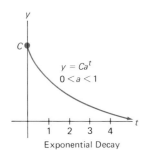

$y = Ca^t$
$0 < a < 1$

Exponential Decay

Figure 9

Problem Set 3-3

1. In each of the following, indicate whether y grows exponentially or decays exponentially with t.

    (a) $y = 128\left(\frac{1}{2}\right)^t$

    (b) $y = 5\left(\frac{5}{3}\right)^t$

    (c) $y = 4(10)^9(1.03)^t$

    (d) $y = 1000(.99)^t$

2. Find the values of y corresponding to $t = 0$, $t = 1$, and $t = 2$ for each case in Problem 1.

3. Use the table at the end of the problem set or a calculator with a $\boxed{y^x}$ key to find each value.
   (a) $(1.08)^{20}$ (b) $(1.12)^{25}$ (c) $1000(1.04)^{40}$ (d) $2000(1.02)^{80}$

4. Evaluate each of the following.
   (a) $(1.01)^{100}$ (b) $(1.02)^{40}$ (c) $100(1.12)^{50}$ (d) $500(1.04)^{30}$

5. Silver City's present population of 1000 is expected to grow exponentially over the next 10 years at 4 percent per year. How many people will it have at the end of that time? *Hint:* $A(t) = A(0)[1 + r]^t$.

6. The value of houses in Longview is said to be growing exponentially at 12 percent per year. What will a house valued at \$100,000 today be worth after 8 years?

7. Under the assumptions concerning world population used in this section, what will be the approximate number of people on earth in each year?
   (a) 1990 (that is, 15 years after 1975)
   (b) 2000
   (c) 2065

8. A certain radioactive substance has a half-life of 40 minutes. What fraction of an initial amount of this substance will remain after 1 hour, 20 minutes (that is, after 2 half-lives)? After 2 hours, 40 minutes?

EXAMPLE A (Compound Interest)   Roger put \$1000 in a money market fund at 15 percent interest compounded annually. How much was it worth after 4 years?

Solution.   We will have to use a calculator since $(1.15)^n$ is not in our table. The answer is

$$1000(1.15)^4 = \$1749.01$$

9. If you put \$100 in the bank for 8 years, how much will it be worth at the end of that time at
   (a) 8 percent compounded annually;
   (b) 12 percent compounded annually?

10. If you invest \$500 in the bank today, how much will it be worth after 25 years at
    (a) 8 percent compounded annually;
    (b) 4 percent compounded annually;
    (c) 12 percent compounded annually?

11. If you put \$3500 in the bank today, how much will it be worth after 40 years at
    (a) 8 percent compounded annually;
    (b) 12 percent compounded annually?

12. Approximately how long will it take for money to accumulate to twice its value if
    (a) it is invested at 8 percent compounded annually;
    (b) it is invested at 12 percent compounded annually?

13. Suppose that you invest P dollars at r percent compounded annually. Write an expression for the amount accumulated after n years.

EXAMPLE B (More on Compound Interest)   If $1000 is invested at 8 percent compounded quarterly, find the accumulated amount after 15 years.

Solution.   Interest calculated at 2 percent ($\frac{1}{4}$ of 8 percent) is converted to principal every 3 months. By the end of the first 3-month period, the account has grown to $1000(1.02) = \$1020$; by the end of the second 3-month period, it has grown to $1000(1.02)^2$; and so on. The accumulated amount after 15 years, or 60 conversion periods, is

$$1000(1.02)^{60} \approx 1000(3.28103)$$

$$= \$3281.03$$

Suppose more generally that P dollars is invested at a rate r (written as a decimal), which is compounded m times per year. Then the accumulated amount A after t years is given by

$$A = P\left(1 + \frac{r}{m}\right)^{tm}$$

In our example

$$A = 1000\left(1 + \frac{.08}{4}\right)^{15 \cdot 4} = 1000(1.02)^{60}$$

Find the accumulated amount for the indicated initial principal, compound interest rate, and total time period.

14. $2000; 8 percent compounded annually; 15 years.
15. $5000; 8 percent compounded semiannually; 5 years.
16. $5000; 12 percent compounded monthly; 5 years.
☐c 17. $3000; 9 percent compounded annually; 10 years.
☐c 18. $3000; 9 percent compounded semiannually; 10 years.
☐c 19. $3000; 9 percent compounded quarterly; 10 years.
☐c 20. $3000; 9 percent compounded monthly; 10 years.
☐c 21. $1000; 8 percent compounded monthly; 10 years.
☐c 22. $1000; 8 percent compounded daily; 10 years.   *Hint:* Assume there are 365 days in a year, so that the interest rate per day is $.08/365$.

MISCELLANEOUS PROBLEMS

23. Let $y = 5400(2/3)^t$. Evaluate y for $t = -1, 0, 1, 2,$ and 3.
24. For what value of t in Problem 23 is $y = 3200/3$?
25. If $(1.023)^T = 2$, find the value of $100(1.023)^{3T}$.
26. If $(.67)^H = \frac{1}{2}$, find the value of $32(.67)^{4H}$.
☐c 27. Suppose the population of a certain city follows the formula

$$p(t) = 4600(1.016)^t$$

where $p(t)$ is the population t years after 1980.

(a) What will the population be in 2020? In 2080?

(b) Experiment with your calculator to find the doubling time for this population.

28. The number of bacteria in a certain culture is known to triple every hour. Suppose the count at 12:00 noon is 162,000. What was the count at 11:00 A.M.? At 8:00 A.M.?

29. If $100 is invested today, how much will it be worth after 5 years at 8 percent interest if interest is:

(a) compounded annually;

(b) compounded quarterly;

[c] (c) compounded monthly;

[c] (d) compounded daily? (There are 365 days in a year.)

30. How long does it take money to double if invested at 12 percent compounded monthly? (Use the compound interest table.)

31. About how long does it take an exponentially growing population to double if its rate of growth is;

(a) 8 percent per year? (Use the compound interest table.)

[c] (b) 6.5 percent per year? (Experiment with your calculator.)

[c] 32. A manufacturer of radial tires found that the percentage P of tires still usable after being driven m miles was given by

$$P = 100(2.71)^{-.000025m}$$

What percentage of tires are still usable at 80,000 miles?

33. A certain radioactive element has a half-life of 1690 years. Starting with 30 milligrams there will be $q(t)$ milligrams left after t years, where $q(t) = 30(1/2)^{kt}$.

(a) Determine the constant k.

[c] (b) How much will be left after 2500 years?

34. One method of depreciation allowed by IRS is the double-declining-balance method. In this method, the original value C of an item is depreciated each year by $100(2/N)$ percent of its value at the beginning of that year, N being the useful life of the item.

(a) Write a formula for the value V of the item after n years.

[c] (b) If an item cost $10,000 and has a useful life of 15 years, calculate its value after 10 years. After 15 years.

(c) Does the value of an item ever become zero by this depreciation method?

[c] 35. (Carbon dating)   All living things contain carbon-12, which is a stable element, and carbon-14, which is radioactive. While a plant or animal is alive, the ratio of these two isotopes of carbon remains unchanged, since carbon-14 is constantly renewed; but after death, no more carbon-14 is absorbed. The half-life of carbon-14 is 5730 years. Bones from a human body were found to contain only 76 percent of the carbon-14 in living bones. How long before did the person die?

[c] 36. Manhattan Island is said to have been bought from the Indians by Peter Minuit in 1626 for $24. If, instead of making this purchase, Minuit had put the money in a savings account drawing interest at 6 percent compounded annually, what would that account be worth in the year 2000?

[c] 37. Hamline University was founded in 1854 with a gift of $25,000 from Bishop Hamline of the Methodist Church. Suppose that Hamline University had wisely put $10,000 of this gift in an endowment drawing 10 percent interest com-

pounded annually, promising not to touch it until 1988 (exactly 134 years later). How much could it then withdraw each year and still maintain this endowment at the 1988 level?

38. **TEASER** Suppose one water lily growing exponentially at the rate of 8 percent per day is able to cover a certain pond in 50 days. How long would it take 10 of these lilies to cover the pond?

TABLE 2   Compound Interest Table

n	$(1.01)^n$	$(1.02)^n$	$(1.04)^n$	$(1.08)^n$	$(1.12)^n$
1	1.01000000	1.02000000	1.04000000	1.08000000	1.12000000
2	1.02010000	1.04040000	1.08160000	1.16640000	1.25440000
3	1.03030100	1.06120800	1.12486400	1.25971200	1.40492800
4	1.04060401	1.08243216	1.16985856	1.36048896	1.57351936
5	1.05101005	1.10408080	1.21665290	1.46932808	1.76234168
6	1.06152015	1.12616242	1.26531902	1.58687432	1.97382269
7	1.07213535	1.14868567	1.31593178	1.71382427	2.21068141
8	1.08285671	1.17165938	1.36856905	1.85093021	2.47596318
9	1.09368527	1.19509257	1.42331181	1.99900463	2.77307876
10	1.10462213	1.21899442	1.48024428	2.15892500	3.10584821
11	1.11566835	1.24337431	1.53945406	2.33163900	3.47854999
12	1.12682503	1.26824179	1.60103222	2.51817012	3.89597599
15	1.16096896	1.34586834	1.80094351	3.17216911	5.47356576
20	1.22019004	1.48594740	2.19112314	4.66095714	9.64629309
25	1.28243200	1.64060599	2.66583633	6.84847520	17.00006441
30	1.34784892	1.81136158	3.24339751	10.06265689	29.95992212
35	1.41660276	1.99988955	3.94608899	14.78534429	52.79961958
40	1.48886373	2.20803966	4.80102063	21.72452150	93.05097044
45	1.56481075	2.43785421	5.84117568	31.92044939	163.98760387
50	1.64463182	2.69158803	7.10668335	46.90161251	289.00218983
55	1.72852457	2.97173067	8.64636692	68.91385611	509.32060567
60	1.81669670	3.28103079	10.51962741	101.25706367	897.59693349
65	1.90936649	3.62252311	12.79873522	148.77984662	1581.87249060
70	2.00676337	3.99955822	15.57161835	218.60640590	2787.79982770
75	2.10912847	4.41583546	18.94525466	321.20452996	4913.05584077
80	2.21671522	4.87543916	23.04979907	471.95483426	8658.48310008
85	2.32978997	5.38287878	28.04360494	693.45648897	15259.20568055
90	2.44863267	5.94313313	34.11933334	1018.91508928	26891.93422336
95	2.57353755	6.56169920	41.51138594	1497.12054855	47392.77662369
100	2.70481383	7.24464612	50.50494818	2199.76125634	83522.26572652

An active participant in the political and religious battles of his day, the Scot. John Napier amused himself by studying mathematics and science. He was interested in reducing the work involved in the calculations of spherical trigonometry, especially as they applied to astronomy. In 1614 he published a book containing the idea that made him famous. He gave it the name *logarithm*.

John Napier
(1550–1617)

3-4　Logarithms and Logarithmic Functions

Napier's approach to logarithms is out of style, but the goal he had in mind is still worth considering. He hoped to replace multiplications by additions. He thought additions were easier to do, and he was right.

Consider the exponential function $f(x) = 2^x$ and recall that

$$2^x \cdot 2^y = 2^{x+y}$$

On the left, we have a multiplication and on the right, an addition. If we are to fulfill Napier's objective, we want logarithms to behave like exponents. That suggests a definition. The logarithm of N to the base 2 is the exponent to which 2 must be raised to yield N. That is,

$$\boxed{\log_2 N = x \quad \text{if and only if} \quad 2^x = N}$$

Thus

$$\log_2 4 = 2 \quad \text{since} \quad 2^2 = 4$$
$$\log_2 8 = 3 \quad \text{since} \quad 2^3 = 8$$
$$\log_2 \sqrt{2} = \tfrac{1}{2} \quad \text{since} \quad 2^{1/2} = \sqrt{2}$$

and, in general,

$$\log_2(2^x) = x \quad \text{since} \quad 2^x = 2^x$$

Has Napier's goal been achieved? Does the logarithm turn a product into a sum? Yes, for note that

$$\log_2(2^x \cdot 2^y) = \log_2(2^{x+y}) \qquad \text{(property of exponents)}$$
$$= x + y \qquad \text{(definition of } \log_2)$$
$$= \log_2(2^x) + \log_2(2^y)$$

Thus

$$\log_2(2^x \cdot 2^y) = \log_2(2^x) + \log_2(2^y)$$

which has the form

$$\log_2(M \cdot N) = \log_2 M + \log_2 N$$

THE GENERAL DEFINITION

What has been done for 2 can be done for any base $a > 1$. The **logarithm of N to the base a** is the exponent x to which a must be raised to yield N. Thus

$$\log_a N = x \quad \text{if and only if} \quad a^x = N$$

Now we can calculate many kinds of logarithms.

$$\log_4 16 = 2 \quad \text{since} \quad 4^2 = 16$$
$$\log_{10} 1000 = 3 \quad \text{since} \quad 10^3 = 1000$$
$$\log_{10}(.001) = -3 \quad \text{since} \quad 10^{-3} = \frac{1}{1000} = .001$$

What is $\log_{10} 7$? We are not ready to answer that yet, except to say it is a number x satisfying $10^x = 7$ (see Section 3-6).

We point out that negative numbers and zero do not have logarithms. Suppose -4 and 0 did have logarithms, that is, suppose

$$\log_a(-4) = m \quad \text{and} \quad \log_a 0 = n$$

Then

$$a^m = -4 \quad \text{and} \quad a^n = 0$$

But that is impossible; we learned earlier that a^x is always positive.

PROPERTIES OF LOGARITHMS

There are three main properties of logarithms.

> **PROPERTIES OF LOGARITHMS**
> 1. $\log_a(M \cdot N) = \log_a M + \log_a N$
> 2. $\log_a(M/N) = \log_a M - \log_a N$
> 3. $\log_a(M^p) = p \log_a M$

To establish Property 1, let

$$x = \log_a M \quad \text{and} \quad y = \log_a N$$

Then, by definition,

$$M = a^x \quad \text{and} \quad N = a^y$$

so that

$$M \cdot N = a^x \cdot a^y = a^{x+y}$$

Thus $x + y$ is the exponent to which a must be raised to yield $M \cdot N$, that is,

$$\log_a(M \cdot N) = x + y = \log_a M + \log_a N$$

Properties 2 and 3 are demonstrated in a similar fashion.

THE LOGARITHMIC FUNCTION

The function determined by

$$g(x) = \log_a x$$

is called the **logarithmic function with base a.** We can get a feeling for the behavior of this function by drawing its graph for $a = 2$ (Figure 10).

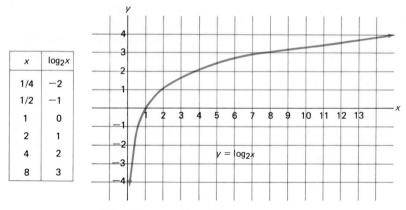

x	$\log_2 x$
1/4	−2
1/2	−1
1	0
2	1
4	2
8	3

Figure 10

Several properties of $y = \log_2 x$ are apparent from this graph. The domain consists of all positive real numbers. If $0 < x < 1$, $\log_2 x$ is negative; if $x > 1$, $\log_2 x$ is positive. The y-axis is a vertical asymptote of the graph since very small positive x values yield large negative y values. Although $\log_2 x$ continues to increase as x increases, even this small part of the complete graph indicates how slowly it grows for large x. In fact, by the time x reaches 1,000,000, $\log_2 x$ is still loafing along at about 20. In this sense, it behaves in a manner opposite to the exponential function 2^x, which grows more and more rapidly as x increases. There is a good reason for this opposite behavior; the two functions are inverses of each other.

INVERSE FUNCTIONS

We begin by emphasizing two facts that you must not forget.

$$a^{\log_a x} = x$$

$$\log_a(a^x) = x$$

For example, $2^{\log_2 7} = 7$ and $\log_2(2^{-19}) = -19$. Both of these facts are direct consequences of the definition of logarithms; the second is also a special case of Property 3, stated earlier. What these facts tell us is that the logarithmic and exponential functions undo each other.

Let us put it in the language of Section 2-7. If $f(x) = a^x$ and $g(x) = \log_a x$, then

$$f(g(x)) = f(\log_a x) = a^{\log_a x} = x$$

and

$$g(f(x)) = g(a^x) = \log_a(a^x) = x$$

Thus g is really f^{-1}. This fact also tells us something about the graphs of g and f: They are simply reflections of each other about the line $y = x$ (Figure 11).

Note finally that $f(x) = a^x$ has the set of all real numbers as its domain and the positive real numbers as its range. Thus its inverse $f^{-1}(x) = \log_a x$ has domain consisting of the positive real numbers and range consisting of all real numbers. We emphasize again a fact that is important to remember. *Negative numbers and zero do not have logarithms.*

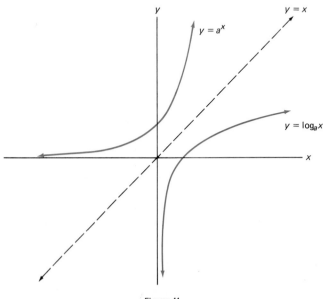

Figure 11

Problem Set 3-4

Write each of the following in logarithmic form. For example, $3^4 = 81$ can be written as $\log_3 81 = 4$.

1. $4^3 = 64$
2. $7^3 = 343$
3. $27^{1/3} = 3$
4. $16^{1/4} = 2$
5. $4^0 = 1$
6. $81^{-1/2} = \frac{1}{9}$
7. $125^{-2/3} = \frac{1}{25}$
8. $2^{9/2} = 16\sqrt{2}$
9. $10^{\sqrt{3}} = a$
10. $5^{\sqrt{2}} = b$
11. $10^a = \sqrt{3}$
12. $b^x = y$

Write each of the following in exponential form. For example, $\log_5 125 = 3$ can be written as $5^3 = 125$.

13. $\log_5 625 = 4$
14. $\log_6 216 = 3$
15. $\log_4 8 = \frac{3}{2}$
16. $\log_{27} 9 = \frac{2}{3}$
17. $\log_{10}(.01) = -2$
18. $\log_3(\frac{1}{27}) = -3$
19. $\log_c c = 1$
20. $\log_b N = x$
21. $\log_c Q = y$

Determine the value of each of the following logarithms.

22. $\log_4 16$
23. $\log_5 25$
24. $\log_7 \frac{1}{7}$
25. $\log_3 \frac{1}{3}$
26. $\log_4 2$
27. $\log_{27} 3$
28. $\log_{10}(10^{-6})$
29. $\log_{10}(.0001)$
30. $\log_8 1$
31. $\log_3 1$
32. $\log_{100} 1000$
33. $\log_8 16$

Find the value of c in each of the following.

34. $\log_c 25 = 2$
35. $\log_c 8 = 3$
36. $\log_4 c = -\frac{1}{2}$
37. $\log_9 c = -\frac{3}{2}$
38. $\log_2(2^{5.6}) = c$
39. $\log_3(3^{-2.9}) = c$
40. $8^{\log_8 11} = c$
41. $5^{2 \log_5 7} = c$
42. $3^{4 \log_3 2} = c$

Given $\log_{10} 2 = .301$ and $\log_{10} 3 = .477$, calculate each of the following without the use of tables. For example, in Problem 43, $\log_{10} 6 = \log_{10} 2 \cdot 3 = \log_{10} 2 + \log_{10} 3$, and in Problem 50, $\log_{10} 54 = \log_{10} 2 \cdot 3^3 = \log_{10} 2 + \log_{10} 3^3 = \log_{10} 2 + 3 \log_{10} 3$.

43. $\log_{10} 6$
44. $\log_{10} \frac{3}{2}$
45. $\log_{10} 16$
46. $\log_{10} 27$
47. $\log_{10} \frac{1}{4}$
48. $\log_{10} \frac{1}{27}$
49. $\log_{10} 24$
50. $\log_{10} 54$
51. $\log_{10} \frac{8}{9}$
52. $\log_{10} \frac{3}{8}$
53. $\log_{10} 5$
54. $\log_{10} \sqrt[3]{3}$

[c] *Your scientific calculator has a $\log_{10}$ key (which may be abbreviated log). Use it to find each of the following.*

55. $\log_{10} 34$
56. $\log_{10} 1417$
57. $\log_{10}(.0123)$
58. $\log_{10}(.3215)$
59. $\log_{10} 9723$
60. $\log_{10}(\frac{21}{312})$

EXAMPLE A (Combining Logarithms) Write the following expression as a single logarithm.

$$2 \log_{10} x + 3 \log_{10}(x + 2) - \log_{10}(x^2 + 5)$$

Solution. We use the properties of logarithms to rewrite this as

$$\log_{10} x^2 + \log_{10}(x + 2)^3 - \log_{10}(x^2 + 5) \qquad \text{(Property 3)}$$

$$= \log_{10} x^2(x + 2)^3 - \log_{10}(x^2 + 5) \qquad \text{(Property 1)}$$

$$= \log_{10}\left[\frac{x^2(x + 2)^3}{x^2 + 5}\right] \qquad \text{(Property 2)}$$

CAUTION

$$\log_b 6 + \log_b 4 = \log_b 10$$
$$\log_b 30 - \log_b 5 = \log_b 25$$

$$\log_b 6 + \log_b 4 = \log_b 24$$
$$\log_b 30 - \log_b 5 = \log_b 6$$

Write each of the following as a single logarithm.

61. $3 \log_{10}(x + 1) + \log_{10}(4x + 7)$
62. $\log_{10}(x^2 + 1) + 5 \log_{10} x$
63. $3 \log_2(x + 2) + \log_2 8x - 2 \log_2(x + 8)$
64. $2 \log_5 x - 3 \log_5(2x + 1) + \log_5(x - 4)$
65. $\frac{1}{2} \log_6 x + \frac{1}{3} \log_6(x^3 + 3)$
66. $-\frac{2}{3} \log_3 x + \frac{5}{2} \log_3(2x^2 + 3)$

EXAMPLE B (Solving Logarithmic Equations) Solve the equation

$$\log_2 x + \log_2(x + 2) = 3$$

Solution. First we note that we must have $x > 0$ so that both logarithms exist. Next we rewrite the equation using the first property of logarithms and then the definition of a logarithm.

$$\log_2 x(x + 2) = 3$$

$$x(x + 2) = 2^3$$

$$x^2 + 2x - 8 = 0$$

$$(x + 4)(x - 2) = 0$$

We reject $x = -4$ (because $-4 < 0$) and keep $x = 2$. To make sure that 2 is a solution, we substitute 2 for x in the original equation.

$$\log_2 2 + \log_2(2 + 2) \stackrel{?}{=} 3$$

$$1 + 2 = 3$$

Solve each of the following equations.

67. $\log_7(x + 2) = 2$
68. $\log_5(3x + 2) = 1$
69. $\log_2(x + 3) = -2$
70. $\log_4(\frac{1}{64}x + 1) = -3$
71. $\log_2 x - \log_2(x - 2) = 3$
72. $\log_3 x - \log_3(2x + 3) = -2$
73. $\log_2(x - 4) + \log_2(x - 3) = 1$
74. $\log_{10} x + \log_{10}(x - 3) = 1$

EXAMPLE C (Change of Base) In Problems 89 and 90, you will be asked to establish the **change-of-base formula**

$$\log_b x = \frac{\log_a x}{\log_a b}$$

Use this formula and the $\boxed{\log_{10}}$ key on a calculator to find $\log_2 13$.

Solution.

$$\log_2 13 = \frac{\log_{10} 13}{\log_{10} 2} \approx \frac{1.1139433}{.30103} \approx 3.7004$$

[c] *Use the method of Example C to find the following.*

75. $\log_2 128$ 76. $\log_3 128$ 77. $\log_3 82$

78. $\log_5 110$ 79. $\log_6 39$ 80. $\log_2(.26)$

MISCELLANEOUS PROBLEMS

81. Find the value of x in each of the following.
    (a) $x = \log_6 36$ (b) $x = \log_4 2$ (c) $\log_{25} x = \frac{3}{2}$
    (d) $\log_4 x = \frac{5}{2}$ (e) $\log_x 10\sqrt{10} = \frac{3}{2}$ (f) $\log_x \frac{1}{8} = -\frac{3}{2}$

82. Write each of the following as a single logarithm.
    (a) $3 \log_2 5 - 2 \log_2 7$
    (b) $\frac{1}{2} \log_5 64 + \frac{1}{3} \log_5 27 - \log_5(x^2 + 4)$
    (c) $\frac{2}{3} \log_{10}(x + 5) + 4 \log_{10} x - 2 \log_{10}(x - 3)$

83. Evaluate $\dfrac{(\log_{27} 3)(\log_{27} 9)(3^{2 \log_3 2})}{\log_3 27 - \log_3 9 + \log_3 1}$

84. Solve for x.
    (a) $2(\log_4 x)^2 + 3 \log_4 x - 2 = 0$
    (b) $(\log_x 8)^2 - \log_x 8 - 6 = 0$
    (c) $\log_x \sqrt{3} + \log_x 3^5 + \log_x(\frac{1}{27}) = \frac{5}{4}$

85. Solve for x.
    (a) $\log_5(2x - 1) = 2$
    (b) $\log_4\left(\dfrac{x - 2}{2x + 3}\right) = 0$
    (c) $\log_4(x - 2) - \log_4(2x + 3) = 0$
    (d) $\log_{10} x + \log_{10}(x - 15) = 2$
    (e) $\dfrac{\log_2(x + 1)}{\log_2(x - 1)} = 2$
    (f) $\log_8[\log_4(\log_2 x)] = 0$

86. Solve for x.
    (a) $2^{\log_2 x} = 16$ (b) $2^{\log_x 2} = 16$ (c) $x^{\log_2 x} = 16$
    (d) $\log_2 x^2 = 2$ (e) $(\log_2 x)^2 = 1$ (f) $x = (\log_2 x)^{\log_2 x}$

87. Solve for y in terms of x.
    (a) $\log_a(x + y) = \log_a x + \log_a y$
    (b) $x = \log_a(y + \sqrt{y^2 - 1})$

88. Show that $f(x) = \log_a(x + \sqrt{1 + x^2})$ is an odd function.

89. Show that $\log_2 x = \log_{10} x \log_2 10$, where $x > 0$. *Hint:* Let $\log_{10} x = c$. Then $x = 10^c$. Next take $\log_2$ of both sides.

90. Use the technique outlined in Problem 89 to show that for $a > 0$, $b > 0$, and $x > 0$

$$\log_a x = \log_b x \log_a b$$

This is equivalent to the change-of-base formula of Example C.

91. Show that $\log_a b = 1/\log_b a$, where a and b are positive.

92. If $\log_b N = 2$, find $\log_{1/b} N$.

93. Graph the equations $y = 3^x$ and $y = \log_3 x$ using the same coordinate axes.

94. Find the solution set for each of the following inequalities.

   (a) $\log_2 x < 0$ (b) $\log_{10} x \geq -1$

   (c) $2 < \log_3 x < 3$ (d) $-2 \leq \log_{10} x \leq -1$

   (e) $2^x > 10$ (f) $2^x < 3^x$

95. Sketch the graph of each of the following functions using the same coordinate axes.

   (a) $f(x) = \log_2 x$ (b) $g(x) = \log_2(x + 1)$ (c) $h(x) = 3 + \log_2 x$

96. **TEASER** Let log represent $\log_{10}$. Evaluate.

   (a) $\log \frac{1}{2} + \log \frac{2}{3} + \log \frac{3}{4} + \cdots + \log \frac{98}{99} + \log \frac{99}{100}$

   (b) $\log \frac{3}{4} + \log \frac{8}{9} + \log \frac{15}{16} + \cdots + \log \frac{99^2 - 1}{99^2} + \log \frac{100^2 - 1}{100^2}$

   (c) $\log_2 3 \cdot \log_3 4 \cdot \log_4 5 \cdot \log_5 6 \cdots \log_{63} 64$

The collected works of this brilliant Swiss mathematician will fill 74 volumes when completed. No other person has written so profusely on mathematical topics. Remarkably, 400 of his research papers were written after he was totally blind. One of his contributions was the introduction of the number $e = 2.71828 \ldots$ as the base for natural logarithms.

Leonhard Euler
1707–1783

3-5 Natural Logarithms and Applications

Napier invented logarithms to simplify arithmetic calculations. Computers and calculators have reduced that application to minor significance, though we shall discuss such a use of logarithms later in this chapter. Here we have in mind deeper applications such as solving exponential equations, defining power functions, and modeling physical phenomena.

Figure 12

In order to make any progress, we shall need an easy way to calculate logarithms. Fortunately, this has been done for us as tables of logarithms to several bases are available. For our purposes in this section, one base is as good as another. Base 10 would be an appropriate choice, but we would rather defer discussion of logarithms to base 10 (common logarithms) until Section 3-6. We have chosen rather to introduce you to the number e (after Euler), which is used as a base of logarithms in all advanced mathematics courses. You will see the importance of logarithms to this base (**natural logarithms**) when you study calculus (also see Figure 12). An approximate value of e is

$$e \approx 2.71828$$

and, like π, e is an irrational number. Table 3 on page 156 shows a table of values of natural logarithms, which we shall denote by ln instead of $\log_e$.

Since ln denotes a genuine logarithm function, we have as in Section 3-4

$$\ln N = x \quad \text{if and only if} \quad e^x = N$$

and consequently

$$\ln e^x = x \quad \text{and} \quad e^{\ln N} = N$$

Moreover the three properties of logarithms hold.

1. $\ln (MN) = \ln M + \ln N$
2. $\ln(M/N) = \ln M - \ln N$
3. $\ln (N^p) = p \ln N$

SOLVING EXPONENTIAL EQUATIONS

Consider first the simple equation

$$5^x = 1.7$$

We call it an *exponential equation* because the unknown is in the exponent. To solve it, we take natural logarithms of both sides.

$$5^x = 1.7$$

$$\ln(5^x) = \ln 1.7$$

$$x \ln 5 = \ln 1.7 \qquad \text{(Property 3)}$$

$$x = \frac{\ln 1.7}{\ln 5}$$

$$x \approx \frac{.531}{1.609} \approx .330 \quad \text{(Table 3)}$$

We point out that the last step can also be done on a scientific calculator, which has a key for calculating natural logarithms.

TABLE 3   Table of Natural Logarithms

x	$\ln x$	x	$\ln x$	x	$\ln x$
		4.0	1.386	8.0	2.079
0.1	-2.303	4.1	1.411	8.1	2.092
0.2	-1.609	4.2	1.435	8.2	2.104
0.3	-1.204	4.3	1.459	8.3	2.116
0.4	-0.916	4.4	1.482	8.4	2.128
0.5	-0.693	4.5	1.504	8.5	2.140
0.6	-0.511	4.6	1.526	8.6	2.152
0.7	-0.357	4.7	1.548	8.7	2.163
0.8	-0.223	4.8	1.569	8.8	2.175
0.9	-0.105	4.9	1.589	8.9	2.186
1.0	0.000	5.0	1.609	9.0	2.197
1.1	0.095	5.1	1.629	9.1	2.208
1.2	0.182	5.2	1.649	9.2	2.219
1.3	0.262	5.3	1.668	9.3	2.230
1.4	0.336	5.4	1.686	9.4	2.241
1.5	0.405	5.5	1.705	9.5	2.251
1.6	0.470	5.6	1.723	9.6	2.262
1.7	0.531	5.7	1.740	9.7	2.272
1.8	0.588	5.8	1.758	9.8	2.282
1.9	0.642	5.9	1.775	9.9	2.293
2.0	0.693	6.0	1.792	10	2.303
2.1	0.742	6.1	1.808	20	2.996
2.2	0.788	6.2	1.825	30	3.401
2.3	0.833	6.3	1.841	40	3.689
2.4	0.875	6.4	1.856	50	3.912
2.5	0.916	6.5	1.872	60	4.094
2.6	0.956	6.6	1.887	70	4.248
2.7	0.993	6.7	1.902	80	4.382
2.8	1.030	6.8	1.917	90	4.500
2.9	1.065	6.9	1.932	100	4.605
3.0	1.099	7.0	1.946		
3.1	1.131	7.1	1.960		
3.2	1.163	7.2	1.974	e	1.000
3.3	1.194	7.3	1.988		
3.4	1.224	7.4	2.001	π	1.145
3.5	1.253	7.5	2.015		
3.6	1.281	7.6	2.028		
3.7	1.308	7.7	2.041		
3.8	1.335	7.8	2.054		
3.9	1.361	7.9	2.067		

To find the natural logarithm of a number N which is either smaller than 0.1 or larger than 10, write N in scientific notation, that is, write $N = c \times 10^k$. Then $\ln N = \ln c + k \ln 10 = \ln c + k(2.303)$. A more complete table of natural logarithms appears as Table A of the Appendix.

Here is a more complicated example.

$$5^{2x-1} = 7^{x+2}$$

Begin by taking natural logarithms of both sides and then solve for x.

$$\ln(5^{2x-1}) = \ln(7^{x+2})$$

$$(2x - 1) \ln 5 = (x + 2) \ln 7 \qquad \text{(Property 3)}$$

$$2x \ln 5 - \ln 5 = x \ln 7 + 2 \ln 7$$

$$2x \ln 5 - x \ln 7 = \ln 5 + 2 \ln 7$$

$$x(2 \ln 5 - \ln 7) = \ln 5 + 2 \ln 7$$

$$x = \frac{\ln 5 + 2 \ln 7}{2 \ln 5 - \ln 7}$$

$$\approx \frac{1.609 + 2(1.946)}{2(1.609) - 1.946} \quad \text{(Table 3 or a calculator)}$$

$$\approx 4.325$$

You get a more accurate answer, 4.321, if you use a calculator all the way.

THE GRAPHS OF ln x AND e^x

We have already pointed out that $\ln e^x = x$ and $e^{\ln x} = x$. Thus $f(x) = \ln x$ and $g(x) = e^x$ are inverse functions, which means that their graphs are reflections of each other across the line $y = x$. They are shown in Figure 13.

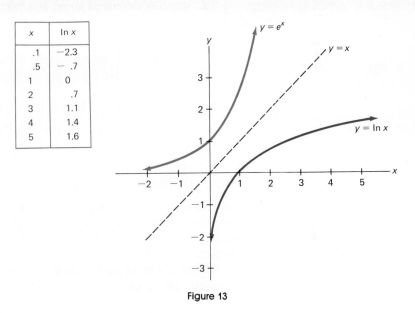

x	ln x
.1	-2.3
.5	$-$ .7
1	0
2	.7
3	1.1
4	1.4
5	1.6

Figure 13

Since $\ln x$ is not defined for $x \leq 0$, it is of some interest to consider $\ln|x|$, which is defined for all x except 0. Its graph is shown in Figure 14 at the top of the next page. Note the symmetry with respect to the y-axis.

EXPONENTIAL FUNCTIONS VERSUS POWER FUNCTIONS

Look closely at the formulas below.

$$f(x) = 2^x \qquad f(x) = x^2$$

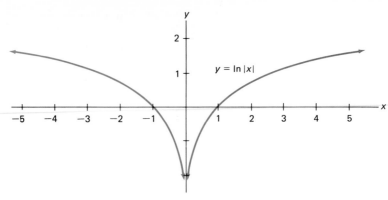

Figure 14

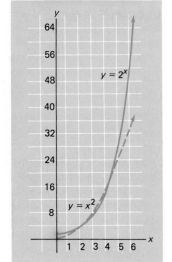

Figure 15

They are very different, yet easily confused. The first is an exponential function, while the second is called a power function. Both grow rapidly for large x, but the exponential function ultimately gets far ahead (see the graphs in Figure 15).

The situation described above is a special instance of two very general classes of functions.

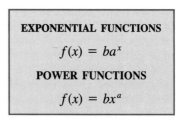

EXPONENTIAL FUNCTIONS

$$f(x) = ba^x$$

POWER FUNCTIONS

$$f(x) = bx^a$$

CURVE FITTING

A recurring theme in science is to fit a mathematical curve to a set of experimental data. Suppose that a scientist, studying the relationship between two variables x and y, obtained the data plotted in Figure 16. In searching for curves to fit these data, the scientist naturally thought of exponential curves and power curves. How did he or she decide if either was appropriate? The scientist took logarithms. Let us see why.

Figure 16

MODEL 1	MODEL 2
$y = ba^x$	$y = bx^a$
$\ln y = \ln b + x \ln a$	$\ln y = \ln b + a \ln x$
$Y = B + Ax$	$Y = B + aX$

Here the scientist made the substitutions $Y = \ln y$, $B = \ln b$, $A = \ln a$, and $X = \ln x$.

In both cases, the final result is a linear equation. But note the difference.

In the first case, ln y is a linear function of x, whereas in the second case, ln y is a linear function of ln x. These considerations suggest the following procedures. Make two additional plots of the data. In the first, plot ln y against x, and in the second, plot ln y against ln x. If the first plotting gives data nearly along a straight line, Model 1 is appropriate; if the second does, then Model 2 is appropriate. If neither plot approximates a straight line, our scientist should look for a different and perhaps more complicated model.

We have used natural logarithms in the discussion above; we could also have used common logarithms (logarithms to the base 10). In the latter case, special kinds of graph paper are available to simplify the curve fitting process. On semilog paper, the vertical axis has a logarithmic scale; on log-log paper, both axes have logarithmic scales. The xy-data can be plotted *directly* on this paper. If semilog paper gives an (approximately) straight line, Model 1 is indicated; if log-log paper does so, then Model 2 is appropriate. You will have ample opportunity to use these special kinds of graph paper in your science courses.

LOGARITHMS AND PHYSIOLOGY

The human body appears to have a built-in logarithmic calculator. What do we mean by this statement?

In 1834, the German physiologist E. Weber noticed an interesting fact. Two heavy objects must differ in weight by considerably more than two light objects if a person is to perceive a difference between them. Other scientists noted the same phenomenon when human subjects tried to differentiate loudness of sounds, pitches of musical tones, brightness of light, and so on. Experiments suggested that people react to stimuli on a logarithmic scale, a result formulated as the Weber-Fechner law (see Figure 17).

$$S = C \ln\left(\frac{R}{r}\right)$$

Here R is the actual intensity of the stimulus, r is the threshold value (smallest value at which the stimulus is observed), C is a constant depending on the type of stimulus, and S is the perceived intensity of the stimulus. Note that a change in R is not as perceptible for large R as for small R because as R increases, the graph of the logarithmic functions gets steadily flatter.

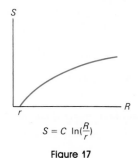

$S = C \ln(\frac{R}{r})$

Figure 17

Problem Set 3-5

In Problems 1–8, find the value of each natural logarithm.

1. ln e 2. $\ln(e^2)$ 3. ln 1   4. $\ln\left(\frac{1}{e}\right)$

5. $\ln\sqrt{e}$ 6. $\ln(e^{1.1})$ 7. $\ln\left(\frac{1}{e^3}\right)$ 8. $\ln(e^n)$

For Problems 9–14, find each value. Assume that ln a = 2.5 and ln b = −.4.

9. ln(ae)

10. $\ln\left(\frac{e}{b}\right)$

11. $\ln\sqrt{b}$

12. ln(a²b¹⁰)

13. $\ln\left(\frac{1}{a^3}\right)$

14. ln(a⁴ᐟ⁵)

Use the table of natural logarithms to calculate the values in Problems 15–20.

15. ln 120 = ln 60 + ln 2   16. ln 150   17. ln 690
18. ln 84   19. ln $\frac{6}{5}$ 20. ln 20,000

In Problems 21–26, use the natural logarithm table to find N.

21. ln N = 2.208   22. ln N = 1.808   23. ln N = −.105
24. ln N = −.916   25. ln N = 4.500   26. ln N = 9.000

© *Use your calculator (ln or logₑ key) to find each of the following.*

27. ln 4.31   28. ln 517   29. ln(.127)

30. ln(.00424)   31. $\ln\left(\frac{6.71}{42.3}\right)$ 32. ln$\sqrt{457}$

33. $\frac{\ln 6.71}{\ln 42.3}$ 34. $\sqrt{\ln 457}$ 35. ln(51.4)³

36. ln(31.2 + 43.1)   37. (ln 51.4)³   38. 3 ln 51.4

© *Use your calculator to find N in each of the following. Hint: ln N = 5.1 if and only if N = e⁵·¹. On some calculators, there is an eˣ key; on others, you use the two keys* INV ln *.*

39. ln N = 2.12   40. ln N = 5.63   41. ln N = −.125
42. ln N = .00257   43. ln$\sqrt{N}$ = 3.41   44. ln N³ = .415

EXAMPLE A (Exponential Equations)   Solve $4^{3x-2} = 15$ for x.

Solution.   Take natural logarithms of both sides and then solve for x. Complete the solution using the ln table or a calculator.

$$\ln 4^{3x-2} = \ln 15$$

$$(3x - 2)\ln 4 = \ln 15$$

$$3x \ln 4 - 2 \ln 4 = \ln 15$$

$$3x \ln 4 = 2 \ln 4 + \ln 15$$

$$x = \frac{2 \ln 4 + \ln 15}{3 \ln 4}$$

$$x \approx 1.32$$

Solve for x using the method above.

45. $3^x = 20$

46. $5^x = 40$

47. $2^{x-1} = .3$

48. $4^x = 3^{2x-1}$

49. $(1.4)^{x+2} = 19.6$

50. $5^x = \frac{1}{2}(4^x)$

EXAMPLE B (Doubling Time)   How long will it take money to double in value if it is invested at 9.5 percent interest compounded annually?

Solution.   For convenience, consider investing $1. From Section 3-3, we know that this dollar will grow to $(1.095)^t$ dollars after t years. Thus we must solve the exponential equation $(1.095)^t = 2$. This we do by the method of Example A, using a calculator. The result is

$$t = \frac{\ln 2}{\ln 1.095} \approx 7.64$$

51. How long would it take money to double at 12 percent compounded annually?

52. How long would it take money to double at 12 percent compounded monthly? *Hint:* After t months, $1 is worth $(1.01)^t$ dollars.

53. How long would it take money to double at 15 percent compounded quarterly?

54. A certain substance decays according to the formula $y = 100e^{-.135t}$, where t is in years. Find its half-life.

55. By finding the natural logarithm of the numbers in each pair, determine which is larger.
    (a) $10^5, 5^{10}$
    (b) $10^9, 9^{10}$
    (c) $10^{20}, 20^{10}$
    (d) $10^{1000}, 1000^{10}$

56. What do your answers in Problem 55 confirm about the growth of 10^x and x^{10} for large x?

57. On the same coordinate plane, graph $y = 3^x$ and $y = x^3$ for $0 \le x \le 4$.

58. By means of a change of variable(s) (as explained in the text), transform each equation below to a linear equation. Find the slope and Y-intercept of the resulting line.
    (a) $y = 3e^{2x}$
    (b) $y = 2x^3$
    (c) $xy = 12$
    (d) $y = x^e$
    (e) $y = 5(3^x)$
    (f) $y = ex^{1.1}$

t	N
0	100
1	700
2	5000
3	40,000

Figure 18

EXAMPLE C (Curve Fitting)   The table in Figure 18 shows the number N of bacteria in a certain culture found after t hours. Which is a better description of these data,

$$N = ba^t \quad \text{or} \quad N = bt^a$$

Find a and b.

Solution.   Following the discussion of curve fitting in the text, we begin by plotting $\ln N$ against t. If the resulting points lie along a line, we choose $N = ba^t$ as the appropriate model. If not, we will plot $\ln N$ against $\ln t$ to check on the second model. Since the fit to a line is quite good (Figure 19 at the top of the next page,), we accept $N = ba^t$ as our model. To find a and b, we write $N = ba^t$ in the form

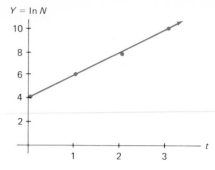

t	$\ln N$
0	4.6
1	6.6
2	8.5
3	10.6

Figure 19

$$\ln N = \ln b + t \ln a \quad \text{or} \quad Y = \ln b + (\ln a)t$$

Examination of the line shows that it has a Y-intercept of 4.6 and a slope of about 2; so for its equation we write $Y = 4.6 + 2t$. Comparing this with $Y = \ln b + (\ln a)t$ gives

$$\ln b = 4.6 \qquad \ln a = 2$$

Finally, we use the natural logarithm table or a calculator to find

$$b \approx 100 \qquad a \approx 7.4$$

Thus the original data are described reasonably well by the equation

$$N = 100(7.4)^t$$

For the data sets below, decide whether $y = ba^x$ or $y = bx^a$ is the better model. Then determine a and b.

59.

x	1	2	3	4
y	96	145	216	325

60.

x	0	1	2	4
y	243	162	108	48

61.

x	1	2	3	5
y	12	190	975	7490

62.

x	1	4	9
y	16	128	432

MISCELLANEOUS PROBLEMS

63. Evaluate without using a calculator or tables.
   (a) $\ln(e^{4.2})$
   (b) $e^{2 \ln 2}$
   (c) $\dfrac{\ln 3e}{2 + \ln 9}$

64. Solve for x.
   (a) $\ln(5 + x) = 1.2$
   (b) $e^{x^2 - x} = 2$
   (c) $e^{2x} = (.6)8^x$

65. Evaluate without use of a calculator or tables.
   (a) $\ln[(e^{3.5})^2]$
   (b) $(\ln e^{3.5})^2$
   (c) $\ln(1/\sqrt{e})$
   (d) $(\ln 1)/(\ln \sqrt{e})$
   (e) $e^{3 \ln 5}$
   (f) $e^{\ln(1/2) + \ln(2/3)}$

66. Since $a^x = e^{\ln a^x} = e^{x \ln a}$, the study of exponential functions can be subsumed under the study of the function e^{kx}. Determine k so that each of the following is true. *Hint:* In (a) rewrite the equation as $3^x = (e^k)^x$ which implies that $3 = e^k$.

Now take natural logarithms.

(a) $3^x = e^{kx}$ (b) $\pi^x = e^{kx}$ (c) $(1/3)^x = e^{kx}$

© 67. Solve for x.

(a) $10^{2x+3} = 200$ (b) $10^{2x} = 8^{x-1}$ (c) $10^{x^2+3x} = 200$

(d) $e^{-.32x} = 1/2$ (e) $x^{\ln x} = 10$ (f) $(\ln x)^{\ln x} = x$

(g) $x^{\ln x} = x$ (h) $\ln x = (\ln x)^{\ln x}$ (i) $(x^2 - 5)^{\ln x} = x$

68. Show that each of the following is an identity. Assume $x > 0$.

(a) $(\sqrt{3})^{3\sqrt{3}} = (3\sqrt{3})^{\sqrt{3}}$ (b) $2.25^{3.375} = 3.375^{2.25}$

(c) $a^{\ln(x^b)} = (a^{\ln x})^b$ (d) $x^x = e^{x \ln x}$

(e) $(\ln x)^x = e^{x \ln(\ln x)}$ (f) $\dfrac{\ln\left(\dfrac{x+1}{x}\right)^x}{\ln\left(\dfrac{x+1}{x}\right)^{x+1}} = \dfrac{\left(\dfrac{x+1}{x}\right)^x}{\left(\dfrac{x+1}{x}\right)^{x+1}}$

© 69. A certain substance decays according to the formula $y = 100e^{-3t}$, where y is the amount present after t years. Find its half-life.

© 70. Suppose the number of bacteria in a certain culture t hours from now will be $200e^{.468t}$. When will the count reach 10,000?

© 71. A radioactive substance decays exponentially with a half-life of 240 years. Determine k in the formula $A = A_0 e^{-kt}$, where A is the amount present after t years and A_0 is the initial amount.

72. By means of a change of variables using natural logarithms, transform each of the following equations to a linear equation. Then find the slope and the Y-intercept of the resulting line.

(a) $xy^2 = 40$ (b) $y = 9e^{2x}$

© 73. Sketch the graph of

$$y = \frac{1}{\sqrt{2\pi}} e^{-(1/2)x^2}$$

This is the famous *normal curve*, so important in statistics.

© 74. In calculus it is shown that

$$e^x \approx 1 + x + \frac{x^2}{2} + \frac{x^3}{6} + \frac{x^4}{24} + \frac{x^5}{120}$$

Use this to approximate e and $e^{-1/2}$.

75. From Problem 74 or from looking at the graph of $y = e^x$, you might guess the true result that $e^x > 1 + x$ for all $x > 0$. Use this and the obvious fact that $(\pi/e) - 1 > 0$ to demonstrate algebraically that $e^\pi > \pi^e$.

© 76. It is important to have a feeling for how various functions grow for large x. Let $\ll$ symbolize the phrase *grows slower than*. Use a calculator to convince yourself that

$$\ln x \ll \sqrt{x} \ll x \ll x^2 \ll e^x \ll x^x$$

© 77. In calculus, it is shown that $(1 + r/m)^m$ gets closer and closer to e^r as m gets larger and larger. Now if P dollars is invested at rate r (written as a decimal) compounded m times per year, it will grow to $P(1 + r/m)^{mt}$ dollars at the end of t years (Example B of Section 3-3). If interest is compounded continuously (see the box e **and interest** on page 155), P dollars will grow to Pe^{rt} dollars at the end of t years. Use these facts to calculate the value of $100 after 10 years if interest is at 12 percent (that is, $r = .12$) and is compounded (a) monthly; (b) daily; (c) hourly; (d) continuously.

78. **TEASER** The *harmonic* sum $S_n = 1 + \frac{1}{2} + \frac{1}{3} + \frac{1}{4} + \cdots + 1/n$ has intrigued both amateur and professional mathematicians since the time of the Greeks. In calculus, it is shown that for $n > 1$

$$\ln n < 1 + \frac{1}{2} + \frac{1}{3} + \frac{1}{4} + \cdots + \frac{1}{n} < 1 + \ln n$$

This shows that S_n grows arbitrarily large but that it grows very very slowly.
(a) About how large would n need to be for $S_n > 100$?
(b) Show how you could stack a pile of identical bricks each of length 1 foot (one brick to a tier as shown in Figure 20) to achieve an overhang of 50 feet. Could you make the overhang 50 million feet? Yes, it does have something to do with part (a).

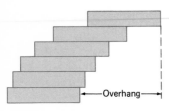

Figure 20

Henry Briggs 1561–1631

When 10 is used as the base for logarithms, some very nice things happen as the display at the right shows. Realizing this, John Napier and his friend Henry Briggs decided to produce a table of these logarithms. The job was accomplished by Briggs after Napier's death and published in 1624 in the famous book *Arithmetica Logarithmica*.

$$\log .0853 = .9309 - 2$$

$$\log .853 \; = .9309 - 1$$

$$\log 8.53 \; = .9309$$

$$\log 85.3 \; = .9309 + 1$$

$$\log 853 \; = .9309 + 2$$

$$\log 8530 = .9309 + 3$$

3-6 Common Logarithms (Optional)

In Section 3-4, we defined the logarithm of a positive number N to the base a as follows.

$$\log_a N = x \quad \text{if and only if} \quad a^x = N$$

Then in Section 3-5, we introduced base e, calling the result natural logarithms. These are the logarithms that are most important in advanced branches of mathematics such as calculus.

In this section, we shall consider logarithms to the base 10, often called **common logarithms,** or Briggsian logarithms. They have been studied by high school and college students for centuries as an aid to computation, a subject we take up in the next section. We shall write $\log N$ instead of $\log_{10} N$ (just as we write $\ln N$ instead of $\log_e N$). Note that

$$\log N = x \quad \text{if and only if} \quad 10^x = N$$

In other words, the common logarithm of any power of 10 is simply its exponent. Thus

$$\log 100 = \log 10^2 = 2$$

$$\log 1 = \log 10^0 = 0$$

$$\log .001 = \log 10^{-3} = -3$$

But how do we find common logarithms of numbers that are not integral powers of 10, such as 8.53 or 14,600? That is the next topic.

FINDING COMMON LOGARITHMS

We know that $\log 1 = 0$ and $\log 10 = 1$. If $1 < N < 10$, we correctly expect $\log N$ to be between 0 and 1. Table B (in the appendix) gives us four-place approximations of the common logarithms of all three-digit numbers between 1 and 10. For example,

$$\log 8.53 = .9309$$

We find this value by locating 8.5 in the left column and then moving right to the entry with 3 as heading. Similarly,

$$\log 1.08 = .0334$$

and

$$\log 9.69 = .9863$$

You should check these values.

You may have noticed in the opening box that log 8.53, log 8530, and log .0853 all have the same positive fractional part, .9309, called the **mantissa** of the logarithm. They differ only in the integer part, called the **characteristic** of the logarithm. To see why this is so, recall that

$$\log(M \cdot N) = \log M + \log N$$

Thus

$$\log 8530 = \log(8.53 \times 10^3) = \log 8.53 + \log 10^3$$

$$= .9309 + 3$$

$$\log .0853 = \log(8.53 \times 10^{-2}) = \log 8.53 + \log 10^{-2}$$

$$= .9309 - 2$$

Clearly the mantissa .9309 is determined by the sequence of digits 8, 5, 3, while the characteristic is determined by the position of the decimal point. Let us say that the decimal point is in **standard position** when it occurs immediately after the first nonzero digit. The characteristic for the logarithm of a number with the decimal point in standard position is 0. The characteristic is k if the decimal point is k places to the right of standard position; it is $-k$ if it is k places to the left of standard position.

8530000.   log 8,530,000 = .9309 + 6

6 places right

.0000853   log .0000853 = .9309 − 5

5 places left

Here is another way of describing the mantissa and characteristic. If $c \times 10^n$ is the scientific notation for N, then $\log c$ is the mantissa and n is the characteristic of $\log N$.

FINDING ANTILOGARITHMS

If we are to make significant use of common logarithms, we must know how to find a number when its logarithm is given. This process is called finding the inverse logarithm, or the **antilogarithm.** The process is simple: Use the mantissa to find the sequence of digits and then let the characteristic tell you where to put the decimal point.

Suppose, for example, that you are given

$$\log N = .4031 - 4$$

Locate .4031 in the body of Table B. You will find it across from 2.5 and below 3. Thus the number N must have 2, 5, 3 as its sequence of digits. Since the characteristic is -4, put the decimal point 4 places to the left of standard position. The result is

$$N = .000253$$

As a second example, let us find antilog 5.9547. The mantissa .9547 gives us the digits 9, 0, 1. Since the characteristic is 5,

$$\text{antilog } 5.9547 = 901,000$$

LINEAR INTERPOLATION

Suppose that for some function f we know $f(a)$ and $f(b)$, but we want $f(c)$, where c is between a and b (see the diagrams in Figure 21). As a reasonable approximation, we may pretend that the graph of f is a straight line between a and b. Then

$$f(c) \approx f(a) + d$$

where, by similarity of triangles (Figure 22),

$$\frac{d}{f(b) - f(a)} = \frac{c - a}{b - a}$$

That is,

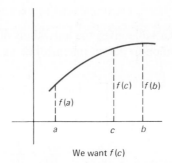

We want $f(c)$

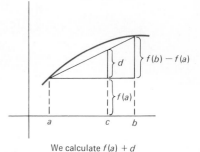

We calculate $f(a) + d$

Figure 21

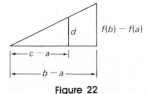

Figure 22

$$d = \frac{f(b) - f(a)}{b - a}(c - a)$$

The process just described is called **linear interpolation.** The process of linear interpolation for the logarithm function is explained in Examples A and B of the problem set, as well as at the beginning of the logarithm tables at the back of the book.

Problem Set 3-6

Find the common logarithm of each of the following numbers.

1. 10,000
2. 1,000,000
3. .01
4. .0001
5. $10^4 10^{3/2}$
6. $10^3 10^{4/3}$
7. $(10^3)^{-5}$
8. $(10^5)^{-1/10}$

Find N in each case.

9. $\log N = 4$
10. $\log N = 6$
11. $\log N = -2$
12. $\log N = -5$
13. $\log N = \frac{3}{2}$
14. $\log N = \frac{1}{3}$
15. $\log N = -\frac{3}{4}$
16. $\log N = -\frac{1}{6}$

Use Table B to find the following logarithms.

17. $\log 4.32$
18. $\log 3.09$
19. $\log 158$
20. $\log 47.3$
21. $\log .0329$
22. $\log .0715$
23. $\log 563,000$
24. $\log 420,000$
25. $\log(9.23 \times 10^8)$
26. $\log(2.83 \times 10^{-11})$

Find N in each case.

27. $\log N = 1.5159$
28. $\log N = 3.9015$
29. $\log N = .0043 - 2$
30. $\log N = .8627 - 4$
31. $\log N = 8.5999$
32. $\log N = 4.7427$

Find the antilogarithm of each number.

33. 2.2201
34. 3.8639
35. $.9232 - 1$
36. $.8500 - 5$

EXAMPLE A (Linear Interpolation in Finding Logarithms)   Find log 34.67.

Solution.   Our table gives the logarithms of 34.6 and 34.7, so we use linear interpolation to get an intermediate value. Here is how we arrange our work.

$$.10\left[.07\begin{bmatrix}\log 34.60 = 1.5391 \\ \log 34.67 = \quad ? \\ \log 34.70 = 1.5403\end{bmatrix}d\right].0012$$

$$\frac{d}{.0012} = \frac{.07}{.10} = \frac{7}{10}$$

$$d = \frac{7}{10}(.0012) \approx .0008$$

$$\log 34.67 \approx \log 34.60 + d \approx 1.5391 + .0008 = 1.5399$$

Use linear interpolation in Table B to find each value.

37. log 5.237   38. log 9.826   39. log 7234

40. log 68.04   41. log .001234   42. log .09876

EXAMPLE B (Interpolation in Finding Antilogarithms)   Find antilog 2.5285.

Solution.   We find .5285 sandwiched between .5276 and .5289 in the body of Table B.

$$.0013\left[.0009\begin{bmatrix}\text{antilog } 2.5276 = 337.0 \\ \text{antilog } 2.5285 = \quad ? \\ \text{antilog } 2.5289 = 338.0\end{bmatrix}d\right]1.0$$

$$\frac{d}{1.0} = \frac{.0009}{.0013} = \frac{9}{13}$$

$$d = \frac{9}{13}(1.0) \approx .7$$

$$\text{antilog } 2.5285 \approx 337.0 + .7 = 337.7$$

Find the antilogarithm of each of the following using linear interpolation.

43. 0.8497   44. 0.8516   45. 3.9130

46. 1.9849   47. .6004 − 2   48. .4946 − 4

MISCELLANEOUS PROBLEMS

49. Without using a calculator, find the common logarithm of each of the following.
   (a) $10^{3/2}\, 10^{-1/4}$ (b) $(.0001)^{1/3}$
   (c) $\sqrt[3]{10}\,\sqrt{10}$ (d) $10^{\log(.001)}$

50. Find N in each case.
   (a) $\log N = 0$ (b) $\log N = -2$
   (c) $\log N = \frac{1}{2}$ (d) $\log N = 10$

51. Use Table B with interpolation to find each of the following.
    (a) log 492.7   (b) log .04705
    (c) antilog 2.9327   (d) antilog($.2698 - 3$)

52. Use Table B to find N. *Hint:* $-2.4473 = .5527 - 3$.
    (a) $\log N = -2.4473$ (b) $\log N = -4.0729$

⟦c⟧ 53. Do Problem 52 using your calculator.

54. Find the characteristic of log N if:
    (a) $10^{11} < N < 10^{12}$;   (b) $.00001 < N < .0001$.

55. Use Table B to find $\log \dfrac{982 - 467}{(982)(267)}$.

56. If $b = .001\,a$ and $\log a = 5.5$, find $\log(a^3/b^4)$.

57. Evaluate antilog$\left[\log\left(\dfrac{999}{4.71 \times 328}\right) - \log 999 + \log 328 \right]$.

58. Find x if $(\log x)^2 + \log(x^2) = 10^{\log 3}$

59. Let $\log N = 15.992$. In the decimal notation for N, how many digits are there before the decimal point?

60. Convince yourself that the number of digits in a positive integer N is $[\log N] + 1$. Then find the number of digits in 50^{50} when expanded out in the usual way in decimal notation.

⟦c⟧ 61. If one can write 6 digits to the inch, about how many miles long would the number $9^{(9^9)}$ be when written in decimal notation?

62. **TEASER** Let $N = 9^{(9^9)}$, A = sum of digits in N, B = sum of digits in A, and C = sum of digits in B. Find C. *Hint:* If a number is divisible by 9, so is the sum of its digits (why?). Since N is divisible by 9, it follows that A, B, and C are all divisible by 9.

"The miraculous powers of modern calculation are due to three inventions: the Arabic Notation, Decimal Fractions, and Logarithms."

F. Cajori, 1897

"Electronic calculators make calculations with logarithms as obsolete as whale oil lamps."

Anonymous Reviewer, 1982

3-7 Calculations with Logarithms (Optional)

For 300 years, scientists depended on logarithms to reduce the drudgery associated with long computations. The invention of electronic computers and calculators has diminished the importance of this long-established technique. Still, we think that any student of algebra should know how products, quotients, powers, and roots can be calculated by means of common logarithms.

About all you need are the three laws stated above and Appendix Table B. A little common sense and the ability to organize your work will help.

PRODUCTS

Suppose you want to calculate $(.00872)(95,300)$. Call this number x. Then by Law 1 and Table B,

$$\log x = \log .00872 + \log 95,300$$
$$= (.9405 - 3) + 4.9791$$
$$= 5.9196 - 3$$
$$= 2.9196$$

Now use Table B backwards to find that antilog $.9196 = 8.31$, so $x =$ antilog $2.9196 = 831$.

Here is a good way to organize your work in a compact systematic way.

$$x = (.00872)(95,300)$$

$$\begin{aligned}
\log .00872 &= .9405 - 3 \\
(+) \underline{\log 95300} &= \underline{4.9791} \\
\log x &= 5.9196 - 3
\end{aligned}$$

$$x = 831$$

QUOTIENTS

Suppose we want to calculate $x = .4362/91.84$. Then by Law 2,

$$\log x = \log .4362 - \log 91.84$$
$$= (.6397 - 1) - 1.9630$$
$$= .6397 - 2.9630$$
$$= -2.3233$$

What we have done is correct; however, it is poor strategy. The result we found for $\log x$ is not in characteristic-mantissa form and therefore is not usable. Remember that the mantissa must be positive. We can bring this about by adding and subtracting 3.

$$-2.3233 = (-2.3233 + 3) - 3 = .6767 - 3$$

Actually it is better to anticipate the need for doing this and arrange the work as follows.

$$x = \frac{.4362}{91.84}$$

$$\begin{array}{ll} \log .4362 = .6397 - 1 = 2.6397 - 3 \\ (-)\ \underline{\log 91.48 =} & \underline{1.9630} \\ \log x = & .6767 - 3 \end{array}$$

$$x = .00475$$

POWERS OR ROOTS

Here the main tool is Law 3. We illustrate with two examples.

$$x = (31.4)^{11}$$
$$\log x = 11 \log 31.4$$
$$\log 31.4 = 1.4969$$
$$11 \log 31.4 = 16.4659$$
$$\log x = 16.4659$$
$$x = 29{,}230{,}000{,}000{,}000{,}000$$
$$= 2.923 \times 10^{16}$$

$$x = \sqrt[4]{.427} = (.427)^{1/4}$$

$$\log x = \frac{1}{4} \log .427$$

$$\log .427 = .6304 - 1$$

$$\frac{1}{4} \log .427 = \frac{1}{4}(.6304 - 1) = \frac{1}{4}(3.6304 - 4) = .9076 - 1$$

$$\log x = .9076 - 1$$

$$x = .8084$$

Notice in the second example that we wrote $3.6304 - 4$ in place of $.6304 - 1$, so that multiplication by $\frac{1}{4}$ gave the logarithm in characteristic-mantissa form.

Problem Set 3-7

Use logarithms and Table B without interpolation to find approximate values for each of the following.

1. $(46.3)(2.76)$
2. $(378)(9.63)$
3. $\dfrac{46.3}{483}$

4. $\dfrac{437}{92300}$
5. $\dfrac{.00912}{.439}$
6. $\dfrac{.0429}{15.7}$

7. $(37.2)^5$
8. $(113)^3$
9. $\sqrt[3]{42.9}$

10. $\sqrt[4]{312}$
11. $\sqrt[5]{.918}$
12. $\sqrt[3]{.0307}$

13. $(14.9)^{2/3}$
14. $(98.6)^{3/4}$

Use logarithms and Table B with interpolation to approximate each of the following.

15. $(31.96)(149)$
16. $(6236)(.00108)$
17. $\dfrac{43.98}{7.16}$

18. $\dfrac{115}{4.623}$
19. $(.1234)^6$
20. $(92.83)^3$

EXAMPLE A (More Complicated Calculations) Use logarithms, without interpolation, to calculate

$$\frac{(31.4)^3(.982)}{(.0463)(824)}$$

Solution. Let N denote the entire numerator, D the entire denominator, and x the fraction N/D. Then

$$\log x = \log N - \log D$$

where

$$\log N = 3 \log 31.4 + \log .982$$

$$\log D = \log .0463 + \log 824$$

Here is a good way to organize the work.

$$\log 31.4 = 1.4969$$
$$3 \log 31.4 = 4.4907$$
$$(+) \quad \underline{\log .982 = .9921 - 1}$$
$$\log N = 5.4828 - 1$$
$$= 4.4828$$

$$\log .0463 = .6656 - 2$$
$$(+) \quad \underline{\log 824 = 2.9159}$$
$$\log D = 3.5815 - 2$$
$$= 1.5815$$

$$\log N = 4.4828$$
$$(-) \underline{\log D = 1.5815}$$
$$\log x = 2.9013$$
$$x = 797$$

Carry out the following calculations using logarithms without interpolation.

21. $\dfrac{(.56)^2(619)}{21.8}$

22. $\dfrac{.413}{(4.9)^2(.724)}$

23. $\dfrac{(14.3)\sqrt{92.3}}{\sqrt[3]{432}}$

24. $\dfrac{(91)(41.3)^{2/3}}{42.6}$

EXAMPLE B (Solving Exponential Equations) Solve the equation

$$2^{2x-1} = 13$$

Solution.   We begin by taking logarithms of both sides and then solving for x.

$$(2x - 1)\log 2 = \log 13$$

$$2x - 1 = \frac{\log 13}{\log 2} = \frac{1.1139}{.3010}$$

$$\log(2x - 1) = \log 1.1139 - \log .3010$$

$$= (1.0469 - 1) - (.4786 - 1)$$

$$= .5683$$

$$2x - 1 = \text{antilog } .5683 = 3.70$$

$$2x = 4.70$$

$$x = 2.35$$

Notice that we did the division of (log 13)/(log 2) by means of logarithms. We could have done it by long division (or on a calculator) if we preferred.

Use logarithms to solve the following exponential equations. You need not interpolate.

25. $3^x = 300$ 26. $5^x = 14$ 27. $10^{2-3x} = 6240$

28. $10^{5x-1} = .00425$ 29. $2^{3x} = 3^{x+2}$ 30. $2^{x^2} = 3^x$

MISCELLANEOUS PROBLEMS

31. Use common logarithms and Table B to calculate each of the following.

(a) $\sqrt[3]{.0427}$ (b) $\dfrac{(42.9)^2(.983)}{\sqrt{323}}$ (c) $\dfrac{10^{6.42}}{8^{7.2}}$

32. Solve for x by using common logarithms.
    (a) $4^{2x} = 150$ (b) $(.975)^x = .5$

33. Solve for x.
    (a) $\log(x + 2) - \log x = 1$ (b) $\log(2x + 1) - \log(x + 3) = 0$
    (c) $\log(x + 3) - \log x + \log 2x^2 = \log 8$
    (d) $\log(x + 3) + \log(x - 1) = \log 4x$

34. Suppose that the amount Q of a radioactive substance (in grams) remaining t years from now will be $Q = (42)2^{-.017t}$. After how many years will the amount remaining be .42 grams?

35. Suppose that the bacteria count in a certain culture t hours from now is $(800)3^t$. When will the count reach 100,000?

36. Assume the population of the earth was 4.19 billion people in 1977 and that the growth rate is 2 percent per year. Then the population t years after 1977 should be $4.19(1.02)^t$ billion.
    (a) What will the population be in the year 2000?
    (b) When will the population reach 8.3 billion?

37. Answer the two questions of Problem 36 assuming the growth rate is only 1.7 percent.

38. **TEASER** Show that log 2 is irrational. For what positive integers n is log n irrational?

Chapter Summary

The symbol $\sqrt[n]{a}$, the principal nth root of a, denotes one of the numbers whose nth power is a. For odd n, that is all that needs to be said. For n even and $a > 0$, we specify that $\sqrt[n]{a}$ signifies the positive nth root. Thus $\sqrt[3]{-8} = -2$ and $\sqrt{16} = \sqrt[2]{16} = 4$. (It is wrong to write $\sqrt{16} = -4$.) These symbols are also called **radicals.** These radicals obey four carefully prescribed rules (page 127). These rules allow us to simplify complicated radical expressions, in particular, to **rationalize denominators.**

The key to understanding **rational exponents** is the definition $a^{1/n} = \sqrt[n]{a}$, which implies $a^{m/n} = (\sqrt[n]{a})^m$. Thus $16^{5/4} = (\sqrt[4]{16})^5 = 2^5 = 32$. The meaning of **real exponents** is determined by considering rational approximations. For example, 2^π is the number that the sequence 2^3, $3^{3.1}$, $2^{3.14}$, . . . approaches. The function $f(x) = a^x$ (and more generally $f(x) = b \cdot a^x$) is called an **exponential function.**

A variable y is **growing exponentially** or **decaying exponentially** according as $a > 1$ or $0 < a < 1$ in the equation $y = b \cdot a^x$. Typical of the former are biological populations; of the latter, radioactive elements. Corresponding key ideas are **doubling times** and **half-lives.**

Logarithms are exponents. In fact, $\log_a N = x$ means $a^x = N$, that is, $a^{\log_a N} = N$. The functions $f(x) = \log_a x$ and $g(x) = a^x$ are **inverses** of each other. Logarithms have three primary properties (page 148). **Natural logarithms** correspond to the choice of base $a = e = 2.71828 \ldots$ and play a fundamental role in advanced courses. **Common logarithms** correspond to base 10 and have historically been used to simplify arithmetic calculations.

Chapter Review Problem Set

1. Simplify, rationalizing all denominators. Assume letters represent positive numbers.

   (a) $\sqrt[3]{\dfrac{-8y^6}{z^{14}}}$ (b) $\sqrt[4]{32x^5y^8}$ (c) $\sqrt{4\sqrt[3]{5}}$

   (d) $\dfrac{2}{\sqrt{x}-\sqrt{y}}$ (e) $\sqrt{50+25x^2}$ (f) $\sqrt{32}+\sqrt{8}$

2. Solve the equations.

   (a) $\sqrt{x-3}=3$ (b) $\sqrt{x}=6-x$

3. Simplify, writing your answer in exponential form with all positive exponents.

   (a) $(25a^2)^{3/2}$ (b) $(a^{-1/2}aa^{-3/4})^2$ (c) $\dfrac{1}{5\sqrt[4]{5^3}}$

   (d) $\dfrac{(3x^{-2}y^{3/4})^2}{3x^2y^{-2/3}}$ (e) $(x^{1/2}-y^{1/2})^2$ (f) $\sqrt[3]{4}\sqrt[4]{4}$

4. Sketch the graph of $y=(\tfrac{3}{2})^x$ and use your graph to estimate the value of $(\tfrac{3}{2})^\pi$.

5. A certain radioactive substance has a half-life of 3 days. What fraction of an initial amount will be left after 243 days?

6. A population grows so that its doubling time is 40 years. If this population is 1 million today, what will it be after 160 years?

7. If $100 is put in the bank at 8 percent interest compounded quarterly, what will it be worth at the end of 10 years?

8. Find x in each of the following.

   (a) $\log_4 64 = x$ (b) $\log_2 x = -3$
   (c) $\log_x 49 = 2$ (d) $\log_4 x = 0$
   (e) $\log_9 27 = x$ (f) $\log_{10} x + \log_{10}(x-3) = 1$
   (g) $a^{\log_a 10} = x$ (h) $x = \log_a(a^{1.14})$

9. Write as a single logarithm.

$$2\log_4(3x+1) - \frac{1}{2}\log_4 x + \log_4(x-1)$$

10. Evaluate.
    (a) $\log_{10}\sqrt{1000}$ (b) $\log_{27} 81$
    (c) $\ln\sqrt{e}$ (d) $\ln(1/2^4)$

11. Use the table of natural logarithms to determine each of the following.
    (a) $\ln(\tfrac{7}{4})^3$ (b) N if $\ln N = 2.230$
    (c) $\ln[(3.4)(9.9)]$ (d) N if $\ln N = -0.105$

12. By taking ln of both sides, solve $2^{x+1}=7$.

13. A certain substance decays according to the formula $y=y_0e^{-.05t}$, where t is measured in years. Find its half-life.

14. A substance initially weighing 100 grams decays exponentially according to the formula $y=100e^{-kt}$. If 30 grams are left after 10 days, determine k.

15. Sketch the graphs of $y=\log_3 x$ and $y=3^x$ using the same coordinate axes.

16. Use common logarithms to calculate

$$\frac{(13.2)^4\sqrt{15.2}}{29.6}$$

The great book of Nature lies open before our eyes and true philosophy is written in it. . . . But we cannot read it unless we have first learned the language and characters in which it is written. . . . It is written in mathematical language and the characters are triangles, circles, and other geometrical figures.

—Galileo

CHAPTER 4

The Trigonometric Functions

Sometime before 100 B.C., the Greeks invented trigonometry to solve problems in astronomy, navigation, and geography. The word "trigonometry" comes from Greek and means "triangle measurement." In its most basic form, trigonometry is the study of relationships between the angles and sides of a right triangle.

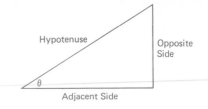

Hypotenuse

Opposite Side

θ

Adjacent Side

4-1   Right-Triangle Trigonometry

A triangle is called a *right triangle* if one of its angles is a right angle, that is, a 90° angle. The other two angles are necessarily acute angles (less than 90°) since the sum of all three angles in a triangle is 180°. Let θ (the Greek letter theta) denote one of these acute angles. We may label the three sides relative to θ: adjacent side, opposite side, and hypotenuse, as shown in the diagram above. In terms of these sides, we introduce the three fundamental ratios of trigonometry, sine θ, cosine θ, and tangent θ. Using obvious abbreviations, we give the following definitions.

$$\sin \theta = \frac{\text{opp}}{\text{hyp}}$$

$$\cos \theta = \frac{\text{adj}}{\text{hyp}}$$

$$\tan \theta = \frac{\text{opp}}{\text{adj}}$$

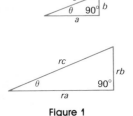

Figure 1

Thus with every acute angle θ, we associate three numbers, sin θ, cos θ, and tan θ. A careful reader might wonder whether these numbers depend only on the size of θ, or if they also depend on the lengths of the sides of the right triangle with which we started. Consider two different right triangles, each with the same angle θ (as in Figure 1). You may think of the lower triangle as a magnification of the upper one. Each of its sides has length r times that of the corresponding side in the upper triangle. If we calculate sin θ from the lower triangle, we get

$$\sin \theta = \frac{\text{opp}}{\text{hyp}} = \frac{rb}{rc} = \frac{b}{c}$$

which is the same result we get using the upper triangle. We conclude that for a given θ, sin θ has the same value no matter which right triangle is used to compute it. So do cos θ and tan θ.

SPECIAL ANGLES

We can use the Pythagorean theorem ($a^2 + b^2 = c^2$) to find the values of sine, cosine, and tangent for the special angles $30°$, $45°$, and $60°$. Consider the two right triangles of Figure 2, which involve these angles.

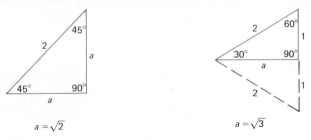

Figure 2

To see that the indicated values of a are correct, note that in the first triangle, $a^2 + a^2 = 2^2$, which gives $a = \sqrt{2}$. In the second, which is half of an equilateral triangle, $a^2 + 1^2 = 2^2$, or $a = \sqrt{3}$.

From these triangles, we obtain the following important facts.

$$\sin 45° = \frac{\sqrt{2}}{2} \qquad \cos 45° = \frac{\sqrt{2}}{2} \qquad \tan 45° = 1$$

$$\sin 30° = \frac{1}{2} \qquad \cos 30° = \frac{\sqrt{3}}{2} \qquad \tan 30° = \frac{1}{\sqrt{3}} = \frac{\sqrt{3}}{3}$$

$$\sin 60° = \frac{\sqrt{3}}{2} \qquad \cos 60° = \frac{1}{2} \qquad \tan 60° = \sqrt{3}$$

OTHER ANGLES

When you need the sine, cosine, or tangent of an angle other than the special ones just considered, you may do one of two things. If you have a scientific calculator, you may simply push two or three keys and have your answer correct to eight or more significant digits. Otherwise, you will need to use Table C in the appendix.

Several facts about Table C should be noted. First, it gives answers usually to four decimal places. Second, angles are measured in degrees and tenths of degrees. By interpolation (see page 166), it is possible to consider angles measured to the nearest hundredth of a degree. Finally, notice that the left column of the table lists angles from $0°$ to $45°$. For angles from $45°$ to $90°$, use the right column; you must then also use the bottom captions. To make sure that you are reading the table (or your calculator) correctly, check that you get each of the following answers.

$$\tan 33.1° = .6519 \qquad \sin 26.9° = .4524$$

$$\cos 54.3° = .5835 \qquad \tan 82° = 7.115$$

APPLICATIONS

Suppose that you wish to measure the distance across a stream but do not want to get your feet wet. Here is how you might proceed.

Pick out a tree at C on the opposite shore and set a stone at B directly across from it on your shore (Figure 3). Set another stone at A, 100 feet up the shore from B. With an angle measuring device (for example, a protractor or a transit), measure angle θ between AB and AC. Then x, the length of BC, satisfies the following equation.

$$\tan \theta = \frac{\text{opp}}{\text{adj}} = \frac{x}{100}$$

or

$$x = 100 \tan \theta$$

For example, if θ measures 29°, you find from your scientific calculator or Table C that $\tan 29° = .5543$. Then $x = 100(.5543) = 55.43$ feet. Since you used stones and trees for points, this suggests that you should not give your answer with such accuracy. It would be better to say that the distance x is approximately 55 feet.

As a more difficult example, consider a church with a steeple, as shown in Figure 4. The problem is to calculate the height of the steeple while standing on the ground. To find the height, mark a point B on the ground directly below the steeple and another point A 200 feet away on the ground. At A, measure the *angles of elevation* α and β to the top and bottom of the steeple. This is all the information you will need, provided you know your trigonometry.

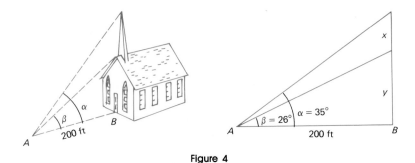

Figure 4

Let x be the height of the steeple and y be the distance from the ground to the bottom of the steeple. Suppose that $\alpha = 35°$ and $\beta = 26°$. Then

$$\tan 35° = \frac{x + y}{200}$$

$$\tan 26° = \frac{y}{200}$$

If you solve for y in the second equation and substitute the value in the first, you will get the following sequence of equations.

$$\tan 35° = \frac{x + 200 \tan 26°}{200}$$

$$200 \tan 35° = x + 200 \tan 26°$$

$$x = 200 \tan 35° - 200 \tan 26°$$

$$= 200(.7002 - .4877)$$

$$x = 42.5 \text{ feet}$$

Problem Set 4-1

In Problems 1–6, use Table C to evaluate each expression. If you have a scientific calculator, use it as a check.

1. $\sin 41.3°$
2. $\tan 54.4°$
3. $\cos 49.2°$
4. $\sin 89.3°$
5. $\tan 72.3°$
6. $\cos 38.7°$

In Problems 7–12, use Table C to find θ. We suggest you also do these problems on a calculator. For example, to do Problem 7 on many calculators, press .2164 $\boxed{\text{INV}}$ $\boxed{\text{sin}}$. This will give the inverse sine of .2164, that is, the angle whose sine is .2164.

7. $\sin \theta = .2164$
8. $\tan \theta = .3096$
9. $\tan \theta = 2.311$
10. $\cos \theta = .9354$
11. $\cos \theta = .3535$
12. $\sin \theta = .7302$

Each of the remaining problems in this problem set involves a considerable amount of arithmetic that you can do by hand (using tables) or by using a calculator. (If you use a calculator to find values for the trigonometric functions, be sure that it is in the degree mode.) In Problems 13–18, find x.

13.
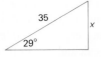

14.

15.
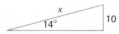

16.

17.
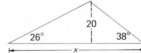

18.
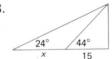

EXAMPLE A (Solving a Right Triangle Given an Angle and a Side) To solve a triangle means to determine all its unknown parts. Solve the right triangle which has hypotenuse of length 14.6 and an angle measuring 33.2°.

Solution. First, we draw the triangle labeling the known parts and assigning letters to the unknown parts. Our convention is to use the first three Greek letters, α, β, and γ (alpha, beta, and gamma) for the angles and a, b, and c for the lengths of the respective sides opposite these angles (see Figure 5). We need to find β, a, and b.

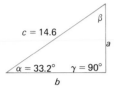

$c = 14.6$

$\alpha = 33.2°$ $\gamma = 90°$

b

Figure 5

(i) $\beta = 90° - 33.2° = 56.8°$

(ii) $\sin 33.2° = a/14.6$, so

$$a = 14.6 \sin 33.2° = (14.6)(.5476) \approx 7.99$$

(iii) $\cos 33.2° = b/14.6$, so

$$b = 14.6 \cos 33.2° = (14.6)(.8368) \approx 12.2$$

Notice that we gave the answers to three significant digits since the given data have three significant digits.

Solve each of the following triangles. First draw the triangle, labeling it as in the example with $\gamma = 90°$.

19. $\alpha = 42°$, $c = 35$ 20. $\beta = 29°$, $c = 50$

21. $\beta = 56.2°$, $c = 91.3$ 22. $\alpha = 69.9°$, $c = 10.6$

23. $\alpha = 39.4°$, $a = 120$ 24. $\alpha = 40.6°$, $b = 163$

EXAMPLE B (Solving a Right Triangle Given Two Sides) Solve the right triangle which has legs $a = 42.8$ and $b = 94.1$.

Solution. First, we draw the triangle and label its parts (Figure 6). We must find α, β, and c.

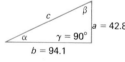

c

α $\gamma = 90°$ $a = 42.8$

$b = 94.1$

Figure 6

(i) $\tan \alpha = \dfrac{42.8}{94.1} \approx .4548$

Now we can find α by using Table C backwards, that is, by looking under tangent in the body of the table for .4548 and determining the corresponding angle. Or better, we can use the $\boxed{\text{INV}}\ \boxed{\text{tan}}$ keys on a scientific calculator. On many calculators, press

$$\boxed{(}\ 42.8\ \boxed{\div}\ 94.1\ \boxed{)}\ \boxed{\text{INV}}\ \boxed{\text{tan}}$$

In either case, the result is $\alpha \approx 24.5°$.

(ii) $\beta = 90° - \alpha \approx 90° - 24.5° = 65.5°$

(iii) We could find c by using $c^2 = a^2 + b^2$. Instead, we use $\sin \alpha$.

$$\sin \alpha = \sin 24.5° = \frac{42.8}{c}$$

$$c = \frac{42.8}{\sin 24.5°} = \frac{42.8}{.4147} \approx 103$$

Solve the right triangles satisfying the given information in Problems 25–32, assuming that c is the hypotenuse. You can do them either with tables or a calculator.

25. $a = 9$, $b = 12$　　　　　　　　　26. $a = 24$, $b = 10$

27. $a = 40$, $c = 50$　　　　　　　　　28. $c = 41$, $a = 40$

29. $a = 14.6$, $c = 32.5$　　　　　　　30. $a = 243$, $c = 419$

31. $a = 9.52$, $b = 14.7$　　　　　　　32. $a = .123$, $b = .456$

33. A straight path leading up a hill rises 26 feet per 100 horizontal feet. What angle does it make with the horizontal?

34. A 20-foot ladder leans against a wall, making an angle of 76° with the level ground. How high up the wall is the upper end of the ladder?

35. Find the angle of elevation of the sun if a woman 5 feet 9 inches tall casts a shadow 46.8 feet long. (The *angle of elevation* is the upward angle made with the horizontal.)

36. A guy wire to a pole makes an angle of 69° with the level ground and is 14 feet from the pole at the ground. How high above the ground is the wire attached to the pole?

37. Suppose that the woman in Problem 35 is walking with her daughter Sue, who is 3 feet 10 inches tall. How long is Sue's shadow?

38. Find the length of the supporting wire in Problem 36.

MISCELLANEOUS PROBLEMS

The following problems can be solved using either a calculator or tables. We recommend using a calculator.

39. Calculate each value.
    (a) tan 14.5°　　　　(b) 24.6 cos 74.3°　　　　(c) 15.6 (sin 14°)²/cos 87°

40. Find θ in each case.
    (a) $\sin \theta = .6691$　　　(b) $\cos \theta = .5519$　　　(c) $\tan \theta = 5.396$

41. From the top of a lighthouse 120 feet above sea level, the *angle of depression* (the downward angle from the horizontal) to a boat adrift on the sea is 9.4° (Figure 7). How far from the foot of the lighthouse is the boat?

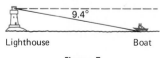

Lighthouse　　　　　　　　　Boat

Figure 7

42. Solve the right triangle in which $b = 67.3$ and $c = 82.9$.

43. Find x in Figure 8.

44. When the *angle of elevation* (the upward angle from the horizontal) of the sun is 28.4°, the Eiffel Tower in Paris casts a horizontal shadow 1822 feet long. How high is the tower?

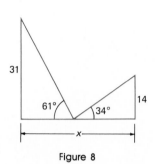

Figure 8

45. With her hands 5 feet above the ground, Sally is pulling on a kite. If the kite is 200 feet above the ground and the kite string makes an angle of 32.4° with the horizontal, how many feet of string are out?

46. A plane is flying directly away from a ground observer at a constant rate, maintaining an elevation of 15,000 feet. At a certain instant, the observer measures the angle of elevation as 44° and 15 seconds later as 31°. How fast is the plane flying in miles per hour?

47. From a window in an office building, I am looking at a television tower that is 600 meters away (horizontally). The angle of elevation of the top of the tower

is 19.6° and the angle of depression of the base of the tower is 21.3°. How tall is the tower?

48. The vertical distance from first to second floor of a certain department store is 28 feet. The escalator, which has a horizontal reach of 96 feet, takes 25 seconds to carry a person between floors. How fast does the escalator travel?

49. The Great Pyramid is about 480 feet high and its square base measures 760 feet on a side. Find the angle of elevation of one of its edges, that is, find β in Figure 9.

50. Find the angle between a principal diagonal and a face diagonal of a cube.

51. A regular hexagon (6 equal sides) is inscribed in a circle of radius 4. Find the perimeter P and area A of this hexagon.

52. A regular decagon (10 equal sides) is inscribed in a circle of radius 12. What percent of the area of the circle is the area of the decagon?

53. Find the area of the regular 6-pointed Star of David that is inscribed in a circle of radius 1 (Figure 10).

54. **TEASER** Find the area of the regular 5-pointed star (the pentagram) that is inscribed in a circle of radius 1 (Figure 11).

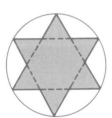

Figure 10

Figure 11

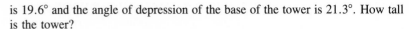

Figure 9

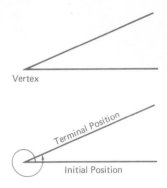

The Dynamic View of Angles

In geometry, we take a static view of angles. An angle is simply the union of two rays with a common endpoint (the vertex). In trigonometry, angles are thought of in a dynamic way. An angle is determined by rotating a ray about its endpoint from an initial position to a terminal position.

4-2 Angles and Arcs

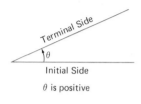

θ is positive

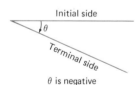

θ is negative

Figure 12

For the solution of right triangles (which involve acute angles), we required only the familiar and simple notion of angle from high-school geometry. But for the broader development of trigonometry, we need the new perspective on angles suggested by our opening display. Not only do we allow arbitrarily large angles, but we also distinguish between positive and negative angles. If an angle is generated by a counterclockwise rotation, it is positive; if generated by a clockwise rotation, it is negative (Figure 12). To know an angle, in trigonometry, is to know how the angle came into being. It is to know the initial side, the terminal side, and the kind of rotation that produced the angle.

DEGREE MEASUREMENT

Take a circle and divide its circumference into 360 equal parts. The angle with vertex at the center determined by one of these parts has measure one **degree** (written 1°). This way of measuring angles is due to the ancient Babylonians and is so familiar that we used it in Section 4-1 without comment. There is a refinement, however, that we avoid. The Babylonians divided each degree into 60 minutes and each minute into 60 seconds; some people still follow this cumbersome practice. If we need to measure angles to finer accuracy than a degree, we will use decimal parts. Thus we write 40.5° rather than 40°30′.

It is important that we be familiar with measuring both positive and negative angles, as well as angles resulting from large rotations. Three angles are shown in Figure 13. Note that all three have the same initial and terminal sides.

RADIAN MEASUREMENT

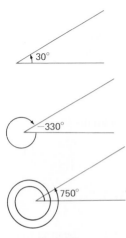

Figure 13

The best way to measure angles is in radians. Take a circle of radius r. The familiar formula $C = 2\pi r$ tells us that the circumference has 2π (about 6.28) arcs of length r around it. The angle with vertex at the center of a circle

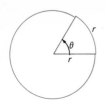

determined by an arc of length equal to its radius measures one **radian** (Figure 14). Thus an angle of size 360° measures 2π radians and an angle of size 180° measures π radians. We abbreviate the latter by writing

$$180° = \pi \text{ radians}$$

θ measures one radian (about 57.3°)

Figure 14

To convert from degrees to radians, all one needs to remember is the result in the box. By dividing by 2, 3, 4, and 6, respectively, we get the conversions for several special angles.

$$90° = \frac{\pi}{2} \text{ radians}$$

$$60° = \frac{\pi}{3} \text{ radians}$$

$$45° = \frac{\pi}{4} \text{ radians}$$

$$30° = \frac{\pi}{6} \text{ radians}$$

If we divide the boxed formula by 180, we get

$$1° = \frac{\pi}{180} \text{ radians}$$

and if we divide by π, we get

$$\frac{180°}{\pi} = 1 \text{ radian}$$

The following rules thus hold.

To convert from degrees to radians, multiply by $\pi/180$.
To convert from radians to degrees, multiply by $180/\pi$.

For example,

$$22° = 22\left(\frac{\pi}{180}\right) \text{ radians} \approx .38397 \text{ radians}$$

and

$$2.3 \text{ radians} = 2.3\left(\frac{180}{\pi}\right)° \approx 131.78°$$

Some scientific calculators have a key that makes these conversions automatically.

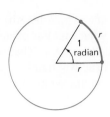

$s = 2r$

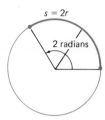

$s = tr$

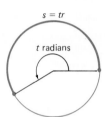

Figure 15

Sector

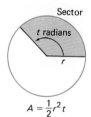

$A = \frac{1}{2}r^2t$

Figure 16

ARC LENGTH AND AREA

Radian measure is almost invariably used in calculus because it is an intrinsic measure. The division of a circle into 360 parts was quite arbitrary; its division into parts of radius length (2π parts) is more natural. Because of this, formulas using radian measure tend to be simple, while those using degree measure are often complicated. As an example, consider arc length. Let t be the radian measure of an angle θ with vertex at the center of a circle of radius r. This angle cuts off an arc of length s which satisfies the simple formula

$$s = rt$$

This follows directly from the fact that an angle of one radian ($t = 1$) cuts off an arc of length r (see Figure 15).

A second nice formula is that for the area of the sector cut off from a circle by a central angle of t radians (Figure 16). Note that the area A of this sector is to the area of the whole circle as t is to 2π, that is, $A/\pi r^2 = t/2\pi$. Thus

$$A = \frac{1}{2}r^2t$$

THE UNIT CIRCLE

The formula for arc length takes a particularly simple form when $r = 1$, namely, $s = t$. We emphasize its meaning. *On the unit circle, the length of an arc is the same as the radian measure of the angle it determines.*

Someone is sure to point out a difficulty in what we have just said. What happens when t is greater than 2π or when t is negative? To understand our meaning, imagine an infinitely long string on which the real number scale has been marked. Think of wrapping this string around the unit circle as shown in Figure 17.

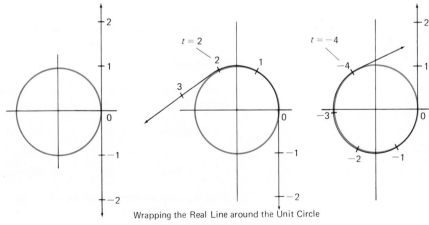

Wrapping the Real Line around the Unit Circle

Figure 17

Now if we think of the directed length (that is, the signed length) of a piece of the string, the formula $s = t$ holds no matter what t is. For example, the length of string corresponding to an angle of 8π radians is 8π. That piece of string wraps counterclockwise around the unit circle exactly 4 times. A piece of string corresponding to an angle of -3π radians would wrap clockwise around the unit circle one and a half times, its directed length being -3π.

Problem Set 4-2

Convert each of the following to radians. You may leave π in your answer.

1. 120°	2. 225°	3. 240°
4. 150°	5. 210°	6. 330°
7. 315°	8. 300°	9. 540°
10. 450°	11. −420°	12. −660°
13. 160°	14. 200°	15. $(20/\pi)°$
16. $(150/\pi)°$		

Convert each of the following to degrees. Give your answer correct to the nearest tenth of a degree.

17. $\dfrac{4}{3}\pi$ radians   18. $\dfrac{5}{6}\pi$ radians   19. $-\dfrac{2\pi}{3}$ radians

20. $-\dfrac{7\pi}{4}$ radians   21. 3π radians   22. 3 radians

ⓒ 23. 4.52 radians   ⓒ 24. $\dfrac{11}{4}$ radians   ⓒ 25. $\dfrac{1}{\pi}$ radians

ⓒ 26. $\dfrac{4}{3\pi}$ radians

27. Find the radian measure of the angle at the center of a circle of radius 6 inches which cuts off an arc of length

 (a) 12 inches;   (b) 18.84 inches.

28. Find the length of the arc cut off on a circle of radius 3 feet by an angle at the center of

 (a) 2 radians;   (b) 5.5 radians;

 (c) $\dfrac{\pi}{4}$ radians;   (d) $\dfrac{5\pi}{6}$ radians.

29. Find the radius r for each of the following.

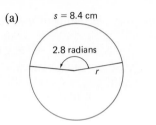

(a) $s = 8.4$ cm

2.8 radians

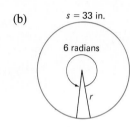

(b) $s = 33$ in.

6 radians

30. Through how many radians does the minute hand of a clock turn in 1 hour? The hour hand in 1 hour? The minute hand in 5 hours?

EXAMPLE A (Locating a Point on the Unit Circle) Figure 18 shows a unit circle with center at the origin. Suppose that a point P moves in a counterclockwise direction around the circle starting at $(1, 0)$. In which quadrant is P when it has traveled a distance of 4 units? Of 40 units?

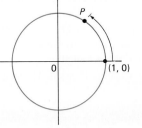

Figure 18

Solution. Keep in mind that the distance P travels equals the radian measure of the angle through which OP turns. A distance of 4 units puts P in quadrant III since $\pi < 4 < 3\pi/2$. Once around the circle is $2\pi \approx 6.28$ units. If you divide 40 by 6.28, you get

$$40 = 6(6.28) + 2.32$$

Since 2.32 is between $\pi/2$ and π, traveling 40 units around the unit circle will put P in quadrant II.

Find the quadrant in which the point P in the example above lies when it has traveled each of the following distances.

31. 3 units   32. 3.2 units   33. 4.7 units

34. 4.8 units   35. $\left(\dfrac{5\pi}{2} + 1\right)$ units   36. $\left(\dfrac{9\pi}{2} - 1\right)$ units

37. 100 units   38. 200 units

EXAMPLE B (Angular Velocity) A formula closely related to the arc length formula $s = rt$ is the formula

$$v = r\omega$$

which connects the speed (velocity) of a point on the rim of a wheel of radius r with the angular velocity ω at which the wheel is turning. Here ω is measured in radians per unit of time. Use this formula to determine the angular velocity in radians per second of a bicycle wheel of radius 16 inches if the bicycle is being ridden down the road at 30 miles per hour.

Solution. We must use consistent units. You can check that the speed of a point on the rim of the wheel (30 miles per hour) translates to 44 feet per second and that the radius of the wheel is $\frac{4}{3}$ feet. Thus

$$44 = \frac{4}{3}\omega$$

or

$$\omega = \frac{3}{4}(44) = 33 \text{ radians per second}$$

39. Sally is pedaling her tricycle so the front wheel (radius 8 inches) turns at 4 revolutions per second. How fast is she moving down the sidewalk in feet per second? *Hint:* Four revolutions per second is 8π radians per second.

40. Suppose that the tire on a car has an outer diameter 2.5 feet. How many revolutions per minute does the tire make when the car is traveling 60 miles per hour?

41. A dead fly is stuck to a belt that passes over two pulleys 6 inches and 8 inches in radius, as shown in Figure 19. Assuming no slippage, how fast is the fly moving when the larger pulley turns at 20 revolutions per minute?

42. How fast (in revolutions per minute) is the smaller wheel in Problem 41 turning?

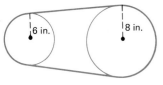

Figure 19

MISCELLANEOUS PROBLEMS

43. Convert to radians.
    (a) $-1440°$ (b) $2\frac{1}{2}$ revolutions   (c) $(60/\pi)°$

44. Convert to degrees.
    (a) $23\pi/36$ radians   (b) -4.63 radians   (c) $3/(2\pi)$ radians

45. Find the length of arc cut off on a circle of radius 4.25 centimeters by each central angle.
    (a) 6 radians   (b) $(18/13\pi)°$ (c) $17\pi/6$ radians

46. The front wheel of Tony's tricycle has a diameter of 20 inches. How far did he travel in pedaling through 60 revolutions?

47. The pedal sprocket of Maria's bicycle has radius 12 centimeters, the rear wheel sprocket has radius 3 centimeters, and the wheels have radius 40 centimeters. How far did Maria travel if she pedaled continuously for 30 revolutions of the pedal sprocket?

48. A belt traveling at the rate of 60 feet per second drives a pulley (a wheel) at the rate of 900 revolutions per minute. Find the radius of the pulley.

49. Assume that the earth is a sphere of radius 3960 miles. How fast (in miles per hour) is a point on the equator moving as a result of the earth's rotation about its axis?

☐ 50. The orbit of the earth about the sun is an ellipse that is nearly circular with radius 93 million miles. Approximately, what is the earth's speed (in miles per hour) in its path around the sun? You will need the fact that a complete orbit takes 365.25 days.

51. The angle subtended by the sun at the earth (93 million miles away) is .0093 radians. Find the diameter of the sun.

☐ 52. A nautical mile is the length of 1 minute ($\frac{1}{60}$ of a degree) of arc on the equator of the earth. How many miles are there in a nautical mile?

53. One of the authors (Dale Varberg) lives at exactly 45° latitude north (see Figure 20). How long would it take him to fly to the North Pole at 600 miles per hour (assuming the earth is a sphere of radius 3960 miles)?

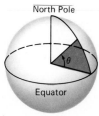

θ measures latitude north

Figure 20

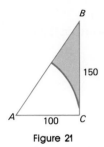

B

150

A   100   C

Figure 21

18

8

10

Figure 22

54. New York City is located at 40.5° latitude north. How far is it from there to the equator?

55. Oslo, Norway, and Leningrad, Russia, are both located at 60° latitude north. Oslo is at longitude 6° east (of the prime meridian) whereas Leningrad is at 30° east. How far apart are these two cities along the 60° parallel?

[c] 56. Find the area of the shaded region of the right triangle *ABC* shown in Figure 21.

57. The minute hand and hour hand of a clock are both 6 inches long and reach to the edge of the dial. Find the area of the pie-shaped region between the two hands at 5:40.

58. A cone has radius of base *R* and slant height *L*. Find the formula for its lateral surface area. *Hint:* Imagine the cone to be made of paper, slit it up the side, and lay it flat in the plane.

59. Find the area of the polar rectangle shown in Figure 22. The two curves are arcs of concentric circles.

60. **TEASER**   Consider two circles both of radius *r* and with the center of each lying on the rim of the other. Find the area of the common part of the two circles.

Definitions of Sine and Cosine

Place an angle θ, whose radian measure is t, in **standard position**, that is, put θ in the coordinate plane so that its vertex is at the origin and its initial side is along the positive x-axis. Let (x,y) be the coordinates of the point of intersection of the terminal side with the unit circle. We define both $\sin \theta$ (sine of θ) and $\sin t$ by

$$\sin \theta = \sin t = y.$$

Similarly,

$$\cos \theta = \cos t = x.$$

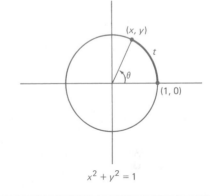

4-3   The Sine and Cosine Functions

In Section 4-1, we defined the sine and cosine for positive acute angles. The definitions in our opening display are more general and hence more widely applicable. They should be studied carefully. Notice that we have defined the sine and cosine for any angle θ and also for the corresponding number t. Both concepts are important. In geometric situations, angles play a central role; thus we are likely to need sines and cosines of angles. But in most of pure mathematics and in many scientific applications, it is the trigonometric functions of numbers that are important. In this connection, we emphasize that the number *t* may be positive or negative, large or small. And we may think of it as the radian measure of an angle, as the directed length of an arc on the unit circle, or simply as a number.

CONSISTENCY WITH EARLIER DEFINITIONS

Do the definitions given in Section 4-1 for the sine and cosine of an acute angle harmonize with those given here? Yes. Take a right triangle ABC with an acute angle θ. Place θ in standard position, thus determining a point $B'(x, y)$ on the unit circle and a point $C'(x, 0)$ directly below it on the x-axis (see Figure 23).

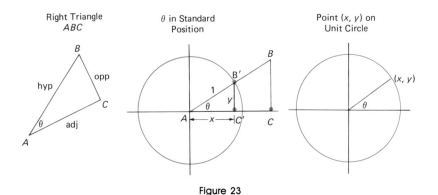

Figure 23

Notice that triangles ABC and $AB'C'$ are similar. It follows that

$$\frac{\text{opp}}{\text{hyp}} = \frac{BC}{AB} = \frac{B'C'}{AB'} = \frac{y}{1} = y$$

$$\frac{\text{adj}}{\text{hyp}} = \frac{AC}{AB} = \frac{AC'}{AB'} = \frac{x}{1} = x$$

On the left are the old definitions of $\sin \theta$ and $\cos \theta$; on the right are the new ones. They are consistent.

SPECIAL ANGLES

In Section 4-1, we learned that

$$\cos 45° = \frac{\sqrt{2}}{2} \qquad \sin 45° = \frac{\sqrt{2}}{2}$$

$$\cos 30° = \frac{\sqrt{3}}{2} \qquad \sin 30° = \frac{1}{2}$$

$$\cos 60° = \frac{1}{2} \qquad \sin 60° = \frac{\sqrt{3}}{2}$$

Making use of the consistency of the old and new definitions of sine and cosine, we conclude that the point on the unit circle corresponding to $\theta = 45° = \pi/4$ radians must have coordinates $(\sqrt{2}/2, \sqrt{2}/2)$. Similarly, the point corresponding to $\theta = 30° = \pi/6$ radians has coordinates $(\sqrt{3}/2, 1/2)$ and the point corresponding to $\theta = 60° = \pi/3$ radians has coordinates $(1/2, \sqrt{3}/2)$.

Now we can make use of obvious symmetries to find the coordinates of many other points on the unit circle. In the two diagrams of Figure 24, we show a number of these points, noting first the radian measure of the angle and then the coordinates of the corresponding point on the unit circle.

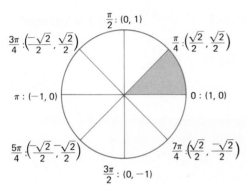
Some multiples of $\pi/4$

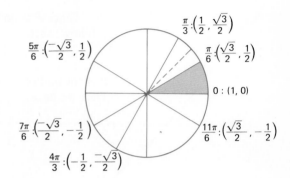
Some multiples of $\pi/6$

Figure 24

Notice, for example, how the coordinates of the points corresponding to $t = 5\pi/6$, $7\pi/6$, and $11\pi/6$ are related to the point corresponding to $t = \pi/6$. You should have no trouble seeing other relationships.

Once we know the coordinates of a point on the unit circle, we can state the sine and cosine of the corresponding angle. In particular, we get the values in the table in Figure 25. They are used so often that you should memorize them.

t	0	$\dfrac{\pi}{6}$	$\dfrac{\pi}{4}$	$\dfrac{\pi}{3}$	$\dfrac{\pi}{2}$	π	$\dfrac{3\pi}{2}$
$\cos t$	1	$\dfrac{\sqrt{3}}{2}$	$\dfrac{\sqrt{2}}{2}$	$\dfrac{1}{2}$	0	-1	0
$\sin t$	0	$\dfrac{1}{2}$	$\dfrac{\sqrt{2}}{2}$	$\dfrac{\sqrt{3}}{2}$	1	0	-1

Figure 25

PROPERTIES OF SINES AND COSINES

Think of what happens to x and y as t increases from 0 to 2π in Figure 26, that is, as P travels all the way around on the unit circle. For example, x steadily decreases until it reaches its smallest value of -1 at $t = \pi$; then it starts to

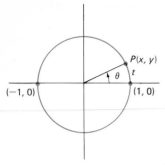

Figure 26

increase until it is back to 1 at $t = 2\pi$. We have just described the behavior of $\cos t$ (or $\cos \theta$) as t increases from 0 to 2π. You should trace the behavior of $\sin t$ in the same way. Notice that both x and y are always between -1 and 1 (inclusive). It follows that

$$-1 \le \sin t \le 1$$
$$-1 \le \cos t \le 1$$

Since P is on the unit circle, $x^2 + y^2 = 1$, and $x = \cos t$ and $y = \sin t$, it follows that

$$(\sin t)^2 + (\cos t)^2 = 1$$

It is conventional to write $\sin^2 t$ instead of $(\sin t)^2$ and $\cos^2 t$ instead of $(\cos t)^2$. Thus we have

$$\sin^2 t + \cos^2 t = 1$$

This is an identity; it is true for all t. Of course we can just as well write

$$\sin^2 \theta + \cos^2 \theta = 1$$

We have established one basic relationship between the sine and the cosine; here are two others, valid for all t.

$$\sin\left(\frac{\pi}{2} - t\right) = \cos t$$
$$\cos\left(\frac{\pi}{2} - t\right) = \sin t$$

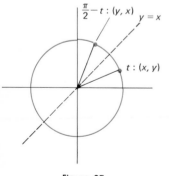

Figure 27

These relationships are easy to see when $0 < t < \pi/2$. Notice that t and $\pi/2 - t$ are measures of complementary angles (two angles with measures totaling 90° or $\pi/2$). That means that t and $\pi/2 - t$ determine points on the unit circle which are reflections of each other about the line $y = x$ (see Figure 27). Thus if one point has coordinates (x, y), the other has coordinates (y, x). The result given above follows from this fact.

Finally, we point out that $t, t \pm 2\pi, t \pm 4\pi, \ldots$ all determine the same point on the unit circle and thus have the same sine and cosine. This repetitive behavior puts the sine and cosine into a special class of functions, for which we give the following definition. A function f is **periodic** if there is a positive number p such that

$$f(t + p) = f(t)$$

for every t in the domain of f. The smallest such p is called the **period** of f. Thus we say that sine and cosine are periodic functions with period 2π and write

$$\sin(t + 2\pi) = \sin t$$
$$\cos(t + 2\pi) = \cos t$$

THE TRIGONOMETRIC POINT $P(t)$

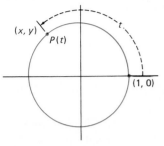

Figure 28

We have introduced $\cos t$ and $\sin t$ as the x- and y-coordinates of the point on the unit circle whose directed distance from $(1, 0)$ along the unit circle is t. This point is called a **trigonometric point** and will be denoted by $P(t)$ (Figure 28). We may regard $P(t)$ to be a function of t, since for each t there is a unique point $P(t)$. This function, moreover, is periodic with period 2π—that is,

$$P(t + 2\pi) = P(t)$$

It follows that

$$P(t + k2\pi) = P(t)$$

for any integer k, a fact that allows us to find the coordinates of $P(t)$ for any t, no matter how large t is. Suppose, for example, that we wish to find the coordinates of $P(16\pi/3)$, shown in Figure 29. Since

$$\frac{16\pi}{3} = \frac{4\pi}{3} + 4\pi$$

it follows that

$$P\left(\frac{16\pi}{3}\right) = P\left(\frac{4\pi}{3}\right)$$

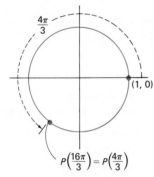

Figure 29

From the special angle diagrams on page 193, $P(4\pi/3)$ has coordinates $(-1/2, -\sqrt{3}/2)$. We conclude that $P(16\pi/3)$ also has these coordinates. This means that

$$\cos\left(\frac{16\pi}{3}\right) = -\frac{1}{2} \qquad \sin\left(\frac{16\pi}{3}\right) = -\frac{\sqrt{3}}{2}$$

Problem Set 4-3

In Problems 1–8, find the coordinates of the trigonometric point P(t) for the indicated value of t. Hint: Begin by drawing a unit circle and locating P(t) on it; then relate P(t) to the diagrams on page 193.

1. $t = \dfrac{13\pi}{6}$

2. $t = \dfrac{19\pi}{6}$

3. $t = \dfrac{19\pi}{4}$

4. $t = \dfrac{15\pi}{4}$

5. $t = 24\pi + \dfrac{5\pi}{4}$

6. $t = 16\pi + \dfrac{5\pi}{6}$

7. $t = -\dfrac{7\pi}{6}$

8. $t = \dfrac{13\pi}{4}$

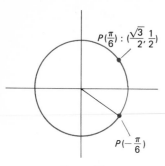

$P(\frac{\pi}{6}) : (\frac{\sqrt{3}}{2}, \frac{1}{2})$

$P(-\frac{\pi}{6})$

Figure 30

EXAMPLE A (Using $P(t)$ to Find Sine and Cosine Values) Find
(a) $\sin(-\pi/6)$; (b) $\cos(29\pi/4)$.

Solution.

(a) We locate $P(-\pi/6)$ on the unit circle (Figure 30) and note that its y-coordinate is $-\frac{1}{2}$ because of its position relative to $P(\pi/6)$. Thus $\sin(-\pi/6) = -\frac{1}{2}$.

(b) We simplify the problem by removing a large multiple of 2π—that is, by noting that

$$\frac{29\pi}{4} = 6\pi + \frac{5\pi}{4}$$

from which we conclude $P(29\pi/4) = P(5\pi/4)$. Then we refer to the diagrams on page 193, or better yet, we simply observe that $P(5\pi/4)$, being diametrically opposite from $P(\pi/4)$ on the unit circle, has coordinates $(-\sqrt{2}/2, -\sqrt{2}/2)$ (see Figure 31). We conclude that

$$\cos\left(\frac{29\pi}{4}\right) = -\frac{\sqrt{2}}{2}$$

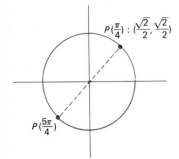

$P(\frac{\pi}{4}) : (\frac{\sqrt{2}}{2}, \frac{\sqrt{2}}{2})$

$P(\frac{5\pi}{4})$

Figure 31

Using the method of Example A, find the value of each of the following.

9. $\sin(-\pi/4)$
10. $\sin(-5\pi/4)$
11. $\sin(9\pi/4)$
12. $\sin(15\pi/4)$
13. $\cos(13\pi/4)$
14. $\cos(-7\pi/4)$
15. $\cos(10\pi/3)$
16. $\cos(25\pi/6)$
17. $\sin(5\pi/2)$
18. $\cos 7\pi$
19. $\sin(-4\pi)$
20. $\cos(7\pi/2)$
21. $\cos(19\pi/6)$
22. $\sin(14\pi/3)$
23. $\cos(-\pi/3)$
24. $\sin(-5\pi/6)$
25. $\cos(125\pi/4)$
26. $\cos(-13\pi/6)$
27. $\sin 510°$
28. $\sin(-390°)$
29. $\cos 840°$
30. $\cos(-720°)$
31. $\cos(-210°)$
32. $\sin 900°$

EXAMPLE B (Sine and Cosine of $-t$) Show that for all t

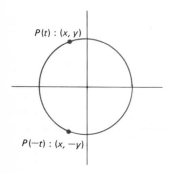

$P(t) : (x, y)$

$P(-t) : (x, -y)$

Figure 32

$$\sin(-t) = -\sin t$$
$$\cos(-t) = \cos t$$

that is, sine is an odd function and cosine is an even function.

Solution. The points $P(-t)$ and $P(t)$ are symmetric with respect to the x-axis (Figure 32). Thus if $P(t)$ has coordinates (x, y), $P(-t)$ has coordinates $(x, -y)$ and so

$$\sin(-t) = -y = -\sin t$$
$$\cos(-t) = x = \cos t$$

33. If $\sin 1.87 = .95557$ and $\cos 1.87 = -0.29476$, find $\sin(-1.87)$ and $\cos(-1.87)$.

34. If $\sin 15.2° = 0.2622$ and $\cos 15.2° = 0.9650$, find $\sin(-15.2°)$ and $\cos(-15.2°)$.

35. Given $P(t)$ with coordinates $(1/\sqrt{5}, -2/\sqrt{5})$.
    (a) What are the coordinates of $P(-t)$?
    (b) What are the values of $\sin(-t)$ and $\cos(-t)$?

36. If t is the radian measure of an angle in quadrant III and $\sin t = -\frac{3}{5}$, evaluate each expression.
    (a) $\sin(-t)$
    (b) $\cos t$ *Hint:* Use the fact that $\sin^2 t + \cos^2 t = 1$.
    (c) $\cos(-t)$

37. Note that $P(t)$ and $P(\pi + t)$ are symmetric with respect to the origin (Figure 33). Use this to show that
    (a) $\sin(\pi + t) = -\sin t$;   (b) $\cos(\pi + t) = -\cos t$.

38. Note that $P(t)$ and $P(\pi - t)$ are symmetric with respect to the y-axis. Use this fact to find identities analogous to those in Problem 37 for $\sin(\pi - t)$ and $\cos(\pi - t)$.

Figure 33

MISCELLANEOUS PROBLEMS

39. Use the unit circle to determine the sign (plus or minus) of each of the following.
    (a) $\cos 2$ (b) $\sin(-3)$ (c) $\cos 428°$
    (d) $\sin 21.4$ (e) $\sin(23\pi/32)$ (f) $\sin(-820°)$

40. Find the coordinates of $P(t)$ for the indicated values of t.
    (a) $t = \pi$ (b) $t = -3\pi/2$ (c) $t = -3\pi/4$
    (d) $t = 5\pi/6$ (e) $t = 44\pi/3$ (f) $t = -93.5\pi$

41. Let $P(t)$ have coordinates $(x, -1/2)$.
    (a) Find the two possible values of x.
    (b) Find the corresponding values of t.

42. With initial point $(0, -1)$, a string of length $4\pi/3$ is wound clockwise around the unit circle. What are the coordinates of the terminal point?

43. For what values of t satisfying $0 \le t < 2\pi$ are the following true?
    (a) $\sin t = \cos t$ (b) $\frac{1}{2} < \sin t < \frac{\sqrt{3}}{2}$
    (c) $\cos^2 t \ge .25$ (d) $\cos^2 t > \sin^2 t$

44. Find the four smallest positive solutions to the following equations.
    (a) $\sin t = 1$ (b) $|\cos t| = \frac{1}{2}$
    (c) $\cos t = -\frac{\sqrt{3}}{2}$ (d) $\sin t = -\frac{\sqrt{2}}{2}$

45. In each case, assume that θ is an angle in standard position with terminal side in the fourth quadrant. Use $\sin^2 \theta + \cos^2 \theta = 1$ to determine the indicated value.
    (a) $\cos \theta$ if $\sin \theta = -\frac{4}{5}$
    (b) $\sin \theta$ if $\cos \theta = \frac{24}{25}$

46. Use the unit circle to find identities for $\sin(2\pi - t)$ and $\cos(2\pi - t)$.

47. If $P(t)$ has coordinates $(\frac{4}{5}, -\frac{3}{5})$, evaluate each of the following. *Hint:* For (e) and (f), see Problems 37 and 38.
    (a) $\sin(-t)$ (b) $\sin(\frac{\pi}{2} - t)$
    (c) $\cos(2\pi + t)$ (d) $\cos(2\pi - t)$
    (e) $\sin(\pi + t)$ (f) $\cos(\pi - t)$

sin t	cos t	sin(t + π)	cos(t + π)	sin(π − t)	sin(2π − t)	Least positive value of t
$\sqrt{3}/2$	$-\frac{1}{2}$					
	$\sqrt{2}/2$	$\sqrt{2}/2$				
$-\frac{1}{2}$			$-\sqrt{3}/2$			
-1						
			$\sqrt{3}/2$		$\frac{1}{2}$	
	0				-1	

48. Fill in all the blanks in the chart above.

49. Recall that [ ] and ( ) denote "the greatest integer in" and "the distance to the nearest integer," respectively. Determine which of the following functions are periodic and, if so, specify the period.
(a) $f(x) = (x)$ (b) $f(x) = (3x)$
(c) $f(x) = [x]$ (d) $f(x) = x - [x]$

50. Suppose that $f(x)$ is periodic with period 2 and that $f(x) = 4 - x^2$ for $0 \le x < 2$. Evaluate each of the following.
(a) $f(2)$ (b) $f(4.5)$
(c) $f(-.5)$ (d) $f(8.8)$

51. Evaluate
$$\sin 1° + \sin 2° + \sin 3° + \cdots + \sin 357° + \sin 358° + \sin 359°$$

52. **TEASER** Evaluate
$$\sin^2 1° + \sin^2 2° + \sin^2 3° + \cdots + \sin^2 357° + \sin^2 358° + \sin^2 359°$$

"Strange as it may sound, the power of mathematics rests on its evasion of all unnecessary thought and on its wonderful saving of mental operations."

Ernst Mach

New Functions from Old Ones

tangent: $\tan t = \dfrac{\sin t}{\cos t}$

cotangent: $\cot t = \dfrac{\cos t}{\sin t}$

secant: $\sec t = \dfrac{1}{\cos t}$

cosecant: $\csc t = \dfrac{1}{\sin t}$

4-4 Four More Trigonometric Functions

Without question, the sine and cosine are the most important of the six trigonometric functions. Not only do they occur most frequently in applications, but the other four functions can be defined in terms of them, as our opening box shows. This means that if you learn all you can about sines and cosines, you will automatically know a great deal about tangents, cotangents, secants, and cosecants. Ernst Mach would say that it is a way to evade unnecessary thought.

Look at the definitions in the opening box again. Naturally, we must rule out any values of t for which a denominator is zero. For example, $\tan t$ is not defined for $t = \pm\pi/2, \pm3\pi/2, \pm5\pi/2$, and so on. Similarly, $\csc t$ is not defined for such values as $t = 0, \pm\pi$, and $\pm2\pi$.

PROPERTIES OF THE NEW FUNCTIONS

The wisdom of the opening paragraph will now be demonstrated. Recall the identity $\sin^2 t + \cos^2 t = 1$. Out of it come two new identities.

$$1 + \tan^2 t = \sec^2 t$$
$$1 + \cot^2 t = \csc^2 t$$

To show that the first identity is correct, we take its left side, express it in terms of sines and cosines, and do a little algebra.

$$1 + \tan^2 t = 1 + \left(\frac{\sin t}{\cos t}\right)^2$$

$$= 1 + \frac{\sin^2 t}{\cos^2 t}$$

$$= \frac{\cos^2 t + \sin^2 t}{\cos^2 t}$$

$$= \frac{1}{\cos^2 t}$$

$$= \left(\frac{1}{\cos t}\right)^2$$

$$= \sec^2 t$$

The second identity is verified in a similar fashion.

Suppose we wanted to know whether cotangent is an even or an odd function (or neither). We simply recall that $\sin(-t) = -\sin t$ and $\cos(-t) = \cos t$ and write

$$\cot(-t) = \frac{\cos(-t)}{\sin(-t)} = \frac{\cos t}{-\sin t} = -\frac{\cos t}{\sin t} = -\cot t$$

Thus cotangent is an odd function.

In a similar vein, recall the identities

(i)
$$\sin\left(\frac{\pi}{2} - t\right) = \cos t$$

(ii)
$$\cos\left(\frac{\pi}{2} - t\right) = \sin t$$

From them, we obtain

$$(iii) \qquad \tan\left(\frac{\pi}{2} - t\right) = \frac{\sin(\pi/2 - t)}{\cos(\pi/2 - t)} = \frac{\cos t}{\sin t} = \cot t$$

These three identities are examples of what are called **cofunction identities.** Sine and cosine are confunctions; so are tangent and cotangent; as are secant and cosecant. Notice that identities (i), (ii), and (iii) all have the form

$$\text{function}\left(\frac{\pi}{2} - t\right) = \text{cofunction}(t)$$

With cosecant as the function, we have

$$\csc\left(\frac{\pi}{2} - t\right) = \sec t$$

ALTERNATIVE DEFINITIONS OF THE TRIGONOMETRIC FUNCTIONS

There is another approach to trigonometry favored by some authors. Let θ be an angle in standard position and suppose that (a, b) is any point on its terminal side at a distance r from the origin (Figure 34). Then

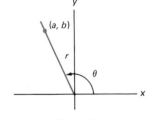

Figure 34

$$\sin \theta = \frac{b}{r} \qquad \cos \theta = \frac{a}{r}$$

$$\tan \theta = \frac{b}{a} \qquad \cot \theta = \frac{a}{b}$$

$$\sec \theta = \frac{r}{a} \qquad \csc \theta = \frac{r}{b}$$

To see that these definitions are equivalent to those we gave earlier, consider first an angle θ with terminal side in quadrant I (see Figure 35).

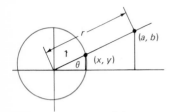

Figure 35

By similar triangles,

$$\frac{b}{r} = \frac{y}{1} \quad \text{and} \quad \frac{a}{r} = \frac{x}{1}$$

Actually these ratios are equal no matter in which quadrant the terminal side of θ is, since b and y always have the same sign, as do a and x. The first two formulas in the box now follow from our original definitions, which say that

$$\sin \theta = y \quad \text{and} \quad \cos \theta = x$$

The others are a consequence of the fact that the remaining four functions can be expressed in terms of sines and cosines.

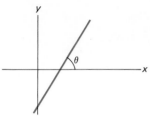

Figure 36

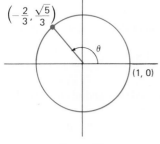

Figure 37

THE TANGENT FUNCTION AND SLOPE

Recall that the slope m of a line is the ratio of rise to run. In particular, if the line goes through the point (a, b) and also the origin, its slope is b/a. But this number b/a is also the tangent of the nonnegative angle θ that the line makes with the positive x-axis (see Figure 34).

In general, the smallest nonnegative angle θ that a line makes with the positive x-axis is called the **angle of inclination** of the line (Figure 36). It follows that for any nonvertical line, the slope m of the line satisfies

$$m = \tan \theta$$

As an example, suppose that a line has angle of inclination $120°$ and goes through the point $(1, 2)$. Then its slope is $m = \tan 120° = -\sqrt{3}$ and the line has equation

$$y - 2 = -\sqrt{3}(x - 1)$$

Problem Set 4-4

1. If $\sin t = \frac{4}{5}$ and $\cos t = -\frac{3}{5}$, evaluate each function.
   (a) $\tan t$ (b) $\cot t$ (c) $\sec t$ (d) $\csc t$

2. If $\sin t = -1/\sqrt{5}$ and $\cos t = 2/\sqrt{5}$, evaluate each function.
   (a) $\tan t$ (b) $\cot t$ (c) $\sec t$ (d) $\csc t$

3. Find the values of $\tan \theta$ and $\csc \theta$ for the angle θ of Figure 37.

4. Find $\cot \alpha$ and $\sec \alpha$ for α as shown in Figure 38.

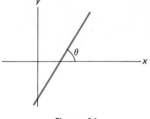

Figure 38

Keeping in mind what you know about the sines and cosines of special angles, find each of the values in Problems 5–22.

5. $\tan(\pi/6)$	6. $\cot(\pi/6)$	7. $\sec(\pi/6)$
8. $\csc(\pi/6)$	9. $\cot(\pi/4)$	10. $\sec(\pi/4)$
11. $\csc(\pi/3)$	12. $\sec(\pi/3)$	13. $\sin(4\pi/3)$
14. $\cos(4\pi/3)$	15. $\tan(4\pi/3)$	16. $\sec(4\pi/3)$
17. $\tan \pi$	18. $\sec \pi$	19. $\tan 330°$
20. $\cot 120°$	21. $\sec 600°$	22. $\csc(-150°)$

23. For what values of t on $0 \le t \le 4\pi$ is each of the following undefined?
    (a) $\sec t$ (b) $\tan t$ (c) $\csc t$ (d) $\cot t$

24. For which values of t on $0 \le t \le 4\pi$ is each of the following equal to 1?
    (a) $\sec t$ (b) $\tan t$ (c) $\csc t$ (d) $\cot t$

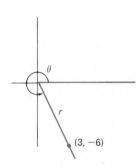

Figure 39

EXAMPLE (Using the a, b, r Definitions) Suppose that the point $(3, -6)$ is on the terminal side of an angle in standard position (Figure 39). Find $\sin \theta$, $\tan \theta$, and $\sec \theta$.

Solution. First we find *r*.

$$r = \sqrt{3^2 + (-6)^2} = \sqrt{45} = 3\sqrt{5}$$

Then

$$\sin \theta = \frac{b}{r} = \frac{-6}{3\sqrt{5}} = -\frac{2}{\sqrt{5}}$$

$$\tan \theta = \frac{b}{a} = \frac{-6}{3} = -2$$

$$\sec \theta = \frac{r}{a} = \frac{3\sqrt{5}}{3} = \sqrt{5}$$

In Problems 25–28, find sin θ, tan θ, and sec θ, assuming that the given point is on the terminal side of θ.

25. $(5, -12)$ 26. $(7, 24)$ 27. $(-1, -2)$

28. $(-3, 2)$

29. If $\tan \theta = \frac{3}{4}$ and θ is an angle in the first quadrant, find $\sin \theta$ and $\sec \theta$. *Hint:* The point $(4, 3)$ is on the terminal side of θ.

30. If $\tan \theta = \frac{3}{4}$ and θ is an angle in the third quadrant, find $\cos \theta$ and $\csc \theta$. *Hint:* The point $(-4, -3)$ is on the terminal side of θ.

31. If $\sin \theta = \frac{5}{13}$ and θ is an angle in the second quadrant, find $\cos \theta$ and $\cot \theta$. *Hint:* A point with y-coordinate 5 and $r = 13$ is on the terminal side of θ. Thus the x-coordinate must be -12.

32. If $\cos \theta = \frac{4}{5}$ and $\sin \theta < 0$, find $\tan \theta$.

33. Where does the line from the origin to $(5, -12)$ intersect the unit circle?

34. Where does the line from the origin to $(-6, 8)$ intersect the unit circle?

35. Find the angle of inclination of the line $5x + 2y = 6$.

36. Find the equation of the line with angle of inclination $75°$ that passes through $(-2, 4)$.

MISCELLANEOUS PROBLEMS

37. Evaluate without use of a calculator.
   (a) $\sec(7\pi/6)$ (b) $\tan(-2\pi/3)$ (c) $\csc(3\pi/4)$
   (d) $\cot(11\pi/4)$ (e) $\csc(570°)$ (f) $\tan(180.045°)$

[c] 38. Calculate.
   (a) $\tan(\sin 2.4)$ (b) $\cot(\tan 1.49)$ (c) $\csc(\sin 11.8°)$
   (d) $\sec^2(\tan 91.2°)$ (e) $\csc(\tan \pi)$ (f) $\tan[\tan(\tan 1.5)]$

39. If $\csc t = 25/24$ and $\cos t < 0$, find each of the following.
   (a) $\sin t$ (b) $\cos t$ (c) $\tan t$
   (d) $\sec(\frac{\pi}{2} - t)$ (e) $\cot(\frac{\pi}{2} - t)$ (f) $\csc(\frac{\pi}{2} - t)$

40. Show that each of the following are identities.
   (a) $\tan(-t) = -\tan t$ (b) $\sec(-t) = \sec t$ (c) $\csc(-t) = -\csc t$

41. Find the two smallest positive values of t that satisfy each of the following.
   (a) $\tan t = -1$ (b) $\sec t = \sqrt{2}$ (c) $|\csc t| = 1$

42. Find the angle of inclination of the line that is perpendicular to the line $4x + 3y = 9$.

43. Write each of the following in terms of sines and cosines and simplify.

(a) $\dfrac{\sec \theta \csc \theta}{\tan \theta + \cot \theta}$

(b) $(\tan \theta)(\cos \theta - \csc \theta)$

(c) $\dfrac{(1 + \tan \theta)^2}{\sec^2 \theta}$

(d) $\dfrac{\sec \theta \cot \theta}{\sec^2 \theta - \tan^2 \theta}$

(e) $\dfrac{\cot \theta - \tan \theta}{\csc \theta - \sec \theta}$

(f) $\tan^4 \theta - \sec^4 \theta$

44. Let θ be a first quadrant angle. Express each of the other five trigonometric functions in terms of $\sin \theta$ alone.

45. Use the identities of Problem 37 in Section 4–3, namely,

$$\sin(t + \pi) = -\sin t \quad \text{and} \quad \cos(t + \pi) = -\cos t$$

to establish each of the following identities.

(a) $\tan(t + \pi) = \tan t$ (b) $\cot(t + \pi) = \cot t$
(c) $\sec(t + \pi) = -\sec t$ (d) $\csc(t + \pi) = -\csc t$

Note: From (a) and (b), we conclude that tangent and cotangent are periodic with period π.

46. Show that $|\sec t| \geq 1$ and $|\csc t| \geq 1$ for all t for which these functions are defined.

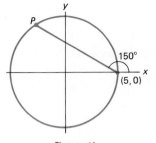

Figure 40

47. If $\tan \theta = \frac{5}{12}$ and $\sin \theta < 0$, evaluate $\cos^2 \theta - \sin^2 \theta$.

48. A wheel of radius 5, centered at the origin, is rotating counterclockwise at a rate of 1 radian per second. At $t = 0$, a speck of dirt on the rim is at $(5, 0)$. What are the coordinates of the speck at time t?

49. At $t = 2\pi/3$, the speck in Problem 44 came loose and flew off along the tangent line. Where did it hit the x-axis?

50. Find the coordinates of P in Figure 40.

51. The face of a clock is in the xy-plane with center at the origin and 12 on the positive y-axis. Both hands of the clock are 5 units long.
(a) Find the slope of the minute hand at 2:24.
(b) Find the slope of the line through the tips of both hands at 12:50.

52. From an airplane h miles above the surface of the earth (a sphere of radius 3960 miles), I can just see a bright light on the horizon d miles away. If I measure the angle of depression of the light as 2.1°, help me determine d and h.

53. A wheel of radius 20 centimeters is used to drive a wheel of radius 50 centimeters by means of a belt that fits around the wheels. How long is the belt if the centers of the two wheels are 100 centimeters apart?

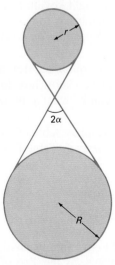

Figure 41

54. **TEASER** Express the length L of the crossed belt that intersects in angle 2α and fits around wheels of radius r and R (Figure 41) in terms of r, R, and α.

t(rad.)	Sin t	Tan t	Cot t	Cos t
.40	.38942	.42279	2.3652	.92106
.41	.39861	.43463	2.3008	.91712
.42	.40776	.44657	2.2393	.91309
.43	.41687	.45862	2.1804	.90897
.44	.42594	.47078	2.1241	.90475
.45	.43497	.48306	2.0702	.90045
.46	.44395	.49545	2.0184	.89605
.47	.45289	.50797	1.9686	.89157
.48	.46178	.52061	1.9208	.88699
.49	.47063	.53339	1.8748	.88233
.50	.47943	.54630	1.8305	.87758

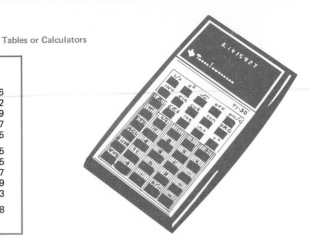

4-5   Finding Values of the Trigonometric Functions

In order to make significant use of the trigonometric functions, we will have to be able to calculate their values for angles other than the special angles we have considered. The simplest procedure is to press the right key on a calculator and read the answer. About the only thing to remember is to make sure the calculator is in the right mode, degree or radian, depending on what we want.

Even though calculators are becoming standard equipment for most mathematics and science students, we think you should also know how to use tables. That is the subject we take up now. We might call it "what to do when your battery goes dead."

The opening display gives a small portion of a five-place table of values for sin t, tan t, cot t, and cos t. (The complete table appears as Table D at the back of the book.) From it we read the following:

$$\sin .44 = .42594 \qquad \tan .44 = .47078$$
$$\cot .44 = 2.1241 \qquad \cos .44 = .90475$$

These results are not exact; they have been rounded off to five significant digits. Keep in mind that you can think of sin .44 in two ways, as the sine of the number .44 or, if you like, as the sine of an angle of radian measure .44.

Table D appears to have two defects. First, t is given only to 2 decimal places. If we need sin .44736, we have to round or perhaps to interpolate (see page 266).

$$\sin .44736 \approx \sin .45 = .43497$$

A more serious defect appears to be the fact that values of t go only to 2.00.

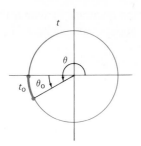

Figure 42

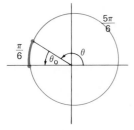

Figure 43

This limitation evaporates once we learn about reference angles and reference numbers, our next topic.

REFERENCE ANGLES AND REFERENCE NUMBERS

Let θ be any angle in standard position and let t be its radian measure. Associated with θ is an acute angle θ_0, called the **reference angle** and defined to be the smallest positive angle between the terminal side of θ and the x-axis (Figure 42). The radian measure t_0 of θ_0 is called the **reference number** corresponding to t. For example, the reference number for $t = 5\pi/6$ is $t_0 = \pi/6$ (Figure 43). Once we know t_0, we can find $\sin t$, $\cos t$, and so on, no matter what t is. Here is how we do it.

Examine the four diagrams in Figure 44.

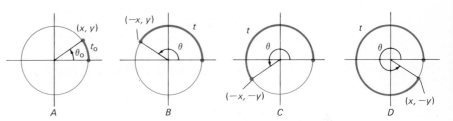

Figure 44

Each angle θ in B, C, and D has θ_0 as its reference angle and, of course, each t has t_0 as its reference number. Now we make a crucial observation. In each case, the point on the unit circle corresponding to t has the same coordinates, except for sign, as the point corresponding to t_0. It follows from this that

$$\sin t = \pm\sin t_0 \qquad \cos t = \pm\cos t_0$$

with the $+$ or $-$ sign being determined by the quadrant in which the terminal side of the angle falls. For example,

$$\sin \frac{5\pi}{6} = \sin \frac{\pi}{6} \qquad \cos \frac{5\pi}{6} = -\cos \frac{\pi}{6}$$

or, in degree notation,

$$\sin 150° = \sin 30° \qquad \cos 150° = -\cos 30°$$

We chose the plus sign for the sine and the minus sign for the cosine because in the second quadrant the sine function is positive, whereas the cosine function is negative.

What we have just said applies to all six trigonometric functions. If T stands for any one of them, then

$$\boxed{T(t) = \pm T(t_0) \quad \text{and} \quad T(\theta) = \pm T(\theta_0)}$$

with the plus or minus sign being determined by the quadrant in which the terminal side of θ lies. Of course $T(t_0)$ itself is always nonnegative since $0 \le t_0 \le \pi/2$.

EXAMPLES

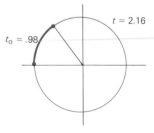

Figure 45

If we wish to calculate cos 2.16 using tables, we must first find the reference number for 2.16. Approximating π by 3.14, we find that (see Figure 45)

$$t_0 = 3.14 - 2.16 = .98$$

and thus, using Table D,

$$\cos 2.16 = -\cos .98 = -.55702$$

Notice we chose the minus sign because the cosine is negative in quadrant II.

To calculate tan 24.95 is slightly more work. First we remove as large a multiple of 2π as possible from 24.95. Using 6.28 for 2π, we get

$$24.95 = 3(6.28) + 6.11$$

The reference number for 6.11 is (see Figure 46)

$$t_0 = 6.28 - 6.11 = .17$$

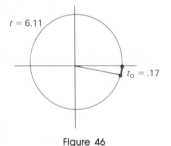

Figure 46

Thus

$$\tan 24.95 = \tan 6.11 = -\tan .17 = -.17166$$

We choose the minus sign because the tangent is negative in quadrant IV.

Now use your pocket calculator to find tan 24.95 the easy way. Be sure you put it in radian mode. You will get $-.18480$ instead of $-.17166$, a rather large discrepancy. Whom should you believe? We suggest that you trust your calculator. The reason we were so far off is that 6.28 is a rather poor approximation for 2π, and multiplying it by 3 made matters worse. Had we used 6.2832 for 2π, we would have obtained $t_0 = .1828$ and tan 24.95 $= -.18486$.

Problem Set 4-5

Find the value of each of the following using Table C or Table D.

1. sin 1.38	2. cos .67	3. cos 42.8°	4. tan 18.0°
5. cot .82	6. tan 1.11	7. sin 68.3°	8. cot 49.6°

EXAMPLE A (Finding Reference Numbers)   Find the reference number t_0 for each of the following values of t.

(a) $t = 20.59$ (b) $t = \dfrac{5\pi}{2} - .92$

$$\begin{array}{r} 3 \\ 6.28\overline{)20.59} \\ \underline{18.84} \\ 1.75 \end{array}$$

Figure 47

Solution.

(a) To get rid of the irrelevant multiples of 2π, we divide 20.59 by 6.28 ($2\pi \approx 6.28$), obtaining 1.75 as remainder (Figure 47). Since 1.75 is between $\pi/2$ and π, we subtract it from π. Thus

$$t_0 \approx \pi - 1.75 \approx 3.14 - 1.75 = 1.39$$

(b) Since $5\pi/2 - .92 = 2\pi + \pi/2 - .92$, it follows that

$$t_0 = \frac{\pi}{2} - .92 \approx 1.57 - .92 = .65$$

Find the reference number t_0 if t has the given value. Use 3.14 for π.

9. 1.84	10. 2.14	11. 3.54	12. 3.74
13. 5.18	14. 6.08	15. 10.48	16. 8.38
17. −1.12	18. −1.86	19. −2.64	20. −4.24

Find the reference number for each of the following. You may leave your answer in terms of π.

21. $13\pi/8$	22. $37\pi/36$	23. $40\pi/3$	24. $-11\pi/5$
25. $3\pi + .24$	26. $3\pi/2 + .17$	27. $3\pi - .24$	28. $3\pi/2 - .17$
29. $11\pi/2$	30. 26π		

Find the value of each of the following using Table D and $\pi = 3.14$. Calculators will give slightly different results because of this crude approximation to π.

31. cos 1.42	32. sin .97	33. tan 1.39	34. cot .08
35. sin 2.14	36. cos 3.08	37. cot 5.62	38. tan 4.11
39. cos(−2.54)	40. sin(−4.18)		

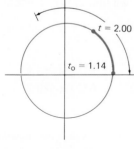

Figure 48

EXAMPLE B (Finding t When sin t or cos t Is Given) Find 2 values of t between 0 and 2π for which (a) $\sin t = .90863$; (b) $\cos t = -.95824$.

Solution.

(a) We get $t = 1.14$ directly from Table D (or using a calculator). Since the sine is also positive in quadrant II, we seek a value of t between $\pi/2$ and π for which 1.14 is the reference number (Figure 48). Only one number fits the bill:

$$\pi - 1.14 \approx 3.14 - 1.14 = 2.00$$

(b) We know that $\cos t_0 = .95824$ and so $t_0 = .29$. Now the cosine is negative in quadrants II and III. Thus we are looking for two numbers between $\pi/2$ and $3\pi/2$ with .29 as reference number (Figure 49). One is $\pi - .29 \approx 3.14 - .29 = 2.85$, and the other $\pi + .29 \approx 3.14 + .29 = 3.43$.

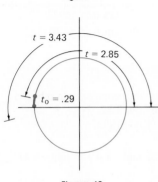

Figure 49

Find two values of t between 0 and 2π for which the given equality holds.

41. $\sin t = .94898$ 42. $\cos t = .72484$ 43. $\cos t = -.08071$

44. $\sin t = -.48818$ 45. $\tan t = 4.9131$ 46. $\cot t = 1.4007$

47. $\tan t = -3.6021$ 48. $\cot t = -.47175$

Find the reference angle (in degrees) for each of the following angles. For example, the reference angle for $\theta = 124.1°$ *is* $\theta_0 = 180° - 124.1° = 55.9°$.

49. $139.6°$ 50. $218.1°$ 51. $348.7°$

52. $375.4°$ 53. $-99.8°$ 54. $-224.4°$

EXAMPLE C (Finding sin θ, cos θ, and so on, When θ Is Any Angle Given in Degrees)   Find the value of each of the following.
(a) $\cos 214.6°$ (b) $\cot 658°$

Solution.   So far, we have used Table C to find the sine, cosine, and so on, of positive angles measuring less than 90°. Here we do this for angles of arbitrary (degree) measure.
(a) The reference angle is

$$214.6° - 180° = 34.6°$$

$$\cos 214.6° = -\cos 34.6° = -.8231$$

We used the minus sign since cosine is negative in quadrant III.
(b) First we reduce our angle by 360°

$$658° = 360° + 298°$$

The reference angle for 298° is $360° - 298°$, or 62°. In the column with cot at the bottom and 62° at the right, we find .5317. Therefore $\cot 658° = -.5317$.

CAUTION

$\cos 99° = \cos 81°$
          $= .1564$

$\cos 99° = -\cos 81°$
          $= -.1564$
Be sure to assign the correct sign.

Find the value of each of the following.

55. $\sin 156.1°$ 56. $\cos 138.7°$ 57. $\tan 348.9°$ 58. $\cot 224.9°$

59. $\cos(-66.1°)$ 60. $\sin 487°$ 61. $\cos 441.3°$ 62. $\sin 180.2°$

63. $\cot(-134°)$ 64. $\tan 311.6°$

Find two different degree values of θ between 0° and 360° for which the given equality holds.

65. $\sin \theta = .3633$ 66. $\cos \theta = .9907$ 67. $\tan \theta = .4942$

68. $\cot \theta = 1.2799$ 69. $\cos \theta = -.9085$ 70. $\sin \theta = -.2045$

MISCELLANEOUS PROBLEMS

71. Use Table C or D to find each of the following. You may approximate π by 3.14.
(a) $\cos 5.63$ (b) $\sin 10.34$ (c) $\tan 8.42$
(d) $\sin 311.3°$ (e) $\tan(-411°)$ (f) $\cos 1989°$

72. Use Tables C *and* D to calculate.
(a) $\sin(\cos 134°)$ (b) $\sin[(\tan 1.5)°]$ (c) $\tan(-5.4°) + \tan(-5.4)$

73. Calculate.
(a) $\cos(\sin 2.42°)$ (b) $\cos^3(\sin^2 2.42)$ (c) $\sqrt{\tan 4.21 + \ln(\sin 7.12)}$

74. Use Table D and $\pi = 3.14$ to find two values of t between 0 and 2π for which each of the following is true.
(a) $\sin t = .62879$ (b) $\cos t = -.90045$ (c) $\tan t = -4.4552$

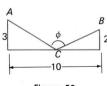

[c] 75. **Find two values** of t between 0 and 2π for which each statement is true, giving your answers correct to 6 decimal places.
(a) $\sin t = .62879$ (b) $\cos t = .34176$ (c) $\tan t = -3.14159$
Note: On many calculators, you would press .62879 [INV] [sin] to get one answer to (a).

76. If $\pi/2 < t < \pi$, then $t_0 = \pi - t$. In a similar manner, express t_0 in terms of t in each case.
(a) $3\pi/2 < t < 2\pi$ (b) $5\pi < t < 11\pi/2$ (c) $-2\pi < t < -3\pi/2$

77. If $0° < \phi < 90°$, express the reference angle θ_0 in terms of ϕ in each case.
(a) $\theta = 180° + \phi$ (b) $\theta = 270° - \phi$ (c) $\theta = \phi - 90°$

78. Without using tables or a calculator, round to the nearest degree the smallest positive angle θ satisfying $\tan \theta = -40,000$.

79. If θ is a fourth quadrant angle whose terminal side coincides with the line $3x + 5y = 0$, find $\sin \theta$.

[c] 80. In calculus, you will learn that

$$\sin t = t - \frac{t^3}{3!} + \frac{t^5}{5!} - \frac{t^7}{7!} + \cdots$$

and

$$\cos t = 1 - \frac{t^2}{2!} + \frac{t^4}{4!} - \frac{t^6}{6!} + \cdots$$

Here, $n! = 1 \cdot 2 \cdot 3 \cdots n$ (for example, $2! = 1 \cdot 2 = 2$ and $3! = 1 \cdot 2 \cdot 3 = 6$). These series are used to construct Tables C and D. If we use just the first three terms in the sine series, we obtain

$$\sin t \approx t - \frac{t^3}{6} + \frac{t^5}{120} = \left[\left(\frac{t^2}{120} - \frac{1}{6} \right) t^2 + 1 \right] t$$

Use the first three terms of these series to approximate each of the following and compare with the corresponding value in Table D.
(a) $\sin(.1)$ (b) $\sin(.4)$ (c) $\cos(.2)$

81. Determine ϕ in Figure 50 so that the path ACB has minimum length.

82. **TEASER**   Let α, β, and γ be acute angles such that $\tan \alpha = 1$, $\tan \beta = 2$, and $\tan \gamma = 3$.
(a) Use your calculator to approximate $\alpha + \beta + \gamma$.
(b) Make a conjecture about the exact value of $\alpha + \beta + \gamma$.
(c) Construct a clever geometric diagram to prove your conjecture.

Figure 50

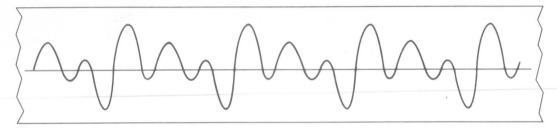

When heart beats, brain activity, or sound waves from a musical instrument are changed into visual images by means of an oscilloscope, they give a regular repetitive pattern which may look something like the diagram above. This repetitive behavior is a characteristic feature of the graphs of the trigonometric functions. In fact, almost any repetitive pattern can be approximated by appropriate combinations of the trigonometric functions.

4-6 Graphs of the Trigonometric Functions

Recall that to graph $y = f(x)$, we first construct a table of values of ordered pairs (x, y), then plot the corresponding points, and finally connect those points with a smooth curve. Here we want to graph $y = \sin t$, $y = \cos t$, and so on, and we will follow a similar procedure. Notice that we use t rather than x as the independent variable because we used t as the variable (radian measure of an angle) in our definition of the trigonometric functions.

We begin with the graphs of the sine and cosine functions. You should become so well acquainted with these two graphs that you can sketch them quickly whenever you need them. This will aid you in two ways. First, these graphs will help you remember many of the important properties of the sine and cosine functions. Second, knowing them will help you graph other more complicated trigonometric functions.

THE GRAPH OF $y = \sin t$

We begin with a table of values (Figure 51).

t	0	$\dfrac{\pi}{6}$	$\dfrac{\pi}{4}$	$\dfrac{\pi}{3}$	$\dfrac{\pi}{2}$	$\dfrac{3\pi}{4}$	π	$\dfrac{5\pi}{4}$	$\dfrac{3\pi}{2}$	$\dfrac{7\pi}{4}$	2π
$y = \sin t$	0	$\dfrac{1}{2}$	$\dfrac{\sqrt{2}}{2}$	$\dfrac{\sqrt{3}}{2}$	1	$\dfrac{\sqrt{2}}{2}$	0	$-\dfrac{\sqrt{2}}{2}$	-1	$-\dfrac{\sqrt{2}}{2}$	0

Figure 51

We have listed values of t between 0 and 2π. That is sufficient to graph one period (shown in Figure 52 as a heavy curve). From there on we can continue the curve indefinitely in either direction in a repetitive fashion, for we learned earlier that $\sin(t + 2\pi) = \sin t$.

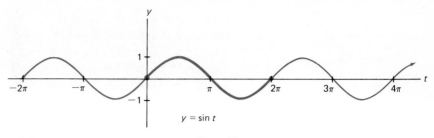

Figure 52

THE GRAPH OF $y = \cos t$

The cosine function is a copycat; its graph is just like that of the sine function but pushed $\pi/2$ units to the left (Figure 53).

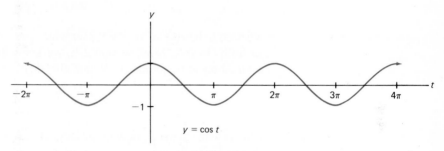

Figure 53

To see that the graph of the cosine function is correct, we might make a table of values and proceed as we did for the sine function. In fact, we ask you to do just that in Problem 1. Alternatively, we can show that

$$\cos t = \sin\left(t + \frac{\pi}{2}\right)$$

This follows directly from identities we have observed earlier.

$$\sin\left(t + \frac{\pi}{2}\right) = \sin\left(\frac{\pi}{2} - (-t)\right)$$

$$= \cos(-t) \qquad \text{(cofunction identity)}$$

$$= \cos t \qquad \text{(cosine is even)}$$

PROPERTIES EASILY OBSERVED FROM THESE GRAPHS

1. Both sine and cosine are periodic with 2π as period.
2. $-1 \leq \sin t \leq 1$ and $-1 \leq \cos t \leq 1$.
3. $\sin t = 0$ if $t = -\pi, 0, \pi, 2\pi$, and so on.
   $\cos t = 0$ if $t = -\pi/2, \pi/2, 3\pi/2$, and so on.

4. $\sin t > 0$ in quadrants I and II.
   $\cos t > 0$ in quadrants I and IV.
5. $\sin(-t) = -\sin t$ and $\cos(-t) = \cos t$.
   The sine is an odd function; its graph is symmetric with respect to the origin. The cosine is an even function; its graph is symmetric with respect to the y-axis.
6. We can see immediately where the sine and cosine functions are increasing and where they are decreasing. For example, the sine function decreases for $\pi/2 \le t \le 3\pi/2$.

THE GRAPH OF $y = \tan t$

Since the tangent function is defined by

$$\tan t = \frac{\sin t}{\cos t}$$

we need to beware of values of t for which $\cos t = 0$: $-\pi/2$, $\pi/2$, $3\pi/2$, and so forth. In fact, from Section 2-5, we know that we should expect vertical asymptotes at these places. Notice also that

$$\tan(-t) = \frac{\sin(-t)}{\cos(-t)} = \frac{-\sin t}{\cos t} = -\tan t$$

which means that the graph of the tangent will be symmetric with respect to the origin. Using these two pieces of information, a small table of values, and the fact that the tangent is periodic, we obtain the graph in Figure 54.

To confirm that the graph is correct near $t = \pi/2$, we suggest looking at Table D. Notice that the tan t steadily increases until at $t = 1.57$, we read tan $t = 1255.8$. But as t takes the short step to 1.58, tan t takes a tremendous plunge to -108.65. In that short space, t has passed through $\pi/2 \approx 1.5708$ and tan t has shot up to celestial heights only to fall to a bottomless pit, from which, however, it manages to escape as t moves to the right.

While we knew the tangent would have to repeat itself every 2π units since the sine and cosine do this, we now notice that it actually repeats itself on intervals of length π. Since the word *period* denotes the length of the shortest interval after which a function repeats itself, the tangent function has period π. For an algebraic demonstration, see Problem 45 of Section 4-4.

THE GRAPH OF $y = \sec t$

Since $\sec t = 1/\cos t$, one way of getting the graph of the secant is by graphing the cosine and then taking reciprocals of the y-coordinates (Figure 55). Note that since $\cos t = 0$ at $t = -\pi/2$, $\pi/2$, $3\pi/2$, and so on, the graph of sec t must have vertical asymptotes at these points.

Just like the cosine, the secant is an even function; that is, $\sec(-t) = \sec t$. And, like the cosine, secant has period 2π. However, notice that if $\cos t$ increases or decreases throughout an interval, sec t does just the opposite. For example, $\cos t$ decreases for $0 < t < \pi/2$, whereas sec t increases there.

t	0	$\frac{\pi}{4}$	$\frac{\pi}{3}$	$\frac{\pi}{2}$	$\frac{2\pi}{3}$	$\frac{3\pi}{4}$	π	$\frac{5\pi}{4}$	$\frac{3\pi}{2}$	$\frac{7\pi}{4}$	2π
$y = \tan t$	0	1	$\sqrt{3}$	undefined	$-\sqrt{3}$	-1	0	1	undefined	-1	0

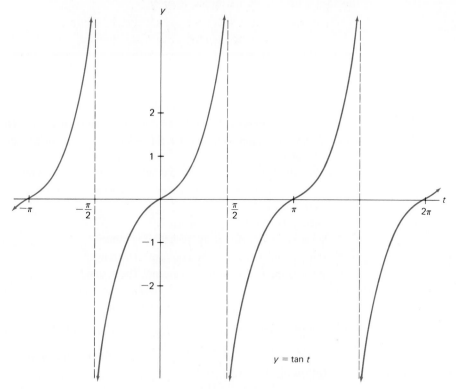

Figure 54

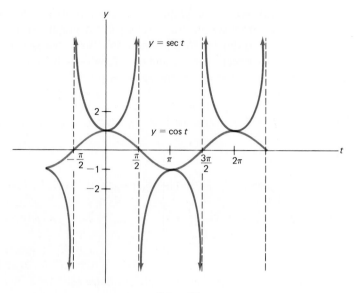

Figure 55

Be sure to study Example A below. It introduces the graph of $y = A \sin Bt$. This important topic is explored more fully in Section 6-5 in connection with simple harmonic motion.

Problem Set 4-6

1. Make a table of values and then sketch the graph of $y = \cos t$.
2. What real numbers constitute the domain of the cosine? The range?
3. Sketch the graph of $y = \cot t$ for $-2\pi \le t \le 2\pi$, being sure to show the asymptotes.
4. What real numbers constitute the entire domain of the cotangent? The range?
5. Using the corresponding fact about the cosine, demonstrate algebraically that $\sec(t + 2\pi) = \sec t$.
6. Sketch the graph of $y = \csc t$.
7. What is the domain of the secant? The range?
8. What is the domain of the cosecant? The range?
9. What is the period of the cotangent? The secant?
10. On the interval $-2\pi \le t \le 2\pi$, where is the cotangent increasing?
11. Which is true: $\cot(-t) = \cot t$ or $\cot(-t) = -\cot t$?
12. Which is true: $\csc(-t) = \csc t$ or $\csc(-t) = -\csc t$?

EXAMPLE A (Some Sine-Related Graphs) Sketch the graph of each of the following for $-2\pi \le t \le 4\pi$.
    (a) $y = 2 \sin t$ (b) $y = \sin 2t$ (c) $y = 3 \sin 4t$

Solution.
    (a) We could graph $y = 2 \sin t$ from a table of values. It is easier, though, to graph $\sin t$ (dotted graph below) and then multiply the ordinates by 2 (Figure 56). Since the graph bobs up and down between $y = -2$ and $y = 2$, we say that it has an **amplitude** of 2. The period is 2π, the same as for $\sin t$.

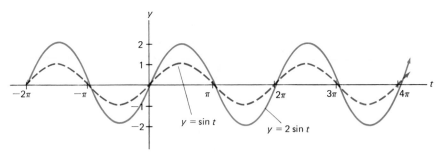

Figure 56

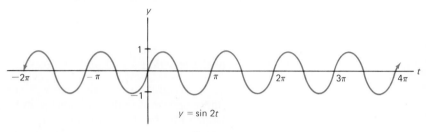

t	$-\pi$	$-\dfrac{3\pi}{4}$	$-\dfrac{\pi}{2}$	$-\dfrac{\pi}{4}$	$-\dfrac{\pi}{12}$	0	$\dfrac{\pi}{12}$	$\dfrac{\pi}{4}$	$\dfrac{\pi}{2}$	$\dfrac{3\pi}{4}$	π
$2t$	-2π	$-\dfrac{3\pi}{2}$	$-\pi$	$-\dfrac{\pi}{2}$	$-\dfrac{\pi}{6}$	0	$\dfrac{\pi}{6}$	$\dfrac{\pi}{2}$	π	$\dfrac{3\pi}{2}$	2π
$\sin 2t$	0	1	0	-1	$-\dfrac{1}{2}$	0	$\dfrac{1}{2}$	1	0	-1	0

y = sin 2t

Figure 57

(b) Here a table of values is advisable, since this is our first example of this type (Figure 57). This graph goes through a complete cycle as t increases from 0 to π; that is, the period of sin $2t$ is π instead of 2π as it was for sin t. The amplitude is 1, just as for sin t.

(c) We can save a lot of work once we recognize how the character of the graph of A sin Bt (and A cos Bt) is determined by the numbers A and B ($B > 0$). The amplitude (which tells how far the graph rises and falls from its median position) is given by $|A|$. The period is given by $2\pi/B$. Thus for $y = 3$ sin $4t$, the amplitude is 3 and the period is $2\pi/4 = \pi/2$. For a quick sketch, we use these two numbers to determine the high and low points and the t-intercepts, connecting these points with a smooth, wavelike curve (Figure 58).

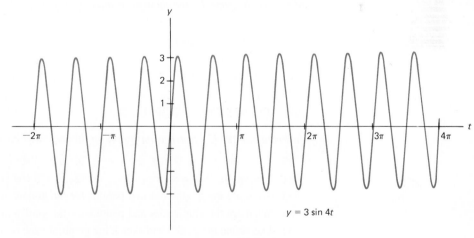

y = 3 sin 4t

Figure 58

In Problems 13–22, determine the amplitude and the period. Then sketch the graph on the indicated interval.

13. $y = 3 \cos t, \; -\pi \le t \le \pi$

14. $y = \frac{1}{2} \cos t, \; -\pi \le t \le \pi$

15. $y = -\sin t, \; -\pi \le t \le \pi$

16. $y = -2 \cos t, \; -\pi \le t \le \pi$

17. $y = \cos 4t, \; -\pi \le t \le \pi$

18. $y = \cos 3t, \; -\pi/2 \le t \le \pi/2$

19. $y = 2 \sin \frac{1}{2} t, \; -2\pi \le t \le 2\pi$

20. $y = 3 \sin \frac{1}{3} t, \; -3\pi \le t \le 3\pi$

21. $y = 2 \cos 3t, \; -\pi \le t \le \pi$

22. $y = 4 \sin 3t, \; -\pi \le t \le \pi$

EXAMPLE B (Graphing Sums of Trigonometric Functions)   Sketch the graph of the equation $y = 2 \sin t + \cos 2t$.

Solution.   We graph $y = 2 \sin t$ and $y = \cos 2t$ on the same coordinate plane (these appear as dotted-line curves in Figure 59) and then add ordinates. Notice that for any t, the ordinates (y-values) of the dotted curves are added to obtain the desired ordinate. The graph of $y = 2 \sin t + \cos 2t$ is quite different from the separate (dotted) graphs but it does repeat itself; it has period 2π.

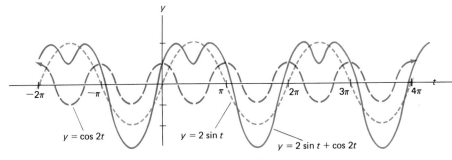

Figure 59

Sketch each graph by the method of adding ordinates. Show at least one complete period.

23. $y = 2 \sin t + \cos t$

24. $y = \sin t + 2 \cos t$

25. $y = \sin 2t + \cos t$

26. $y = \sin t + \cos 2t$

27. $y = \sin \frac{1}{2} t + \frac{1}{2} \sin t$

28. $y = \cos \frac{1}{2} t + \frac{1}{2} \cos t$

MISCELLANEOUS PROBLEMS

In Problems 29–32, sketch each graph on the indicated interval.

29. $y = -\cos t, \; -\pi \le t \le \pi$

30. $y = 3 \sin t, \; -\pi \le t \le \pi$

31. $y = \sin 4t, \; 0 \le t \le \pi$

32. $y = 3 \cos \frac{1}{2} t, \; -2\pi \le t \le 2\pi$

33. What are the amplitudes and periods for the graphs in Problems 29 and 31?

34. What are the amplitudes and periods for the graphs in Problems 30 and 32?

35. Determine the period and sketch the graph of each of the following, showing at least three periods.

(a) $y = \tan 2t$

(b) $y = 3 \tan(t/2)$

36. Follow the directions of Problem 35.
    (a) $y = 2 \cot 2t$ (b) $y = \sec 3t$

37. Sketch, using the same axes, the graphs of
    (a) $f(t) = \sin t$;   (b) $g(t) = 3 + \sin t$;   (c) $h(t) = \sin(t - \pi/4)$.

38. Sketch, using the same axes, the graphs of
    (a) $f(t) = \cos t$;   (b) $g(t) = -2 + \cos t$;   (c) $h(t) = \cos(t + \pi/3)$.

39. Sketch the graph of $y = \cos 3t + 2 \sin t$ for $-\pi \leq t \leq \pi$. Use the method of adding ordinates.

40. Sketch the graph of $y = t + \sin t$ on $-4\pi \leq t \leq 4\pi$. Use the method of adding ordinates.

☐ 41. **Sketch the graph** of $y = t - \cos t$ for $0 \leq t \leq 6$ by actually calculating the y values corresponding to $t = 0, .5, 1, 1.5, 2, 2.5, \ldots, 6$.

42. By sketching the graphs of $y = t$ and $y = 3 \sin t$ on the same coordinate axes, determine approximately all solutions of $t = 3 \sin t$.

43. The strength I of current (in amperes) in a wire of an alternating current circuit might satisfy

$$I = 30 \sin(120\pi t)$$

where time t is measured in seconds.
    (a) What is the period?
    (b) How many cycles (periods) are there in one second?
    (c) What is the maximum strength of the current?

☐ 44. Sketch the graph of $y = (\sin t)/t$ on $-3\pi \leq t \leq 3\pi$. Be sure to plot several points for t near 0 (for example, $t = -.5, -.2, -.1, .1, .2, .5$). What value does y seem to approach as t approaches 0?

45. Consider $y = \sin(1/t)$ on the interval $0 < t \leq 1$.
    (a) Where does its graph cross the t-axis?
    (b) Evaluate y for $t = 2/\pi, 2/3\pi, 2/5\pi, 2/7\pi, \ldots$.
    (c) Sketch the graph as best you can, using a large unit on the t-axis.

46. TEASER   How many solutions does each equation have on the indicated interval for t?
    (a) $\sin t = t/60, t \geq 0$
    (b) $\sin(1/t) = t/60, t \geq .06$
    (c) $\sin(1/t) = t/60, t > 0$

Chapter Summary

The word **trigonometry** means triangle measurement. In its elementary historical form, it is the study of how to **solve** right triangles when appropriate information is given. The main tools are the three trigonometric ratios $\sin \theta$, $\cos \theta$, and $\tan \theta$, which were first defined only for acute angles θ.

In order to give the subject its modern general form, we first generalized the notion of an angle θ, allowing θ to have arbitrary size and measuring it either in **degrees** or **radians.** Such an angle θ can be placed in **standard position** in a coordinate system, where it will cut off an arc of directed length t (the radian measure of θ) stretching from $(1, 0)$ to (x, y) on the unit circle. This allowed us to make the key definitions

$$\sin \theta = \sin t = y \qquad \cos \theta = \cos t = x$$

on which all of modern trigonometry rests.

From the above definitions, we derived several identities, of which the most important is

$$\sin^2 t + \cos^2 t = 1$$

We also defined four additional functions

$$\tan t = \frac{\sin t}{\cos t} \qquad \cot t = \frac{\cos t}{\sin t}$$

$$\sec t = \frac{1}{\cos t} \qquad \csc t = \frac{1}{\sin t}$$

To evaluate the trigonometric functions, we may use either a scientific calculator or Tables C and D in the Appendix. If the tables are used, the notions of **reference angle** and **reference number** become important. Finally we graphed several of the trigonometric functions, noting especially their **periodic** behavior.

Chapter Review Problem Set

1. Solve the following right triangles ($\gamma = 90°$).
   (a) $\alpha = 47.1°$, $c = 36.9$ (b) $a = 417$, $c = 573$

2. At a distance of 10 feet from a wall, the angle of elevation of the top of a mural with respect to eye level is $18°$ and the corresponding angle of depression of the bottom is $10°$. How high is the mural?

3. Change $33°$ to radians. Change $9\pi/4$ radians to degrees.

4. How far does a wheel of radius 30 centimeters roll along level ground in making 100 revolutions?

5. Calculate each of the following without use of tables or a calculator.
   (a) $\sin(7\pi/6)$ (b) $\cos(11\pi/6)$
   (c) $\tan(13\pi/4)$ (d) $\sin(41\pi/6)$

6 Evaluate.
   (a) $\sin 411°$ (b) $\cos 1312°$
   (c) $\tan 5.77$ (d) $\sin 13.12$

7. Write in terms of $\sin t$.
   (a) $\sin(-t)$ (b) $\sin(t + 4\pi)$
   (c) $\sin(\pi + t)$ (d) $\cos\left(\dfrac{\pi}{2} - t\right)$

8. For what values of t between 0 and 2π is
   (a) $\cos t > 0$ (b) $\cos 2t > 0$?

9. If $(-5, -12)$ is on the terminal side of an angle θ in standard position, find
   (a) $\cot \theta$;   (b) $\sec \theta$.

10. If $\sin \theta = \frac{2}{3}$ and θ is a second quadrant angle, find $\tan \theta$.

11. Sketch the graph of $y = 3 \cos 2t$ for $-\pi \leq t \leq 2\pi$.

12. Sketch the graph of $y = \sin t + \sin 2t$ using the method of adding ordinates.

13. What is the range of the sine function? Of the cosecant function?

14. Using the facts that the sine function is odd and the cosine function is even, show that cotangent is an odd function.

15. Give the general definition of $\cos t$ based on the unit circle.

For just as in nature itself there is no middle ground between truth and falsehood, so in rigorous proofs one must either establish his point beyond doubt, or else beg the question inexcusably. There is no chance of keeping one's feet by invoking limitations, distinctions, verbal distortions, or other mental acrobatics. One must with a few words and at the first assault become Caesar or nothing at all.

Galileo

CHAPTER 5

Trigonometric Identities and Equations

5-1 · Identities

Complicated combinations of the six trigonometric functions occur often in mathematics. It is important that we, like the professor above, be able to write a complicated trigonometric expression in a simpler or more convenient form. To do this requires two things. We must be good at algebra and we must know the fundamental identities of trigonometry.

THE FUNDAMENTAL IDENTITIES

We list eleven fundamental identities, which should be memorized.

1. $\tan t = \dfrac{\sin t}{\cos t}$

2. $\cot t = \dfrac{\cos t}{\sin t} = \dfrac{1}{\tan t}$

3. $\sec t = \dfrac{1}{\cos t}$

4. $\csc t = \dfrac{1}{\sin t}$

5. $\sin^2 t + \cos^2 t = 1$

6. $1 + \tan^2 t = \sec^2 t$

7. $1 + \cot^2 t = \csc^2 t$

8. $\sin\left(\dfrac{\pi}{2} - t\right) = \cos t$

9. $\cos\left(\dfrac{\pi}{2} - t\right) = \sin t$

10. $\sin(-t) = -\sin t$

11. $\cos(-t) = \cos t$

We have seen all these identities before. The first four are actually definitions; the others were established either in the text or the problem sets of Sections 4-3 and 4-4.

PROVING NEW IDENTITIES

The professor's work in our opening cartoon can be viewed in two ways. The more likely way of looking at it is that she wanted to simplify the complicated expression

$$(\sec t + \tan t)(1 - \sin t)$$

But it could be that someone had conjectured that

$$(\sec t + \tan t)(1 - \sin t) = \cos t$$

is an identity and that the professor was trying to prove it. It is this second concept we want to discuss now.

Suppose someone claims that a certain equation is an identity—that is, true for all values of the variable for which both sides make sense. How can you check on such a claim? The procedure used by the professor is one we urge you to follow. Start with the more complicated looking side and try to use a chain of equalities to produce the other side.

Suppose we wish to prove that

$$\sin t + \cos t \cot t = \csc t$$

is an identity. We begin with the left side and rewrite it step by step, using algebra and the fundamental identities, until we get the right side.

$$\sin t + \cos t \cot t = \sin t + \cos t \left(\frac{\cos t}{\sin t}\right)$$

$$= \frac{\sin^2 t + \cos^2 t}{\sin t}$$

$$= \frac{1}{\sin t}$$

$$= \csc t$$

When proving that an equation is an identity, it pays to look before you leap. Changing the more complicated side to sines and cosines, as in the above example, is often the best thing to do. But not always. Sometimes the simpler side gives us a clue as to how we should reshape the other side. For example, the left side of

$$\tan t = \frac{(\sec t - 1)(\sec t + 1)}{\tan t}$$

suggests that we try to rewrite the right side in terms of $\tan t$. This can be done by multiplying out the numerator and making use of the fundamental identity $\sec^2 t = 1 + \tan^2 t$.

$$\frac{(\sec t - 1)(\sec t + 1)}{\tan t} = \frac{\sec^2 t - 1}{\tan t} = \frac{\tan^2 t}{\tan t} = \tan t$$

Proving an identity is something like a game in that it requires a strategy. If one strategy does not work, try another, and still another, until you succeed.

A POINT OF LOGIC

Why all the fuss about working with just one side of a conjectured identity? First of all, it offers good practice in manipulating trigonometric expressions. But there is also a point of logic. If you operate on both sides simultaneously, you are in effect assuming that you already have an identity. That is bad logic and it can be corrected only by carefully checking that each step is reversible. To make this point clear, consider the equation

$$1 - x = x - 1$$

which is certainly not an identity. Yet when we square both sides we get

$$1 - 2x + x^2 = x^2 - 2x + 1$$

which is an identity. The trouble here is that squaring both sides is not a reversible operation.

The situation contrasts sharply with our procedure for solving conditional equations, in which we often perform an operation on both sides. For example, in the case of the equation

$$\sqrt{2x + 1} = 1 - x$$

we even square both sides. We are protected from error here by checking our solutions in the original equation.

Problem Set 5-1

1. Express entirely in terms of $\sin t$.
   (a) $\cos^2 t$ (b) $\tan t \cos t$
   (c) $\dfrac{3}{\csc^2 t} + 2\cos^2 t - 2$ (d) $\cot^2 t$

2. Express entirely in terms of $\cos t$.
   (a) $\sin^2 t$ (b) $\tan^2 t$
   (c) $\csc^2 t$ (d) $(1 + \sin t)^2 - 2\sin t$

3. Express entirely in terms of $\tan t$.
   (a) $\cot^2 t$ (b) $\sec^2 t$
   (c) $\sin t \sec t$ (d) $2\sec^2 t - 2\tan^2 t + 1$

4. Express entirely in terms of $\sec t$.
   (a) $\cos^4 t$ (b) $\tan^2 t$
   (c) $\tan t \csc t$ (d) $\tan^2 t - 2\sec^2 t + 5$

EXAMPLE A (Proving Identities) Prove that the following is an identity.

$$\csc \theta - \sin \theta = \cot \theta \cos \theta$$

Solution.   The left side looks inviting, as $\csc \theta = 1/\sin \theta$. We rewrite it a step at a time.

$$\csc \theta - \sin \theta = \frac{1}{\sin \theta} - \sin \theta$$

$$= \frac{1 - \sin^2 \theta}{\sin \theta}$$

$$= \frac{\cos^2 \theta}{\sin \theta}$$

$$= \frac{\cos \theta}{\sin \theta} \cdot \cos \theta$$

$$= \cot \theta \cos \theta$$

Prove that each of the following is an identity.

5. $\cos t \sec t = 1$

6. $\sin t \csc t = 1$

7. $\tan x \cot x = 1$

8. $\sin x \sec x = \tan x$

9. $\cos y \csc y = \cot y$

10. $\tan y \cos y = \sin y$

11. $\cot \theta \sin \theta = \cos \theta$

12. $\dfrac{\sec \theta}{\csc \theta} = \tan \theta$

13. $\dfrac{\tan u}{\sin u} = \dfrac{1}{\cos u}$

14. $\dfrac{\sin u}{\csc u} + \dfrac{\cos u}{\sec u} = 1$

15. $(1 + \sin z)(1 - \sin z) = \dfrac{1}{\sec^2 z}$

16. $(\sec z - 1)(\sec z + 1) = \tan^2 z$

17. $(1 - \sin^2 x)(1 + \tan^2 x) = 1$

18. $(1 - \cos^2 x)(1 + \cot^2 x) = 1$

19. $\sec t - \sin t \tan t = \cos t$

20. $\sin t(\csc t - \sin t) = \cos^2 t$

21. $\dfrac{\sec^2 t - 1}{\sec^2 t} = \sin^2 t$

22. $\dfrac{1 - \csc^2 t}{\csc^2 t} = \dfrac{-1}{\sec^2 t}$

23. $\cos t(\tan t + \cot t) = \csc t$

24. $\dfrac{1}{\sin t \cos t} - \dfrac{\cos t}{\sin t} = \tan t$

EXAMPLE B (Expressing All Trigonometric Functions in Terms of One of Them)   If $\pi/2 < t < \pi$, express $\cos t$, $\tan t$, $\cot t$, $\sec t$, and $\csc t$ in terms of $\sin t$.

Solution.   Since $\cos^2 t = 1 - \sin^2 t$ and cosine is negative in quadrant II,

$$\cos t = -\sqrt{1 - \sin^2 t}$$

Also

$$\tan t = \frac{\sin t}{\cos t} = -\frac{\sin t}{\sqrt{1 - \sin^2 t}}$$

$$\cot t = \frac{1}{\tan t} = -\frac{\sqrt{1 - \sin^2 t}}{\sin t}$$

$$\sec t = \frac{1}{\cos t} = -\frac{1}{\sqrt{1 - \sin^2 t}}$$

$$\csc t = \frac{1}{\sin t}$$

25. If $\pi/2 < t < \pi$, express $\sin t$, $\tan t$, $\cot t$, $\sec t$, and $\csc t$ in terms of $\cos t$.

26. If $\pi < t < 3\pi/2$, express $\sin t$, $\cos t$, $\cot t$, $\sec t$, and $\csc t$ in terms of $\tan t$.

27. If $\pi/2 < t < \pi$ and $\sin t = \frac{4}{5}$, find the values of the other five functions for the same value of t. *Hint:* Use the results of Example B.

28. If $\pi < t < 3\pi/2$ and $\tan t = 2$, find $\sin t$, $\cos t$, $\cot t$, $\sec t$, and $\csc t$.

EXAMPLE C (How to Proceed When Neither Side Is Simple) Prove that

$$\frac{\sin t}{1 - \cos t} = \frac{1 + \cos t}{\sin t}$$

is an identity.

Solution. Since both sides are equally complicated, it would seem to make no difference which side we choose to manipulate. We will try to transform the left side into the right side. Seeing $1 + \cos t$ in the numerator of the right side suggests multiplying the left side by $(1 + \cos t)/(1 + \cos t)$.

$$\frac{\sin t}{1 - \cos t} = \frac{\sin t}{1 - \cos t} \cdot \frac{1 + \cos t}{1 + \cos t} = \frac{\sin t(1 + \cos t)}{1 - \cos^2 t}$$

$$= \frac{\sin t(1 + \cos t)}{\sin^2 t}$$

$$= \frac{1 + \cos t}{\sin t}$$

Prove that each of the following is an identity.

29. $\dfrac{\sec t - 1}{\tan t} = \dfrac{\tan t}{\sec t + 1}$

30. $\dfrac{1 - \tan \theta}{1 + \tan \theta} = \dfrac{\cot \theta - 1}{\cot \theta + 1}$

Hint: In Problem 30, multiply numerator and denominator of the left side by $\cot \theta$.

31. $\dfrac{\tan^2 x}{\sec x + 1} = \dfrac{1 - \cos x}{\cos x}$

32. $\dfrac{\cot x}{\csc x + 1} = \dfrac{\csc x - 1}{\cot x}$

33. $\dfrac{\sin t + \cos t}{\tan^2 t - 1} = \dfrac{\cos^2 t}{\sin t - \cos t}$

34. $\dfrac{\sec t - \cos t}{1 + \cos t} = \sec t - 1$

MISCELLANEOUS PROBLEMS

35. Express $[(\sin x + \cos x)^2 - 1]\sec x \csc^3 x$ as follows.
    (a) Entirely in terms of $\sin x$.
    (b) Entirely in terms of $\tan x$.

36. If $\sec t = 8$, find the values of (a) $\cos t$; (b) $\cot^2 t$; (c) $\csc^2 t$.

In Problems 37–56, prove that each equation is an identity. Do this by taking one side and showing by a chain of equalities that it is equal to the other side.

37. $(1 + \tan^2 t)(\cos t + \sin t) = (1 + \tan t)\sec t$

38. $1 - (\cos t + \sin t)(\cos t - \sin t) = 2 \sin^2 t$

39. $2 \sec^2 y - 1 = \dfrac{1 + \sin^2 y}{\cos^2 y}$

40. $(\sin x + \cos x)(\sec x + \csc x) = 2 + \tan x + \cot x$

41. $\dfrac{\cos z}{1 + \cos z} = \dfrac{\sin z}{\sin z + \tan z}$

42. $2 \sin^2 t + 3 \cos^2 t + \sec^2 t = (\sec t + \cos t)^2$

43. $(\csc t + \cot t)^2 = \dfrac{1 + \cos t}{1 - \cos t}$

44. $\sec^4 y - \tan^4 y = \dfrac{1 + \sin^2 y}{\cos^2 y}$

45. $\dfrac{\cos x + \sin x}{\cos x - \sin x} = \dfrac{1 + \tan x}{1 - \tan x}$

46. $\dfrac{1 + \cos x}{1 - \cos x} - \dfrac{1 - \cos x}{1 + \cos x} = 4 \cot x \csc x$

47. $(\sec t + \tan t)(\csc t - 1) = \cot t$

48. $\sec t + \cos t = \sin t \tan t + 2 \cos t$

49. $\dfrac{\cos^3 t + \sin^3 t}{\cos t + \sin t} = 1 - \sin t \cos t$

50. $\dfrac{\tan x}{1 + \tan x} + \dfrac{\cot x}{1 - \cot x} = \dfrac{\tan x + \cot x}{\tan x - \cot x}$

51. $\dfrac{1 - \cos \theta}{1 + \cos \theta} = \left(\dfrac{1 - \cos \theta}{\sin \theta}\right)^2$

52. $\dfrac{(\sec^2 \theta + \tan^2 \theta)^2}{\sec^4 \theta - \tan^4 \theta} = \sec^2 \theta + \tan^2 \theta$

53. $(\csc t - \cot t)^4 (\csc t + \cot t)^4 = 1$

54. $(\sec t + \tan t)^5 (\sec t - \tan t)^6 = \dfrac{1 - \sin t}{\cos t}$

55. $\sin^6 u + \cos^6 u = 1 - 3 \sin^2 u \cos^2 u$

56. $\dfrac{\cos^2 x - \cos^2 y}{\cot^2 x - \cot^2 y} = \sin^2 x \sin^2 y$

57. In a later section, we will learn that

$$\tan 3x = \dfrac{3 \tan x - \tan^3 x}{1 - 3 \tan^2 x}$$

Taking this for granted, show that

$$\cot 3x = \dfrac{3 \cot x - \cot^3 x}{1 - 3 \cot^2 x}$$

Note the similarity in form of these two identities.

58. **TEASER** Generalize Problem 57 by showing that if $\tan kx = f(\tan x)$ and if k is an odd number, then $\cot kx = f(\cot x)$. *Hint:* Let $x = \pi/2 - y$.

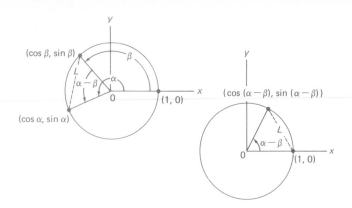

Equal Chords

In the two unit circles at the right, the two dotted chords have the same length, being chords for angles of the same size, namely, $\alpha - \beta$. Out of this simple observation, we can extract several of the most important identities of trigonometry.

5-2 Addition Laws

When you study calculus, you will meet expressions like $\cos(\alpha + \beta)$ and $\sin(\alpha - \beta)$. It will be very important to rewrite these expressions directly in terms of $\sin \alpha$, $\cos \alpha$, $\sin \beta$, and $\cos \beta$. It might be tempting to replace $\cos(\alpha + \beta)$ by $\cos \alpha + \cos \beta$ and $\sin(\alpha - \beta)$ by $\sin \alpha - \sin \beta$, but that would be terribly wrong. To see this, let's try $\alpha = \pi/3$ and $\beta = \pi/6$.

$$\cos\left(\frac{\pi}{3} + \frac{\pi}{6}\right) = \cos\frac{\pi}{2} = 0 \qquad \cos\frac{\pi}{3} + \cos\frac{\pi}{6} = \frac{1}{2} + \frac{\sqrt{3}}{2} \approx 1.4$$

$$\sin\left(\frac{\pi}{3} - \frac{\pi}{6}\right) = \sin\frac{\pi}{6} = .5 \qquad \sin\frac{\pi}{3} - \sin\frac{\pi}{6} = \frac{\sqrt{3}}{2} - \frac{1}{2} \approx .4$$

To obtain correct expressions is the goal of this section.

A KEY IDENTITY

The opening display shows two chords of equal length L. Using the formula for the distance between two points (Figure 1) and the identity $\sin^2 \theta + \cos^2 \theta = 1$, we have the following expression for the square of the chord on the right.

$$L^2 = [\cos(\alpha - \beta) - 1]^2 + \sin^2(\alpha - \beta)$$

$$= \cos^2(\alpha - \beta) - 2\cos(\alpha - \beta) + 1 + \sin^2(\alpha - \beta)$$

$$= [\cos^2(\alpha - \beta) + \sin^2(\alpha - \beta)] + 1 - 2\cos(\alpha - \beta)$$

$$= 2 - 2\cos(\alpha - \cdot\beta)$$

A similar calculation for the square of the chord on the left gives

$$L^2 = (\cos \alpha - \cos \beta)^2 + (\sin \alpha - \sin \beta)^2$$

> **Distance Formula**
>
> The distance between (x_1, y_1) and (x_2, y_2) is
> $$\sqrt{(x_2 - x_1)^2 + (y_2 - y_1)^2}$$

Figure 1

$$= \cos^2 \alpha - 2 \cos \alpha \cos \beta + \cos^2 \beta + \sin^2 \alpha - 2 \sin \alpha \sin \beta + \sin^2 \beta$$

$$= 1 - 2 \cos \alpha \cos \beta - 2 \sin \alpha \sin \beta + 1$$

$$= 2 - 2(\cos \alpha \cos \beta + \sin \alpha \sin \beta)$$

When we equate these two expressions for L^2, we get our key identity

$$\cos(\alpha - \beta) = \cos \alpha \cos \beta + \sin \alpha \sin \beta$$

Our derivation is based on a picture in which α and β are positive angles with $\alpha > \beta$. Minor modifications would establish the identity for arbitrary angles α and β and hence also for their radian measures s and t. Thus for all real numbers s and t,

$$\boxed{\cos(s - t) = \cos s \cos t + \sin s \sin t}$$

We can use this identity to calculate $\cos(\pi/12)$ by thinking of $\pi/12$ as $\pi/3 - \pi/4$.

$$\cos \frac{\pi}{12} = \cos\left(\frac{\pi}{3} - \frac{\pi}{4}\right) = \cos \frac{\pi}{3} \cos \frac{\pi}{4} + \sin \frac{\pi}{3} \sin \frac{\pi}{4}$$

$$= \frac{1}{2} \cdot \frac{\sqrt{2}}{2} + \frac{\sqrt{3}}{2} \cdot \frac{\sqrt{2}}{2} = \frac{\sqrt{2} + \sqrt{6}}{4} \approx .9659$$

In words, this identity says: *The cosine of a difference is the cosine of the first times the cosine of the second plus the sine of the first times the sine of the second*. It is important to memorize this identity in words so you can easily apply it to $\cos(3u - v)$, $\cos[s - (-t)]$, or even $\cos[(\pi/2 - s) - t]$, as we shall have to do soon.

RELATED IDENTITIES

In the boxed identity above, we replace t by $-t$ and use the fundamental identities $\cos(-t) = \cos t$ and $\sin(-t) = -\sin t$ to get

$$\cos[s - (-t)] = \cos s \cos(-t) + \sin s \sin(-t)$$

$$= \cos s \cos t + (\sin s)(-\sin t)$$

This gives us the **addition law for cosines.**

$$\boxed{\cos(s + t) = \cos s \cos t - \sin s \sin t}$$

We illustrate this law by calculating $\cos(13\pi/12)$.

$$\cos \frac{13\pi}{12} = \cos\left(\frac{3\pi}{4} + \frac{\pi}{3}\right) = \cos \frac{3\pi}{4} \cos \frac{\pi}{3} - \sin \frac{3\pi}{4} \sin \frac{\pi}{3}$$

$$= \frac{-\sqrt{2}}{2} \cdot \frac{1}{2} - \frac{\sqrt{2}}{2} \cdot \frac{\sqrt{3}}{2} = -\frac{\sqrt{2} + \sqrt{6}}{4} \approx -.9659$$

There is also an identity involving $\sin(s + t)$. To derive this identity, we use the cofunction identity $\sin u = \cos(\pi/2 - u)$ to write

$$\sin(s + t) = \cos\left[\frac{\pi}{2} - (s + t)\right] = \cos\left[\left(\frac{\pi}{2} - s\right) - t\right]$$

Then we use our key identity for the cosine of a difference to obtain

$$\cos\left(\frac{\pi}{2} - s\right)\cos t + \sin\left(\frac{\pi}{2} - s\right)\sin t$$

Two applications of cofunction identities give us the result we want, the **addition law for sines.**

$$\boxed{\sin(s + t) = \sin s \cos t + \cos s \sin t}$$

Finally, replacing t by $-t$ in this last result leads to

$$\boxed{\sin(s - t) = \sin s \cos t - \cos s \sin t}$$

If we let $s = \pi/2$ in the latter, we get another important identity—but it is one we already know.

$$\sin\left(\frac{\pi}{2} - t\right) = \sin\frac{\pi}{2}\cos t - \cos\frac{\pi}{2}\sin t = 1 \cdot \cos t - 0 \cdot \sin t = \cos t$$

Problem Set 5-2

Find the value of each expression. Note that in each case, the answers to parts (a) and (b) are different.

1. (a) $\sin\dfrac{\pi}{4} + \sin\dfrac{\pi}{6}$ (b) $\sin\left(\dfrac{\pi}{4} + \dfrac{\pi}{6}\right)$

2. (a) $\cos\dfrac{\pi}{4} + \cos\dfrac{\pi}{6}$ (b) $\cos\left(\dfrac{\pi}{4} + \dfrac{\pi}{6}\right)$

3. (a) $\cos\dfrac{\pi}{4} - \cos\dfrac{\pi}{6}$ (b) $\cos\left(\dfrac{\pi}{4} - \dfrac{\pi}{6}\right)$

4. (a) $\sin\dfrac{\pi}{4} - \sin\dfrac{\pi}{6}$ (b) $\sin\left(\dfrac{\pi}{4} - \dfrac{\pi}{6}\right)$

Use the identities derived in this section to show that the equalities in Problems 5–12 are identities.

5. $\sin(t + \pi) = -\sin t$

6. $\cos(t + \pi) = -\cos t$

7. $\sin\left(t + \dfrac{3\pi}{2}\right) = -\cos t$

8. $\cos\left(t + \dfrac{3\pi}{2}\right) = \sin t$

9. $\sin\left(t - \dfrac{\pi}{2}\right) = -\cos t$

10. $\cos\left(t - \dfrac{\pi}{2}\right) = \sin t$

11. $\cos\left(t + \dfrac{\pi}{3}\right) = \dfrac{1}{2}\cos t - \dfrac{\sqrt{3}}{2}\sin t$

12. $\sin\left(t + \dfrac{\pi}{3}\right) = \dfrac{1}{2}\sin t + \dfrac{\sqrt{3}}{2}\cos t$

EXAMPLE A (Recognizing Expressions as Single Sines or Cosines)   Write as a single sine or cosine.
(a) $\sin \frac{7}{6} \cos \frac{1}{6} + \cos \frac{7}{6} \sin \frac{1}{6}$
(b) $\cos(x + h)\cos h + \sin(x + h)\sin h$

Solution.

(a) By the addition law for sines,
$$\sin \tfrac{7}{6} \cos \tfrac{1}{6} + \cos \tfrac{7}{6} \sin \tfrac{1}{6} = \sin(\tfrac{7}{6} + \tfrac{1}{6}) = \sin \tfrac{4}{3}$$

(b) This we recognize as the cosine of a difference
$$\cos(x + h)\cos h + \sin(x + h)\sin h = \cos[(x + h) - h] = \cos x$$

Write each of the following as a single sine or cosine.

13. $\cos \frac{1}{2} \cos \frac{3}{2} - \sin \frac{1}{2} \sin \frac{3}{2}$

14. $\cos 2 \cos 3 + \sin 2 \sin 3$

15. $\sin \dfrac{7\pi}{8} \cos \dfrac{\pi}{8} + \cos \dfrac{7\pi}{8} \sin \dfrac{\pi}{8}$

16. $\sin \dfrac{5\pi}{16} \cos \dfrac{\pi}{16} - \cos \dfrac{5\pi}{16} \sin \dfrac{\pi}{16}$

17. $\cos 33° \cos 27° - \sin 33° \sin 27°$

18. $\sin 49° \cos 41° + \cos 49° \sin 41°$

19. $\sin(\alpha + \beta) \cos \beta - \cos(\alpha + \beta) \sin \beta$

20. $\cos(\alpha + \beta) \cos(\alpha - \beta) - \sin(\alpha + \beta) \sin(\alpha - \beta)$

EXAMPLE B (Using the Addition Laws)   Suppose that α is a first quadrant angle with $\cos \alpha = \frac{4}{5}$ and β is a second quadrant angle with $\sin \beta = \frac{12}{13}$. Evaluate $\sin(\alpha + \beta)$ and $\cos(\alpha + \beta)$ and then determine the quadrant for $\alpha + \beta$.

Solution.   We are going to need $\sin \alpha$ and $\cos \beta$. We can find them by using the identity $\sin^2 \theta + \cos^2 \theta = 1$, but we have to be careful about signs.
$$\sin \alpha = \sqrt{1 - \cos^2 \alpha} = \sqrt{1 - \tfrac{16}{25}} = \tfrac{3}{5}$$
$$\cos \beta = -\sqrt{1 - \sin^2 \beta} = -\sqrt{1 - \tfrac{144}{169}} = -\tfrac{5}{13}$$

We chose the plus sign in the first case because α is a first quadrant angle and the minus sign in the second because β is a second quadrant angle, where the cosine is negative. Then

$$\sin(\alpha + \beta) = \sin\alpha\cos\beta + \cos\alpha\sin\beta$$

$$= \left(\frac{3}{5}\right)\left(\frac{-5}{13}\right) + \left(\frac{4}{5}\right)\left(\frac{12}{13}\right) = \frac{33}{65}$$

$$\cos(\alpha + \beta) = \cos\alpha\cos\beta - \sin\alpha\sin\beta$$

$$= \left(\frac{4}{5}\right)\left(\frac{-5}{13}\right) - \left(\frac{3}{5}\right)\left(\frac{12}{13}\right) = \frac{-56}{65}$$

Since $\sin(\alpha + \beta)$ is positive and $\cos(\alpha + \beta)$ is negative, $\alpha + \beta$ is a second quadrant angle.

21. If α and β are third quadrant angles with $\sin\alpha = -\frac{4}{5}$ and $\cos\beta = -\frac{5}{13}$, find $\sin(\alpha + \beta)$ and $\cos(\alpha + \beta)$. In what quadrant does the terminal side of $\alpha + \beta$ lie?

22. Let α and β be second quadrant angles with $\sin\alpha = \frac{2}{3}$ and $\sin\beta = \frac{3}{4}$. Find $\sin(\alpha + \beta)$ and $\cos(\alpha + \beta)$ and determine the quadrant for $\alpha + \beta$.

23. Let α be a first quadrant angle with $\sin\alpha = 1/\sqrt{10}$ and β be a second quadrant angle with $\cos\beta = -\frac{1}{2}$. Find $\sin(\alpha - \beta)$ and $\cos(\alpha - \beta)$ and determine the quadrant for $\alpha - \beta$.

24. Let α and β be second and third quadrant angles, respectively, with $\cos\alpha = \cos\beta = -\frac{3}{7}$. Find $\sin(\alpha - \beta)$ and $\cos(\alpha - \beta)$ and determine the quadrant for $\alpha - \beta$.

EXAMPLE C (Tangent Identities)   Verify the **addition law for tangents.**

$$\tan(s + t) = \frac{\tan s + \tan t}{1 - \tan s \tan t}$$

Solution.

$$\tan(s + t) = \frac{\sin(s + t)}{\cos(s + t)}$$

$$= \frac{\sin s \cos t + \cos s \sin t}{\cos s \cos t - \sin s \sin t}$$

$$= \frac{\dfrac{\sin s \cos t}{\cos s \cos t} + \dfrac{\cos s \sin t}{\cos s \cos t}}{\dfrac{\cos s \cos t}{\cos s \cos t} - \dfrac{\sin s \sin t}{\cos s \cos t}}$$

$$= \frac{\tan s + \tan t}{1 - \tan s \tan t}$$

The key step was the third one, in which we divided both the numerator and the denominator by $\cos s \cos t$.

Establish that each equation in Problems 25–28 is an identity.

25. $\tan(s - t) = \dfrac{\tan s - \tan t}{1 + \tan s \tan t}$

26. $\tan(s + \pi) = \tan s$

27. $\tan\left(t + \dfrac{\pi}{4}\right) = \dfrac{1 + \tan t}{1 - \tan t}$

28. $\tan\left(t - \dfrac{\pi}{3}\right) = \dfrac{\tan t - \sqrt{3}}{1 + \sqrt{3}\tan t}$

MISCELLANEOUS PROBLEMS

29. Express in terms of $\sin t$ and $\cos t$.
    (a) $\sin(t - \tfrac{5}{6}\pi)$ (b) $\cos(\tfrac{\pi}{6} - t)$

30. Express $\tan(\theta + \tfrac{3}{4}\pi)$ in terms of $\tan \theta$.

31. Let α and β be first and third quadrant angles, respectively, with $\sin \alpha = \tfrac{2}{3}$ and $\cos \beta = -\tfrac{1}{3}$. Evaluate each of the following exactly.
    (a) $\cos \alpha$ (b) $\sin \beta$ (c) $\cos(\alpha + \beta)$
    (d) $\sin(\alpha - \beta)$ (e) $\tan(\alpha + \beta)$ (f) $\sin(2\beta)$

32. If $0 \le t \le \pi/2$ and $\cos(t + \pi/6) = .8$, find the exact value of $\sin t$ and $\cos t$.
    Hint: $t = (t + \pi/6) - \pi/6$.

33. Evaluate each of the following (the easy way).
    (a) $\sin(t + \pi/3)\cos t - \cos(t + \pi/3)\sin t$
    (b) $\cos 175° \cos 25° + \sin 175° \sin 25°$
    (c) $\sin t \cos(1 - t) + \cos t \sin(1 - t)$

34. Find the exact value of $\cos 85° \cos 40° + \cos 5° \cos 50°$.

35. Show that each of the following is an identity.
    (a) $\sin(x + y)\sin(x - y) = \sin^2 x - \sin^2 y$
    (b) $\dfrac{\sin(x + y)}{\cos(x - y)} = \dfrac{\tan x + \tan y}{1 + \tan x \tan y}$
    (c) $\dfrac{\cos 5t}{\sin t} - \dfrac{\sin 5t}{\cos t} = \dfrac{\cos 6t}{\sin t \cos t}$

36. Show that the following are identities.
    (a) $\cot(u + v) = \dfrac{\cot u \cot v - 1}{\cot u + \cot v}$
    (b) $\dfrac{\sin(u + v)}{\sin(u - v)} = \dfrac{\tan u + \tan v}{\tan u - \tan v}$
    (c) $\dfrac{\cos 2t}{\sin t} + \dfrac{\sin 2t}{\cos t} = \csc t$

37. Let θ be the smallest counterclockwise angle from the line $y = m_1 x + b_1$ to the line $y = m_2 x + b_2$, where $m_1 m_2 \ne -1$. Show that

$$\tan \theta = \dfrac{m_2 - m_1}{1 + m_1 m_2}$$

38. Find the counterclockwise angle θ from the line $3x - 4y = 1$ to the line $2x + 6y = 3$. (See Problem 37.)

39. Use the addition and subtraction laws (the four boxed formulas of this section) to prove the following **product identities**.
    (a) $\cos s \cos t = \tfrac{1}{2}[\cos(s + t) + \cos(s - t)]$
    (b) $\sin s \sin t = -\tfrac{1}{2}[\cos(s + t) - \cos(s - t)]$
    (c) $\sin s \cos t = \tfrac{1}{2}[\sin(s + t) + \sin(s - t)]$
    (d) $\cos s \sin t = \tfrac{1}{2}[\sin(s + t) - \sin(s - t)]$

40. Use the identities of Problem 39 to prove the following **factoring identities**.
    Hint: Let $u = s + t$ and $v = s - t$.

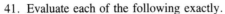

(a) $\cos u + \cos v = 2 \cos \dfrac{u + v}{2} \cos \dfrac{u - v}{2}$

(b) $\cos u - \cos v = -2 \sin \dfrac{u + v}{2} \sin \dfrac{u - v}{2}$

(c) $\sin u + \sin v = 2 \sin \dfrac{u + v}{2} \cos \dfrac{u - v}{2}$

(d) $\sin u - \sin v = 2 \cos \dfrac{u + v}{2} \sin \dfrac{u - v}{2}$

Figure 2

41. Evaluate each of the following exactly.
   (a) $\cos 105° \cos 45°$ (b) $\sin 15° - \sin 75°$
   (c) $\cos 15° + \cos 30° + \cos 45° + \cos 60° + \cos 75°$

42. Show that each of the following is an identity.
   (a) $\dfrac{\cos 9t + \cos 3t}{\sin 9t - \sin 3t} = \cot 3t$ (b) $\dfrac{\sin 3u + \sin 7u}{\cos 3u + \cos 7u} = \tan 5u$
   (c) $\cos 10\beta + \cos 2\beta + 2 \cos 8\beta \cos 6\beta = 4 \cos^2 6\beta \cos 2\beta$

43. Stack three identical squares and consider angles α, β, and γ as shown in Figure 2. Prove that $\alpha + \beta = \gamma$.

44. **TEASER** Consider an oblique triangle (no right angles) with angles α, β, and γ. Prove that

$$\tan \alpha + \tan \beta + \tan \gamma = \tan \alpha \tan \beta \tan \gamma$$

I could do this problem if $\sin (2t) = 2 \sin t$

Wishful Thinking

"Wishful thinking is imagining good things you don't have . . . . [It] may be bad as too much salt is bad in the soup and even a little garlic is bad in the chocolate pudding. I mean, wishful thinking may be bad if there is too much of it or in the wrong place, but it is good in itself and may be a great help in life and in problem solving."

George Polya
in *Mathematical Discovery*

5-3 Double-Angle and Half-Angle Formulas

George Polya would agree that the student in our opening panel is wishing for too much. And there is a better way than wishing to get formulas for $\sin 2t$ and $\cos 2t$. All we have to do is to think of $2t$ as $t + t$ and apply the addition laws of the previous section.

$$\sin(t + t) = \sin t \cos t + \cos t \sin t = 2 \sin t \cos t$$
$$\cos(t + t) = \cos t \cos t - \sin t \sin t = \cos^2 t - \sin^2 t$$

DOUBLE-ANGLE FORMULAS

We have just derived two very important results. They are called *double-angle formulas*, though double-number formulas would perhaps be more appropriate.

$$\sin 2t = 2 \sin t \cos t$$
$$\cos 2t = \cos^2 t - \sin^2 t$$

Suppose $\sin t = \frac{2}{5}$ and $\pi/2 < t < \pi$. Then we can calculate both $\sin 2t$ and $\cos 2t$, but we must first find $\cos t$. Since $\pi/2 < t < \pi$, the cosine is negative, and therefore

$$\cos t = -\sqrt{1 - \sin^2 t} = -\sqrt{1 - \frac{4}{25}} = -\frac{\sqrt{21}}{5}$$

The double-angle formulas now give

$$\sin 2t = 2 \sin t \cos t = 2\left(\frac{2}{5}\right)\left(\frac{-\sqrt{21}}{5}\right) = \frac{-4\sqrt{21}}{25} \approx -.73$$

$$\cos 2t = \cos^2 t - \sin^2 t = \left(\frac{-\sqrt{21}}{5}\right)^2 - \left(\frac{2}{5}\right)^2 = \frac{17}{25} \approx .68$$

CAUTION

$\cos 6\theta = 6 \cos \theta$ ~~(crossed out)~~

$\cos 6\theta = 2 \cos^2 3\theta - 1$

There are two other forms of the cosine double-angle formula that are often useful. If, in the expression $\cos^2 t - \sin^2 t$, we replace $\cos^2 t$ by $1 - \sin^2 t$, we obtain

$$\cos 2t = 1 - 2 \sin^2 t$$

and, alternatively, if we replace $\sin^2 t$ by $1 - \cos^2 t$, we have

$$\cos 2t = 2 \cos^2 t - 1$$

Of course, in all that we have done, we may replace the number t by the angle θ; hence the name double-angle formulas.

Once we grasp the generality of the four boxed formulas, we can write numerous others that follow from them. For example,

$$\sin 6\theta = 2 \sin 3\theta \cos 3\theta$$

$$\cos 4u = \cos^2 2u - \sin^2 2u$$

$$\cos t = 1 - 2 \sin^2\left(\frac{t}{2}\right)$$

$$\cos t = 2 \cos^2\left(\frac{t}{2}\right) - 1$$

The last two of these identities lead us directly to the half-angle formulas.

HALF-ANGLE FORMULAS

In the identity $\cos t = 1 - 2 \sin^2(t/2)$, we solve for $\sin(t/2)$.

$$2 \sin^2\left(\frac{t}{2}\right) = 1 - \cos t$$

$$\sin^2\left(\frac{t}{2}\right) = \frac{1 - \cos t}{2}$$

CAUTION

$$\sin 75° = \tfrac{1}{2} \sin 150° = \tfrac{1}{4}$$

$$\sin 75° = \sqrt{\frac{1 - \cos 150°}{2}}$$

$$\boxed{\sin\left(\frac{t}{2}\right) = \pm\sqrt{\frac{1 - \cos t}{2}}}$$

Similarly, if we solve $\cos t = 2 \cos^2(t/2) - 1$ for $\cos(t/2)$, the result is

$$\boxed{\cos\left(\frac{t}{2}\right) = \pm\sqrt{\frac{1 + \cos t}{2}}}$$

In both of these formulas, the choice of the plus or minus sign is determined by the interval on which $t/2$ lies. For example,

$$\cos\left(\frac{5\pi}{8}\right) = \cos\left(\frac{5\pi/4}{2}\right) = -\sqrt{\frac{1 + \cos(5\pi/4)}{2}}$$

$$= -\sqrt{\frac{1 - \sqrt{2}/2}{2}} = -\frac{\sqrt{2 - \sqrt{2}}}{2}$$

We chose the minus sign because $5\pi/8$ corresponds to an angle in quadrant II, where the cosine is negative.

As a second example, suppose that $\cos\theta = .4$, where θ is a fourth quadrant angle. Then we can calculate $\sin(\theta/2)$ and $\cos(\theta/2)$, observing first that $\theta/2$ is necessarily a second quadrant angle.

$$\sin\left(\frac{\theta}{2}\right) = \sqrt{\frac{1 - \cos\theta}{2}} = \sqrt{\frac{1 - .4}{2}} \approx .548$$

$$\cos\left(\frac{\theta}{2}\right) = -\sqrt{\frac{1 + \cos\theta}{2}} = -\sqrt{\frac{1 + .4}{2}} \approx -.837$$

Problem Set 5-3

1. If $\cos t = \frac{4}{5}$ with $0 < t < \pi/2$, show that $\sin t = \frac{3}{5}$. Then use formulas from this section to calculate
   (a) $\sin 2t$;   (b) $\cos 2t$;
   (c) $\cos(t/2)$;   (d) $\sin(t/2)$.

2. If $\sin t = -\frac{2}{3}$ with $3\pi/2 < t < 2\pi$, show that $\cos t = \sqrt{5}/3$. Then calculate
   (a) $\sin 2t$;   (b) $\cos 2t$;
   (c) $\cos(t/2)$;   (d) $\sin(t/2)$.

Use formulas from this section to simplify the expressions in Problems 3–16. For example, $2\sin(.5)\cos(.5) = \sin 1$.

3. $2\sin 5t\cos 5t$ 4. $2\sin 3\theta\cos 3\theta$

5. $\cos^2(3t/2) - \sin^2(3t/2)$ 6. $\cos^2(7\pi/8) - \sin^2(7\pi/8)$

7. $2\cos^2(y/4) - 1$ 8. $2\cos^2(\alpha/3) - 1$

9. $1 - 2\sin^2(.6t)$ 10. $2\sin^2(\pi/8) - 1$

11. $\sin^2(\pi/8) - \cos^2(\pi/8)$ 12. $2\sin(.3)\cos(.3)$

13. $\dfrac{1 + \cos x}{2}$ 14. $\dfrac{1 - \cos y}{2}$

15. $\dfrac{1 - \cos 4\theta}{2}$ 16. $\dfrac{1 + \cos 8u}{2}$

17. Use the half-angle formulas to calculate
    (a) $\sin(\pi/8)$;   (b) $\cos(112.5°)$.

18. Calculate, using half-angle formulas,
    (a) $\cos 67.5°$;   (b) $\sin(\pi/12)$.

19. Use the addition law for tangents (Example C of Section 5-2) to show that
$$\tan 2t = \frac{2\tan t}{1 - \tan^2 t}$$

20. Use the identity of Problem 19 to evaluate $\tan 2t$ given that
    (a) $\tan t = 3$;
    (b) $\cos t = \frac{4}{5}$ and $0 < t < \pi/2$.

21. Use the half-angle formulas for sine and cosine to show that
$$\tan\left(\frac{t}{2}\right) = \pm\sqrt{\frac{1 - \cos t}{1 + \cos t}}$$

22. Use the identity of Problem 21 to evaluate
    (a) $\tan(\pi/8)$;   (b) $\tan 112.5°$.

EXAMPLE (Using Double-Angle and Half-Angle Formulas to Prove New Identities) Prove that the following are identities.

(a) $\sin 3t = 3\sin t - 4\sin^3 t$

(b) $\tan\dfrac{t}{2} = \dfrac{\sin t}{1 + \cos t}$

Solution.

(a) We think of $3t$ as $2t + t$ and use the addition law for sines and then double-angle formulas.

$$\sin 3t = \sin(2t + t)$$
$$= \sin 2t \cos t + \cos 2t \sin t$$
$$= (2\sin t \cos t)\cos t + (1 - 2\sin^2 t)\sin t$$
$$= 2\sin t(1 - \sin^2 t) + \sin t - 2\sin^3 t$$
$$= 2\sin t - 2\sin^3 t + \sin t - 2\sin^3 t$$
$$= 3\sin t - 4\sin^3 t$$

(b) This is the unambiguous form for $\tan(t/2)$ (see Problem 21). To prove it, think of t as $2(t/2)$ and apply double-angle formulas to the right side.

$$\frac{\sin t}{1 + \cos t} = \frac{\sin(2(t/2))}{1 + \cos(2(t/2))}$$
$$= \frac{2\sin(t/2)\cos(t/2)}{1 + 2\cos^2(t/2) - 1}$$
$$= \frac{\sin(t/2)}{\cos(t/2)}$$
$$= \tan\frac{t}{2}$$

Now prove that each of the following is an identity.

23. $\cos 3t = 4\cos^3 t - 3\cos t$

24. $(\sin t + \cos t)^2 = 1 + \sin 2t$

25. $\csc 2t + \cot 2t = \cot t$

26. $\sin^2 t \cos^2 t = \frac{1}{8}(1 - \cos 4t)$

27. $\dfrac{\sin\theta}{1 - \cos\theta} = \cot\dfrac{\theta}{2}$

28. $1 - 2\sin^2\theta = 2\cot 2\theta \sin\theta \cos\theta$

29. $\dfrac{2\tan\alpha}{1 + \tan^2\alpha} = \sin 2\alpha$

30. $\dfrac{1 - \tan^2\alpha}{1 + \tan^2\alpha} = \cos 2\alpha$

31. $\sin 4\theta = 4\sin\theta(2\cos^3\theta - \cos\theta)$

Hint: $4\theta = 2(2\theta)$.

32. $\cos 4\theta = 8\cos^4\theta - 8\cos^2\theta + 1$

MISCELLANEOUS PROBLEMS

33. Write a simple expression for each of the following.
   (a) $2 \sin(x/2) \cos(x/2)$
   (b) $\cos^2 3t - \sin^2 3t$
   (c) $2 \sin^2(y/4) - 1$
   (d) $(\cos 4t - 1)/2$
   (e) $(1 - \cos 4t)/(1 + \cos 4t)$
   (f) $(\sin 6y)/(1 + \cos 6y)$

34. Find the exact value of each of the following.
   (a) $\sin 15° \cos 15°$
   (b) $\cos^2 105° - \sin^2 105°$
   (c) $\sin 15°$
   (d) $\cos 105°$

35. If $\pi < t < 3\pi/2$ and $\cos t = -5/13$, find each value.
   (a) $\sin 2t$
   (b) $\cos(t/2)$
   (c) $\tan(t/2)$

36. If the trigonometric point $P(t)$ on the unit circle has coordinates $(-\frac{3}{5}, \frac{4}{5})$, find the coordinates for each of the following points.
   (a) $P(2t)$
   (b) $P(t/2)$
   (c) $P(4t)$

In Problems 37–52, prove that each equation is an identity.

37. $\cos^4 z - \sin^4 z = \cos 2z$

38. $(1 - \cos 4x)/\tan^2 2x = 2 \cos^2 2x$

39. $1 + (1 - \cos 8t)/(1 + \cos 8t) = \sec^2 4t$

40. $\sec 2t = (\sec^2 t)/(2 - \sec^2 t)$

41. $\tan(\theta/2) - \sin \theta = (-\sin \theta)/(1 + \sec \theta)$

42. $(2 - \sec^2 2\theta) \tan 4\theta = 2 \tan 2\theta$

43. $3 \cos 2t + 4 \sin 2t = (3 \cos t - \sin t)(\cos t + 3 \sin t)$

44. $\csc 2x - \cot 2x = \tan x$

45. $2(\cos 3x \cos x + \sin 3x \sin x)^2 = 1 + \cos 4x$

46. $\dfrac{1 + \sin 2x + \cos 2x}{1 + \sin 2x - \cos 2x} = \cot x$

47. $\tan 3t = \dfrac{3 \tan t - \tan^3 t}{1 - 3 \tan^2 t}$

48. $\cos^4 u = \frac{3}{8} + \frac{1}{2} \cos 2u + \frac{1}{8} \cos 4u$

49. $\sin^4 u + \cos^4 u = \frac{3}{4} + \frac{1}{4} \cos 4u$

50. $\cos^6 u - \sin^6 u = \cos 2u - \frac{1}{4} \sin^2 2u \cos 2u$

51. Prove that $\cos^2 x + \cos^2 2x + \cos^2 3x = 1 + 2 \cos x \cos 2x \cos 3x$ is an identity. *Hint:* Use half-angle formulas and factoring identities.

52. [c] Calculate $(\sin 2t)[3 + (16 \sin^2 t - 16)\sin^2 t] - \sin 6t$ for $t = 1, 2,$ and 3. Guess at an identity and then prove it.

53. If α, β, and γ are the three angles of a triangle, prove that
$$\sin 2\alpha + \sin 2\beta + \sin 2\gamma = 4 \sin \alpha \sin \beta \sin \gamma$$

54. Show that $\cos x \cos 2x \cos 4x \cos 8x \cos 16x = (\sin 32x)/(32 \sin x)$

55. Figure 3 shows two abutting circles of radius 1, one centered at the origin, the other at $(2, 0)$. Find the exact coordinates of P, the point where the line through $(-1, 0)$ and tangent to the second circle meets the first circle.

56. **TEASER** Determine the exact value of
$$\sin 1° \sin 3° \sin 5° \sin 7° \cdots \sin 175° \sin 177° \sin 179°$$

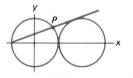

Figure 3

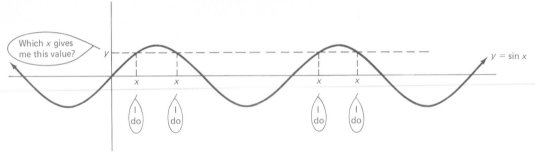

The sine function is not one–to–one.

5-4 Inverse Trigonometric Functions

The diagram above shows why the ordinary sine function does not have an inverse. We learned in Section 2-7 (a section worth reviewing now) that only one-to-one functions have inverses. To be one-to-one means that for each y there is at most one x that corresponds to it. The sine function is about as far from being one-to-one as possible. For each y between -1 and 1, there are infinitely many x's giving that y-value. To make the sine function have an inverse, we will have to restrict its domain drastically.

THE INVERSE SINE

Consider the graph of the sine function again (Figure 4). We want to restrict its domain in such a way that the sine assumes its full range of values but takes on each value only once. There are many possible choices, but the one commonly used is $-\pi/2 \le x \le \pi/2$. Notice the corresponding part of the sine graph below. From now on, whenever we need an inverse sine function, we always assume the domain of the sine has been restricted to $-\pi/2 \le x \le \pi/2$.

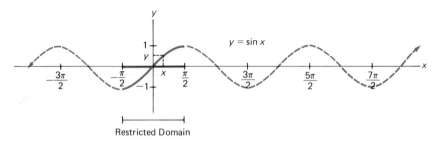

Restricted Domain

Figure 4

Having done this, we see that each y corresponds to exactly one x. We write $x = \sin^{-1} y$ (x is the inverse sine of y). Thus

$$\sin^{-1}\left(\frac{1}{2}\right) = \frac{\pi}{6}$$

$$\sin^{-1}(1) = \frac{\pi}{2}$$

$$\sin^{-1}(-1) = -\frac{\pi}{2}$$

$$\sin^{-1}\left(\frac{-\sqrt{2}}{2}\right) = -\frac{\pi}{4}$$

Please note that $\sin^{-1}y$ does not mean $1/(\sin y)$; you should not think of -1 as an exponent when used as a superscript on a function.

An alternate notation for $x = \sin^{-1}y$ is $x = \arcsin y$ (x is the arcsine of y). This is appropriate notation, since $\pi/6 = \arcsin\frac{1}{2}$ could be interpreted as saying that $\pi/6$ is the arc (on the unit circle) whose sine is $\frac{1}{2}$.

Recall from Section 2-7 that if f is a one-to-one function, then

$$x = f^{-1}(y) \quad \text{if and only if} \quad y = f(x)$$

Here the corresponding statement is

$$x = \sin^{-1}y \quad \text{if and only if} \quad y = \sin x \quad \text{and} \quad -\frac{\pi}{2} \le x \le \frac{\pi}{2}$$

Moreover

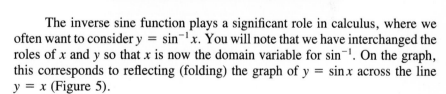

$$\sin(\sin^{-1}y) = y \quad \text{for} \quad -1 \le y \le 1$$

$$\sin^{-1}(\sin x) = x \quad \text{for} \quad -\frac{\pi}{2} \le x \le \frac{\pi}{2}$$

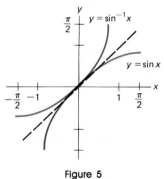

Figure 5

The inverse sine function plays a significant role in calculus, where we often want to consider $y = \sin^{-1}x$. You will note that we have interchanged the roles of x and y so that x is now the domain variable for $\sin^{-1}$. On the graph, this corresponds to reflecting (folding) the graph of $y = \sin x$ across the line $y = x$ (Figure 5).

THE INVERSE COSINE

One look at the graph of $y = \cos x$ should convince you that we cannot restrict the domain of the cosine to the same interval as that for the sine (Figure 6). We choose rather to use the interval $0 \le x \le \pi$, in which the cosine is one-to-one.

Having made the needed restriction, we may reasonably talk about $\cos^{-1}$. Moreover,

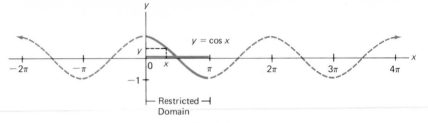

Figure 6

$$x = \cos^{-1}y \quad \text{if and only if} \quad y = \cos x \quad \text{and} \quad 0 \le x \le \pi$$

In particular,

$$\cos^{-1} 1 = 0$$

$$\cos^{-1} \frac{\sqrt{3}}{2} = \frac{\pi}{6}$$

$$\cos^{-1} 0 = \frac{\pi}{2}$$

$$\cos^{-1}(-1) = \pi$$

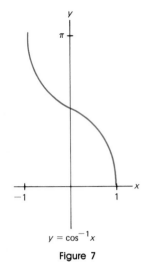

$y = \cos^{-1}x$

Figure 7

The graph of $y = \cos^{-1} x$ is shown in Figure 7. It is the graph of $y = \cos x$ reflected across the line $y = x$.

THE INVERSE TANGENT

To make $y = \tan x$ have an inverse, we restrict x to $-\pi/2 < x < \pi/2$. Thus

$$x = \tan^{-1}y \quad \text{if and only if} \quad y = \tan x \quad \text{and} \quad -\frac{\pi}{2} < x < \frac{\pi}{2}$$

The graphs of the tangent function and its inverse are shown in Figure 8.
Notice that the graph of $y = \tan^{-1}x$ has horizontal asymptotes at $y = \pi/2$ and $y = -\pi/2$.

THE INVERSE SECANT

The secant function has an inverse, provided we restrict its domain to $0 \le x \le \pi$, excluding $\pi/2$. Thus

$$x = \sec^{-1}y \quad \text{if and only if} \quad y = \sec x \quad \text{and} \quad 0 \le x \le \pi, x \ne \frac{\pi}{2}$$

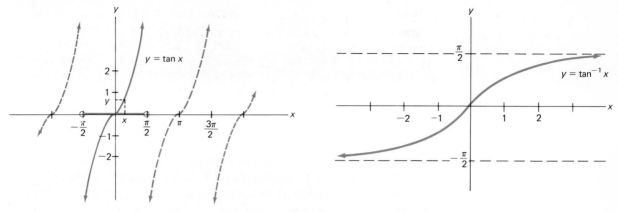

Figure 8

$$\sec^{-1} 2 = \frac{1}{\cos 2}$$

$$\sec^{-1} 2 = \cos^{-1} \frac{1}{2} = \frac{\pi}{3}$$

(Some authors choose to restrict the domain of the secant to $\{x : 0 \le x < \pi/2$ or $\pi \le x < 3\pi/2\}$. For this reason, check an author's definition before using any stated fact about the inverse secant.) The graphs of $y = \sec x$ and $y = \sec^{-1} x$ are shown in Figure 9.

Since the secant and cosine are reciprocals of each other, it is not surprising that $\sec^{-1} x$ is related to $\cos^{-1} x$. In fact, for every x in the domain of $\sec^{-1} x$, we have

$$\sec^{-1} x = \cos^{-1} \left(\frac{1}{x} \right)$$

This follows from the fact that each side is limited in value to the interval 0 to π and has the same cosine.

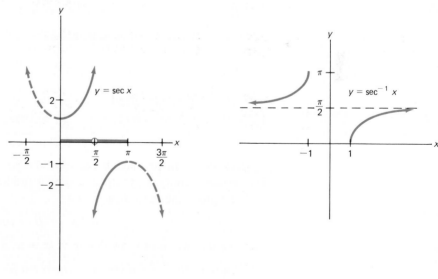

Figure 9

$$\cos(\sec^{-1} x) = \frac{1}{\sec(\sec^{-1} x)} = \frac{1}{x}$$

The two other inverse trigonometric functions, $\cot^{-1} x$ and $\csc^{-1} x$, are of less importance. They are introduced in Problem 66.

INVERSE TRIGONOMETRIC FUNCTIONS AND CALCULATORS

Most scientific calculators have been programmed to give values of $\sin^{-1}$, $\cos^{-1}$, and $\tan^{-1}$ that are consistent with the definitions we have given. For example, to obtain $\sin^{-1}(.32)$ on many calculators, press the buttons .32 $\boxed{\text{INV}}$ $\boxed{\text{sin}}$; on other calculators, there is a button marked $\sin^{-1}$ and you simply press .32 $\boxed{\sin^{-1}}$. Normally, you cannot get $\sec^{-1}$ directly; instead you must use the identity $\sec^{-1} x = \cos^{-1}(1/x)$. Of course, in every case you must put your calculator in the appropriate mode, depending on whether you want the answer in degrees or radians.

Here is a problem that requires thinking in addition to use of a calculator. Find all values of t between 0 and 2π for which $\tan t = 2.12345$. A calculator (in radian mode) immediately gives the solution

$$t_0 = \tan^{-1}(2.12345) = 1.13067$$

However, there is another solution having t_0 as reference number, namely,

$$t = \pi + t_0 = 4.27227$$

THREE IDENTITIES

Here are three identities connecting sines, cosines, and their inverses.

(i) $\qquad\qquad\qquad\qquad \cos(\sin^{-1} x) = \sqrt{1 - x^2}$

(ii) $\qquad\qquad\qquad\qquad \sin(\cos^{-1} x) = \sqrt{1 - x^2}$

(iii) $\qquad\qquad\qquad \sin^{-1} x + \cos^{-1} x = \dfrac{\pi}{2}$

To prove the first identity, we let $\theta = \sin^{-1} x$. Remember that this means that $x = \sin \theta$, with $-\pi/2 \le \theta \le \pi/2$. Then

$$\cos(\sin^{-1} x) = \cos \theta = \pm\sqrt{1 - \sin^2 \theta} = \pm\sqrt{1 - x^2}$$

Finally, choose the plus sign because $\cos \theta$ is positive for $-\pi/2 \le \theta \le \pi/2$. The second identity is proved in a similar fashion.

To prove the third identity, let

$$\alpha = \sin^{-1} x \qquad \beta = \cos^{-1} x$$

and note that we must show that $\alpha + \beta = \pi/2$. Now

$$\sin(\alpha + \beta) = \sin \alpha \cos \beta + \cos \alpha \sin \beta$$

$$= \sin(\sin^{-1} x) \cos(\cos^{-1} x) + \cos(\sin^{-1} x) \sin(\cos^{-1} x)$$

$$= x \cdot x + \sqrt{1 - x^2} \cdot \sqrt{1 - x^2}$$
$$= x^2 + 1 - x^2$$
$$= 1$$

From this, we conclude that $\alpha + \beta$ is either $\pi/2$ or some number that differs from $\pi/2$ by a multiple of 2π. But since $-\pi/2 \le \alpha \le \pi/2$ and $0 \le \beta \le \pi$, it follows that

$$-\frac{\pi}{2} \le \alpha + \beta \le \frac{3\pi}{2}$$

The only possibility on this interval is $\alpha + \beta = \pi/2$.

Problem Set 5-4

Find the exact value of each of the following (without using a calculator).

1. $\sin^{-1}(\sqrt{3}/2)$ 2. $\cos^{-1} \frac{1}{2}$ 3. $\arcsin(\sqrt{2}/2)$

4. $\arccos(\sqrt{2}/2)$ 5. $\tan^{-1} 0$ 6. $\tan^{-1} 1$

7. $\tan^{-1} \sqrt{3}$ 8. $\tan^{-1}(\sqrt{3}/3)$ 9. $\arccos(-\frac{1}{2})$

10. $\arcsin(-\frac{1}{2})$ 11. $\sec^{-1} \sqrt{2}$ 12. $\sec^{-1}(-2/\sqrt{3})$

Use a calculator or Table D to find each value (in radians) in Problems 13–18.

13. $\sin^{-1} .21823$ 14. $\cos^{-1} .30582$ 15. $\sin^{-1}(-0.21823)$

16. $\cos^{-1}(-0.30582)$ 17. $\tan^{-1} .20660$ 18. $\tan^{-1}(1.2602)$

© 19. Calculate, using $\sec^{-1} x = \cos^{-1}(1/x)$.
     (a) $\sec^{-1} 1.4263$ (b) $\sec^{-1}(-2.6715)$

© 20. Calculate.
     (a) $\sec^{-1}(\pi + 1)$ (b) $\sec^{-1}(-\sqrt{5}/2)$

Solve for t, where $0 \le t < 2\pi$. Use a calculator if you have one.

21. $\sin t = .3416$ 22. $\cos t = .9812$

23. $\tan t = 3.345$ 24. $\sec t = 1.342$

Find the following without the use of tables or a calculator.

25. $\sin(\sin^{-1} \frac{2}{3})$ 26. $\cos(\cos^{-1}(-\frac{1}{4}))$

27. $\tan(\tan^{-1} 10)$ 28. $\cos^{-1}(\cos(\pi/2))$

29. $\sin^{-1}(\sin(\pi/3))$ 30. $\tan^{-1}(\tan(\pi/4))$

31. $\sin^{-1}(\cos(\pi/4))$ 32. $\cos^{-1}(\sin(-\pi/6))$

33. $\cos(\sin^{-1} \frac{4}{5})$ 34. $\sin(\cos^{-1} \frac{3}{5})$

     Hint: Use the identities established on page 244.

35. $\cos(\tan^{-1} \frac{1}{2})$ 36. $\cos(\tan^{-1}(-\frac{3}{4}))$

37. $\cos(\sec^{-1} 3)$ 38. $\sec(\cos^{-1}(-.4))$

39. $\sec^{-1}(\sec(2\pi/3))$ 40. $\sec(\sec^{-1} 2.56)$

□ *Use a calculator to find each value.*

41. $\cos(\sin^{-1}(-.2564))$

42. $\tan^{-1}(\sin 14.1)$

43. $\sin^{-1}(\cos 1.12)$

44. $\cos^{-1}(\cos^{-1} .91)$

45. $\tan(\sec^{-1} 2.5)$

46. $\sec^{-1}(\sin 1.67)$

EXAMPLE A (Complicated Evaluations Involving Inverses) Evaluate
(a) $\sin(2 \cos^{-1} \frac{2}{3})$;   (b) $\tan(\tan^{-1} 2 + \sin^{-1} \frac{4}{5})$.

Solution.

(a) Let $\theta = \cos^{-1}(\frac{2}{3})$ so that $\cos \theta = \frac{2}{3}$ and

$$\sin \theta = \sqrt{1 - \cos^2 \theta} = \sqrt{1 - \frac{4}{9}} = \frac{\sqrt{5}}{3}$$

Then apply the double-angle formula for $\sin 2\theta$ as indicated below.

$$\sin\left(2 \cos^{-1} \frac{2}{3}\right) = \sin 2\theta$$

$$= 2 \sin \theta \cos \theta$$

$$= 2 \frac{\sqrt{5}}{3} \cdot \frac{2}{3}$$

$$= \frac{4}{9}\sqrt{5}$$

(b) Let $\alpha = \tan^{-1} 2$ and $\beta = \sin^{-1}(\frac{4}{5})$ and apply the identity

$$\tan(\alpha + \beta) = \frac{\tan \alpha + \tan \beta}{1 - \tan \alpha \tan \beta}$$

Now $\tan \alpha = 2$ and

$$\tan \beta = \frac{\sin \beta}{\cos \beta} = \frac{\frac{4}{5}}{\sqrt{1 - \left(\frac{4}{5}\right)^2}} = \frac{\frac{4}{5}}{\frac{3}{5}} = \frac{4}{3}$$

Therefore

$$\tan(\alpha + \beta) = \frac{2 + \frac{4}{3}}{1 - 2 \cdot \frac{4}{3}} = -2$$

Evaluate by using the method of Example A, not by using a calculator.

47. $\sin(2 \cos^{-1} \frac{3}{5})$

48. $\sin(2 \cos^{-1} \frac{1}{2})$

49. $\cos(2 \sin^{-1}(-\frac{2}{3}))$

50. $\tan(2 \tan^{-1} \frac{1}{3})$

51. $\sin(\cos^{-1} \frac{3}{5} + \cos^{-1} \frac{5}{13})$

52. $\tan(\tan^{-1} \frac{1}{2} + \tan^{-1}(-3))$

53. $\cos(\sec^{-1} \frac{3}{2} - \sec^{-1} \frac{4}{3})$

54. $\sin(\sin^{-1} \frac{4}{5} + \sec^{-1} 3)$

EXAMPLE B (More Identities)   Show that

$$\cos(2\ \tan^{-1} x) = \frac{1 - x^2}{1 + x^2}$$

Solution.   We will apply the double-angle formula

$$\cos 2\theta = 2 \cos^2 \theta - 1$$

Here $\theta = \tan^{-1} x$, so that $x = \tan \theta$. Then

$$\cos(2\ \tan^{-1} x) = \cos(2\theta)$$

$$= 2 \cos^2 \theta - 1$$

$$= \frac{2}{\sec^2 \theta} - 1$$

$$= \frac{2}{1 + \tan^2 \theta} - 1$$

$$= \frac{2}{1 + x^2} - 1$$

$$= \frac{1 - x^2}{1 + x^2}$$

Show that each of the following is an identity.

55. $\tan(\sin^{-1} x) = \dfrac{x}{\sqrt{1 - x^2}}$

56. $\sin(\tan^{-1} x) = \dfrac{x}{\sqrt{1 + x^2}}$

57. $\tan(2\ \tan^{-1} x) = \dfrac{2x}{1 - x^2}$

58. $\cos(2\ \sin^{-1} x) = 1 - 2x^2$

59. $\cos(2\ \sec^{-1} x) = \dfrac{2}{x^2} - 1$

60. $\sec(2\ \tan^{-1} x) = \dfrac{1 + x^2}{1 - x^2}$

MISCELLANEOUS PROBLEMS

61. Without using tables or a calculator, find each value (in radians).
    (a) $\arcsin(-\sqrt{3}/2)$ (b) $\tan^{-1}(-\sqrt{3})$ (c) $\sec^{-1}(-2)$

[c] 62. Calculate each of the following (radian mode).
    (a) $\dfrac{2 \arccos(.956)}{3 \arcsin(-.846)}$ (b) $.3624\ \sec^{-1}(4.193)$
    (c) $\cos^{-1}(2 \sin .1234)$ (d) $\sin[\arctan(4.62) - \arccos(-.48)]$

63. Without using tables or a calculator, find each value. Then check using your calculator.
    (a) $\tan[\tan^{-1}(43)]$ (b) $\cos[\sin^{-1}(\frac{5}{13})]$
    (c) $\sin[\frac{\pi}{4} + \sin^{-1}(.8)]$ (d) $\cos[\sin^{-1}(.6) + \sec^{-1}(3)]$

[c] 64. Try to calculate each of the following and then explain why your calculator gives you an error message.
    (a) $\cos[\sin^{-1}(2)]$ (b) $\cos^{-1}(\tan 2)$ (c) $\tan[\arctan 3 + \arctan(\frac{1}{3})]$

65. Solve for x.
   (a) $\cos(\sin^{-1}x) = \frac{3}{4}$ (b) $\sin(\cos^{-1}x) = \sqrt{.19}$
   (c) $\sin^{-1}(3x - 5) = \frac{\pi}{6}$ (d) $\tan^{-1}(x^2 - 3x + 3) = \frac{\pi}{4}$

66. To determine inverses for cotangent and cosecant, we restrict their domains to $0 < x < \pi$ and $-\pi/2 \le x \le \pi/2$, $x \ne 0$, respectively. With these restrictions understood, find each value.
   (a) $\cot^{-1}(\sqrt{3})$ (b) $\cot^{-1}(-1/\sqrt{3})$ (c) $\cot^{-1}(0)$
   (d) $\csc^{-1}(2)$ (e) $\csc^{-1}(-1)$ (f) $\csc^{-1}(-2/\sqrt{3})$

67. It is always true that $\sin(\sin^{-1}x) = x$, but it is not always true that $\sin^{-1}(\sin x) = x$. For example, $\sin^{-1}(\sin \pi) \ne \pi$. Instead,

$$\sin^{-1}(\sin \pi) = \sin^{-1}(0) = 0$$

Find each value.
   (a) $\sin^{-1}[\sin(\pi/2)]$ (b) $\sin^{-1}[\sin(3\pi/4)]$
   (c) $\sin^{-1}[\sin(5\pi/4)]$ (d) $\sin^{-1}[\sin(3\pi/2)]$
   (e) $\cos^{-1}[\cos(3\pi)]$ (f) $\tan^{-1}[\tan(13\pi/4)]$

68. Sketch the graph of $y = \sin^{-1}(\sin x)$ for $-2\pi \le x \le 4\pi$. *Hint:* See Problem 67.

69. For each of the following right triangles, write θ explicitly in terms of x.

(a)
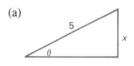

(b)
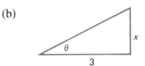

(c)

(d)
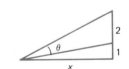

70. In some computer languages (for example, BASIC and FORTRAN), the only built-in inverse trigonometric function is $\tan^{-1}$. Establish the following identities, which show why this is sufficient.
   (a) $\sin^{-1} x = \tan^{-1}\left(\dfrac{x}{\sqrt{1 - x^2}}\right)$

   (b) $\cos^{-1} x = \dfrac{\pi}{2} - \sin^{-1} x = \dfrac{\pi}{2} - \tan^{-1}\left(\dfrac{x}{\sqrt{1 - x^2}}\right)$

71. Assume that your calculator's $\tan^{-1}$ button is working but not the $\sin^{-1}$ or $\cos^{-1}$ buttons. Use the results in Problem 70 to calculate each of the following.
   (a) $\sin^{-1}(.6)$ (b) $\sin^{-1}(-.3)$
   (c) $\cos^{-1}(.8)$ (d) $\cos^{-1}(-.9)$

72. Show that $\arctan(\frac{1}{4}) + \arctan(\frac{3}{5}) = \pi/4$. *Hint:* Show that both sides have the same tangent, using the formula for $\tan(\alpha + \beta)$.

73. In 1706, John Machin used the following formula to calculate π to 100 decimal places, a tremendous feat for its day. Establish this formula. *Hint:* Apply the addition formula for the tangent to the left side. Think of the right side as $4\theta = 2(2\theta)$ and use the tangent double angle formula twice.

$$\frac{\pi}{4} + \arctan\left(\frac{1}{239}\right) = 4\arctan\left(\frac{1}{5}\right)$$

74. Show that

$$\arctan\left(\frac{1}{3}\right) + \arctan\left(\frac{1}{5}\right) + \arctan\left(\frac{1}{7}\right) + \arctan\left(\frac{1}{8}\right) = \frac{\pi}{4}$$

75. A picture 4 feet high is hung on a museum wall so that its bottom is 7 feet above the floor. A viewer whose eye level is 5 feet above the floor stands b feet from the wall.
    (a) Express θ, the vertical angle subtended by the picture at her eye, explicitly in terms of b.
    (b) Calculate θ when $b = 8$.
    (c) Determine b so $\theta = 30°$.

76. **TEASER**   A goat is tethered to a stake at the edge of a circular pond of radius r by means of a rope of length kr, $0 < k \leq 2$. Find an explicit formula for its grazing area in terms of r and k.

Two Bugs on a Circle

Two bugs crawl around the unit circle starting together at $(1, 0)$, one moving at one unit per second, the other moving twice as fast. When will one bug be directly above the other bug?

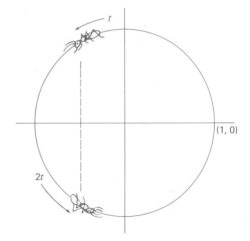

5-5   Trigonometric Equations

What does the bug problem have to do with trigonometric equations? Well, you should agree that after t seconds the slow bug, having traveled t units along the unit circle, is at $(\cos t, \sin t)$. The fast bug is at $(\cos 2t, \sin 2t)$. One bug will be directly above the other bug when their two x-coordinates are equal. This means we must solve the equation

$$\cos 2t = \cos t$$

Specifically, we must find the first $t > 0$ that makes this equality true. We shall solve this equation in due time, but first we ought to solve some simpler trigonometric equations.

SIMPLE EQUATIONS

Suppose we are asked to solve the equation

$$\sin t = \frac{1}{2}$$

for t. The number $t = \pi/6$ occurs to us right away. But that is not the only answer. All numbers that measure angles in the first or second quadrant and have $\pi/6$ as their reference number are solutions. Thus,

$$\ldots, -\frac{11\pi}{6}, -\frac{7\pi}{6}, \frac{\pi}{6}, \frac{5\pi}{6}, \frac{13\pi}{6}, \ldots$$

all work. In fact, one characteristic of trigonometric equations is that, if they have one solution, they have infinitely many solutions.

Let us alter the problem. Suppose we wish to solve $\sin t = \frac{1}{2}$ for $0 \le t < 2\pi$. Then the answers are $\pi/6$ and $5\pi/6$. In the following pages, we shall assume that, unless otherwise specified, we are to find only those solutions on the interval $0 \le t < 2\pi$.

For a second example, let us solve $\sin t = -.5234$ for $0 \le t < 2\pi$. Our unthinking action is to use a calculator to find $\sin^{-1}(-.5234)$, but note that this gives us the value $-.55084$, which is not on the required interval. The better way to proceed is to recognize that the given equation has two solutions but that they both have t_0 as their reference number, where

$$t_0 = \sin^{-1}(.5234) = .55084$$

The two solutions we seek are $\pi + t_0$ and $2\pi - t_0$, that is, 3.69243 and 5.73235.

Do you remember how we solved the equation $x^2 = 4x$? We rewrote it with 0 on one side, factored the other side, and then set each factor equal to 0 (see Figure 10). We follow exactly the same procedure with our next trigonometric equation. To solve

$$\cos t \cot t = -\cos t$$

use the following steps:

$$\cos t \cot t + \cos t = 0$$

$$\cos t (\cot t + 1) = 0$$

$$\cos t = 0, \cot t + 1 = 0$$

$$\cos t = 0, \qquad \cot t = -1$$

Thus our problem is reduced to solving two simple equations. The first has the two solutions, $\pi/2$ and $3\pi/2$; the second has solutions $3\pi/4$ and $7\pi/4$. Thus the set of all solutions of

$$x^2 = 4x$$
$$x^2 - 4x = 0$$
$$x(x - 4) = 0$$
$$x = 0, x - 4 = 0$$
$$x = 0, 4$$

Figure 10

$$\cos t \cot t = -\cos t$$

on the interval $0 \le t < 2\pi$ is

$$\left\{ \frac{\pi}{2}, \frac{3\pi}{4}, \frac{3\pi}{2}, \frac{7\pi}{4} \right\}$$

EQUATIONS OF QUADRATIC FORM

In Section 1-4, we solved quadratic equations by a number of techniques (factoring, taking square roots, and using the quadratic formula). We use the same techniques here. For example,

$$\cos^2 t = \frac{3}{4}$$

is analogous to $x^2 = \frac{3}{4}$. We solve such an equation by taking square roots.

$$\cos t = \pm \frac{\sqrt{3}}{2}$$

The set of solutions on $0 \le t < 2\pi$ is

$$\left\{ \frac{\pi}{6}, \frac{5\pi}{6}, \frac{7\pi}{6}, \frac{11\pi}{6} \right\}$$

As a second example, consider the equation

$$2 \sin^2 t - \sin t - 1 = 0$$

Think of it as being like

$$2x^2 - x - 1 = 0$$

Now

$$2x^2 - x - 1 = (2x + 1)(x - 1)$$

and so

$$2 \sin^2 t - \sin t - 1 = (2 \sin t + 1)(\sin t - 1)$$

When we set each factor equal to zero and solve, we get

$$2 \sin t + 1 = 0 \qquad\qquad \sin t - 1 = 0$$
$$\sin t = -\tfrac{1}{2} \qquad\qquad \sin t = 1$$
$$t = \frac{7\pi}{6}, \frac{11\pi}{6} \qquad\qquad t = \frac{\pi}{2}$$

The set of all solutions on $0 \le t < 2\pi$ is

$$\left\{ \frac{\pi}{2}, \frac{7\pi}{6}, \frac{11\pi}{6} \right\}$$

USING IDENTITIES TO SOLVE EQUATIONS

Consider the equation

$$\tan^2 x = \sec x + 1$$

The identity $\sec^2 x = \tan^2 x + 1$ suggests writing everything in terms of $\sec x$.

$$\sec^2 x - 1 = \sec x + 1$$

$$\sec^2 x - \sec x - 2 = 0$$

$$(\sec x + 1)(\sec x - 2) = 0$$

$$\sec x + 1 = 0 \qquad \sec x - 2 = 0$$

$$\sec x = -1 \qquad \sec x = 2$$

$$x = \pi \qquad x = \frac{\pi}{3}, \frac{5\pi}{3}$$

Thus, the set of solutions on $0 \le t < 2\pi$ is $\{\pi/3,\ \pi,\ 5\pi/3\}$. Unfamiliarity with the secant may hinder you at the last step. If so, use $\sec x = 1/\cos x$ to write the equations in terms of cosines and solve the equations

$$\cos x = -1 \qquad \cos x = \frac{1}{2}$$

SOLUTION TO THE TWO-BUG PROBLEM

Our opening display asked when one bug would first be directly above the other. We reduced that problem to solving

$$\cos 2t = \cos t$$

for t. Using a double-angle formula, we may write

$$2 \cos^2 t - 1 = \cos t$$

$$2 \cos^2 t - \cos t - 1 = 0$$

$$(2 \cos t + 1)(\cos t - 1) = 0$$

$$\cos t = -\frac{1}{2} \qquad \cos t = 1$$

$$t = \frac{2\pi}{3}, \frac{4\pi}{3} \qquad t = 0$$

The smallest positive solution is $t = 2\pi/3$. After a little over 2 seconds, the slow bug will be directly above the fast bug.

Problem Set 5-5

Solve each of the following, finding all solutions on the interval 0 to 2π, excluding 2π.

1. $\sin t = 0$
2. $\cos t = 1$
3. $\sin t = -1$
4. $\tan t = -\sqrt{3}$
5. $\sin t = 2$
6. $\sec t = \frac{1}{2}$
7. $2 \cos x + \sqrt{3} = 0$
8. $2 \sin x + 1 = 0$
9. $\tan^2 x = 1$
10. $4 \sin^2 \theta - 3 = 0$
11. $(2 \cos \theta + 1)(2 \sin \theta - \sqrt{2}) = 0$
12. $(\sin \theta - 1)(\tan \theta + 1) = 0$
13. $\sin^2 x + \sin x = 0$
14. $2 \cos^2 x - \cos x = 0$
15. $\tan^2 \theta = \sqrt{3} \tan \theta$
16. $\cot^2 \theta = -\cot \theta$
17. $2 \sin^2 x = 1 + \cos x$
18. $\sec^2 x = 1 + \tan x$

19. $\boxed{c}$ $\tan^2 x - 3 \tan x + 1 = 0$
20. $\boxed{c}$ $\cos 2t = 3 \sin t$

CAUTION

$\begin{array}{l}\text{tan}^2\,\theta = \sqrt{3}\ \text{tan}\ \theta \\ \text{tan}\ \theta = \sqrt{3}\end{array}$

$\tan^2 \theta = \sqrt{3} \tan \theta$
$\tan \theta (\tan \theta - \sqrt{3}) = 0$
$\tan \theta = 0 \qquad \tan \theta = \sqrt{3}$

EXAMPLE A (Solving by Squaring Both Sides) Solve

$$1 - \cos t = \sqrt{3} \sin t$$

Solution.   Since the identity relating sines and cosines involves their squares, we begin by squaring both sides. Then we express everything in terms of $\cos t$ and solve.

$$(1 - \cos t)^2 = 3 \sin^2 t$$

$$1 - 2 \cos t + \cos^2 t = 3(1 - \cos^2 t)$$

$$\cos^2 t - 2 \cos t + 1 = 3 - 3 \cos^2 t$$

$$4 \cos^2 t - 2 \cos t - 2 = 0$$

$$2 \cos^2 t - \cos t - 1 = 0$$

$$(2 \cos t + 1)(\cos t - 1) = 0$$

$$\cos t = -\frac{1}{2} \qquad \cos t = 1$$

$$t = \frac{2\pi}{3}, \frac{4\pi}{3} \qquad t = 0$$

Since squaring may introduce extraneous solutions, it is important to check our answers. We find that $4\pi/3$ is extraneous, since substituting $4\pi/3$ for t in the original equation gives us $1 + \frac{1}{2} = -\frac{3}{2}$. However, 0 and $2\pi/3$ are solutions, as you should verify.

Solve each of the following equations on the interval $0 \le t < 2\pi$; check your answers.

21. $\sin t + \cos t = 1$
22. $\sin t - \cos t = 1$
23. $\sqrt{3}(1 - \sin t) = \cos t$
24. $1 + \sin t = \sqrt{3} \cos t$
25. $\sec t + \tan t = 1$
26. $\tan t - \sec t = 1$

EXAMPLE B (Finding All of the Solutions) Find the entire set of solutions of the equation $\cos 2t = \cos t$.

Solution. In the text, we found 0, $2\pi/3$, and $4\pi/3$ to be the solutions for $0 \le t < 2\pi$. Clearly we get new solutions by adding 2π again and again to any of these numbers. The same holds true for subtracting 2π. In fact, the entire solution set consists of all those numbers of the form $2\pi k$, $2\pi/3 + 2\pi k$, or $4\pi/3 + 2\pi k$, where k is any integer.

Find the entire solution set of each of the following equations.

27. $\sin t = \frac{1}{2}$ 28. $\cos t = -\frac{1}{2}$ 29. $\tan t = 0$

30. $\tan t = -\sqrt{3}$ 31. $\sin^2 t = \frac{1}{4}$ 32. $\cos^2 t = 1$

EXAMPLE C (Multiple-Angle Equations) Find all solutions of $\cos 4t = \frac{1}{2}$ on the interval $0 \le t < 2\pi$.

Solution. There will be more answers than you think. We know that $\cos 4t$ equals $\frac{1}{2}$ when

$$4t = \frac{\pi}{3}, \frac{5\pi}{3}, \frac{7\pi}{3}, \frac{11\pi}{3}, \frac{13\pi}{3}, \frac{17\pi}{3}, \frac{19\pi}{3}, \frac{23\pi}{3}$$

that is, when

$$t = \frac{\pi}{12}, \frac{5\pi}{12}, \frac{7\pi}{12}, \frac{11\pi}{12}, \frac{13\pi}{12}, \frac{17\pi}{12}, \frac{19\pi}{12}, \frac{23\pi}{12}$$

The reason that there are 8 solutions instead of 2 is that $\cos 4t$ completes 4 periods on the interval $0 \le t < 2\pi$.

Solve each of the following equations, finding all solutions on the interval $0 \le t < 2\pi$.

33. $\sin 2t = 0$ 34. $\cos 2t = 0$ 35. $\sin 4t = 1$

36. $\cos 4t = 1$ 37. $\tan 2t = -1$ 38. $\tan 3t = 0$

MISCELLANEOUS PROBLEMS

In Problems 39–56, find all solutions to the given equation on $0 \le x < 2\pi$.

39. $2 \sin^2 x = \sin x$ 40. $2 \cos x \sin x + \cos x = 0$

© 41. $\cos^2 x = \frac{1}{3}$ © 42. $\tan^2 x + 2 \tan x = 0$

43. $2 \tan x - \sec^2 x = 0$ 44. $\tan^2 x = 1 + \sec x$

45. $\tan 2x = 3 \tan x$ 46. $\cos(x/2) - \cos x = 1$

© 47. $\sin^2 x + 3 \sin x - 1 = 0$ © 48. $\tan^2 x - 2 \tan x - 10 = 0$

49. $\sin 2x + \sin x + 4 \cos x = -2$

50. $\cos x + \sin x = \sec x + \sec x \tan x$

51. $\sin x \cos x = -\sqrt{3}/4$ 52. $4 \sin x - 4 \sin^3 x + \cos x = 0$

© 53. $\cos x - 2 \sin x = 2$ © 54. $\sin x + \cos x = \frac{1}{3}$

55. $\cos^8 x - \sin^8 x = 0$ 56. $\cos^6 x + \sin^6 x = \frac{13}{16}$

57. A ray of light from the lamp L in Figure 11 reflects off a mirror to the object O.

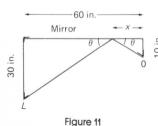

Figure 11

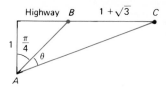

Figure 12

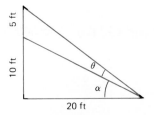

Figure 13

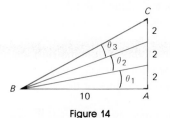

Figure 14

(a) Find the distance x.

(b) Write an equation for θ.

(c) Solve this equation.

58. Tom and John are lost in a desert 1 mile from a highway, at point A in Figure 12. Each strikes out in a different direction to get to the highway. Tom gets to the highway at point B and John arrives at point C, $1 + \sqrt{3}$ miles farther down the road. Write an equation for θ and solve it.

59. Mr. Quincy built a slide with a 10-foot rise and 20-foot base (Figure 13). (a) Find the angle α in degrees. (b) By how much (θ in Figure 13) would the angle of the slide increase if he made the rise 15 feet, keeping the base at 20 feet?

60. Find the angles θ_1, θ_2, and θ_3 shown in Figure 14. Your answers should convince you that the angle ABC is not trisected.

61. Solve the equation

$$\sin 4t + \sin 3t + \sin 2t = 0$$

Hint: Use the identity $\sin u + \sin v = 2 \sin((u + v)/2) \cos((u - v)/2)$.

62. Solve the equation

$$\cos 5t + \cos 3t - 2 \cos t = 0$$

Hint: Use the identity $\cos u + \cos v = 2 \cos((u + v)/2) \cos((u - v)/2)$.

63. Solve $\cos^8 u + \sin^8 u = \frac{41}{128}$ for $0 \leq u \leq \pi$. *Hint:* Begin by using half-angle formulas.

64. **TEASER** Show that $t = \pi/4$ is the only solution on $0 \leq t \leq \pi$ to the equation

$$\frac{a + b \cos t}{b + a \sin t} = \frac{a + b \sin t}{b + a \cos t}$$

Chapter Summary

An **identity** is an equality that is true for all values of the unknown for which both sides of the equality make sense. Our first task was to establish the fundamental identities of trigonometry, here arranged by category.

Basic Identities

1. $\tan t = \dfrac{\sin t}{\cos t}$

2. $\cot t = \dfrac{\cos t}{\sin t} = \dfrac{1}{\tan t}$

3. $\sec t = \dfrac{1}{\cos t}$

4. $\csc t = \dfrac{1}{\sin t}$

5. $\sin^2 t + \cos^2 t = 1$

6. $1 + \tan^2 t = \sec^2 t$

7. $1 + \cot^2 t = \csc^2 t$

Cofunction Identities

8. $\sin\left(\dfrac{\pi}{2} - t\right) = \cos t$

9. $\cos\left(\dfrac{\pi}{2} - t\right) = \sin t$

Odd-Even Identities

10. $\sin(-t) = -\sin t$ 11. $\cos(-t) = \cos t$

Addition Formulas

12. $\sin(s + t) = \sin s \cos t + \cos s \sin t$
13. $\sin(s - t) = \sin s \cos t - \cos s \sin t$
14. $\cos(s + t) = \cos s \cos t - \sin s \sin t$
15. $\cos(s - t) = \cos s \cos t + \sin s \sin t$

Double-Angle Formulas

16. $\sin 2t = 2 \sin t \cos t$
17. $\cos 2t = \cos^2 t - \sin^2 t = 1 - 2 \sin^2 t = 2 \cos^2 t - 1$

Half-Angle Formulas

18. $\sin \dfrac{t}{2} = \pm \sqrt{\dfrac{1 - \cos t}{2}}$ 19. $\cos \dfrac{t}{2} = \pm \sqrt{\dfrac{1 + \cos t}{2}}$

Once we have memorized the fundamental identities, we can use them to prove thousands of other identities. The suggested technique is to take one side of a proposed identity and show by a chain of equalities that it is equal to the other.

A **trigonometric equation** is an equality involving trigonometric functions that is true only for some values of the unknown (for example, $\sin 2t = \frac{1}{2}$). Here our job is to solve the equation, that is, to find the values of the unknown that make it true.

With their natural domains, the trigonometric functions are not one-to-one and therefore do not have inverses. However, there are standard ways to restrict the domains so that inverses exist. Here are the results.

$$x = \sin^{-1} y \quad \text{means} \quad y = \sin x \quad \text{and} \quad \frac{-\pi}{2} \le x \le \frac{\pi}{2}$$

$$x = \cos^{-1} y \quad \text{means} \quad y = \cos x \quad \text{and} \quad 0 \le x \le \pi$$

$$x = \tan^{-1} y \quad \text{means} \quad y = \tan x \quad \text{and} \quad \frac{-\pi}{2} < x < \frac{\pi}{2}$$

$$x = \sec^{-1} y \quad \text{means} \quad y = \sec x \quad \text{and} \quad 0 \le x \le \pi, x \ne \frac{\pi}{2}$$

Chapter Review Problem Set

1. Prove that the following are identities.
   (a) $\cot \theta \cos \theta = \csc \theta - \sin \theta$
   (b) $\dfrac{\cos x \tan^2 x}{\sec x + 1} = 1 - \cos x$

2. Express each of the following in terms of $\sin x$ and simplify.

(a) $\dfrac{(\cos^2 x - 1)(1 + \tan^2 x)}{\csc x}$

(b) $\dfrac{\cos^2 x \csc x}{1 + \csc x}$

3. Use appropriate identities to simplify and then calculate each of the following.

(a) $2 \cos^2 22.5° - 1$

(b) $\sin 37° \cos 53° + \cos 37° \sin 53°$

(c) $\cos 108° \cos 63° + \sin 108° \sin 63°$

4. If $\cos t = -\frac{4}{5}$ and $\pi < t < 3\pi/2$, calculate

(a) $\sin 2t$; (b) $\sin(t/2)$.

5. Prove that the following are identities.

(a) $\sin 2t \cos t - \cos 2t \sin t = \sin t$

(b) $\sec 2t + \tan 2t = \dfrac{\cos t + \sin t}{\cos t - \sin t}$

(c) $\dfrac{\cos(\alpha + \beta)}{\cos \alpha \cos \beta} = \tan \alpha (\cot \alpha - \tan \beta)$

6. Solve the following trigonometric equations for t, $0 \le t < 2\pi$.

(a) $\cos t = -\sqrt{3}/2$

(b) $(2 \sin t + 1) \tan t = 0$

(c) $\cos^2 t + 2 \cos t - 3 = 0$

(d) $\sin t - \cos t = 1$

(e) $\sin 3t = 1$

7. What is the standard way to restrict the domain of sine, cosine, and tangent so that they have inverses?

8. Calculate each of the following without the help of a calculator.

(a) $\sin^{-1}(-\sqrt{3}/2)$

(b) $\cos^{-1}(-\sqrt{3}/2)$

(c) $\tan^{-1}(-\sqrt{3})$

(d) $\tan(\tan^{-1} 6)$

(e) $\cos^{-1}(\cos 3\pi)$

(f) $\sin(\cos^{-1} \frac{2}{3})$

(g) $\cos(2 \cos^{-1} .7)$

(h) $\sin(2 \cos^{-1} \frac{5}{13})$

9. Sketch the graph of $y = \tan^{-1} x$.

10. Find an approximate value for $\tan^{-1}(-1000)$.

11. Show that

$$\frac{\pi}{4} = \arctan \frac{1}{2} + \arctan \frac{1}{3}$$

*Thus one sees in the sciences
many brilliant theories which
have remained unapplied for
a long time suddenly be-
coming the foundation of most
important applications, and
likewise applications very simple in appearance giving birth
to ideas of the most abstract
theories.*

Marquis de Condorcet

CHAPTER 6

Applications of Trigonometry

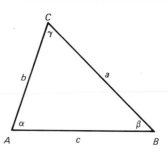

6-1 Oblique Triangles: Law of Sines

We learned in Section 4-1 how to solve a right triangle. But can we solve an oblique triangle—that is, one without a 90° angle? One valuable tool is the **law of sines,** stated above. It is valid for any triangle whatever, but we initially establish it for the case where all angles are acute.

PROOF OF THE LAW OF SINES

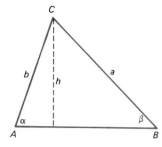

Figure 1

Consider a triangle with all acute angles, labeled as in Figure 1. Drop a perpendicular of length h from vertex C to the opposite side. Then by right-triangle trigonometry,

$$\sin \alpha = \frac{h}{b} \qquad \sin \beta = \frac{h}{a}$$

If we solve for h in these two equations and equate the results, we obtain

$$b \sin \alpha = a \sin \beta$$

Finally, dividing both sides by ab yields

$$\frac{\sin \alpha}{a} = \frac{\sin \beta}{b}$$

Since the roles of β and γ can be interchanged, the same reasoning gives

$$\frac{\sin \alpha}{a} = \frac{\sin \gamma}{c}$$

Next consider a triangle with an obtuse angle α ($90° < \alpha < 180°$). Drop a perpendicular of length h from vertex C to the extension of AB (see Figure 2). Notice that angle α' is the reference angle for α and so $\sin \alpha = \sin \alpha'$. It follows from right-triangle trigonometry that

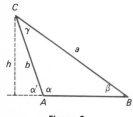

Figure 2

$$\sin \alpha = \sin \alpha' = \frac{h}{b} \qquad \sin \beta = \frac{h}{a}$$

just as in the acute case. The rest of the argument is identical with that case.

SOLVING A TRIANGLE (AAS)

Suppose that we know two angles and any side of a triangle. For example, suppose that in triangle ABC, $\alpha = 103.5°$, $\beta = 27.5°$, and $c = 45.3$ (Figure 3). Our task is to find γ, a, and b.

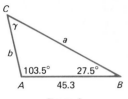

Figure 3

1. Since $\alpha + \beta + \gamma = 180°$, $\gamma = 180° - (103.5° + 27.5°) = 49°$.
2. By the law of sines,

$$\frac{a}{\sin 103.5°} = \frac{45.3}{\sin 49°}$$

$$a = \frac{(45.3)(\sin 103.5°)}{\sin 49°}$$

$$= \frac{(45.3)(\sin 76.5°)}{\sin 49°}$$

$$= \frac{(45.3)(.9724)}{.7547}$$

$$\approx 58.4$$

3. Also by the law of sines,

$$\frac{b}{\sin 27.5°} = \frac{45.3}{\sin 49°}$$

$$b = \frac{(45.3)(\sin 27.5°)}{\sin 49°}$$

$$= \frac{(45.3)(.4617)}{.7547}$$

$$\approx 27.7$$

SOLVING A TRIANGLE (SSA)

Suppose that two sides and the angle opposite one of them are given. This is called the **ambiguous case** because the given information may not determine a unique triangle.

   If α, a, and b are given, we consider trying to construct a triangle fitting these data by first drawing angle α, then marking off b on one of its sides thus determining vertex C. Finally, we attempt to locate vertex B by striking off a circular arc of radius a with center at C. If $a \geq b$, this can always be done in a unique way. Figure 4 on the next page illustrates this both for α acute and α obtuse. If $a < b$, there are several possibilities (Figure 5).

   Fortunately, we are able to decide which of these possibilities is the case if we draw an approximate picture and then attempt to apply the law of sines. First, note that if $a \geq b$ there is one triangle corresponding to the data and for it β is an acute angle. Application of the law of sines will give $\sin \beta$, which allows determination of β.

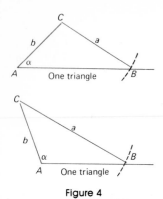

Figure 4

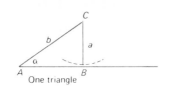

One triangle

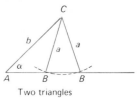

Two triangles

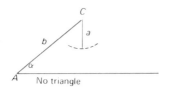

One triangle

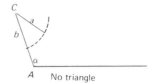

No triangle

Figure 5

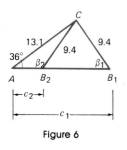

Figure 6

If $a < b$, we may attempt to apply the law of sines. If it yields $\sin \beta = 1$, we have a unique right triangle. If it yields $\sin \beta < 1$, we have two triangles corresponding to the two angles β_1 and β_2 (one acute, the other obtuse) with this sine. If it yields $\sin \beta > 1$, we have an inconsistency in the data; no triangle satisfying the data exists.

Suppose, for example, that we are given $\alpha = 36°$, $a = 9.4$, and $b = 13.1$ (Figure 6). We must find β, γ, and c. Since $a < b$, there may be zero, one, or two triangles. We proceed to compute $\sin \beta$.

1. $$\frac{\sin \beta}{13.1} = \frac{\sin 36°}{9.4}$$

$$\sin \beta = \frac{(13.1) \sin 36°}{9.4} = \frac{(13.1)(.5878)}{9.4} \approx .8191$$

Since $\sin \beta < 1$, there are two triangles.

$$\beta_1 = 55° \qquad \beta_2 = 125°$$

2. $\gamma_1 = 180° - (36° + 55°) = 89°$
   $\gamma_2 = 180° - (36° + 125°) = 19°$

3. $$\frac{c_1}{\sin 89°} = \frac{9.4}{\sin 36°}$$

$$c_1 = \frac{9.4}{\sin 36°} (\sin 89°)$$

$$= \frac{9.4}{.5878}(.9998) \approx 16.0$$

$$\frac{c_2}{\sin 19°} = \frac{9.4}{\sin 36°}$$

$$c_2 = \frac{9.4}{\sin 36°}(\sin 19°)$$

$$= \frac{9.4}{.5878}(.3256) \approx 5.2$$

Problem Set 6-1

Solve the triangles of Problems 1–10 using either Table C or a calculator.

1. $\alpha = 42.6°$, $\beta = 81.9°$, $a = 14.3$
2. $\beta = 123°$, $\gamma = 14.2°$, $a = 295$
3. $\alpha = \gamma = 62°$, $b = 50$
4. $\alpha = \beta = 14°$, $c = 30$
5. $\alpha = 115°$, $a = 46$, $b = 34$
6. $\beta = 143°$, $a = 46$, $b = 84$
7. $\alpha = 30°$, $a = 8$, $b = 5$
8. $\beta = 60°$, $a = 11$, $b = 12$
9. $\alpha = 30°$, $a = 5$, $b = 8$
10. $\beta = 60°$, $a = 12$, $b = 11$

11. Two observers stationed 110 meters apart at A and B on the bank of a river are looking at a tower situated at a point C on the opposite bank. They measure angles CAB and CBA to be 43° and 57°, respectively. How far is the first observer from the tower?

12. A telegraph pole leans away from the sun at an angle of 11° to the vertical. The pole casts a shadow 96 feet long on horizontal ground when the angle of elevation of the sun is 23°. Find the length of the pole (see Figure 7).

13. A vertical pole 60 feet long is standing by the side of an inclined road. It casts a shadow 138 feet long directly downhill along the road when the angle of elevation of the sun is 58°. Find the angle of inclination θ of the road (see Figure 8).

14. Two forest rangers 15 miles apart at points A and B observe a fire at a point C. The ranger at A measures angle CAB as 43.6° and the one at B measures angle CBA as 79.3°. How far is the fire from each ranger? How far is the fire from a straight road that goes from A to B?

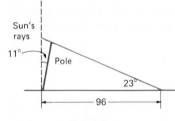

Figure 7

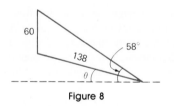

Figure 8

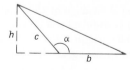

Figure 9

EXAMPLE (An Important Area Formula)   Consider a triangle with two sides b and c and included angle α. Show that the area A of the triangle is

$$A = \tfrac{1}{2}bc \sin \alpha$$

Solution.   Let h denote the altitude of the triangle as shown in the diagrams of Figure 9. Whether α is acute or obtuse, we have $\sin \alpha = h/c$, that is,

$h = c \sin \alpha$. We conclude that

$$A = \frac{1}{2}bh = \frac{1}{2}bc \sin \alpha$$

15. Find the area of the triangle with sides $b = 20$, $c = 30$, and included angle $\alpha = 40°$.
16. Find the area of the triangle with $a = 14.6$, $b = 31.7$, and $\gamma = 130.2°$.
17. Find the area of the triangle with $c = 30.1$, $\alpha = 25.3°$, and $\beta = 112.2°$.
18. Find the area of the triangle with $a = 20$, $\alpha = 29°$, and $\gamma = 46°$.

MISCELLANEOUS PROBLEMS

19. The children's slide at the park is 30 feet long and inclines 36° from the horizontal. The ladder to the top is 18 feet long. How steep is the ladder, that is, what angle does it make with the horizontal? Assume the slide is straight and that the bottom end of the slide is at the same level as the bottom end of the ladder.

20. Prevailing winds have caused an old tree to incline 11° eastward from the vertical. The sun in the west is 32° above the horizontal. How long a shadow is cast by the tree if the tree measures 114 feet from top to bottom?

21. A rectangular room, 16 feet by 30 feet, has an open beam ceiling. The two parts of the ceiling make angles of 65° and 32° with the horizontal (an end view is shown in Figure 10). Find the total area of the ceiling.

22. Sheila Sather, traveling north on a straight road at a constant rate of 60 miles per hour, sighted flames shooting up into the air at a point 20° west of north. Exactly 1 hour later, the fire was 59° west of south. Determine the shortest distance from the road to the fire.

23. A lighthouse stands at a certain distance out from a straight shoreline. It throws a beam of light that revolves at a constant rate of one revolution per minute. A short time after shining on the nearest point on the shore, the beam reaches a point on the shore that is 2640 feet from the lighthouse, and 3 seconds later it reaches a point 2000 feet farther along the shore. How far is the lighthouse from the shore?

24. In Figure 11, AC is 10 meters longer than CB. Determine the length of CD.

25. Four line segments of lengths 3, 4, 5, and 6 radiate like spokes from a common point. Their outer ends are the vertices of a quadrilateral Q. Determine the maximum possible area of Q.

26. Figure 12 illustrates the Pythagorean theorem ($a^2 + b^2 = c^2$). A rubber band is stretched around this figure. Show that the area of the region enclosed by the rubber band is $2(ab + c^2)$.

27. Let 2ϕ denote the angle at a point of the regular 6-pointed star shown in Figure 13. Express the area A of this star in terms of ϕ and the edge length r.

28. **TEASER** Figure 14 shows two mirrors intersecting at an angle of 15°. A light ray from S is reflected at P and again at Q and then is absorbed at R. It is given that $ST = RU = 5$, $OT = 50$, and $OU = 20$. Find the length $x + y + z$ of the path of the light ray. As indicated, the angle of incidence equals the angle of reflection.

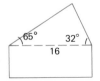

Figure 10

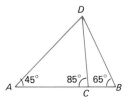

Figure 11

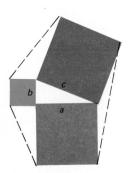

Figure 12

Figure 13

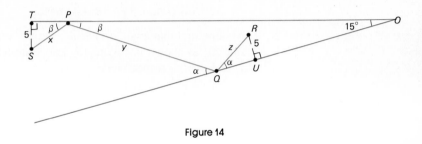

Figure 14

The Law of Cosines

Consider an arbitrary triangle with angles α, β, γ and corresponding opposite sides $a, b, c,$ respectively. Then

$$a^2 = b^2 + c^2 - 2bc \cos \alpha$$

$$b^2 = a^2 + c^2 - 2ac \cos \beta$$

$$c^2 = a^2 + b^2 - 2ab \cos \gamma$$

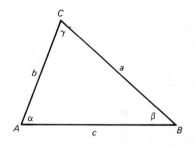

6-2   Oblique Triangles: Law of Cosines

When two sides and the included angle (SAS) or three sides (SSS) of a triangle are given, we cannot apply the law of sines to solve the triangle. Rather, we need the law of cosines, stated above in symbols. Actually it is wise to learn the law in words.

> *The square of any side is equal to the sum of the squares of the other two sides minus twice the product of those sides times the cosine of the angle between them.*

Notice what happens when $\gamma = 90°$ so that $\cos \gamma = 0$. The law of cosines

$$c^2 = a^2 + b^2 - 2ab \cos \gamma$$

becomes

$$c^2 = a^2 + b^2$$

which is just the Pythagorean theorem. In fact, you should think of the law of cosines as a generalization of the Pythagorean theorem, with the term $-2ab \cos \gamma$ acting as a correction term when γ is not $90°$.

PROOF OF THE LAW OF COSINES

Assume first that angle α is acute. Drop a perpendicular CD from vertex C to side AB as shown in Figure 15. Label the lengths of CD, AD, and DB by h, x, and $c - x$, respectively.

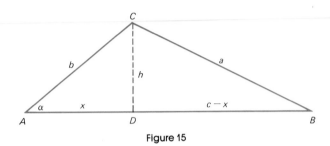

Figure 15

Consider the two right triangles ADC and BDC. By the Pythagorean theorem,

$$h^2 = b^2 - x^2 \quad \text{and} \quad h^2 = a^2 - (c - x)^2$$

Equating these two expressions for h^2 gives

$$a^2 - (c - x)^2 = b^2 - x^2$$
$$a^2 = b^2 - x^2 + (c - x)^2$$
$$a^2 = b^2 - x^2 + c^2 - 2cx + x^2$$
$$a^2 = b^2 + c^2 - 2cx$$

Now $\cos \alpha = x/b$, and so $x = b \cos \alpha$. Thus

$$a^2 = b^2 + c^2 - 2cb \cos \alpha$$

which is the result we wanted.

Next we give the proof of the law of cosines for the obtuse angle case. Again drop a perpendicular from vertex C to side AB extended and label the resulting diagram as shown in Figure 16.

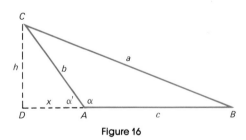

Figure 16

From consideration of triangles ADC and BDC and the Pythagorean theorem, we obtain

$$h^2 = b^2 - x^2 \quad \text{and} \quad h^2 = a^2 - (c + x)^2$$

Algebra analogous to that used in the acute angle case yields

$$a^2 = b^2 + c^2 + 2cx$$

Now α' is the reference angle for α, and so $\cos \alpha = -\cos \alpha'$. Also $\cos \alpha' = x/b$. Therefore,

$$x = b \cos \alpha' = -b \cos \alpha$$

When we substitute this expression for x in the equation above, we get

$$a^2 = b^2 + c^2 - 2cb \cos \alpha$$

SOLVING A TRIANGLE (SAS)

Consider a triangle with $b = 18.1$, $c = 12.3$, and $\alpha = 115°$ (Figure 17). We want to determine a, β, and γ.

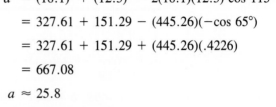

1. By the law of cosines,

$$a^2 = (18.1)^2 + (12.3)^2 - 2(18.1)(12.3) \cos 115°$$
$$= 327.61 + 151.29 - (445.26)(-\cos 65°)$$
$$= 327.61 + 151.29 + (445.26)(.4226)$$
$$= 667.08$$
$$a \approx 25.8$$

2. Now we can use the law of sines.

$$\frac{\sin \beta}{18.1} = \frac{\sin 115°}{25.8}$$

$$\sin \beta = \frac{(18.1) \sin 115°}{25.8} = \frac{(1.81)(\sin 65°)}{25.8}$$

$$= \frac{(18.1)(.9063)}{25.8} = .6358$$

$$\beta \approx 39.5°$$

3. $\gamma \approx 180° - (115° + 39.5°) = 25.5°$.

SOLVING A TRIANGLE (SSS)

If $a = 13.1$, $b = 15.5$, and $c = 17.2$, then we must determine the three angles.

1. By the law of cosines,

$$a^2 = b^2 + c^2 - 2bc \cos \alpha$$

Thus

$$\cos \alpha = \frac{b^2 + c^2 - a^2}{2bc}$$

$$= \frac{(15.5)^2 + (17.2)^2 - (13.1)^2}{2(15.5)(17.2)} = .6836$$

$$\alpha \approx 46.9°$$

2. By the law of sines,

$$\frac{\sin \beta}{15.5} = \frac{\sin 46.9°}{13.1}$$

$$\sin \beta = \frac{(15.5)(\sin 46.9°)}{13.1} = \frac{(15.5)(.7302)}{13.1} = .8640$$

$$\beta \approx 59.8°$$

3. $\gamma = 180° - (46.9° + 59.8°) = 73.3°$.

Problem Set 6-2

In Problems 1–8, solve the triangles satisfying the given data. Use either Table C or a calculator.

1. $\alpha = 60°$, $b = 14$, $c = 10$
2. $\beta = 60°$, $a = c = 8$
3. $\gamma = 120°$, $a = 8$, $b = 10$
4. $\alpha = 150°$, $b = 35$, $c = 40$
5. $a = 5$, $b = 6$, $c = 7$
6. $a = 10$, $b = 20$, $c = 25$
7. $a = 12.2$, $b = 19.1$, $c = 23.8$
8. $a = .11$, $b = .21$, $c = .31$

9. At one corner of a triangular field, the angle measures 52.4°. The sides that meet at this corner are 100 meters and 120 meters long. How long is the third side?

10. To approximate the distance between two points A and B on opposite sides of a swamp, a surveyor selects a point C and measures it to be 140 meters from A and 260 meters from B. Then she measures the angle ACB, which turns out to be 49°. What is the calculated distance from A to B?

11. Two runners start from the same point at 12:00 noon, one of them heading north at 6 miles per hour and the other heading 68° east of north at 8 miles per hour (Figure 18). What is the distance between them at 3:00 that afternoon?

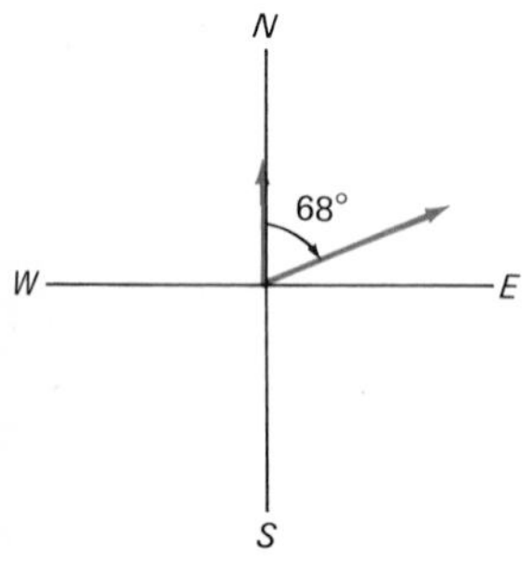

12. A 50-foot pole stands on top of a hill which slants 20° from the horizontal. How long must a rope be to reach from the top of the pole to a point 88 feet directly downhill (that is, on the slant) from the base of the pole?

13. A triangular garden plot has sides of length 35 meters, 40 meters, and 60 meters. Find the largest angle of the triangle.

14. A piece of wire 60 inches long is bent into the shape of a triangle. Find the angles of the triangle if two of the sides have lengths 24 inches and 20 inches.

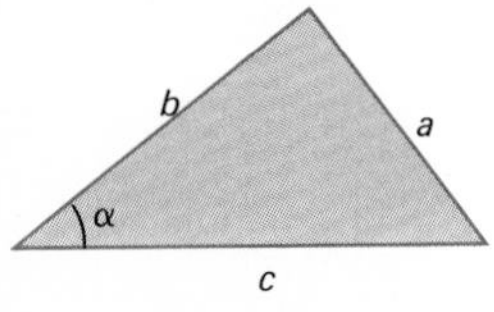

EXAMPLE (Heron's Area Formula)   Show that a triangle with sides a, b, and c (Figure 19) and semiperimeter $s = (a + b + c)/2$ has area A given by

Figure 18

Figure 19

$$A = \sqrt{s(s-a)(s-b)(s-c)}$$

Solution. The proof is subtle, depending on the clever matching of the area formula from the last section with the law of cosines. Begin by writing the law of cosines in the form

$$2bc \cos \alpha = b^2 + c^2 - a^2$$

a formula we will use shortly. Next, take the area formula $A = \frac{1}{2} bc \sin \alpha$ in its squared form and manipulate it very carefully.

$$A^2 = \frac{1}{4}b^2c^2 \sin^2 \alpha = \frac{1}{4}b^2c^2(1 - \cos^2 \alpha)$$

$$= \frac{1}{16}(2bc)(1 + \cos \alpha)(2bc)(1 - \cos \alpha)$$

$$= \frac{1}{16}(2bc + 2bc \cos \alpha)(2bc - 2bc \cos \alpha)$$

$$= \frac{1}{16}(2bc + b^2 + c^2 - a^2)(2bc - b^2 - c^2 + a^2)$$

$$= \frac{1}{16}[(b + c)^2 - a^2][a^2 - (b - c)^2]$$

$$= \frac{(b + c + a)(b + c - a)(a - b + c)(a + b - c)}{2 \quad\quad 2 \quad\quad 2 \quad\quad 2}$$

$$= \left[\frac{a + b + c}{2}\right]\left[\frac{a + b + c}{2} - a\right]\left[\frac{a + b + c}{2} - b\right]\left[\frac{a + b + c}{2} - c\right]$$

$$= s(s - a)(s - b)(s - c)$$

15. The area of the right triangle with sides 3, 4, and 5 is 6. Confirm that Heron's formula gives the same answer.

16. Find the area of the triangle with sides 31, 42, and 53.

17. Find the area of the triangle with sides 5.9, 6.7, and 10.3.

18. Use the answer you got to Problem 16 to find the length h of the shortest altitude of the triangle with sides 31, 42, and 53.

MISCELLANEOUS PROBLEMS

19. A triangular garden plot has sides measuring 42 meters, 50 meters, and 63 meters. Find the measure of the smallest angle.

20. A diagonal and a side of a parallelogram measure 80 centimeters and 25 centimeters, respectively, and the angle between them measures 47°. Find the length of the other diagonal. Recall that the diagonals of a parallelogram bisect each other.

21. Two cars, starting from the intersection of two straight highways, travel along

the highways at speeds of 55 miles per hour and 65 miles per hour, respectively. If the angle of intersection of the highways measures 72°, how far apart are the cars after 36 minutes?

22. Buoys A, B, and C mark the vertices of a triangular racing course on a lake. Buoys A and B are 4200 feet apart, buoys A and C are 3800 feet apart, and angle CAB measures 100°. If the winning boat in a race covered the course in 6.4 minutes, what was its average speed in miles per hour?

23. A quadrilateral Q has sides of length 1, 2, 3, and 4, respectively. The angle between the first pair of sides is 120°. Find the angle between the other pair of sides and also the exact area of Q.

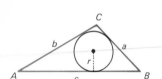

Figure 20

24. For the triangle ABC in Figure 20, let r be the radius of the inscribed circle and let $s = (a + b + c)/2$ be its semiperimeter.
    (a) Show that the area of the triangle is rs.
    (b) Show that $r = \sqrt{(s - a)(s - b)(s - c)/s}$.
    (c) Find r for a triangle with sides 5, 6, and 7.

25. Consider a triangle with sides of length 4, 5, and 6. Show that one of its angles is twice another. *Hint:* Show that the cosine of twice one angle is equal to the cosine of another angle.

26. In the triangle with sides of length a, b, and c, let a_1, b_1, and c_1 denote the lengths of the corresponding medians to these sides from the opposite vertices. Show that

$$a_1^2 + b_1^2 + c_1^2 = \tfrac{3}{4}(a^2 + b^2 + c^2)$$

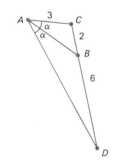

Figure 21

27. Determine the length of AB in Figure 21.

28. **TEASER** The two hands of a clock are 4 and 5 inches long, respectively. At some time between 1:45 and 2, the tips of the hands are 8 inches apart. What time is it then?

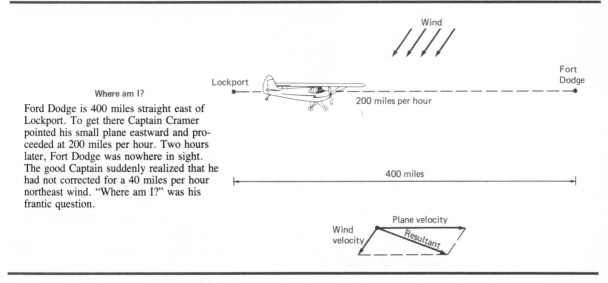

Where am I?

Ford Dodge is 400 miles straight east of Lockport. To get there Captain Cramer pointed his small plane eastward and proceeded at 200 miles per hour. Two hours later, Fort Dodge was nowhere in sight. The good Captain suddenly realized that he had not corrected for a 40 miles per hour northeast wind. "Where am I?" was his frantic question.

6-3 Vectors and Scalars

Many quantities that occur in science (for example, length, mass, volume, and electric charge) can be specified by giving a single number. We call these quantities *scalar quantities;* the numbers that measure their magnitudes are called **scalars.** Other quantities, such as velocity, force, torque, and displacement, must be specified by giving both a magnitude and a direction. We call such quantities **vectors** and represent them by arrows (directed line segments). The length of the arrow is the **magnitude** of the vector; its direction is the **direction** of the vector. Thus, in our opening diagram, the plane's velocity appears as an arrow 200 units long pointing eastward, while the wind velocity is shown as an arrow 40 units long pointing southwest. But how shall we put these two vectors together, that is, find their resultant? Before we try to answer, we introduce more terminology.

Arrows that we draw, like those shot from a bow, have two ends. There is the initial or feather end, which we shall call the **tail**, and the pointed or terminal end, which we shall call the **head** (Figure 22). Two vectors are considered to be **equivalent** if they have the same magnitude and direction (Figure 23). We shall symbolize vectors by boldface letters, such as **u** and **v**. (Since this is hard to accomplish in normal writing, you might use $\vec{u}$ and $\vec{v}$.)

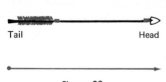

Figure 22

Equivalent vectors

Figure 23

ADDITION OF VECTORS

To find the **sum,** or resultant, of **u** and **v**, move **v** without changing its magnitude or direction until its tail coincides with the head of **u**. Then **u** + **v** is the vector connecting the tail of **u** to the head of **v** (see the left diagram in Figure 24).

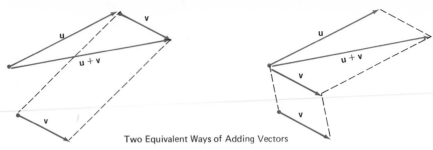

Two Equivalent Ways of Adding Vectors

Figure 24

As an alternative way to find $\mathbf{u} + \mathbf{v}$, move $\mathbf{v}$ so that its tail coincides with that of $\mathbf{u}$. Then $\mathbf{u} + \mathbf{v}$ is the vector with this common tail that is the diagonal of the parallelogram with $\mathbf{u}$ and $\mathbf{v}$ as sides. This method (called the *parallelogram law*) is illustrated on the right in Figure 24.

You should convince yourself that addition is commutative and associative, that is, $\mathbf{u} + \mathbf{v} = \mathbf{v} + \mathbf{u}$ and $(\mathbf{u} + \mathbf{v}) + \mathbf{w} = \mathbf{u} + (\mathbf{v} + \mathbf{w})$.

SCALAR MULTIPLICATION AND SUBTRACTION

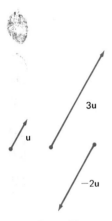

Figure 25

If $\mathbf{u}$ is a vector, then $3\mathbf{u}$ is the vector with the same direction as $\mathbf{u}$ but three times as long; $-2\mathbf{u}$ is twice as long as $\mathbf{u}$ and oppositely directed (Figure 25). More generally, $c\mathbf{u}$ has magnitude $|c|$ times that of $\mathbf{u}$ and is similarly or oppositely directed, depending on whether c is positive or negative. In particular, $(-1)\mathbf{u}$ (usually written $-\mathbf{u}$) has the same length as $\mathbf{u}$ but the opposite direction. It is called the **negative** of $\mathbf{u}$ because when we add it to $\mathbf{u}$, the result is a vector that has shriveled to a point. This special vector (the only vector without direction) is called the **zero vector** and is denoted by $\mathbf{0}$. It is the identity element for addition; that is, $\mathbf{u} + \mathbf{0} = \mathbf{0} + \mathbf{u} = \mathbf{u}$. Finally, subtraction is defined by

$$\mathbf{u} - \mathbf{v} = \mathbf{u} + (-\mathbf{v})$$

CAPTAIN CRAMER'S QUESTION

Consider again the problem posed in the display that opens this section. Physicists tell us that velocities add as vectors. Consequently, our first problem is to add $\mathbf{u}$, which points east and is 200 units long, to a vector $\mathbf{v}$, which points southwest and is 40 units long. Specifically, our aim is to find the length of $\mathbf{u} + \mathbf{v}$, denoted by $\|\mathbf{u} + \mathbf{v}\|$, and the angle α that $\mathbf{u} + \mathbf{v}$ makes with $\mathbf{u}$. The situation is shown in Figure 26. Note that we interpret a northeast wind to mean that $\beta = 45°$.

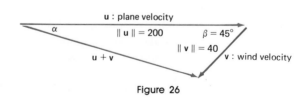

Figure 26

Now, by the law of cosines,

$$\|\mathbf{u} + \mathbf{v}\|^2 = (200)^2 + (40)^2 - 2(200)(40) \cos 45°$$

$$\approx 30{,}300$$

Thus

$$\|\mathbf{u} + \mathbf{v}\| \approx 174$$

Next, by the law of sines,

$$\frac{\sin \alpha}{40} = \frac{\sin 45°}{174}$$

or

$$\sin \alpha = (40 \sin 45°)/174 \approx .1626$$

$$\alpha \approx 9.4°$$

Where is Captain Cramer? Since his true velocity is the vector $\mathbf{u} + \mathbf{v}$ and since he flew at this velocity for 2 hours, he is at a point $2(174) = 348$ miles from Lockport along the line that makes an angle of $9.4°$ with the line between Lockport and Fort Dodge. Of course, he is also 80 miles southwest of Fort Dodge.

Problem Set 6-3

In Problems 1–4, draw the vector **w** *so that its tail is at the heavy dot.*

1. $\mathbf{w} = \mathbf{u} + \mathbf{v}$

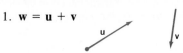

2. $\mathbf{w} = \mathbf{u} - \mathbf{v}$

3. $\mathbf{w} = -2\mathbf{u} + \frac{1}{2}\mathbf{v}$

4. $\mathbf{w} = \mathbf{u} - 3\mathbf{v}$

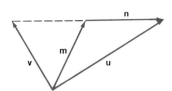

Figure 27

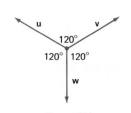

Figure 28

Figure 29

5. Figure 27 shows a parallelogram. Express **w** in terms of **u** and **v**.
6. In the large triangle of Figure 28, **m** is a median (it bisects the side to which it is drawn). Express **m** and **n** in terms of **u** and **v**.
7. In Figure 29, $\mathbf{w} = -(\mathbf{u} + \mathbf{v})$ and $\|\mathbf{u}\| = \|\mathbf{v}\| = 1$. Find $\|\mathbf{w}\|$.
8. Do Problem 7 if the top angle is $90°$ and the two side angles are $135°$.

EXAMPLE A (Displacements Are Vectors) In navigation, directions are specified by giving an angle, called the **bearing,** with respect to a north-south line. Thus a bearing of N35°E denotes an angle whose initial side points north and whose terminal side is 35° east of north. If a ship sails 70 miles in the direction N35°E and then 90 miles straight east, what is its distance and bearing with respect to its starting point?

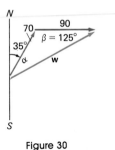

Figure 30

Solution. Our job is to determine the length and bearing of **w** (see Figure 30). We first use a little geometry to determine that $\beta = 125°$. Then, by the law of cosines,

$$\|\mathbf{w}\|^2 = (70)^2 + (90)^2 - 2(70)(90) \cos 125° \approx 20{,}227$$

$$\|\mathbf{w}\| \approx 142$$

By the law of sines,

$$\frac{\sin \alpha}{90} = \frac{\sin 125°}{142}$$

$$\sin \alpha = \frac{90 \sin 125°}{142} \approx .5192$$

$$\alpha \approx 31°$$

Thus the bearing of **w** is N66°E.

9. If I walk 10 miles N45°E and then 10 miles straight north, how far am I from my starting point?

10. In Problem 9, what is the bearing of my final position with respect to my starting point?

11. An airplane flew 100 kilometers in the direction S51°W and then 145 kilometers S39°W. What was the airplane's distance and bearing with respect to its starting point?

12. A ship sailed 11.2 miles straight north and then 48.3 miles N13.2°W. Find its distance and bearing with respect to the starting point.

EXAMPLE B (Velocities Are Vectors) The river is flowing at 6 miles per hour and Jane's boat travels at 20 miles per hour in still water. In what direction should she head her boat if she wants to go straight across the river?

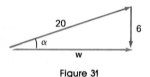

Figure 31

Solution. Our job is simply to determine α in Figure 31.

$$\sin \alpha = \tfrac{6}{20} = .3000$$

$$\alpha \approx 17°$$

13. If the river (see Example B) is $\frac{1}{2}$ mile wide, how long will it take Jane to get across? *Hint:* First determine $\|\mathbf{w}\|$, which is her actual speed with respect to the shore.

14. If Jane (see Example B) had not corrected for the current (that is, if she had

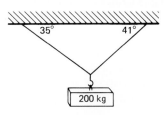

Figure 32

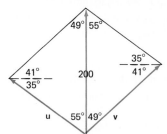

Figure 33

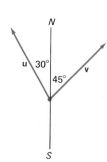

Figure 34

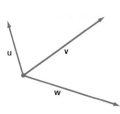

Figure 35

pointed her boat straight across), where would she have landed on the opposite shore?

[c] 15. A wind with velocity 58 miles per hour is blowing in the direction N20°W. An airplane which flies at 425 miles per hour in still air is supposed to fly straight north. How should the airplane be headed and how fast will it then be flying with respect to the ground?

16. A ship is sailing due south at 20 miles per hour. A man walks west (that is, at right angles to the side of the ship) across the deck at 3 miles per hour. What is the magnitude and direction of his velocity relative to the surface of the water?

EXAMPLE C (Forces Are Vectors)   A weight of 200 kilograms is supported by two wires, as shown in Figure 32. Find the magnitude of the tension in each wire.

Solution.   The weight and the two tensions are forces which behave as vectors. These three forces must balance; that is, the two forces exerted by the wires must add together and cancel the downward force of the weight. This will happen if their sum is a vector of magnitude 200 pointing upward, as shown in Figure 33. This figure is a parallelogram composed of two congruent triangles. Using the given 35° and 41° angles and the fact that the angles of a triangle have a sum of 180°, we can find all the angles of the figure, as shown. By the law of sines,

$$\frac{\|\mathbf{u}\|}{\sin 49°} = \frac{200}{\sin 76°}$$

$$\|\mathbf{u}\| = \frac{200 \sin 49°}{\sin 76°} \approx 156$$

Similarly,

$$\|\mathbf{v}\| = \frac{200 \sin 55°}{\sin 76°} \approx 169$$

17. In Figure 34, $\|\mathbf{u}\| = \|\mathbf{v}\| = 10$. Find the magnitude and direction of a force $\mathbf{w}$ needed to counterbalance $\mathbf{u}$ and $\mathbf{v}$.

18. John pushes on a post from the direction S30°E with a force of 50 pounds. Wayne pushes on the same post from the direction S60°W with a force of 40 pounds. What is the magnitude and direction of the resultant force?

[c] 19. A body weighing 237.5 pounds is held in equilibrium by two ropes that make angles of 27.34° and 39.22°, respectively, with the vertical. Find the magnitude of the force exerted on the body by each rope.

20. A 250-kilogram weight rests on a smooth (friction negligible) inclined plane that makes an angle of 30° with the horizontal. What force parallel to the plane will just keep the weight from sliding down the plane? *Hint:* Consider the downward force of 250 kilograms to be the sum of two forces, one parallel to the plane and one perpendicular to it.

MISCELLANEOUS PROBLEMS

21. Draw the sum of the three vectors shown in Figure 35.
22. Draw $\mathbf{u} - \mathbf{v} + \frac{1}{2}\mathbf{w}$ for Figure 35.

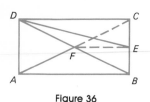

Figure 36

Figure 37

23. Three vectors form the edges of a triangle and are oriented clockwise around it. What is their sum?

24. Four vectors each of length 1 point in the directions N, N30°E, N60°E, and E, respectively. Find the exact length and direction of their sum.

25. Refer to Figure 36. Express each of the following in terms of $\overrightarrow{AD}$ and $\overrightarrow{AB}$.
   (a) $\overrightarrow{BD}$ (b) $\overrightarrow{AF}$ (c) $\overrightarrow{DE}$ (d) $\overrightarrow{AF} - \overrightarrow{DE}$

26. Let **u**, **v**, and **w** be the vectors from the vertices of a triangle to the midpoints of the opposite edges (the medians). Show that $\mathbf{u} + \mathbf{v} + \mathbf{w} = \mathbf{0}$.

27. Alice and Bette left point P at the same time and met at point Q two hours later. To get there Alice walked a straight path, but Bette first walked 1 mile south and then 2 miles in the direction S60°E. How fast did Alice walk, assuming that she walked at a constant rate?

[c] 28. Suppose that **u**, **v**, and **w** of Figure 37 point in the directions N60°E, S45°E, N25°W and have lengths 60, $30\sqrt{2}$, 100, respectively. Find the length of $\mathbf{u} + \mathbf{v} + \mathbf{w}$.

[c] 29. Two men are pushing an object along the ground. One is pushing with a force of 50 pounds in the direction N32°W and the other is pushing with a force of 100 pounds in the direction N30°E. In what direction is the object moving?

[c] 30. A pilot, flying in a wind which is blowing 80 miles per hour due south, discovers that she is heading due east when she points her plane in the direction N60°E. Find the air speed (speed in still air) of the plane.

[c] 31. What heading and air speed are required for a plane to fly 600 miles per hour due north if a wind of 56 miles per hour is blowing in the direction S12°E?

[c] 32. A spacecraft designed to softland on the moon has three legs whose feet form the vertices of an equilateral triangle on the ground. Each leg makes an angle of 35° with the vertical. If the impact force of 9000 pounds is evenly distributed, find the compression force on each leg.

[c] 33. What is the smallest force needed to keep a car weighing 3625 pounds from rolling down a hill that makes an angle of 10.35° with the horizontal?

34. Work Example C a different way as follows. For equilibrium, the magnitude $\|\mathbf{u}\| \sin 55°$ of the leftward force must equal the magnitude $\|\mathbf{v}\| \sin 49°$ of the rightward force. Similarly, the downward force of 200 must just balance the upward force of $\|\mathbf{u}\| \cos 55° + \|\mathbf{v}\| \cos 49°$. Solve the resulting pair of equations for $\|\mathbf{u}\|$ and $\|\mathbf{v}\|$ and confirm that you get the same answers we got earlier.

35. Suppose as in Example C that a weight w is supported by two wires making angles α and $\beta = 60°$ with the ceiling and creating tensions of 90 pounds in the first wire and 75 pounds in the second wire. Determine α and w by reasoning as in Problem 34.

36. **TEASER** Consider a horizontal triangular table (Figure 38) with each vertex angle being less than 120°. Three strings are knotted together at P and pass over frictionless pulleys at the vertices. Identical weights w are attached to the free ends of the strings. Show that at equilibrium, the angles between the strings at P are equal, that is, show that $\alpha + \beta = \alpha + \gamma = \beta + \gamma$.

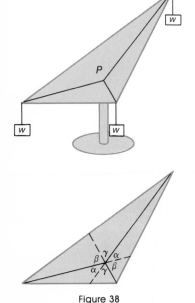

Figure 38

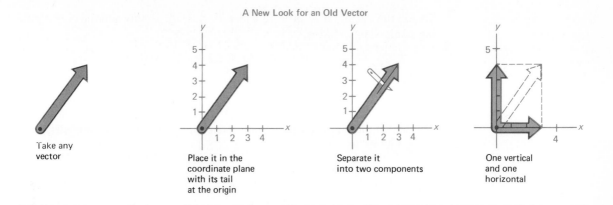

A New Look for an Old Vector

Take any vector

Place it in the coordinate plane with its tail at the origin

Separate it into two components

One vertical and one horizontal

6-4   The Algebra of Vectors

Our treatment of vectors in Section 6-3 was mainly geometric. To give the subject an algebraic appearance, we first suppose that all vectors have been placed in the ordinary cartesian coordinate plane with their tails attached to the origin, (0, 0). In this case, both the magnitude and the direction of a vector are completely determined by the position of its head.

Next we select two vectors to play a permanent and special role. The first, called **i**, is the vector from (0, 0) to (1, 0); the second, called **j**, is the vector from (0, 0) to (0, 1). Then, as Figure 39 makes clear, an arbitrary vector **u** with its head at (a, b) can be expressed uniquely in the form

$$\mathbf{u} = a\mathbf{i} + b\mathbf{j}$$

The vectors $a\mathbf{i}$ and $b\mathbf{j}$ are called the **horizontal** and **vertical vector components** of **u**, while a and b are called its **scalar components.** Notice that the length of **u** is easily expressed in terms of its scalar components: $\|\mathbf{u}\| = \sqrt{a^2 + b^2}$.

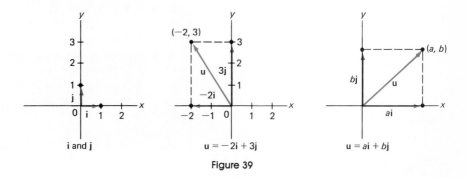

i and j

$u = -2i + 3j$

$u = ai + bj$

Figure 39

ALGEBRAIC OPERATIONS

To add the vectors $\mathbf{u} = a\mathbf{i} + b\mathbf{j}$ and $\mathbf{v} = c\mathbf{i} + d\mathbf{j}$, simply add the corresponding components; that is,

$$\mathbf{u} + \mathbf{v} = (a + c)\mathbf{i} + (b + d)\mathbf{j}$$

Similarly, to multiply $\mathbf{u}$ by the scalar k, multiply each component by k. Thus

$$k\mathbf{u} = (ka)\mathbf{i} + (kb)\mathbf{j}$$

To see that these new algebraic rules are equivalent to the old geometric ones, study Figure 40.

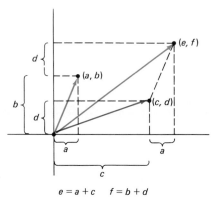
$$e = a + c \qquad f = b + d$$

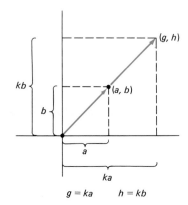
$$g = ka \qquad h = kb$$

Figure 40

Once the rules for addition and scalar multiplication are established, the rule for subtraction follows easily.

$$\mathbf{u} - \mathbf{v} = \mathbf{u} + (-1)\mathbf{v} = a\mathbf{i} + b\mathbf{j} + (-1)(c\mathbf{i} + d\mathbf{j}) = (a - c)\mathbf{i} + (b - d)\mathbf{j}$$

Moreover, with vectors written in component form, it is a simple matter to establish all of the following properties. Keep in mind that $\mathbf{0} = 0\mathbf{i} + 0\mathbf{j}$.

ALGEBRAIC PROPERTIES OF VECTORS

1. $\mathbf{u} + \mathbf{v} = \mathbf{v} + \mathbf{u}$
2. $\mathbf{u} + (\mathbf{v} + \mathbf{w}) = (\mathbf{u} + \mathbf{v}) + \mathbf{w}$
3. $\mathbf{u} + \mathbf{0} = \mathbf{0} + \mathbf{u} = \mathbf{u}$
4. $\mathbf{u} + (-\mathbf{u}) = -\mathbf{u} + \mathbf{u} = \mathbf{0}$
5. $k(\mathbf{u} + \mathbf{v}) = k\mathbf{u} + k\mathbf{v}$
6. $(k + l)\mathbf{u} = k\mathbf{u} + l\mathbf{u}$
7. $(kl)\mathbf{u} = k(l\mathbf{u}) = l(k\mathbf{u})$
8. $1\mathbf{u} = \mathbf{u}$
9. $0\mathbf{u} = \mathbf{0} = k\mathbf{0}$

THE DOT PRODUCT

Is there a sensible way to multiply two vectors together? Yes; in fact, there are two kinds of products. One, called the vector product, requires three-dimensional space and therefore falls outside of the scope of this course. The other, called the **dot product** or **scalar product**, can be introduced now. If $\mathbf{u} = a\mathbf{i} + b\mathbf{j}$ and $\mathbf{v} = c\mathbf{i} + d\mathbf{j}$, then the dot product of $\mathbf{u}$ and $\mathbf{v}$ is the scalar given by

$$\mathbf{u} \cdot \mathbf{v} = ac + bd$$

Why would anyone be interested in the dot product? To answer this, we need its geometric interpretation. Suppose the heads of $\mathbf{u} = a\mathbf{i} + b\mathbf{j}$ and $\mathbf{v} = c\mathbf{i} + d\mathbf{j}$ are at (a, b) and (c, d), as shown in Figure 41. Then we may think of $\mathbf{v} - \mathbf{u}$ as the vector from (a, b) to (c, d). Let θ denote the smallest positive angle between $\mathbf{u}$ and $\mathbf{v}$. By the law of cosines,

$$\|\mathbf{v} - \mathbf{u}\|^2 = \|\mathbf{u}\|^2 + \|\mathbf{v}\|^2 - 2\|\mathbf{u}\| \, \|\mathbf{v}\| \cos \theta$$
$$(a - c)^2 + (b - d)^2 = a^2 + b^2 + c^2 + d^2 - 2\|\mathbf{u}\| \, \|\mathbf{v}\| \cos \theta$$
$$-2ac - 2bd = -2\|\mathbf{u}\| \, \|\mathbf{v}\| \cos \theta$$
$$ac + bd = \|\mathbf{u}\| \, \|\mathbf{v}\| \cos \theta$$

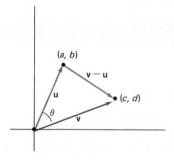

Figure 41

The last equality gives us a geometric formula for the dot product.

$$\mathbf{u} \cdot \mathbf{v} = \|\mathbf{u}\| \, \|\mathbf{v}\| \cos \theta$$

Of what use is this formula? For one thing, it gives us an easy way to tell when two vectors are perpendicular. Since $\cos \theta$ is zero if and only if θ is 90°, we see that:

Two vectors are perpendicular if and only if their dot product is zero.

More generally, we can use the formula to find the angle between any two vectors. For example, if $\mathbf{u} = 3\mathbf{i} + 4\mathbf{j}$ and $\mathbf{v} = -2\mathbf{i} + 3\mathbf{j}$, then

$$\cos \theta = \frac{\mathbf{u} \cdot \mathbf{v}}{\|\mathbf{u}\| \, \|\mathbf{v}\|} = \frac{(3)(-2) + (4)(3)}{\sqrt{9 + 16}\sqrt{4 + 9}} = \frac{6}{5\sqrt{13}} \approx .3328$$

We conclude that $\theta \approx 70.6°$.

Important applications of the dot product are to projections of one vector on another (see Example B) and to the concept of work from physics (see Example C).

Problem Set 6-4

In Problems 1–4, find $3\mathbf{u} - \mathbf{v}$, $\mathbf{u} \cdot \mathbf{v}$, *and* $\cos \theta$ *for the given vectors* $\mathbf{u}$ *and* $\mathbf{v}$.

1. $\mathbf{u} = 3\mathbf{i} - 4\mathbf{j}$, $\mathbf{v} = 5\mathbf{i} + 12\mathbf{j}$
2. $\mathbf{u} = \mathbf{i} + \sqrt{3}\mathbf{j}$, $\mathbf{v} = 6\mathbf{i} - 8\mathbf{j}$
3. $\mathbf{u} = 2\mathbf{i} - \mathbf{j}$, $\mathbf{v} = 3\mathbf{i} - 4\mathbf{j}$
4. $\mathbf{u} = \mathbf{i} + \mathbf{j}$, $\mathbf{v} = \mathbf{i} - \mathbf{j}$

ⓒ 5. If $\mathbf{u} = 14.1\mathbf{i} + 32.7\mathbf{j}$ and $\mathbf{v} = 19.2\mathbf{i} - 13.3\mathbf{j}$, find θ, the smallest positive angle between $\mathbf{u}$ and $\mathbf{v}$.

ⓒ 6. Determine the length of $2\mathbf{u} - 3\mathbf{v}$, where $\mathbf{u}$ and $\mathbf{v}$ are the vectors in Problem 5.

EXAMPLE A (Writing Vectors in the Form $a\mathbf{i} + b\mathbf{j}$**)** Write the vector $\mathbf{w}$ from the point $P(1, 5)$ to the point $Q(6, 2)$ in the form $a\mathbf{i} + b\mathbf{j}$.

Solution. We need the horizontal and vertical components of $\mathbf{w}$. They are obtained by subtracting the coordinates of the tail from those of the head. This gives $(6 - 1, 2 - 5)$ or $(5, -3)$, which are the coordinates of the head of $\mathbf{w}$ translated so that its tail is at the origin (Figure 42). Thus $\mathbf{w} = 5\mathbf{i} - 3\mathbf{j}$.

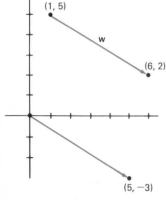

Figure 42

In Problems 7–10, let $\mathbf{u}$ *be the vector from P to Q and* $\mathbf{v}$ *be the vector from P to R. Write both vectors in the form* $a\mathbf{i} + b\mathbf{j}$ *and then find* $\mathbf{u} \cdot \mathbf{v}$.

7. $P(1, 1)$, $Q(6, 3)$, $R(5, -2)$
8. $P(-1, 2)$, $Q(-3, 6)$, $R(0, -5)$
9. $P(1, 1)$, $Q(-3, -4)$, $R(-5, 6)$
10. $P(-1, -1)$, $Q(3, -5)$, $R(2, 4)$
11. If $\mathbf{u}$ is a vector 10 units long pointing in the direction N30°W, write $\mathbf{u}$ in the form $a\mathbf{i} + b\mathbf{j}$.
12. If $\mathbf{u}$ is a vector 9 units long pointing in the direction S21°W, write $\mathbf{u}$ in the form $a\mathbf{i} + b\mathbf{j}$.
13. Determine x so that $x\mathbf{i} + \mathbf{j}$ is perpendicular to $3\mathbf{i} - 4\mathbf{j}$.
14. Determine two vectors that are perpendicular to $2\mathbf{i} + 5\mathbf{j}$. *Hint:* Try $x\mathbf{i} + \mathbf{j}$ and $x\mathbf{i} - \mathbf{j}$.
15. Find a vector of unit length that has the same direction as $\mathbf{u} = 3\mathbf{i} - 4\mathbf{j}$. *Hint:* Try $\mathbf{u}/\|\mathbf{u}\|$.
16. Find two vectors of unit length that are perpendicular to $2\mathbf{i} + 3\mathbf{j}$. *Hint:* See Problems 14 and 15.

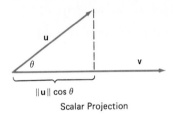

$\|\mathbf{u}\| \cos \theta$

Scalar Projection

Figure 43

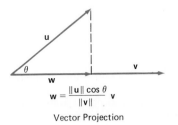

$\mathbf{w} = \dfrac{\|\mathbf{u}\| \cos \theta}{\|\mathbf{v}\|} \mathbf{v}$

Vector Projection

Figure 44

17. Find a vector twice as long as $\mathbf{u} = 2\mathbf{i} - 5\mathbf{j}$ and with opposite direction.

18. For any angle θ, show that $\mathbf{u} = (\cos \theta)\mathbf{i} + (\sin \theta)\mathbf{j}$ and $\mathbf{v} = (\sin \theta)\mathbf{i} - (\cos \theta)\mathbf{j}$ are perpendicular unit vectors.

19. Show that $\mathbf{u} \cdot \mathbf{u} = \|\mathbf{u}\|^2$ for any vector $\mathbf{u} = a\mathbf{i} + b\mathbf{j}$.

20. Show that $(k\mathbf{u}) \cdot \mathbf{v} = k(\mathbf{u} \cdot \mathbf{v})$.

EXAMPLE B (The Projection of One Vector on Another)   The scalar $\|\mathbf{u}\| \cos \theta$ is called the **scalar projection of u on v** for reasons that should be apparent from Figure 43. It is positive, zero, or negative, depending on whether θ is acute, right, or obtuse. If we multiply this scalar times $\mathbf{v}/\|\mathbf{v}\|$, we get the **vector projection of u on v.** Its geometric interpretation is shown in Figure 44. Derive formulas for these projections in terms of the dot product.

Solution.   Since

$$\|\mathbf{u}\| \cos \theta = \frac{\|\mathbf{u}\| \, \|\mathbf{v}\| \cos \theta}{\|\mathbf{v}\|}$$

it follows that

$$\text{scalar proj. } \mathbf{u} \text{ on } \mathbf{v} = \frac{\mathbf{u} \cdot \mathbf{v}}{\|\mathbf{v}\|}$$

$$\text{vector proj. } \mathbf{u} \text{ on } \mathbf{v} = \frac{\mathbf{u} \cdot \mathbf{v}}{\|\mathbf{v}\|^2} \mathbf{v} = \frac{\mathbf{u} \cdot \mathbf{v}}{\mathbf{v} \cdot \mathbf{v}} \mathbf{v}$$

In Problems 21–26, let $\mathbf{u} = 2\mathbf{i} + 9\mathbf{j}$, $\mathbf{v} = 4\mathbf{i} + 3\mathbf{j}$, *and* $\mathbf{w} = -5\mathbf{i} - 12\mathbf{j}$. *In each case, sketch the appropriate vectors and then find the indicated quantity.*

21. Scalar projection of $\mathbf{u}$ on $\mathbf{v}$.

22. Vector projection of $\mathbf{u}$ on $\mathbf{v}$.

23. Vector projection of $\mathbf{u}$ on $\mathbf{w}$.

24. Scalar projection of $\mathbf{v}$ on $\mathbf{w}$.

25. Scalar projection of $\mathbf{w}$ on $\mathbf{v}$.

26. Vector projection of $\mathbf{v}$ on $\mathbf{w}$.

EXAMPLE C (Work Done by a Force)   In physics, the **work** done by a force $\mathbf{F}$ in moving an object from P to Q is defined to be the product of the magnitude of that force times the distance from P to Q. This assumes that the force is in the direction of the motion. In the more general case where the force $\mathbf{F}$ is at an angle to the motion, we must replace the magnitude of $\mathbf{F}$ by its scalar projection in the direction of the motion. If both $\mathbf{F}$ and the displacement $\mathbf{D}$ are treated as vectors, the work done is

$$(\text{scalar proj. } \mathbf{F} \text{ on } \mathbf{D})\|\mathbf{D}\| = \frac{\mathbf{F} \cdot \mathbf{D}}{\|\mathbf{D}\|} \, \|\mathbf{D}\|$$

Thus

$$\boxed{\text{Work} = \mathbf{F} \cdot \mathbf{D}}$$

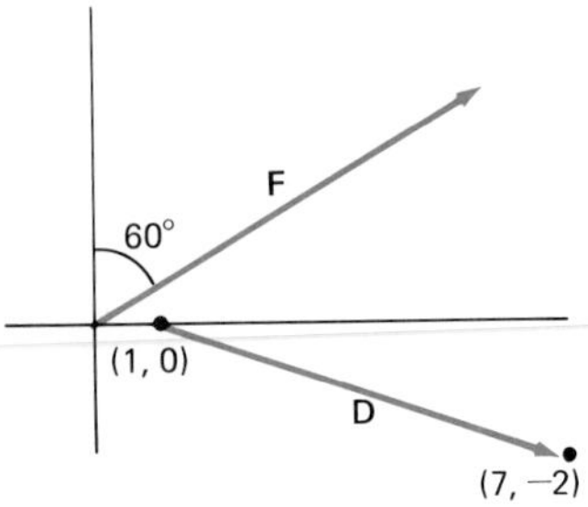

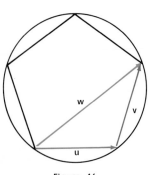

Figure 45

Use this to find the work done by a force of 80 pounds in the direction N60°E in moving an object from (1, 0) to (7, −2) as in Figure 45.

Solution. It is simply a matter of writing **F** and **D** in the form $a\mathbf{i} + b\mathbf{j}$ and taking their dot product.

$$\mathbf{F} = 80\cos 30°\mathbf{i} + 80\sin 30°\mathbf{j}$$
$$= 40\sqrt{3}\mathbf{i} + 40\mathbf{j}$$
$$\mathbf{D} = 6\mathbf{i} - 2\mathbf{j}$$
$$\mathbf{F} \cdot \mathbf{D} = 240\sqrt{3} - 80 \approx 336$$

If the units of distance are feet, the work done is 336 foot-pounds.

27. Find the work done by the force $\mathbf{F} = 3\mathbf{i} + 10\mathbf{j}$ in moving an object north 10 units.

28. Find the work done by a S70°E force of 100 dynes in moving an object 50 centimeters east.

29. Find the work done by a N45°E force of 50 dynes in moving an object from (1, 1) to (6, 9), with the distance measured in centimeters.

30. Find the work done by $\mathbf{F} = 3\mathbf{i} + 4\mathbf{j}$ in moving an object from (0, 0) to (−6, 0). Interpret the negative answer.

MISCELLANEOUS PROBLEMS

31. If $\mathbf{u} = 2\mathbf{i} + 3\mathbf{j}$ and $\mathbf{v} = 3\mathbf{i} + 4\mathbf{j}$, find $\|3\mathbf{u} - \mathbf{v}\|$.

32. For **u** and **v** in Problem 31, find $\|\mathbf{u}\|$, $\|\mathbf{v}\|$, $\mathbf{u} \cdot \mathbf{v}$, and θ (the angle between **u** and **v**).

33. Find two vectors of length 1 that are perpendicular to $3\mathbf{i} - 4\mathbf{j}$.

34. Find the vector $a\mathbf{i} + b\mathbf{j}$ that is 12 units long with the same direction as $3\mathbf{i} - 4\mathbf{j}$.

35. Find the vector projection of $5\mathbf{i} + 3\mathbf{j}$ on $3\mathbf{i} - 4\mathbf{j}$. Also find the angle between these two vectors.

36. Find the work done by a force of 100 pounds directed N45° E in moving an object along the line from (1, 1) to (7, 5), distances measured in feet.

37. Show that for any two vectors **u** and **v**,

$$|\mathbf{u} \cdot \mathbf{v}| \le \|\mathbf{u}\| \|\mathbf{v}\|$$

When will equality hold?

38. Prove that $\mathbf{u} \cdot \mathbf{v} = \mathbf{v} \cdot \mathbf{u}$ and that $\mathbf{u} \cdot (\mathbf{v} + \mathbf{w}) = \mathbf{u} \cdot \mathbf{v} + \mathbf{u} \cdot \mathbf{w}$.

39. If $\mathbf{u} + \mathbf{v}$ and $\mathbf{u} - \mathbf{v}$ are perpendicular, what can we conclude about $\|\mathbf{u}\|$ and $\|\mathbf{v}\|$?

40. Show that $\|\mathbf{u} + \mathbf{v}\|^2 + \|\mathbf{u} - \mathbf{v}\|^2 = 2(\|\mathbf{u}\|^2 + \|\mathbf{v}\|^2)$.

41. Find the exact value of sin 18°, which may be needed to complete Problem 42. *Hint:* Let $\theta = 18°$ and note that $\cos 3\theta = \sin 2\theta = 2\sin\theta\cos\theta$ and that $\cos 3\theta = \cos(2\theta + \theta) = (1 - 4\sin^2\theta)\cos\theta$.

42. **TEASER** Let **u** and **v** denote adjacent edges of a regular pentagon that is inscribed in a circle of radius 1 (Figure 46). Let $\mathbf{w} = \mathbf{u} + \mathbf{v}$.
    (a) Express $\mathbf{u} \cdot \mathbf{w}$ and $\|\mathbf{u}\|\|\mathbf{w}\|$ in terms of cos 36°.
    (b) Use your calculator to guess the exact value of $(\|\mathbf{u}\|\|\mathbf{w}\|)^2$ and then prove that this is the correct value.

Figure 46

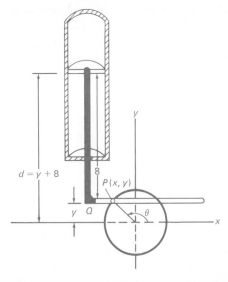

A Piston Problem

One end of an 8-foot shaft is attached to a piston that moves up and down. The other end is attached to a wheel by means of a horizontal slotted arm which fits over a peg P on the rim. Starting at an initial position of $\theta = \pi/4$, the wheel of radius 2 feet rotates at a rate of 3 radians per second. Find a formula for d, the vertical distance from the piston to the wheel center, after t seconds.

6-5 Simple Harmonic Motion

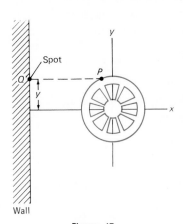

Figure 47

The up-and-down motion of the piston above is an example of what is called simple harmonic motion. Notice right away that the motion of the piston is essentially the same as that of the point Q. That means we want to find y; and y is just the y-coordinate of the peg P.

Perhaps the piston-wheel device seems complicated, so let's consider another version of the same problem. Imagine the wheel shown in Figure 47 to be turning at a uniform rate in the counterclockwise direction. Emanating from P, a point attached to the rim, is a horizontal beam of light, which projects a bright spot at Q on a nearby vertical wall. As the wheel turns, the spot at Q moves up and down. Our task is to express the y-coordinate of Q (which is also the y-coordinate of P) in terms of the elapsed time t.

The solution to this problem depends on a number of factors (the rate at which the wheel turns, the radius of the wheel, and the location of P at $t = 0$). We think it wise to begin with a simple case and gradually extend to more general situations.

Case 1 Suppose the wheel has radius 1, that it turns at 1 radian per second, and that it starts at $\theta = 0$. Then at time t, θ will measure t radians and P will have y-coordinate

$$y = \sin t$$

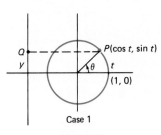

Case 1

Figure 48

(see Figure 48). Keep in mind that this equation describes the up-and-down motion of Q.

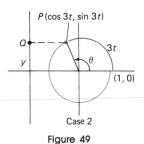

P(cos 3t, sin 3t)

Q

y

3t

θ

(1, 0)

Case 2

Figure 49

Case 2   Let everything be as in the first case, but now let the wheel turn at 3 radians per second (Figure 49). Then at time t, θ will measure $3t$ radians and both P and Q will have y-coordinate.

$$y = \sin 3t$$

Case 3   Next increase the radius of the wheel to 2 feet, but leave the other information as in Case 2 (Figure 50). Now the coordinates of P are $(2 \cos 3t,\ 2 \sin 3t)$ and

$$y = 2 \sin 3t$$

Case 4   Finally, let the wheel start at $\theta = \pi/4$ rather than $\theta = 0$. With the help of Figure 51, we see that

$$y = 2 \sin\left(3t + \frac{\pi}{4}\right)$$

Case 4 describes the wheel of the original piston-wheel problem. The number y measures the distance between Q and the x-axis, and $d = y + 8$ is the distance from the piston to the x-axis. Thus the answer to the question first posed is

$$d = 8 + 2 \sin\left(3t + \frac{\pi}{4}\right)$$

The number 8 does not interest us; it is the sine expression that is significant. As a matter of fact, equations of the form

$$y = A \sin(Bt + C) \quad \text{and} \quad y = A \cos(Bt + C)$$

with $B > 0$ arise often in physics. Any straight-line motion which can be described by one of these formulas is called **simple harmonic motion.** Cases

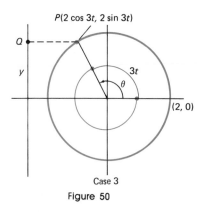

P(2 cos 3t, 2 sin 3t)

Q

y

3t

θ

(2, 0)

Case 3

Figure 50

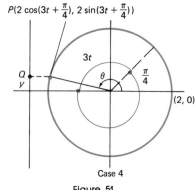

P(2 cos(3t + $\frac{\pi}{4}$), 2 sin(3t + $\frac{\pi}{4}$))

3t

Q

y

θ

$\frac{\pi}{4}$

(2, 0)

Case 4

Figure 51

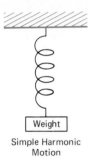

Simple Harmonic
Motion

Figure 52

1–4 are examples of this motion. Other examples from physics occur in connection with the motion of a weight attached to a vibrating spring (Figure 52) and the motion of a water molecule in an ocean wave. Voltage in an alternating current, although it does not involve motion, is given by the same kind of sine (or cosine) equation.

GRAPHS

The graphs of the four boxed equations given on pages 283-84 are worthy of study. They are shown in Figure 53. Note how the graph of $y = \sin t$ is progressively modified as we move from Case 1 to Case 4.

Under each graph are listed three important numbers, numbers that identify the critical features of the graph. The **period** is the length of the shortest interval after which the graph repeats itself. The **amplitude** is the maximum distance of the graph from its median position (the t-axis). The **phase shift** measures the distance the graph is shifted horizontally from its normal position.

You might have expected a phase shift of $-\pi/4$ in Case 4, since the initial angle of the wheel measured $\pi/4$ radians. But, note that factoring 3 from $3t + \pi/4$ gives

$$y = 2 \sin\left(3t + \frac{\pi}{4}\right) = 2 \sin 3\left(t + \frac{\pi}{12}\right)$$

If you recall our discussion of translations (see Section 2-6), you see why the graph is shifted $\pi/12$ units to the left. Note in particular that $y = 0$ when $t = -\pi/12$ instead of when $t = 0$.

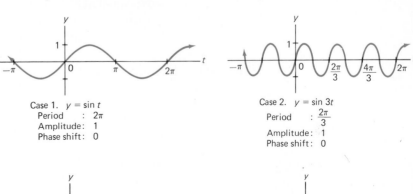

Case 1. $y = \sin t$
Period   : 2π
Amplitude: 1
Phase shift: 0

Case 2. $y = \sin 3t$
Period   : $\dfrac{2\pi}{3}$
Amplitude: 1
Phase shift: 0

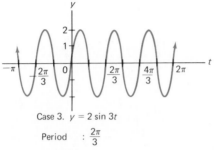

Case 3. $y = 2 \sin 3t$

Period   : $\dfrac{2\pi}{3}$

Amplitude : 2
Phase shift : 0

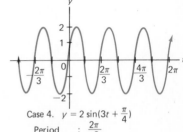

Case 4. $y = 2 \sin(3t + \frac{\pi}{4})$
Period   : $\dfrac{2\pi}{3}$
Amplitude: 2
Phase shift: $-\dfrac{\pi}{12}$

Figure 53

GRAPHING IN THE GENERAL CASE

If

$$y = A \sin(Bt + C) \quad \text{or} \quad y = A \cos(Bt + C)$$

with $B > 0$, all three concepts (period, amplitude, phase shift) make good sense. We have the following formulas.

Period:	$\dfrac{2\pi}{B}$
Amplitude:	$\|A\|$
Phase shift:	$\dfrac{-C}{B}$

Knowing these three numbers is a great aid in graphing. For example, to graph

$$y = 3 \cos\left(4t - \frac{\pi}{4}\right)$$

we recall the graph of $y = \cos t$ and then modify it using the three numbers.

$$\text{Period:} \quad \frac{2\pi}{B} = \frac{2\pi}{4} = \frac{\pi}{2}$$

$$\text{Amplitude:} \quad |A| = |3| = 3$$

$$\text{Phase shift:} \quad -\frac{C}{B} = \frac{\pi/4}{4} = \frac{\pi}{16}$$

The result is shown in Figure 54.

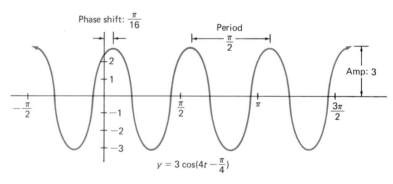

$$y = 3 \cos\left(4t - \frac{\pi}{4}\right)$$

Figure 54

Problem Set 6-5

1. Sketch the graphs of the following equations in the order given. Use the interval $-2\pi \leq t \leq 2\pi$.
   (a) $y = \cos t$ (b) $y = \cos 2t$
   (c) $y = 4 \cos 2t$ (d) $y = 4 \cos(2t + \pi/3)$

2. Sketch the graphs of the following on $-2\pi \leq t \leq 4\pi$.
   (a) $y = \sin t$ (b) $y = \sin \frac{1}{2} t$
   (c) $y = 3 \sin \frac{1}{2} t$ (d) $y = 3 \sin\left(\frac{1}{2} t + \pi/2\right)$

EXAMPLE A (More Graphing) Sketch the graph of $y = 3 \sin(\frac{1}{2}t + \pi/8)$.

Solution. We begin by finding the three key numbers.

$$\text{Period:} \qquad \frac{2\pi}{B} = \frac{2\pi}{\frac{1}{2}} = 4\pi$$

$$\text{Amplitude:} \qquad |A| = |3| = 3$$

$$\text{Phase shift:} \qquad \frac{-C}{B} = -\frac{\pi/8}{\frac{1}{2}} = -\frac{\pi}{4}$$

Then we draw the graph of Figure 55.

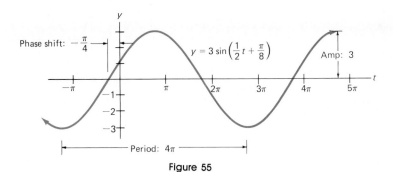

Figure 55

In Problems 3–6, find the period, amplitude, and phase shift. Then sketch the graph.

3. (a) $y = 4 \sin 2t$ (b) $y = 3 \cos\left(t + \frac{\pi}{8}\right)$

   (c) $y = \sin\left(4t + \frac{\pi}{8}\right)$ (d) $y = 3 \cos\left(3t - \frac{\pi}{2}\right)$

4. (a) $y = \frac{1}{2} \cos 3t$ (b) $y = 3 \sin\left(t - \frac{\pi}{6}\right)$

   (c) $y = 2 \sin\left(\frac{1}{2} t + \frac{\pi}{8}\right)$ (d) $y = \frac{1}{2} \sin(2t - 1)$

5. $y = 3 + 2 \cos\left(\frac{1}{2}t - \pi/16\right)$. *Hint:* The number 3 lifts the graph of $y = 2\cos\left(\frac{1}{2}t - \pi/16\right)$ up 3 units.

6. $y = 4 + 3 \sin(2t + \pi/16)$

EXAMPLE B (*Negative A*) Sketch the graph of $y = -3 \cos 2t$.

Solution. We begin by asking how the graph of $y = -3 \cos 2t$ relates to that of $y = 3 \cos 2t$. Clearly, every y value has the opposite sign, which has the effect of reflecting the graph about the t-axis. Then we calculate the three crucial numbers.

$$\text{Period:} \qquad \frac{2\pi}{B} = \frac{2\pi}{2} = \pi$$

$$\text{Amplitude:} \qquad |A| = |-3| = 3$$

$$\text{Phase shift:} \qquad \frac{-C}{B} = \frac{0}{2} = 0$$

Finally we sketch the graph (Figure 56).

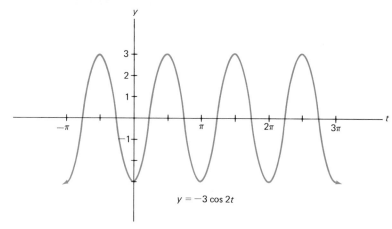

$y = -3 \cos 2t$

Figure 56

Now sketch the graphs of the equations in Problems 7–12.

7. $y = -2 \sin 3t$ 8. $y = -4 \cos \frac{1}{2}t$

9. $y = \sin(2t - \pi/3)$ 10. $y = -\cos(3t + \pi)$

11. $y = -2 \cos\left(t - \frac{1}{6}\right)$ 12. $y = -3 \sin(3t + 3)$

13. A wheel with center at the origin is rotating counterclockwise at 4 radians per second. There is a small hole in the wheel 5 centimeters from the center. If that hole has initial coordinates $(5, 0)$, what will its coordinates be after t seconds?

14. Answer Problem 13 if the hole is initially at $(0, 5)$.

15. A free-hanging shaft, 8 centimeters long, is attached to the wheel of Problem 13 by putting a bolt through the hole (Figure 57). What are the coordinates of P, the bottom point of the shaft, at time t (assuming the shaft continues to hang vertically)?

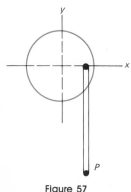

Figure 57

16. Suppose the wheel of Problem 13 rotates at 3 revolutions per second. What are the coordinates of the hole after t seconds?

MISCELLANEOUS PROBLEMS

17. Find the period, amplitude, and phase shift for each graph.
    (a) $y = \sin 5t$ (b) $y = \frac{3}{2} \cos(\frac{1}{2}t)$
    (c) $y = 2 \cos(4t - \pi)$ (d) $y = -4 \sin(3t + 3\pi/4)$

18. Sketch the graphs of the equations in Problem 17 on the interval $-\pi \le t \le 2\pi$.

19. The weight attached to a spring (Figure 58) is bobbing up and down so that

$$y = 8 + 4 \cos\left(\frac{\pi}{2}t + \frac{\pi}{4}\right)$$

where y and t are measured in feet and seconds, respectively. What is the closest the weight gets to the ceiling and when does this first happen for $t > 0$?

20. The equations $x = 2 + 2 \cos 4t$ and $y = 6 + 2 \sin 4t$ give the coordinates of a point moving along the circumference of a circle. Determine the center and radius of the circle. How long does it take for the point to make a complete revolution?

21. Consider the wheel-piston device shown in Figure 59 (which is analogous to the crankshaft and piston in an automobile engine). The wheel has a radius of 1 foot and rotates counterclockwise at 1 radian per second; the connecting rod is 5 feet long. If the point P is initially at $(1, 0)$, find the y-coordinate of Q after t seconds. Assume the x-coordinate is always zero.

22. Redo problem 21, but assume the wheel has radius 2 feet and rotates at 60 revolutions per second and that P is initially at $(2, 0)$. Is Q executing simple harmonic motion in either of these problems?

23. The voltage drop E across the terminals in a certain alternating current circuit is approximately $E = 156 \sin(110\pi t)$, where t is in seconds. What is the maximum voltage drop and what is the **frequency** (number of cycles per second) for this circuit?

24. The carrier wave for the radio wave of a certain FM station has the form $y = A \sin(2\pi \cdot 10^8 t)$, where t is measured in seconds. What is the frequency for this wave?

25. The AM radio wave for a certain station has the form

$$y = 55[1 + .02 \sin(2400\pi t)] \sin(2 \times 10^5 \pi t)$$

    (a) Find y when $t = 3$.
    (b) Find y when $t = .03216$.
    [c] (c) Find y when $t = .0000321$.

26. In predator-prey systems, the number of predators and the number of prey tend to vary periodically. In a certain region with coyotes as predators and rabbits as prey, the rabbit population R varied according to the formula

$$R = 1000 + 150 \sin 2t$$

where t was measured in years after January 1, 1950.
    (a) What was the maximum rabbit population?
    (b) When was it first reached?
    [c] (c) What was the population on January 1, 1953?

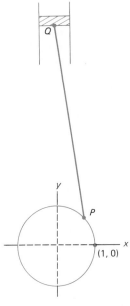

Ceiling

Weight

Figure 58

Q

P

$(1, 0)$

Figure 59

27. The number of coyotes C in Problem 26 satisfied

$$C = 200 + 50 \sin(2t - .7)$$

Sketch the graphs of C and R using the same coordinate system and attempt to explain the phase shift in C.

[c] 28. Sketch the graph of $y = 2^{-t} \cos 2t$ for $0 \le t \le 3\pi$. This is an example of damped harmonic motion, which is typical of harmonic motion where there is friction.

29. Use addition laws to write each of the following in the form $A_1 \sin Bt + A_2 \cos Bt$
    (a) $4 \sin(2t - \frac{\pi}{4})$ (b) $3 \cos(3t + \frac{\pi}{3})$

    Note: The same idea would work on any expression of the form $A \sin(Bt + C)$ or $A \cos(Bt + C)$.

30. Determine C so that

$$5 \sin 4t + 12 \cos 4t = 13 \sin(4t + C)$$

31. Suppose that A_1 and A_2 are both positive. Show that

$$A_1 \sin Bt + A_2 \cos Bt = A \sin(Bt + C)$$

    where $A = \sqrt{A_1^2 + A_2^2}$ and $C = \tan^{-1}(A_2/A_1)$.

32. Generalize Problem 31 by showing that $A_1 \sin Bt + A_2 \cos Bt$ can always be written in the form $A \sin(Bt + C)$. *Hint:* Choose A as in Problem 31 and let C be the radian measure of an angle that has (A_1, A_2) on its terminal side.

33. Use the result in Problems 31 and 32 to write each of the following in the form $A \sin(Bt + C)$.
    (a) $4 \cos 2t + 3 \sin 2t$ (b) $3 \sin 4t - \sqrt{3} \cos 4t$

34. Give an argument to show that

$$A_1 \sin Bt + A_2 \cos Bt, \qquad A_1 A_2 \ne 0, B \ne 0$$

    is not a polynomial in t for any choices of A_1, A_2, and B.

35. Find the maximum and minimum values of $\cos t \pm \sin t$.

36. **TEASER** Prove that $\sin(\cos t) < \cos(\sin t)$ for all t. *Hint:* First show that

$$-\frac{\pi}{2} < \cos t < \frac{\pi}{2} - |\sin t|$$

Chapter Summary

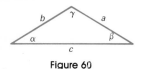

Figure 60

A triangle like the one in Figure 60 that has no right angle is called an oblique triangle. If any three of the six parts α, β, γ, a, b, and c—including at least one side—are given, we can find the remaining parts by using the **law of sines**

$$\frac{a}{\sin \alpha} = \frac{b}{\sin \beta} = \frac{c}{\sin \gamma}$$

and the **law of cosines**

$$a^2 = b^2 + c^2 - 2bc \cos \alpha \qquad \text{(one of three forms)}$$

There is, however, one case (given two sides and an angle opposite one of them) in which there might be no solution or two solutions. We call it the ambiguous case.

Vectors, represented by arrows, play an important role in science because they have both **magnitude** (length) and **direction.** We call two vectors **equivalent** if they have the same magnitude and direction. Vectors can be added (using the parallelogram law) and multiplied by **scalars,** which are real numbers. The special vectors **i** and **j** allow us to write any vector in the form $a\mathbf{i} + b\mathbf{j}$, where a and b are scalars. If $\mathbf{u} = a\mathbf{i} + b\mathbf{j}$ is a vector, its **length** $\|\mathbf{u}\|$ is given by

$$\|\mathbf{u}\| = \sqrt{a^2 + b^2}$$

If $\mathbf{v} = c\mathbf{i} + d\mathbf{j}$ is another such vector, the **dot product** $\mathbf{u} \cdot \mathbf{v}$ of $\mathbf{u}$ and $\mathbf{v}$ is

$$\mathbf{u} \cdot \mathbf{v} = ac + bd = \|\mathbf{u}\| \|\mathbf{v}\| \cos \theta$$

where θ is the smallest positive angle between $\mathbf{u}$ and $\mathbf{v}$. Two vectors are perpendicular if and only if their dot product is zero.

The equations $y = A \sin(Bt + C)$ and $y = A \cos(Bt + C)$ with $B > 0$ describe a common phenomenon known as **simple harmonic motion.** We can quickly draw the graphs of these equations by making use of three key numbers: the **amplitude,** $|A|$, the **period,** $2\pi/B$, and the **phase shift,** $-C/B$.

Chapter Review Problem Set

1. Solve each of the following triangles using Table C or a calculator.
   (a) $\alpha = 104.9°$, $\gamma = 36°$, $b = 149$
   (b) $a = 14.6$, $b = 89.2$, $c = 75.8$
   (c) $\gamma = 35°$, $a = 14$, $b = 22$
   (d) $\beta = 48.6°$, $c = 39.2$, $b = 57.6$

2. For the triangle in Figure 61, find x and the area of the triangle.

3. If $\mathbf{u}$ is a vector 10 units long pointing in the direction N30°E and $\mathbf{v} = 3\mathbf{i} - 9\mathbf{j}$, calculate $\mathbf{u} + \frac{1}{3}\mathbf{v}$ and write it in the form $a\mathbf{i} + b\mathbf{j}$.

4. If $\mathbf{u} = 3\mathbf{i} - 4\mathbf{j}$ and $\mathbf{v} = 5\mathbf{i} + 12\mathbf{j}$, calculate each of the following.
   (a) $\|\mathbf{u}\|$ (b) $\|\mathbf{v}\|$ (c) $\mathbf{u} \cdot \mathbf{v}$
   (d) The angle θ between $\mathbf{u}$ and $\mathbf{v}$.
   (e) The scalar projection of $\mathbf{u}$ on $\mathbf{v}$.
   (f) The vector projection of $\mathbf{u}$ on $\mathbf{v}$.

5. Two men, A and B, are pushing an object along the ground. A pushes with a force $\mathbf{u}$ of 100 pounds in the direction N(arctan $\frac{4}{3}$) E; B pushes with a force $\mathbf{v}$ of 40 pounds straight east.
   (a) Find the resultant force $\mathbf{w} = \mathbf{u} + \mathbf{v}$ and write it in the form $a\mathbf{i} + b\mathbf{j}$.
   (b) Calculate the work done by A when the object is moved a distance of 2 feet in the direction of $\mathbf{w}$.

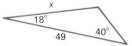

x

$18°$

49 $40°$

Figure 61

6. Find the period, amplitude, and phase shift for each of the following.
   (a) $y = \cos 2t$ (b) $y = 3 \cos 4t$
   (c) $y = 2 \sin(3t - (\pi/2))$ (d) $y = -2 \sin(\frac{1}{2}t + \pi)$

7. Sketch the graphs of the equations in Problem 6 on $-\pi \le t \le \pi$.

8. A wheel of radius 4 feet with center at the origin is rotating counterclockwise at $3\pi/4$ radians per second (Figure 62). If a paint speck P has coordinates $(-4, 0)$ initially, what will be its coordinates after t seconds?

9. A point is moving in a straight line according to the equation $x = 3 \cos(5t + 3\pi)$, where x is in feet and t in seconds.
   (a) What is the period of the motion?
   (b) Find the initial position of the point relative to the point $x = 0$.
   (c) When is $x = 3$ for the first time?

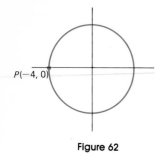

$P(-4, 0)$

Figure 62

If it is true . . . that the essence of the triumphs of science and its progress consists in that it enables us to consider evident and necessary, views which our ancestors held to be unintelligible and were unable to comprehend, then the extension of the number concept to include the irrational and, we will at once add, the imaginary, is the greatest step which pure mathematics has ever taken.

Hermann Hankel

CHAPTER 7

The Complex Number System

$$i = \sqrt{-1}$$

"The Divine Spirit found a sublime outlet in that wonder of analysis, that portent of the ideal world, that amphibian between being and not-being, which we call the imaginary root of negative unity."

Gottfried Wilhelm Leibniz
1646–1716

7-1   The Complex Numbers

As early as 1550, the Italian mathematician Raffael Bombelli had introduced numbers like $\sqrt{-1}$, $\sqrt{-2}$, and so on, to solve certain equations. But mathematicians had a hard time deciding whether they should be considered legitimate numbers. Even Leibniz, who ranks with Newton as the greatest mathematician of the late seventeenth century, called them amphibians between being and not being. He wrote them down, he used them in calculations, but he carefully covered his tracks by calling them imaginary numbers. Unfortunately that name (which was actually first used by Descartes) has stuck, though these numbers are now well accepted by all mathematicians and have numerous applications in science. Let us see what these new numbers are and why they are needed.

Go back to the whole numbers 0, 1, 2, 3, ... We can easily solve the equation $x + 3 = 7$ within this system ($x = 4$). On the other hand, the equation $x + 7 = 3$ has no whole number solution. To solve it, we need the negative integer -4. Similarly, we cannot solve $3x = 2$ in the integers. We can say that the solution is $\frac{2}{3}$ only after the rational numbers have been introduced. To their dismay, the Greeks discovered that $x^2 - 2 = 0$ had no rational solution. We conquered that problem by enlarging our family of numbers to the real numbers. But there are still simple equations without solutions. Consider $x^2 + 1 = 0$. Try as you will, you will never solve it within the real number system.

By now, the procedure is well established. When we need new numbers, we invent them. This time we invent a number denoted by i (or by $\sqrt{-1}$) which satisfies $i^2 = -1$. However, we cannot get by with just one number. For after we have adjoined it to the real numbers, we still must be able to multiply and add. Thus with i, we also need numbers such as

$$2i \qquad -4i \qquad (\tfrac{3}{2})i, \ldots$$

which are called pure imaginary numbers. We also need

$$3 + 2i \qquad 11 + (-4i) \qquad \tfrac{3}{4} + \tfrac{3}{2}i, \ldots$$

and it appears that we need even more complicated things such as

$$(3 + 8i + 2i^2 + 6i^3)(5 + 2i)$$

Actually, this last number can be simplified to $1 + 12i$ as we shall see later. In fact, no matter how many additions and multiplications we do, after the expressions are simplified we shall never have anything more complicated than a number of the form $a + bi$ (a fact that Figure 1 is meant to illustrate). Such numbers, that is, numbers of the form $a + bi$, where a and b are real, are called **complex numbers**. We refer to a as the **real part** and b as the **imaginary part** of $a + bi$. Since we shall agree that $0 \cdot i = 0$, it follows that $a + 0i = a$, and so every real number is automatically a complex number. If $b \neq 0$, then $a + bi$ is nonreal, and in this case $a + bi$ is said to be an imaginary number.

ADDITION AND MULTIPLICATION

We cannot say anything sensible about operations for complex numbers until we agree on the meaning of equality. The definition that seems most natural is this:

$$\boxed{a + bi = c + di \quad \text{means} \quad a = c \quad \text{and} \quad b = d}$$

That is, two complex numbers are equal if and only if their real parts and their imaginary parts are equal. As an example, suppose $x^2 + yi = 4 - 3i$. Then we know that $x = \pm 2$ and $y = -3$.

Now we can consider addition. Actually we have already used the plus sign in $a + bi$. That was like trying to add apples and bananas. The addition can be indicated but no further simplification is possible. We do not get apples and we do not get bananas; we get fruit salad.

When we have two numbers of the form $a + bi$, we can actually perform an addition. We just add the real parts and the imaginary parts separately—that is, we add the apples and we add the bananas. Thus

$$(3 + 2i) + (6 + 5i) = 9 + 7i$$

and, more generally,

$$\boxed{(a + bi) + (c + di) = (a + c) + (b + d)i}$$

When we consider multiplication, our desire to maintain the standard properties of numbers leads to a definition that looks complicated. Thus let us first look at some examples.

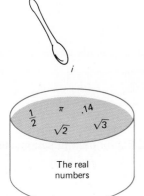

The real numbers

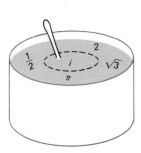

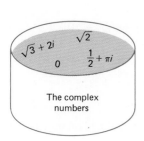

The complex numbers

Figure 1

(i)	$2(3 + 4i) = 6 + 8i$	(distributive property)
(ii)	$(3i)(-4i) = -12i^2 = 12$	(commutative and associative properties and $i^2 = -1$)

(iii)	$(3 + 2i)(6 + 5i) = (3 + 2i)6 + (3 + 2i)5i$	(distributive property)
	$= 18 + 12i + 15i + 10i^2$	(distributive, commutative, and associative properties)
	$= 18 + (12 + 15)i - 10$	(distributive property)
	$= 8 + 27i$	(commutative property)

The same kind of reasoning applied to the general case leads to

$$(a + bi)(c + di) = (ac - bd) + (ad + bc)i$$

which we take as the definition of multiplication for complex numbers.

Actually there is no need to memorize the formula for multiplication. Just do what comes naturally (that is, use familiar properties) and then replace i^2 by -1 wherever it arises, as in the following example.

$$(2 - 3i)(5 + 4i) = (10 - 12i^2) + (8i - 15i)$$

$$= (10 + 12) + (-7i)$$

$$= 22 - 7i$$

Consider the more complicated expression mentioned earlier. After noting that $i^3 = i^2 i = -i$, we have

$$(3 + 8i + 2i^2 + 6i^3)(5 + 2i) = (3 + 8i - 2 - 6i)(5 + 2i)$$

$$= (1 + 2i)(5 + 2i)$$

$$= (5 + 4i^2) + (2i + 10i)$$

$$= 1 + 12i$$

SUBTRACTION AND DIVISION

Subtraction is easy; we simply subtract corresponding real and imaginary parts. For example,

$$(3 + 6i) - (5 + 2i) = (3 - 5) + (6i - 2i)$$
$$= -2 + 4i$$

and

$$(5 + 2i) - (3 + 7i) = (5 - 3) + (2i - 7i)$$
$$= 2 + (-5i)$$
$$= 2 - 5i$$

Division is somewhat more difficult. We first note that $a - bi$ is called the **conjugate** of $a + bi$. Thus $2 - 3i$ is the conjugate of $2 + 3i$ and $-2 + 5i$ is the conjugate of $-2 - 5i$. Next, we observe that a complex number times its conjugate is a real number. For example,

$$(3 + 4i)(3 - 4i) = 9 + 16 = 25$$

and in general

$$(a + bi)(a - bi) = a^2 + b^2$$

To simplify the quotient of two complex numbers, multiply the numerator and denominator by the conjugate of the denominator, as illustrated below.

$$\frac{2 + 3i}{3 + 4i} = \frac{(2 + 3i)(3 - 4i)}{(3 + 4i)(3 - 4i)} = \frac{18 + i}{9 + 16} = \frac{18}{25} + \frac{1}{25}i$$

The effect of this multiplication is to replace a complex denominator by a real one. Notice that the result of the division is a number in the form $a + bi$.

A GENUINE EXTENSION

We assert that the complex numbers constitute a genuine enlargement of the real numbers. This means first of all that they include the real numbers, since any real number a can be written as $a + 0i$. Second, the complex numbers satisfy all the standard properties of the real numbers (associative, commutative, and distributive laws and existence of inverses). However, we do not introduce the order relation $<$ for the complex numbers.

Figure 2 on the next page indicates how the number systems of mathematics are built up. Is there any need to enlarge the number system again? The answer is no, and for a good reason. With the complex numbers, we can solve any equation that arises in algebra. Right now, we expect you to take this statement on faith. Another look at the quadratic formula provides some evidence.

THE QUADRATIC FORMULA REVISITED

In Section 1-4, we completed the square to solve

$$ax^2 + bx + c = 0$$

where a, b, and c are arbitrary real numbers ($a \neq 0$). We learned that solutions

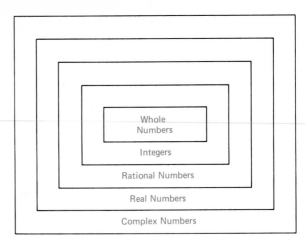

Whole
Numbers

Integers

Rational Numbers

Real Numbers

Complex Numbers

Figure 2

are given by the quadratic formula

$$x = \frac{-b \pm \sqrt{b^2 - 4ac}}{2a}$$

If $b^2 - 4ac \geq 0$, this formula yields real number answers; but if $b^2 - 4ac < 0$, there are no real solutions, since a negative number cannot have a real square root. That is all that we could say until now. However, with the introduction of the complex numbers, we can make perfectly good sense of $\sqrt{b^2 - 4ac}$ even if $b^2 - 4ac < 0$. Here is how we do it.

If $A > 0$, we define $\sqrt{-A}$ by

$$\sqrt{-A} = \sqrt{A(-1)} = \sqrt{A}\sqrt{-1} = \sqrt{A}\,i$$

For example,

$$\sqrt{-36} = \sqrt{36}\,i = 6i$$

and

$$\sqrt{-27} = \sqrt{27}\,i = 3\sqrt{3}\,i$$

With this understanding, the quadratic formula yields solutions to $ax^2 + bx + c = 0$ even when $b^2 - 4ac < 0$. In that case, the two solutions are complex numbers; in fact, they are conjugates of each other.

Here are two illustrations. The equation

$$x^2 - 2x + 10 = 0$$

has solutions

$$x = \frac{2 \pm \sqrt{4 - 40}}{2} = \frac{2 \pm \sqrt{-36}}{2} = \frac{2 \pm 6i}{2} = 1 \pm 3i$$

Similarly,

$$9x^2 + 3x + 1 = 0$$

has solutions

$$x = \frac{-3 \pm \sqrt{9 - 36}}{18} = \frac{-3 \pm \sqrt{-27}}{18} = \frac{-3 \pm 3\sqrt{3}\,i}{18} = -\frac{1}{6} \pm \frac{\sqrt{3}}{6}i$$

Given a quadratic equation with real coefficients a, b, and c, we can describe its solutions in terms of the number $b^2 - 4ac$, called the **discriminant**.

1. If $b^2 - 4ac > 0$, there are two real solutions.
2. If $b^2 - 4ac = 0$, there is one real solution.
3. If $b^2 - 4ac < 0$, there are two complex (nonreal) solutions, which are conjugates of each other.

Problem Set 7-1

Carry out the indicated operations and write the answer in the form $a + bi$.

1. $(2 + 3i) + (-4 + 5i)$
2. $(3 - 4i) + (-5 - 6i)$
3. $5i - (4 + 6i)$
4. $(-3i + 4) + 8i$
5. $(3i - 6) + (3i + 6)$
6. $(6 + 3i) + (6 - 3i)$
7. $4i^2 + 7i$
8. $i^3 + 2i$
9. $i(4 - 11i)$
10. $(3 + 5i)i$
11. $(3i + 5)(2i + 4)$
12. $(2i + 3)(3i - 7)$
13. $(3i + 5)^2$
14. $(-3i - 5)^2$
15. $(5 + 6i)(5 - 6i)$
16. $(\sqrt{3} + \sqrt{2}\,i)(\sqrt{3} - \sqrt{2}\,i)$
17. $\dfrac{5 + 2i}{1 - i}$
18. $\dfrac{4 + 9i}{2 + 3i}$
19. $\dfrac{5 + 2i}{i}$
20. $\dfrac{-4 + 11i}{-i}$
21. $\dfrac{(2 + i)(3 + 2i)}{1 + i}$
22. $\dfrac{(4 - i)(5 - 2i)}{4 + i}$
23. $\dfrac{4 + 2i}{4 + 3i}$
24. $\dfrac{-3 + 2i}{5 - 3i}$
25. $\dfrac{2(2 + i)^2}{i}$
26. $\dfrac{3(1 - 3i)^2}{1 + i}$

EXAMPLE A (Multiplicative Inverses) Find $(5 + 4i)^{-1}$, the multiplicative inverse of $5 + 4i$.

Solution. The multiplicative inverse is just the reciprocal. Thus

$$(5 + 4i)^{-1} = \frac{1}{5 + 4i} = \frac{1(5 - 4i)}{(5 + 4i)(5 - 4i)} = \frac{5 - 4i}{25 - 16i^2}$$

$$= \frac{5 - 4i}{25 + 16} = \frac{5 - 4i}{41} = \frac{5}{41} - \frac{4}{41}i$$

Find each inverse.

27. $(2 - i)^{-1}$　　　　　　　　　　28. $(3 + 2i)^{-1}$

29. $(\sqrt{3} + i)^{-1}$　　　　　　　　　30. $(\sqrt{2} - \sqrt{2}\,i)^{-1}$

Use the results above to perform the following divisions. For example, division by $2 - i$ is the same as multiplication by $(2 - i)^{-1}$.

31. $\dfrac{2 + 3i}{2 - i}$　　　　　　　　　32. $\dfrac{1 + 2i}{3 + 2i}$

33. $\dfrac{4 - i}{\sqrt{3} + i}$　　　　　　　　34. $\dfrac{\sqrt{2} + \sqrt{2}\,i}{\sqrt{2} - \sqrt{2}\,i}$

EXAMPLE B (The Quadratic Formula)　Solve the equation

$$2x^2 + 3x + 4 = 0$$

Solution.

$$x = \frac{-3 \pm \sqrt{9 - 32}}{4} = \frac{-3 \pm \sqrt{-23}}{4} = \frac{-3 \pm \sqrt{23}\,i}{4} = -\frac{3}{4} \pm \frac{\sqrt{23}}{4}\,i$$

Solve each of the following equations.

35. $x^2 - x + 1 = 0$　　　　　　　　36. $x^2 + x + 2 = 0$

37. $4x^2 - 8x + 3 = 0$　　　　　　　38. $9x^2 + 12x + 4 = 0$

39. $x^2 - \sqrt{3}\,x + 7 = 0$　　　　　　40. $2x^2 + 2\sqrt{2}\,x + 3 = 0$

41. $5x^2 + 6x + 4 = 0$　　　　　　　42. $5x^2 + 2x + 2 = 0$

EXAMPLE C (High Powers of i)　Simplify (a) i^{51}; (b) $(2i)^6\,i^{19}$.

Solution.　Keep in mind that $i^2 = -1$, $i^3 = i^2 i = -i$ and $i^4 = i^2 i^2 = (-1)(-1) = 1$. Then use the usual rules of exponents.

(a) $i^{51} = i^{48}\,i^3 = (i^4)^{12}\,i^3 = (1)^{12}(-i) = -i$

(b) $(2i)^6\,i^{19} = 2^6\,i^6\,i^{19} = 64i^{25} = 64i^{24}\,i = 64i$

Simplify each of the following.

43. i^{94}　　　　44. i^{39}　　　　45. $(-i)^{17}$　　　　46. $(2i)^3\,i^{12}$

47. $\dfrac{(3i)^{16}}{(9i)^5}$　　　48. $\dfrac{2^8\,i^{19}}{(-2i)^{11}}$　　　49. $(1 + i)^3$　　　50. $(2 - i)^4$

EXAMPLE D (Complex Roots)　We say that a is a 4th root of b if $a \cdot a \cdot a \cdot a = b$ (more briefly, $a^4 = b$). Thus 3 is a 4th root of 81 because $3 \cdot 3 \cdot 3 \cdot 3 = 81$. Show that $1 + i$ is a 4th root of -4.

Solution.

$$(1 + i)(1 + i)(1 + i)(1 + i) = (1 + i)^2\,(1 + i)^2$$

$$= (1 + 2i - 1)(1 + 2i - 1)$$

$$= (2i)(2i)$$

$$= 4i^2$$

$$= -4$$

51. Show that i is a 4th root of 1. Can you find three other 4th roots of 1?
52. Show that $-1 - i$ is a 4th root of -4.
53. Show that $1 - i$ is a 4th root of -4.
54. Show that $-1 + i$ is a 4th root of -4.

MISCELLANEOUS PROBLEMS

55. Perform the indicated operations and simplify

    (a) $2 + 3i - i(4 - 3i)$ (b) $\dfrac{i^7 + 2i^2}{i^3}$

    (c) $3 + 4i + (3 + 4i)(-1 + 2i)$ (d) $i^{14} + \dfrac{5 + 2i}{5 - 2i}$

    (e) $\dfrac{5 - 2i}{3 + 4i} + \dfrac{5 + 2i}{3 - 4i}$ (f) $\dfrac{(3 + 2i)(3 - 2i)}{2\sqrt{3} + i}$

56. Simplify.
    (a) $1 + i + i^2 + i^3 + \cdots + i^{16}$

    (b) $1 + \dfrac{1}{i} + \dfrac{1}{i^2} + \dfrac{1}{i^3} + \cdots + \dfrac{1}{i^{16}}$

    (c) $(1 - i)^{16}$

57. Find a and b so that each is true.
    (a) $(2 + i)(2 - i)(a + bi) = 10 - 4i$
    (b) $(2 + i)(a - bi) = 8 - i$
58. Show that -2, $1 + \sqrt{3}i$, and $1 - \sqrt{3}i$ are each cube roots of -8.
59. Let $x = -\dfrac{1}{2} + \dfrac{\sqrt{3}}{2}i$. Find and simplify.

    (a) x^3
    (b) $1 + x + x^2$
    (c) $(1 - x)(1 - x^2)$
60. Show that $2 + i$ and $-2 - i$ are both solutions to $x^2 - 3 - 4i = 0$.
61. Solve each of the following quadratic equations.
    (a) $x^2 - 8x + 25 = 0$ (b) $ix^2 + 9x - 20i = 0$

62. Is $\dfrac{1}{a + bi}$ ever equal to $\dfrac{1}{a} + \dfrac{1}{bi}$?

63. A standard notation for the conjugate of the complex number x is $\bar{x}$. Thus, $\overline{a + bi} = a - bi$. Show that each of the following is true in general.
    (a) $\overline{x + y} = \bar{x} + \bar{y}$ (b) $\overline{xy} = \bar{x}\,\bar{y}$ (c) $\overline{x^{-1}} = \bar{x}^{-1}$ (d) $\overline{x/y} = \bar{x}/\bar{y}$

64. **TEASER** The absolute value of a complex number x is defined by $|x| = \sqrt{x\bar{x}}$. Note that this is consistent with the meaning of absolute value when x is a real number. Prove the triangle inequality for complex numbers, that is, show $|x + y| \leq |x| + |y|$.

Jean–Robert Argand (1768–1822)

Though several mathematicians (for example, De Moivre, Euler, Gauss) had thought of complex numbers as points in the plane before Argand, this obscure Swiss bookkeeper gets credit for the idea. In 1806 he wrote a small book on the geometric representation of complex numbers. It was his only contribution to mathematics.

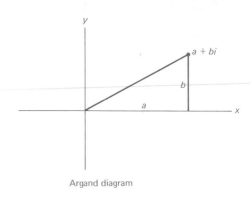

Argand diagram

7-2 Polar Representation of Complex Numbers

Throughout this book we have used the fact that a real number can be thought of as a point on a line. Now we are going to learn that a complex number can be represented as a point in the plane. This simple idea leads rather quickly to the fruitful notion of the polar form for a complex number. This in turn aids in the multiplication and division of complex numbers and greatly facilitates the calculation of powers and roots of complex numbers.

COMPLEX NUMBERS AS POINTS IN THE PLANE

Consider a complex number $a + bi$. It is determined by the two real numbers a and b, that is, by the ordered pair (a, b). But (a, b), in turn, determines a

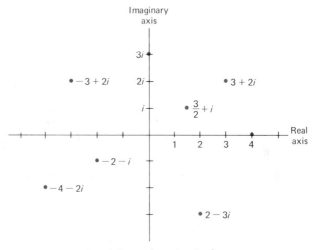

Argand diagram (complex plane)

Figure 3

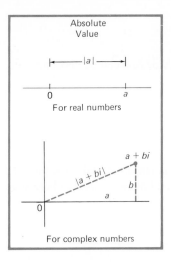

Absolute Value

For real numbers

For complex numbers

Figure 4

point in the plane. That point we now label with the complex number $a + bi$. Thus $2 + 4i$, $2 - 4i$, $-3 + 2i$, and all other complex numbers may be used as labels for points in the plane (Figure 3). The plane with points labeled this way is called the **Argand diagram** or **complex plane.** Note that $3i = 0 + 3i$ labels a point on the y-axis, which we now call the **imaginary axis,** while $4 = 4 + 0i$ corresponds to a point on the x-axis (called the **real axis**).

Recall that the absolute value of a real number a (written $|a|$) is its distance from the origin on the real line. The concept of absolute value is extended to a complex number $a + bi$ by defining

$$\left| a + bi \right| = \sqrt{a^2 + b^2}$$

which is also its distance from the origin (Figure 4). Thus while there are only two real numbers with absolute value of 5, namely -5 and 5, there are infinitely many complex numbers with absolute value 5. They include 5, -5, $5i$, $3 + 4i$, $3 - 4i$, $-\sqrt{21} + 2i$, and, in fact, all complex numbers on a circle of radius 5 centered at the origin (Figure 5).

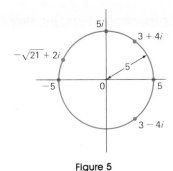

Figure 5

POLAR FORM

There is another geometric way to describe complex numbers, a way that will prove very useful to us. For the complex number $a + bi$, which we have already identified with the point (a, b) in the plane, let r denote its distance from the origin and let θ be one of the angles that a ray from the origin through the point makes with the positive x-axis. Then from Figure 6 (or one of its analogues in another quadrant), we see that

$$a = r \cos \theta \qquad b = r \sin \theta$$

This means that we can write

$$a + bi = r \cos \theta + (r \sin \theta)i$$

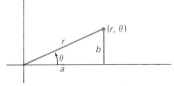

Figure 6

or

$$a + bi = r(\cos \theta + i \sin \theta)$$

The boxed expression gives the **polar form** of $a + bi$. Notice that r is just the absolute value of $a + bi$; we shall refer to θ as its angle.

To put a number $a + bi$ in polar form, we use the formulas

$$r = \sqrt{a^2 + b^2} \qquad \cos \theta = \frac{a}{r}$$

For example, for $2\sqrt{3} - 2i$,

$$r = \sqrt{(2\sqrt{3})^2 + (-2)^2} = \sqrt{12 + 4} = 4$$

$$\text{cos } \theta = \frac{2\sqrt{3}}{4} = \frac{\sqrt{3}}{2}$$

Since $2\sqrt{3} - 2i$ is in quadrant IV and $\cos \theta = \sqrt{3}/2$, θ can be chosen as an angle of $11\pi/6$ radians or $330°$. Thus

$$2\sqrt{3} - 2i = 4\left(\cos \frac{11\pi}{6} + i \sin \frac{11\pi}{6}\right)$$

$$= 4(\cos 330° + i \sin 330°)$$

For some numbers, finding the polar form is almost trivial. Just picture in your mind (Figure 7) where -6 and $4i$ are located in the complex plane and you will know that

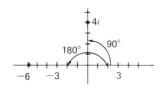

Figure 7

$$-6 = 6(\cos 180° + i \sin 180°)$$

$$4i = 4(\cos 90° + i \sin 90°)$$

Changing from the Cartesian form $a + bi$ to polar form is what we have just illustrated. Going in the opposite direction is much easier. For example, to change the polar form $3(\cos 240° + i \sin 240°)$ to Cartesian form, we simply calculate the sine and cosine of $240°$ and remove the parentheses.

$$3(\cos 240° + i \sin 240°) = 3\left(-\frac{1}{2} + i\frac{-\sqrt{3}}{2}\right)$$

$$= -\frac{3}{2} - \frac{3\sqrt{3}}{2}i$$

MULTIPLICATION AND DIVISION

The polar form is ideally suited for multiplying and dividing complex numbers. Let U and V be complex numbers given in polar form by

$$U = r(\cos \alpha + i \sin \alpha)$$

$$V = s(\cos \beta + i \sin \beta)$$

Then

$$U \cdot V = rs[\cos(\alpha + \beta) + i \sin(\alpha + \beta)]$$

$$\frac{U}{V} = \frac{r}{s}[\cos(\alpha - \beta) + i \sin(\alpha - \beta)]$$

In words, to multiply two complex numbers, we multiply their absolute values and add their angles. To divide two complex numbers, we divide their absolute values and subtract their angles (in the correct order). Thus if

$$U = 4(\cos 75° + i \sin 75°)$$

$$V = 3(\cos 60° + i \sin 60°)$$

then

$$U \cdot V = 12(\cos 135° + i \sin 135°)$$

$$\frac{U}{V} = \frac{4}{3} (\cos 15° + i \sin 15°)$$

To establish the multiplication formula we use a bit of trigonometry.

$$U \cdot V = r(\cos \alpha + i \sin \alpha)s(\cos \beta + i \sin \beta)$$

$$= rs(\cos \alpha \cos \beta + i \cos \alpha \sin \beta + i \sin \alpha \cos \beta + i^2 \sin \alpha \sin \beta)$$

$$= rs[(\cos \alpha \cos \beta - \sin \alpha \sin \beta) + i(\sin \alpha \cos \beta + \cos \alpha \sin \beta)]$$

$$= rs[\cos(\alpha + \beta) + i \sin(\alpha + \beta)]$$

The key step was the last one, where we used the addition laws for the cosine and the sine.

You will be asked to establish the division formula in Problem 56.

GEOMETRIC ADDITION AND MULTIPLICATION

Having learned that the complex numbers can be thought of as points in a plane, we should not be surprised that the operations of addition and multiplication have a geometric interpretation. Let U and V be any two complex numbers; that is, let

$$U = a + bi = r(\cos \alpha + i \sin \alpha)$$

$$V = c + di = s(\cos \beta + i \sin \beta)$$

Addition is accomplished algebraically by adding the real parts and imaginary parts separately.

$$U + V = (a + c) + (b + d)i$$

To accomplish the same thing geometrically, we construct the parallelogram that has O, U, and V as three of its vertices (see Figure 8). Then $U + V$ corresponds to the vertex opposite the origin, as you should be able to show by finding the coordinates of this vertex.

To multiply algebraically, we use the polar forms of U and V, adding the angles and multiplying the absolute values.

$$U \cdot V = rs[\cos(\alpha + \beta) + i \sin(\alpha + \beta)]$$

To interpret this geometrically (for the case where α and β are between $0°$ and $180°$), first draw triangle OAU, where A is the point $1 + 0i$. Then construct triangle OVW similar to triangle OAU in the manner indicated in Figure 9. We claim that $W = U \cdot V$. Certainly W has the correct angle, namely, $\alpha + \beta$. Moreover, by similarity of triangles,

$$\frac{\overline{OW}}{\overline{OV}} = \frac{\overline{OU}}{\overline{OA}} = \frac{\overline{OU}}{1}$$

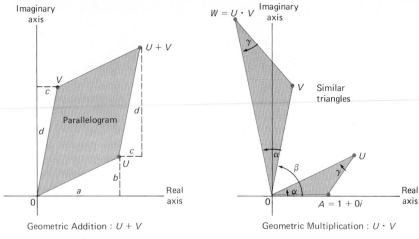

Geometric Addition : $U + V$

Figure 8

Geometric Multiplication : $U \cdot V$

Figure 9

(Here we are using $\overline{OW}$ for the length of the line segment from O to W.) Thus

$$|W| = \overline{OW} = \overline{OU} \cdot \overline{OV} = |U| \cdot |V|$$

so W also has the correct absolute value.

Problem Set 7-2

In Problems 1–12, plot the given numbers in the complex plane.

1. $2 + 3i$

2. $2 - 3i$

3. $-2 - 3i$

4. $-2 + 3i$

5. 5

6. -6

7. $-4i$

8. $6i$

9. $\frac{3}{5} - \frac{4}{5}i$

10. $-\frac{5}{13} + \frac{12}{13}i$

11. $2\left(\cos \dfrac{\pi}{4} + i \sin \dfrac{\pi}{4}\right)$

12. $3\left(\cos \dfrac{7\pi}{6} + i \sin \dfrac{7\pi}{6}\right)$

13. Find the absolute values of the numbers in Problems 1, 3, 5, 7, 9, and 11.

14. Find the absolute values of the numbers in Problems 2, 4, 6, 8, and 10.

Express each of the following in the form $a + bi$.

15. $4\left(\cos \dfrac{3\pi}{2} + i \sin \dfrac{3\pi}{2}\right)$

16. $5(\cos \pi + i \sin \pi)$

17. $2(\cos 225° + i \sin 225°)$

18. $\frac{3}{2}(\cos 300° + i \sin 300°)$

Express each of the following in polar form. For example, $1 + i = \sqrt{2} \, (\cos 45° + i \sin 45°)$

19. -4

20. 9

21. $-5i$

22. $4i$

23. $2 - 2i$ 24. $-5 - 5i$ 25. $2\sqrt{3} + 2i$ 26. $-4\sqrt{3} + 4i$

© 27. $5 + 4i$ © 28. $3 + 2i$

For Problems 29–36 let $u = 2(\cos 140° + i \sin 140°)$, $v = 3(\cos 70° + i \sin 70°)$ *and* $w = \frac{1}{2}(\cos 55° + i \sin 55°)$. *Calculate each product or quotient, leaving your answer in polar form. For example,*

$$\frac{u^2}{w} = \frac{u \cdot u}{w} = \frac{4(\cos 280° + i \sin 280°)}{\frac{1}{2}(\cos 55° + i \sin 55°)} = 8(\cos 225° + i \sin 225°)$$

29. uv 30. uw 31. vw 32. uvw

33. u/v 34. uv/w 35. $1/w$ 36. $1/v$

EXAMPLE (Finding Products in Two Ways)   Find the product

$$(\sqrt{3} + i)(-4 - 4\sqrt{3}\,i)$$

directly and then by using the polar form.

Solution.   *Method 1*   We use the definition of multiplication given in Section 7-1 to get

$$(\sqrt{3} + i)(-4 - 4\sqrt{3}\,i) = (-4\sqrt{3} + 4\sqrt{3}) + (-4 - 12)i$$
$$= -16i$$

Method 2   We change both numbers to polar form, multiply by the method of this section, and finally change to $a + bi$ form.

$$\sqrt{3} + i = 2(\cos 30° + i \sin 30°)$$
$$-4 - 4\sqrt{3}i = 8(\cos 240° + i \sin 240°)$$
$$(\sqrt{3} + i)(-4 - 4\sqrt{3}i) = 16(\cos 270° + i \sin 270°)$$
$$= 16(0 - i)$$
$$= -16i$$

Find each of the following products in two ways, giving your final answer in $a + bi$ *form.*

37. $(4 - 4i)(2 + 2i)$ 38. $(\sqrt{3} + i)(2 - 2\sqrt{3}i)$

39. $(1 + \sqrt{3}i)(1 + \sqrt{3}i)$ 40. $(\sqrt{2} + \sqrt{2}i)(\sqrt{2} + \sqrt{2}i)$

Find the following products and quotients, giving your answers in polar form. Start by changing each of the given complex numbers to polar form.

41. $4i(2\sqrt{3} - 2i)$ 42. $(-2i)(5 + 5i)$

43. $\dfrac{4i}{2\sqrt{3} - 2i}$ 44. $\dfrac{-2i}{5 + 5i}$

45. $(2\sqrt{2} - 2\sqrt{2}i)(2\sqrt{2} - 2\sqrt{2}i)$ 46. $(1 - \sqrt{3}i)(1 - \sqrt{3}i)$

MISCELLANEOUS PROBLEMS

47. Plot the given number in the complex plane and find its absolute value.
    (a) $-5 + 12i$ (b) $-4i$ (c) $5(\cos 60° + i \sin 60°)$

48. Express in the form $a + bi$.
    (a) $5[\cos(3\pi/2) + i \sin(3\pi/2)]$ (b) $4(\cos 180° + i \sin 180°)$
    (c) $2(\cos 315° + i \sin 315°)$ (d) $3[\cos(-2\pi/3) + i \sin(-2\pi/3)]$

49. Express in polar form
    (a) 12 (b) $-\sqrt{2} + \sqrt{2}i$ (c) $-3i$
    (d) $2 - 2\sqrt{3}i$ (e) $4\sqrt{3} + 4i$ (f) $2(\cos 45° - i \sin 45°)$

50. Write in the form $a + bi$.
    (a) $2(\cos 37° + i \sin 37°)8(\cos 113° + i \sin 113°)$
    (b) $6(\cos 123° + i \sin 123°)/[3(\cos 33° + i \sin 33°)]$

51. Perform the indicated operations and write your answer in polar form.
    (a) $1.5(\cos 110° + i \sin 110°)4(\cos 30° + i \sin 30°)2(\cos 20° + i \sin 20°)$
    (b) $\dfrac{12(\cos 115° + i \sin 115°)}{4(\cos 55° + i \sin 55°)(\cos 20° + i \sin 20°)}$
    (c) $\dfrac{(-\sqrt{2} + \sqrt{2}i)(2 - 2\sqrt{3}i)}{4\sqrt{3} + 4i}$ (See Problem 49.)

52. Calculate and write your answer in the form $a + bi$.
    (a) $\dfrac{(-\sqrt{2} + \sqrt{2}i)^2(2 - 2\sqrt{3}i)^3}{4\sqrt{3} + 4i}$ (See Problem 49.)
    (b) $(-1 + \sqrt{3}i)^5(-1 - \sqrt{3}i)^{-4}$

53. In each case, find two values for $z = a + bi$.
    (a) The imaginary part of z is 5 and $|z| = 13$.
    (b) The number z lies on the line $y = x$ and $|z| = 8$.

54. Find four complex numbers $z = a + bi$ that are located on the hyperbola $y^2 - x^2 = 2$ and satisfy $|z| = \sqrt{10}$.

55. Let $u = r(\cos \theta + i \sin \theta)$. Write each of the following in polar form.
    (a) u^3 (b) $\bar{u}$ ($\bar{u}$ is the conjugate of u)
    (c) $u\bar{u}$ (d) $1/u$
    (e) u^{-2} (f) $-u$

56. Prove the division formula: If $U = r(\cos \alpha + i \sin \alpha)$ and $V = s(\cos \beta + i \sin \beta)$, then

$$\frac{U}{V} = \frac{r}{s}[\cos(\alpha - \beta) + i \sin(\alpha - \beta)]$$

57. Let U and V be complex numbers. Give a geometric interpretation for
    (a) $|U - V|$ and (b) the angle of $U - V$.

58. By expanding $(\cos \theta + i \sin \theta)^3$ in two different ways, derive the formulas.
    (a) $\cos 3\theta = 4 \cos^3 \theta - 3 \cos \theta$
    (b) $\sin 3\theta = -4 \sin^3 \theta + 3 \sin \theta$

59. Let $z_k = 2[\cos(k\pi/4) + i \sin(k\pi/4)]$. Find the exact value of each of the following.
    (a) $z_1 z_2 z_3 \cdots z_8$
    (b) $z_1 + z_2 + z_3 + \cdots + z_8$ (Think geometrically.)

60. TEASER Let $z_k = \dfrac{k}{k+1}(\cos k° + i \sin k°)$. Find the exact value of the product

$$z_1 z_2 z_3 \cdots z_{179} z_{180}$$

Abraham De Moivre (1667–1754)

Though he was a Frenchman, De Moivre spent most of his life in London. There he became an intimate friend of the great Isaac Newton, inventor of calculus. De Moivre made many contributions to mathematics but his reputation rests most securely on the theorem that bears his name.

$$(\cos \theta + i \sin \theta)^n$$
$$= \cos n\theta + i \sin n\theta$$

De Moivre's Theorem

7-3 Powers and Roots of Complex Numbers

De Moivre's theorem tells us how to raise a complex number of absolute value 1 to an integral power. We can easily extend it to cover the case of any complex number, no matter what its absolute value. Then with a little work, we can use it to find roots of complex numbers. Here we are in for a surprise. Take the number $8i$, for example. After some fumbling around, we find that one of its cube roots is $-2i$, because $(-2i)^3 = -8i^3 = 8i$. We shall find that it has two other cube roots (both nonreal numbers). In fact, we shall see that every number has exactly three cube roots, four 4th roots, five 5th roots, and so on. To put it in a spectacular way, we claim that any number, for example $37 + 3.5i$, has 1,000,000 millionth roots.

POWERS OF COMPLEX NUMBERS

To raise the complex number $r(\cos \theta + i \sin \theta)$ to the nth power, n a positive integer, we simply find the product of n factors of $r(\cos \theta + i \sin \theta)$. But from Section 7-2, we know that we multiply complex numbers by multiplying their absolute values and adding their angles. Thus,

$$[r(\cos \theta + i \sin \theta)]^n$$
$$= \underbrace{r \cdot r \cdots r}_{n \text{ factors}}[\cos \underbrace{(\theta + \theta + \cdots + \theta)}_{n \text{ terms}} + i \sin(\theta + \theta + \cdots + \theta)]$$

In short,

$$[r(\cos\theta + i\sin\theta)]^n = r^n(\cos n\theta + i\sin n\theta)$$

When $r = 1$, this is De Moivre's theorem.

As a first illustration, let us find the 6th power of a complex number that is already in polar form.

$$\left[2\left(\cos\frac{\pi}{6} + i\sin\frac{\pi}{6}\right)\right]^6 = 2^6\left[\cos\left(6\cdot\frac{\pi}{6}\right) + i\sin\left(6\cdot\frac{\pi}{6}\right)\right]$$

$$= 64(\cos\pi + i\sin\pi)$$

$$= 64(-1 + i\cdot 0)$$

$$= -64$$

To find $(1 - \sqrt{3}i)^5$, we could use repeated multiplication of $1 - \sqrt{3}i$ by itself. But how much better to change $1 - \sqrt{3}i$ to polar form and use the boxed formula above.

$$1 - \sqrt{3}i = 2(\cos 300° + i\sin 300°)$$

Then

$$(1 - \sqrt{3}i)^5 = 2^5(\cos 1500° + i\sin 1500°)$$

$$= 32(\cos 60° + i\sin 60°)$$

$$= 32\left(\frac{1}{2} + i\frac{\sqrt{3}}{2}\right)$$

$$= 16 + 16\sqrt{3}i$$

THE THREE CUBE ROOTS OF 8*i*

Because finding roots is tricky, we begin with an example before attempting the general case. We have already noted that $-2i$ is one cube root of $8i$, but now we claim there are two others. How shall we find them? We begin by writing $8i$ in polar form.

$$8i = 8(\cos 90° + i\sin 90°)$$

Finding cube roots is the opposite of cubing. This suggests that we take the real cube root (rather than the cube) of 8 and divide (rather than multiply) the angle 90° by 3. This would give us one cube root

$$2(\cos 30° + i\sin 30°)$$

which reduces to

$$2\left(\frac{\sqrt{3}}{2} + \frac{1}{2}i\right) = \sqrt{3} + i$$

Is this really a cube root of $8i$? For fear that you might be suspicious of the polar form, we will cube it the old-fashioned way and check.

$$(\sqrt{3} + i)^3 = (\sqrt{3} + i)(\sqrt{3} + i)(\sqrt{3} + i)$$
$$= [(3 - 1) + 2\sqrt{3}i](\sqrt{3} + i)$$
$$= 2(1 + \sqrt{3}i)(\sqrt{3} + i)$$
$$= 2(0 + 4i)$$
$$= 8i$$

Of course, the check using polar form is more direct.

$$[2(\cos 30° + i \sin 30°)]^3 = 2^3(\cos 90° + i \sin 90°)$$
$$= 8(0 + i)$$
$$= 8i$$

The process described above yielded one cube root of $8i$ (namely, $\sqrt{3} + i$); there are two others. Let us go back to our representation of $8i$ in polar form. We used the angle $90°$; we could as well have used $90° + 360° = 450°$.

$$8i = 8(\cos 450° + i \sin 450°)$$

Now if we take the real cube root of 8 and divide $450°$ by 3 we get

$$2(\cos 150° + i \sin 150°) = 2\left(-\frac{\sqrt{3}}{2} + \frac{1}{2}i\right) = -\sqrt{3} + i$$

We could again check that this is indeed a cube root of $8i$.

What worked once might work twice. Let us write $8i$ in polar form in a third way, this time adding $2(360°)$ to its angle of $90°$.

$$8i = 8(\cos 810° + i \sin 810°)$$

The corresponding cube root is

$$2(\cos 270° + i \sin 270°) = 2(0 - i) = -2i$$

This does not come as a surprise, since we knew that $-2i$ was one of the cube roots of $8i$.

If we add $3(360°)$(that is, $1080°$) to $90°$, do we get still another cube root of $8i$? No, for if we write

$$8i = 8(\cos 1170° + i \sin 1170°)$$

the corresponding cube root of $8i$ would be

$$2(\cos 390° + i \sin 390°) = 2(\cos 30° + i \sin 30°)$$

But this is the same as the first cube root we found. The truth is that we have found all the cube roots of $8i$, namely, $\sqrt{3} + i$, $-\sqrt{3} + i$, and $-2i$.

Let us summarize. The number $8i$ has three cube roots given by

$$2\left[\cos\left(\frac{90°}{3}\right) + i \sin\left(\frac{90°}{3}\right)\right]$$

$$2\left[\cos\left(\frac{90° + 360°}{3}\right) + i \sin\left(\frac{90° + 360°}{3}\right)\right]$$

$$2\left[\cos\left(\frac{90° + 720°}{3}\right) + i \sin\left(\frac{90° + 720°}{3}\right)\right]$$

We can say the same thing in a shorter way by writing

$$2\left[\cos\left(\frac{90° + k \cdot 360°}{3}\right) + i \sin\left(\frac{90° + k \cdot 360°}{3}\right)\right] \qquad k = 0, 1, 2$$

ROOTS OF COMPLEX NUMBERS

We are ready to generalize. If $u \neq 0$, then

$$u = r(\cos \theta + i \sin \theta)$$

has n distinct nth roots $u_0, u_1, \ldots, u_{n-1}$ given by

$$u_k = \sqrt[n]{r}\left[\cos\left(\frac{\theta + k \cdot 360°}{n}\right) + i \sin\left(\frac{\theta + k \cdot 360°}{n}\right)\right]$$

$$k = 0, 1, 2, \ldots, n - 1$$

Recall that $\sqrt[n]{r}$ denotes the positive real nth root of $r = |u|$. In our example, it was $\sqrt[3]{|8i|} = \sqrt[3]{8} = 2$. To see that each value of u_k is an nth root, simply raise it to the nth power. In each case, you should get u.

The boxed formula assumes that θ is given in degrees. If θ is in radians, the formula takes the following form.

$$u_k = \sqrt[n]{r}\left[\cos\left(\frac{\theta + 2k\pi}{n}\right) + i \sin\left(\frac{\theta + 2k\pi}{n}\right)\right]$$

$$k = 0, 1, 2, \ldots, n - 1$$

A REAL EXAMPLE

Let us use the boxed formula to find the six 6th roots of 64. (Keep in mind that a real number is a special kind of complex number.) Changing to polar form, we write

$$64 = 64(\cos 0° + i \sin 0°)$$

Applying the formula with $r = |64| = 64$, $\theta = 0°$, and $n = 6$ gives

$$u_0 = 2(\cos 0° + i \sin 0°) = 2$$

$$u_1 = 2(\cos 60° + i \sin 60°) = 1 + \sqrt{3}\,i$$

$$u_2 = 2(\cos 120° + i \sin 120°) = -1 + \sqrt{3}\,i$$

$$u_3 = 2(\cos 180° + i \sin 180°) = -2$$

$$u_4 = 2(\cos 240° + i \sin 240°) = -1 - \sqrt{3}\,i$$

$$u_5 = 2(\cos 300° + i \sin 300°) = 1 - \sqrt{3}\,i$$

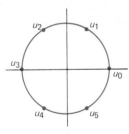

Figure 10

Notice that two of the roots, 2 and -2, are real; the other four are not real.

If you plot these six numbers (Figure 10) you will find that they lie on a circle of radius 2 centered at the origin and that they are equally spaced around the circle. This is typical of what happens in general.

Problem Set 7-3

Find each of the following, leaving your answer in polar form.

1. $\left[2\left(\cos \dfrac{\pi}{4} + i \sin \dfrac{\pi}{4}\right)\right]^3$

2. $\left[3\left(\cos \dfrac{5\pi}{6} + i \sin \dfrac{5\pi}{6}\right)\right]^2$

3. $[\sqrt{5}(\cos 11° + i \sin 11°)]^6$

4. $[\frac{1}{3}(\cos 12.5° + i \sin 12.5°)]^4$

5. $(1 + i)^8$

6. $(1 - i)^4$

Find each of the following powers. Write your answer in $a + bi$ form.

7. $(\cos 36° + i \sin 36°)^{10}$

8. $(\cos 27° + i \sin 27°)^{10}$

9. $(\sqrt{3} + i)^5$

10. $(2 - 2\sqrt{3}i)^4$

Find the nth roots of u for the given u and n, leaving your answers in polar form. Plot these roots in the complex plane.

11. $u = 125(\cos 45° + i \sin 45°); n = 3$

12. $u = 81(\cos 80° + i \sin 80°); n = 4$

13. $u = 64\left(\cos \dfrac{\pi}{2} + i \sin \dfrac{\pi}{2}\right); n = 6$

14. $u = 3^8\left(\cos \dfrac{2\pi}{3} + i \sin \dfrac{2\pi}{3}\right); n = 8$

15. $u = 4(\cos 112° + i \sin 112°); n = 4$

16. $u = 7(\cos 200° + i \sin 200°); n = 5$

Find the nth roots of u for the given u and n. Write your answers in the $a + bi$ form.

17. $u = 16, n = 4$

18. $u = -16, n = 4$

19. $u = 4i, n = 2$

20. $u = -27i, n = 3$

21. $u = -4 + 4\sqrt{3}i, n = 2$

22. $u = -2 - 2\sqrt{3}i, n = 4$

EXAMPLE (Roots of Unity) The *n*th roots of 1, called the **nth roots of unity,** play an important role in advanced algebra. Find the five 5th roots of unity, plot them, and show that four of the roots are powers of the 5th root.

Solution. First we represent 1 in polar form.

$$1 = 1(\cos 0° + i \sin 0°)$$

The five 5th roots are (according to the formula developed in this section)

$$u_0 = \cos 0° + i \sin 0° = 1$$

$$u_1 = \cos 72° + i \sin 72°$$

$$u_2 = \cos 144° + i \sin 144°$$

$$u_3 = \cos 216° + i \sin 216°$$

$$u_4 = \cos 288° + i \sin 288°$$

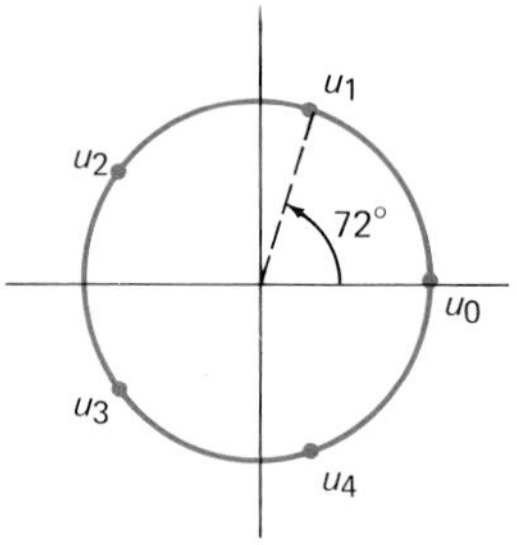

These roots are plotted in Figure 11. They lie on the unit circle and are equally spaced around it. Finally notice that

$$u_1 = u_1$$

$$u_2 = u_1^2$$

$$u_3 = u_1^3$$

$$u_4 = u_1^4$$

$$u_0 = u_1^5$$

Figure 11

Thus all the roots are powers of u_1. These powers of u_1 repeat in cycles of 5. For example, note that

$$u_1^6 = u_1^5 u_1 = u_1$$

$$u_1^7 = u_1^5 u_1^2 = u_1^2$$

In each of the following, find all the nth roots of unity for the given n and plot them in the complex plane.

23. $n = 4$ 24. $n = 6$ 25. $n = 10$ 26. $n = 12$

MISCELLANEOUS PROBLEMS

27. Calculate each of the following, leaving your answer in polar form.
    (a) $[3(\cos 20° + i \sin 20°)]^4$
    © (b) $[2.46(\cos 1.54 + i \sin 1.54)]^5$
    (c) $[2(\cos 50° + i \sin 50°)(\cos 30° + i \sin 30°)]^3$
    (d) $\left(\dfrac{8[\cos(2\pi/3) + i \sin(2\pi/3)]}{4[\cos(\pi/4) + i \sin(\pi/4)]} \right)^4$

28. Change to polar form, calculate, and then change back to $a + bi$ form.
    (a) $(1 - \sqrt{3}i)^5$
    (b) $[(\sqrt{3} + i)(2 - 2i)/(-1 + \sqrt{3}i)]^3$

29. Find the five 5th roots of $32(\cos 255° + i \sin 255°)$, giving your answers in polar form.

30. Find the three cube roots of $-4\sqrt{2} - 4\sqrt{2}i$, giving your answers in polar form.

31. Write the eight 8th roots of 1 in $a + bi$ form and calculate their sum and product.

32. Solve the equation $x^3 - 4 - 4\sqrt{3}i = 0$. You may give your answers in polar form.

33. Find the solution to $x^5 + \sqrt{2} - \sqrt{2}i = 0$ with the largest real part. Write your answer in the form $a + bi$.

34. Solve the equation $x^3 + 27 = 0$ in two ways.
    (a) By finding the three cube roots of -27.
    (b) By writing $x^3 + 27 = (x + 3)(x^2 - 3x + 9)$ and using the quadratic formula.

35. Find the six solutions to $x^6 - 1 = 0$ by two different methods.

36. Show that $\cos(\pi/3) + i \sin(\pi/3)$ is a solution to $2x^4 + x^2 + x + 1 = 0$.

37. Find all six solutions to $x^6 + x^4 + x^2 + 1 = 0$. *Hint*: The left side can be factored as $(x^2 + 1)(x^4 + 1)$.

38. Show that DeMoivre's theorem is valid when n is a negative integer.

39. If A is a nonreal number, we agree that $\sqrt{A}$ stands for the one of the two square roots with nonnegative real part. For example, the two square roots of $-4 + 4\sqrt{3}i$ are $\sqrt{2} + \sqrt{6}i$ and $-\sqrt{2} - \sqrt{6}i$, but we agree that

$$\sqrt{-4 + 4\sqrt{3}i} = \sqrt{2} + \sqrt{6}i$$

Evaluate.
    (a) $\sqrt{1 + \sqrt{3}i}$ (b) $\sqrt{-1 + \sqrt{3}i}$

40. The quadratic formula is valid even for quadratic equations with nonreal coefficients if we follow the agreement of Problem 39. Solve the following equations.
    (a) $x^2 - 2x + \sqrt{3}i = 0$
    (b) $x^2 - 4ix - 5 + \sqrt{3}i = 0$

41. Let n be an integer that is not divisible by 3. Simplify

$$(-1 + \sqrt{3}i)^n + (-1 - \sqrt{3}i)^n$$

as much as possible.

42. **TEASER** Let $1, u, u^2, u^3, \cdots, u^{15}$ be the sixteen 16th roots of unity. Calculate each of the following. Look for a simple way in each case.
    (a) $1 + u + u^2 + u^3 + \cdots + u^{15}$
    (b) $1 \cdot u \cdot u^2 \cdot u^3 \cdots u^{15}$
    (c) $(1 - u)(1 - u^2)(1 - u^3) \cdots (1 - u^{15})$
    (d) $(1 + u)(1 + u^2)(1 + u^4)(1 + u^8)(1 + u^{16})$

Chapter Summary

With just the real numbers, we cannot solve certain quadratic equations, such as the simple equation $x^2 + 1 = 0$. To solve such equations, we introduce first the number $i = \sqrt{-1}$ and then all numbers of the form $a + bi$, where a and

b are real numbers. We call these numbers **complex numbers.** We can add, subtract, multiply, and divide these numbers; the result is always a complex number. Division is most easily accomplished via the notion of the **conjugate** $a - bi$ of the number $a + bi$. With the complex numbers at our disposal, we can solve any quadratic equation (with real coefficients) using the familiar **quadratic formula.**

A complex number $a + bi$ can be represented geometrically as a point (a, b) in a plane called the **complex plane,** or **Argand diagram.** The horizontal and vertical axes are known as the **real axis** and **imaginary axis,** respectively. The distance from the origin to (a, b) is $\sqrt{a^2 + b^2}$; it is also the absolute value of $a + bi$, denoted by $|a + bi|$. If we let $r = \sqrt{a^2 + b^2}$ and θ be the angle that the ray from the origin through (a, b) makes with the positive x-axis, we obtain the **polar form** of $a + bi$, namely,

$$r(\cos \theta + i \sin \theta)$$

This form facilitates multiplication and division and is especially useful in finding powers and roots of complex numbers. A significant result is the formula

$$[r(\cos \theta + i \sin \theta)]^n = r^n(\cos n\theta + i \sin n\theta)$$

Chapter Review Problem Set

1. Write each of the following in the form $a + bi$.
   (a) $(3 - 2i) + (-7 + 6i) - (2 - 3i)$
   (b) $(3 - 2i)(3 + 2i) - 4i^2 + (2i)^3$
   (c) $(3 + 4i)(-2 + i)$
   (d) $\dfrac{3 - i}{2 + 3i}$
   (e) $(2 + i)^3 + i^{73}$
   (f) $(5 - 3i)^{-1} + i^{-1}$

2. Solve the following quadratic equations.
   (a) $x^2 + 4 = 0$
   (b) $x^2 + 2x + 4 = 0$
   (c) $3x^2 - 2x + 4 = 0$
   (d) $x^2 + 2ix + 4 = 0$

3. Plot the following numbers in the complex plane.
   (a) $3 - 4i$ (b) -6
   (c) $5i$ (d) $3(\cos(3\pi/4) + i \sin(3\pi/4))$
   (e) $4(\cos 300° + i \sin 300°)$

4. Find the absolute value of each number in Problem 3.

5. Express $4(\cos 150° + i \sin 150°)$ in the form $a + bi$.

6. Express in polar form.
   (a) $3i$ (b) -6
   (c) $-1 - i$ (d) $2\sqrt{3} - 2i$

7. Let $u = 8(\cos 105° + i \sin 105°)$ and $v = 4(\cos 40° + i \sin 40°)$. Calculate each of the following, leaving your answer in polar form.

    (a) uv (b) u/v (c) u^3 (d) $u^2 v^3$

8. Find all the 6th roots of $2^6(\cos 120° + i \sin 120°)$, leaving your answers in polar form.

9. Find the five solutions to $x^5 - 1 = 0$.

He who loves practice without theory is like the sailor who boards ship without a rudder and compass and never knows where he may cast.

Leonardo da Vinci

CHAPTER 8

Theory of Polynomials

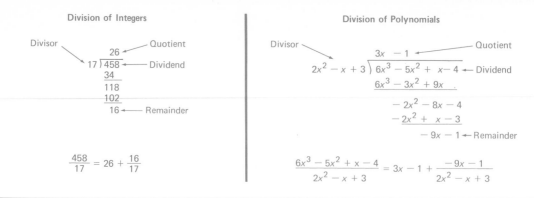

8-1   Division of Polynomials

In Section 1-3, we learned that polynomials in x can always be added, subtracted, and multiplied; the result in every case is a polynomial. For example,

$$(3x^2 + x - 2) + (3x - 2) = 3x^2 + 4x - 4$$

$$(3x^2 + x - 2) - (3x - 2) = 3x^2 - 2x$$

$$(3x^2 + x - 2) \cdot (3x - 2) = 9x^3 - 3x^2 - 8x + 4$$

Now we are going to study division of polynomials. Occasionally the division is exact.

$$(3x^2 + x - 2) \div (3x - 2) = \frac{(3x - 2)(x + 1)}{3x - 2} = x + 1$$

We say $3x - 2$ is an exact divisor, or factor, of $3x^2 + x - 2$. More often than not, the division is inexact and there is a nonzero remainder.

THE DIVISION ALGORITHM

The opening display shows that the process of division for polynomials is much the same as for integers. Both processes (we call them *algorithms*) involve subtraction. The first one shows how many times 17 can be subtracted from 458 before obtaining a remainder less than 17. The answer is 26 times, with a remainder of 16. The second algorithm shows how many times $2x^2 - x + 3$ can be subtracted from $6x^3 - 5x^2 + x - 4$ before obtaining a remainder of lower degree than $2x^2 - x + 3$. The answer is $3x - 1$ times, with a remainder of $-9x - 1$. Thus in these two examples we have

$$458 - 26(17) = 16$$

$$(6x^3 - 5x^2 + x - 4) - (3x - 1)(2x^2 - x + 3) = -9x - 1$$

Of course, we can also write these equalities as

$$458 = (17)(26) + 16$$

$$6x^3 - 5x^2 + x - 4 = (2x^2 - x + 3)(3x - 1) + (-9x - 1)$$

Notice that both of them can be summarized in the words

$$\text{Dividend} = (\text{divisor}) \cdot (\text{quotient}) + \text{remainder}$$

This statement is very important and is worth stating again for polynomials in very precise language.

THE DIVISION LAW FOR POLYNOMIALS

If $P(x)$ and $D(x)$ are any two nonconstant polynomials, then there are unique polynomials $Q(x)$ and $R(x)$ such that

$$P(x) = D(x)Q(x) + R(x)$$

where $R(x)$ is either zero or it is of lower degree than $D(x)$.

Here you should think of $P(x)$ as the dividend, $D(x)$ as the divisor, $Q(x)$ as the quotient, and $R(x)$ as the remainder.

The algorithm we use to find $Q(x)$ and $R(x)$ was illustrated in the opening display. Here is another illustration, this time for polynomials with some nonreal coefficients. Notice that we always arrange the divisor and dividend in descending powers of x before we start the division process.

$$
\begin{array}{r}
x - i \\
3x + i \overline{\smash{\big)}\ 3x^2 - 2ix + 7} \\
\underline{3x^2 + ix} \\
-3ix + 7 \\
\underline{-3ix + 1} \\
6
\end{array}
\qquad
\begin{array}{l}
Q(x) = x - i \\
R(x) = 6
\end{array}
$$

SYNTHETIC DIVISION

It is often necessary to divide by polynomials of the form $x - c$. For such a division, there is a shortcut called *synthetic division*. We illustrate how it works for

$$(2x^3 - x^2 + 5) \div (x - 2)$$

Certainly the result depends on the coefficients; the powers of x serve mainly to determine the placement of the coefficients. Below, we show the division in its usual form and then in a skeletal form with the x's omitted. Note that we leave a blank space for the missing first degree term in the long division but indicate it with a 0 in the skeletal form.

LONG DIVISION

$$
\begin{array}{r}
2x^2 + 3x\ + 6 \\
x - 2\ \overline{\smash{\big)}\ 2x^3 - \ x^2 \qquad\ + 5} \\
\underline{2x^3 - 4x^2} \\
3x^2 \\
\underline{3x^2 - 6x} \\
6x + \ 5 \\
\underline{6x - 12} \\
17
\end{array}
$$

FIRST CONDENSATION

$$
\begin{array}{r}
②\qquad③\qquad⑥ \\
①\ -2\ \overline{\smash{\big)}\ 2 \quad -1 \quad 0 \quad 5} \\
2 \quad -4 \\
3 \quad ⓪ \\
③ \quad -6 \\
6 \quad ⑤ \\
⑥ \quad -12 \\
17
\end{array}
$$

We can condense things still more by discarding all of the circled digits. The coefficients of the quotient, 2, 3, and 6, the remainder, 17, and the numbers from which they were calculated remain. All of the important numbers appear in the diagram below, on the left. On the right, we show the final modification. There we have changed the divisor from -2 to 2 to allow us to do addition rather than subtraction at each stage.

SECOND CONDENSATION

$$
\begin{array}{r}
-2\ \underline{\big|\ 2 \quad -1 \quad\ \ 0 \quad\ \ 5} \\
-4 \quad -6\ \ -12 \\
\hline
2 \quad\ \ 3 \quad\ \ 6 \quad\ \ 17
\end{array}
$$

SYNTHETIC DIVISION

$$
\begin{array}{r}
2\ \underline{\big|\ 2 \quad -1 \quad\ \ 0 \quad\ \ 5} \\
4 \quad\ \ 6 \quad\ \ 12 \\
\hline
2 \quad\ \ 3 \quad\ \ 6 \quad\ \ 17
\end{array}
$$

The process shown in the final format is called **synthetic division.** We can describe it by a series of steps.

1. To divide by $x - 2$, use 2 as the synthetic divisor.
2. Write down the coefficients of the dividend. Be sure to write zeros for missing powers.
3. Bring down the first coefficient.
4. Follow the arrows, first multiplying by 2 (the divisor), then adding, multiplying the sum by 2, adding, and so on.
5. The last number in the third row is the remainder and the others are the coefficients of the quotient.

Here is another example. We use synthetic division to divide $3x^3 + x^2 - 15x - 5$ by $x + \frac{1}{3}$, that is, by $x - (-\frac{1}{3})$.

$$
\begin{array}{r}
-\frac{1}{3}\ \underline{\big|\ 3 \quad\ \ 1 \quad -15 \quad -5} \\
-1 \quad\ \ 0 \quad\ \ 5 \\
\hline
3 \quad\ \ 0 \quad -15 \quad\ \ 0
\end{array}
$$

Since the remainder is 0, the division is exact. We conclude that

$$3x^3 + x^2 - 15x - 5 = (x + \tfrac{1}{3})(3x^2 - 15)$$

PROPER AND IMPROPER RATIONAL EXPRESSIONS

A rational expression (ratio of two polynomials) is said to be **proper** if the degree of its numerator is smaller than that of its denominator. Thus

$$\frac{x + 1}{x^2 - 3x + 2}$$

is a proper rational expression, but

$$\frac{2x^3}{x^2 - 3}$$

is improper. The division law, $P(x) = D(x)Q(x) + R(x)$, can be written as

$$\frac{P(x)}{D(x)} = Q(x) + \frac{R(x)}{D(x)}$$

It implies that any improper rational expression can be rewritten as the sum of a polynomial and a proper rational expression. For example,

$$\frac{2x^3}{x^2 - 3} = 2x + \frac{6x}{x^2 - 3}$$

a result we obtained by dividing $x^2 - 3$ into $2x^3$.

Problem Set 8-1

In Problems 1–8, find the quotient and the remainder if the first polynomial is divided by the second.

1. $x^3 - x^2 + x + 3;\ x^2 - 2x + 3$
2. $x^3 - 2x^2 - 7x - 4;\ x^2 + 2x + 1$
3. $6x^3 + 7x^2 - 18x + 15;\ 2x^2 + 3x - 5$
4. $10x^3 + 13x^2 + 5x + 12;\ 5x^2 - x + 4$
5. $4x^4 - x^2 - 6x - 9;\ 2x^2 - x - 3$
6. $25x^4 - 20x^3 + 4x^2 - 4;\ 5x^2 - 2x + 2$
7. $2x^5 - 2x^4 + 9x^3 - 12x^2 + 4x - 16;\ 2x^3 - 2x^2 + x - 4$
8. $3x^5 - x^4 - 8x^3 - x^2 - 3x + 12;\ 3x^3 - x^2 + x - 4$

In Problems 9–14, write each rational expression as the sum of a polynomial and a proper rational expression.

9. $\dfrac{x^3 + 2x^2 + 5}{x^2}$

10. $\dfrac{2x^3 - 4x^2 - 3}{x^2 + 1}$

11. $\dfrac{x^3 - 4x + 5}{x^2 + x - 2}$

12. $\dfrac{2x^3 + x - 8}{x^2 - x + 4}$ 13. $\dfrac{2x^2 - 4x + 5}{x^2 + 1}$ 14. $\dfrac{5x^3 - 6x + 11}{x^3 - x}$

In Problems 15–24, use synthetic division to find the quotient and remainder if the first polynomial is divided by the second.

15. $2x^3 - x^2 + x - 4; x - 1$ 16. $3x^3 + 2x^2 - 4x + 5; x - 2$

17. $3x^3 + 5x^2 + 2x - 10; x - 1$ 18. $2x^3 - 5x^2 + 4x - 4; x - 2$

19. $x^4 - 2x^2 - 1; x - 3$ 20. $x^4 + 3x^2 - 340; x - 4$

21. $x^3 + 2x^2 - 3x + 2; x + 1$ 22. $x^3 - x^2 + 11x - 1; x + 1$

23. $2x^4 + x^3 + 4x^2 + 7x + 4; x + \frac{1}{2}$

24. $2x^4 + x^3 + x^2 + 10x - 8; x - \frac{1}{2}$

EXAMPLE (Division by $x - (a + bi)$)   Find the quotient and remainder when $x^3 - 2x^2 - 6ix + 18$ is divided by $x - 2 - 3i$.

Solution.   Synthetic division works just fine even when some or all of the coefficients are nonreal. Since $x - 2 - 3i = x - (2 + 3i)$, we use $2 + 3i$ as the synthetic divisor.

$$
\begin{array}{r|rrrr}
2 + 3i & 1 & -2 & -6i & 18 \\
 & & 2 + 3i & -9 + 6i & -18 - 27i \\
\hline
 & 1 & 3i & -9 & -27i
\end{array}
$$

Quotient: $x^2 + 3ix - 9$; remainder: $-27i$.

Use synthetic division to find the quotient and remainder when the first polynomial is divided by the second.

▣ 25. $x^3 - 2x^2 + 5x + 30; x - 2 + 3i$

▣ 26. $2x^3 - 11x^2 + 44x + 35; x - 3 - 4i$

▣ 27. $x^4 - 17; x - 2i$

▣ 28. $x^4 + 18x^2 + 90; x - 3i$

MISCELLANEOUS PROBLEMS

29. Express $(2x^4 - x^3 - x^2 - 2)/(x^3 + 1)$ as a polynomial plus a proper rational expression.

30. Find the quotient and the remainder when $x^4 + 6x^3 - 2x^2 + 4x - 15$ is divided by $x^2 - 2x + 3$.

31. Find by inspection the quotient and the remainder when the first polynomial is divided by the second.
    (a) $2x^3 + 3x^2 - 11x + 9; x^2$
    (b) $2(x + 3)^2 + 10(x + 3) - 14; x + 3$
    (c) $(x - 4)^5 + x^2 + x + 1; (x - 4)^3$
    (d) $(x^2 + 3)^3 + 2x(x^2 + 3) + 4x - 1; x^2 + 3$

32. Use synthetic division to find the quotient and the remainder when the first polynomial is divided by the second.
    (a) $x^4 - 4x^3 + 29; x - 3$
    (b) $2x^4 - x^3 + 2x - 4; x + \frac{1}{2}$

(c) $x^4 + 4x^3 + 4\sqrt{3}x^2 + 3\sqrt{3}x + 3\sqrt{3}; x + \sqrt{3}$

☐ (d) $x^3 - (3 + 2i)x^2 + 10ix + 20 - 12i; x - 3$

33. Show that the second polynomial is a factor of the first and determine the other factor.

(a) $x^5 + x^4 - 16x - 16; x - 2$

(b) $x^5 + 32; x + 2$

(c) $x^4 - \frac{3}{2}x^3 + 3x^2 + 6x + 2; x + \frac{1}{2}$

☐ (d) $x^3 - 2ix^2 + x - 2i; x - 2i$

34. Use synthetic division to show that the second polynomial is a factor of the first and determine the other factor. *Hint:* In (a), you will want to use the fact that $x^2 - 1 = (x - 1)(x + 1)$.

(a) $x^4 + x^3 - x - 1; x^2 - 1$

(b) $x^4 - x^3 + 2x^2 - 4x - 8; x^2 - x - 2$

(c) $x^4 + 2x^3 - 4x - 4; x^2 - 2$

☐ (d) $x^5 + x^4 - 16x - 16; x^2 + 4$

35. Find k so that the second polynomial is a factor of the first.

(a) $x^3 + x^2 - 10x + k; x - 4$

(b) $x^4 + kx + 10; x + 2$

(c) $k^2x^3 - 4kx + 4; x - 1$

36. Determine h and k so that both $x - 3$ and $x + 2$ are factors of $x^4 - x^3 + hx^2 + kx - 6$.

37. Determine a, b, and c so that $(x - 1)^3$ is a factor of $x^4 + ax^3 + bx^2 + cx - 4$.

38. **TEASER** Let a, b, c, and d be distinct integers and suppose that $x - a$, $x - b$, $x - c$, and $x - d$ are factors of the polynomial $p(x)$, which has integral coefficients. Prove that $p(n)$ is not a prime number for any integer n.

Young Scholar: And does $x^4 + 99x^3 + 21$ have a zero?

Carl Gauss: Yes.

Young Scholar: How about $\pi x^{67} - \sqrt{3x^{19}} + 4i$?

Carl Gauss: It does.

Young Scholar: How can you be sure?

Carl Gauss: When I was young, 22 I think, I proved that every nonconstant polynomial, no matter how complicated, has at least one zero. Many people call it the *Fundamental Theorem of Algebra.*

Carl F. Gauss (1777–1855)
"The Prince of Mathematicians"

8-2 Factorization Theory for Polynomials

Our young scholar could have asked a harder question. How do you find the zeros of a polynomial? Even the eminent Gauss would have had trouble with that question. You see, it is one thing to know a polynomial has zeros; it is quite another thing to find them.

Even though it is a difficult task and one at which we will have only limited success, our goal for this and the next two sections is to develop methods for finding zeros of polynomials. Remember that a polynomial is an expression of the form

$$P(x) = a_n x^n + a_{n-1} x^{n-1} + \cdots + a_1 x + a_0$$

Unless otherwise specified, the coefficients (the a_i's) are allowed to be *complex* numbers. And by a **zero** of $P(x)$, we mean any complex number c (real or nonreal) such that $P(c) = 0$. The number c is also called a **solution,** or a **root,** of the equation $P(x) = 0$. Note the use of words: Polynomials have zeros, but polynomial equations have solutions.

THE REMAINDER AND FACTOR THEOREMS

Recall the division law from Section 8-1, which had as its conclusion

$$P(x) = D(x)Q(x) + R(x)$$

If $D(x)$ has the form $x - c$, this becomes

$$P(x) = (x - c)Q(x) + R$$

where R, which is of lower degree than $x - c$, must be a constant. This last equation is an identity; it is true for all values of x, including $x = c$. Thus

$$P(c) = (c - c)Q(c) + R = 0 + R$$

We have just proved an important result.

REMAINDER THEOREM

If a polynomial $P(x)$ is divided by $x - c$, then the constant remainder R is given by $R = P(c)$.

Here is a nice example. Suppose we want to know the remainder R when $P(x) = x^{1000} + x^{22} - 15$ is divided by $x - 1$. We could, of course, divide it out—but what a waste of energy, especially since we know the remainder theorem. From it we learn that

$$R = P(1) = 1^{1000} + 1^{22} - 15 = -13$$

Much more important than the mere calculation of remainders is a consequence called the *factor theorem*. Since $R = P(c)$, as we have just seen, we may rewrite the division law as

$$P(x) = (x - c)Q(x) + P(c)$$

It is plain to see that $P(c) = 0$ if and only if the division of $P(x)$ by $x - c$ is exact; that is, if and only if $x - c$ is a factor of $P(x)$.

FACTOR THEOREM

A polynomial $P(x)$ has c as a zero if and only if it has $x - c$ as a factor.

Sometimes it is easy to spot one zero of a polynomial. If so, the factor theorem may help us find the other zeros. Consider the polynomial

$$P(x) = 3x^3 - 8x^2 + 3x + 2$$

Notice that

$$P(1) = 3 - 8 + 3 + 2 = 0$$

so 1 is a zero. By the factor theorem, $x - 1$ is a factor of $P(x)$. We can use synthetic division to find the other factor.

$$
\begin{array}{r|rrrr}
1 & 3 & -8 & 3 & 2 \\
  &   & 3 & -5 & -2 \\
\hline
  & 3 & -5 & -2 & 0
\end{array}
$$

The remainder is 0 as we expected, and

$$P(x) = (x - 1)(3x^2 - 5x - 2)$$

Using the quadratic formula, we find the zeros of $3x^2 - 5x - 2$ to be $(5 \pm \sqrt{49})/6$, which simplify to 2 and $-\frac{1}{3}$. Thus $P(x)$ has 1, 2, and $-\frac{1}{3}$ as its three zeros.

COMPLETE FACTORIZATION OF POLYNOMIALS

In the example above, we did not really need the quadratic formula. If we had been clever, we would have factored $3x^2 - 5x - 2$.

$$3x^2 - 5x - 2 = (3x + 1)(x - 2)$$
$$= 3(x + \tfrac{1}{3})(x - 2)$$

Thus $P(x)$, our original polynomial, may be written as

$$P(x) = 3(x - 1)(x + \tfrac{1}{3})(x - 2)$$

from which all three of the zeros are immediately evident.

But now we make another key observation. Notice that $P(x)$ can be factored as a product of its leading coefficient and three factors of the form $(x - c)$, where the c's are the zeros of $P(x)$. This holds true in general.

COMPLETE FACTORIZATION THEOREM

If

$$P(x) = a_n x^n + a_{n-1} x^{n-1} + \cdots + a_1 x + a_0$$

is an nth degree polynomial with $n > 0$, then there are n numbers c_1, c_2, . . . , c_n, not necessarily distinct, such that

$$P(x) = a_n(x - c_1)(x - c_2) \cdots (x - c_n)$$

The c's are the zeros of $P(x)$; they may or may not be real numbers.

To prove the complete factorization theorem, we must go back to Carl Gauss and our opening display. In his doctoral dissertation in 1799, Gauss gave a proof of the following important theorem, a proof that unfortunately is beyond the scope of this book.

FUNDAMENTAL THEOREM OF ALGEBRA

Every nonconstant polynomial has at least one zero.

Now let $P(x)$ be any polynomial of degree $n > 0$. By the fundamental theorem, it has a zero, which we may call c_1. By the factor theorem, $x - c_1$ is a factor of $P(x)$; that is,

$$P(x) = (x - c_1)P_1(x)$$

where $P_1(x)$ is a polynomial of degree $n - 1$ and with the same leading coefficient as $P(x)$—namely, a_n.

If $n - 1 > 0$, we may repeat the argument on $P_1(x)$. It has a zero c_2 and hence a factor $x - c_2$; that is,

$$P_1(x) = (x - c_2)P_2(x)$$

where $P_2(x)$ has degree $n - 2$. For our original polynomial $P(x)$, we may now write

$$P(x) = (x - c_1)(x - c_2)P_2(x)$$

Continuing in the pattern now established, we eventually get

$$P(x) = (x - c_1)(x - c_2) \cdots (x - c_n)P_n$$

where P_n has degree zero; that is, P_n is a constant. In fact, $P_n = a_n$, since the leading coefficient stayed the same at each step. This establishes the complete factorization theorem.

ABOUT THE NUMBER OF ZEROS

Each of the numbers c_i in

$$P(x) = a_n(x - c_1)(x - c_2) \cdots (x - c_n)$$

is a zero of $P(x)$. Are there any other zeros? No, for if d is any number different from each of the c_i's, then

$$P(d) = a_n(d - c_1)(d - c_2) \cdots (d - c_n) \neq 0$$

All of this tempts us to say that a polynomial of degree n has exactly n zeros. But hold on! The numbers $c_1, c_2, \ldots, c_n$ need not all be different. For example, the sixth degree polynomial

$$P(x) = 4(x - 2)^3(x + 1)(x - 4)^2$$

has only three distinct zeros, 2, -1, and 4. We have to settle for the following statement.

An nth degree polynomial has at most n distinct zeros.

There is a way in which we can say that there are exactly n zeros. Call c a **zero of multiplicity k** of $P(x)$ if $x - c$ appears k times in its complete factorization. For example, in

$$P(x) = 4(x - 2)^3(x + 1)(x - 4)^2$$

the zeros 2, -1, and 4 have multiplicities 3, 1, and 2, respectively. A zero of multiplicity 1 is called a **simple zero.** Notice in our example that the multiplicities add to 6, the degree of the polynomial. In general, we may say this:

An nth degree polynomial has exactly n zeros provided we count a zero of multiplicity k as k zeros.

Problem Set 8-2

Use the remainder theorem to find P(c). Check your answer by substituting c for x.

1. $P(x) = 2x^3 - 5x^2 + 3x - 4; c = 2$
2. $P(x) = x^3 + 4x^2 - 11x - 5; c = 3$

3. $P(x) = 8x^4 - 3x^2 - 2; c = \frac{1}{2}$

4. $P(x) = 2x^4 + \frac{3}{4}x + \frac{3}{2}; c = -\frac{1}{2}$

Find the remainder if the first polynomial is divided by the second. Do it without actually dividing.

5. $x^{10} - 15x + 8; x - 1$

6. $2x^{20} + 5; x + 1$

7. $64x^6 + 13; x + \frac{1}{2}$

8. $81x^3 + 9x^2 - 2; x - \frac{1}{3}$

Find all of the zeros of the given polynomial and give their multiplicities.

9. $(x - 1)(x + 2)(x - 3)$

10. $(x + 2)(x + 5)(x - 7)$

11. $(2x - 1)(x - 2)^2 x^3$

12. $(3x + 1)(x + 1)^3 x^2$

☐ 13. $3(x - 1 - 2i)(x + \frac{2}{3})$

14. $5(x - 2 + \sqrt{5})(x - \frac{4}{5})$

In Problems 15–18, show that $x - c$ is a factor of $P(x)$.

15. $P(x) = 2x^3 - 7x^2 + 9x - 4; c = 1$

16. $P(x) = 3x^3 + 4x^2 - 6x - 1; c = 1$

17. $P(x) = x^3 - 7x^2 + 16x - 12; c = 3$

18. $P(x) = x^3 - 8x^2 + 13x + 10; c = 5$

19. In Problem 15, you know that $P(x)$ has 1 as a zero. Find the other zeros.

20. Find all of the zeros of $P(x)$ in Problem 16. Then factor $P(x)$ completely.

21. Find all of the zeros of $P(x)$ in Problem 17.

22. Find all of the zeros of $P(x)$ in Problem 18.

In Problems 23–26, factor the given polynomial into linear factors. You should be able to do it by inspection.

23. $x^2 - 5x + 6$

24. $2x^2 - 14x + 24$

25. $x^4 - 5x^2 + 4$

26. $x^4 - 13x^2 + 36$

In Problems 27–30, factor $P(x)$ into linear factors given that c is a zero of $P(x)$.

27. $P(x) = x^3 - 3x^2 - 28x + 60; c = 2$

28. $P(x) = x^3 - 2x^2 - 29x - 42; c = -2$

29. $P(x) = x^3 + 3x^2 - 10x - 12; c = -1$

30. $P(x) = x^3 + 11x^2 - 5x - 55; c = -11$

EXAMPLE A (Finding a Polynomial from Its Zeros)

(a) Find a cubic polynomial having simple zeros 3, $2i$, and $-2i$.

(b) Find a polynomial $P(x)$ with integral coefficients and having $\frac{1}{2}$ and $-\frac{2}{3}$ as simple zeros and 1 as a zero of multiplicity 2.

Solution.

(a) Let us call the required polynomial $P(x)$. Then

$$P(x) = a(x - 3)(x - 2i)(x + 2i)$$

where a can be any nonzero number. Choosing $a = 1$ and multiplying, we have

$$P(x) = (x - 3)(x^2 + 4) = x^3 - 3x^2 + 4x - 12$$

(b) $P(x) = a(x - \tfrac{1}{2})(x + \tfrac{2}{3})(x - 1)^2$

We choose $a = 6$ to eliminate fractions.

$$P(x) = 6(x - \tfrac{1}{2})(x + \tfrac{2}{3})(x - 1)^2$$
$$= 2(x - \tfrac{1}{2})3(x + \tfrac{2}{3})(x - 1)^2$$
$$= (2x - 1)(3x + 2)(x^2 - 2x + 1)$$
$$= 6x^4 - 11x^3 + 2x^2 + 5x - 2$$

In Problems 31–38, find a polynomial $P(x)$ with integral coefficients having the given zeros. Assume each zero to be simple (multiplicity 1) unless otherwise indicated.

31. 2, 1, and -4
32. 3, -2, and 5
33. $\tfrac{1}{2}$, $-\tfrac{5}{6}$
34. $\tfrac{3}{7}$, $\tfrac{3}{4}$
35. 2, $\sqrt{5}$, $-\sqrt{5}$
36. -3, $\sqrt{7}$, $-\sqrt{7}$
37. $\tfrac{1}{2}$ (multiplicity 2), -2 (multiplicity 3)
38. 0, -2, $\tfrac{3}{4}$ (multiplicity 3)

In Problems 39–42, find a polynomial $P(x)$ having only the indicated simple zeros.

39. 2, -2, i, $-i$
40. $2i$, $-2i$, $3i$, $-3i$
41. 2, -5, $2 + 3i$, $2 - 3i$
42. -3, 2, $1 - 4i$, $1 + 4i$

EXAMPLE B (More on Zeros of Polynomials) Show that 2 is a zero of multiplicity 3 of the polynomial

$$P(x) = 2x^5 - 17x^4 + 51x^3 - 58x^2 + 4x + 24$$

and find the remaining zeros.

Solution. We must show that $x - 2$ appears as a factor 3 times in the factored form of $P(x)$. Synthetic division can be used successively, as shown next.

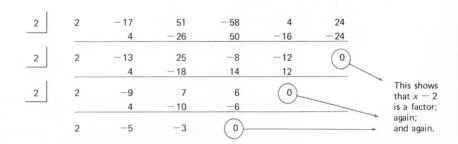

The final quotient $2x^2 - 5x - 3$ factors as $(2x + 1)(x - 3)$. Therefore, the remaining two zeros are $-\tfrac{1}{2}$ and 3. The factored form of $P(x)$ is

$$2(x - 2)^3(x + \tfrac{1}{2})(x - 3)$$

43. Show that 1 is a zero of multiplicity 3 of the polynomial $x^5 + 2x^4 - 6x^3 - 4x^2 + 13x - 6$, find the remaining zeros, and factor completely.

44. Show that the polynomial $x^5 - 11x^4 + 46x^3 - 90x^2 + 81x - 27$ has 3 as a zero of multiplicity 3, find the remaining zeros, and factor completely.

45. Show that the polynomial

$$x^6 - 8x^5 + 7x^4 + 32x^3 + 31x^2 + 40x + 25$$

has -1 and 5 as zeros of multiplicity 2 and find the remaining zeros.

46. Show that the polynomial

$$x^6 + 3x^5 - 9x^4 - 50x^3 - 84x^2 - 72x - 32$$

has 4 as a simple zero and -2 as a zero of multiplicity 3. Find the remaining zeros.

EXAMPLE C (Factoring a Polynomial with a Given Nonreal Zero)   Show that $1 + 2i$ is a zero of $P(x) = x^3 - (1 + 2i)x^2 - 4x + 4 + 8i$. Then factor $P(x)$ into linear factors.

Solution.   We start by using synthetic division.

$$\begin{array}{r|rrrr} 1 + 2i & 1 & -1 - 2i & -4 & 4 + 8i \\ & & 1 + 2i & 0 & -4 - 8i \\ \hline & 1 & 0 & -4 & 0 \end{array}$$

Therefore, $1 + 2i$ is a zero and $x - 1 - 2i$ is a factor of $P(x)$, and

$$P(x) = (x - 1 - 2i)(x^2 - 4)$$
$$= (x - 1 - 2i)(x + 2)(x - 2)$$

In each of the following, factor P(x) into linear factors given that c is a zero of P(x).

47. $P(x) = x^3 - 2ix^2 - 9x + 18i; c = 2i$
48. $P(x) = x^3 + 3ix^2 - 4x - 12i; c = -3i$
49. $P(x) = x^3 + (1 - i)x^2 - (1 + 2i)x - 1 - i; c = 1 + i$
50. $P(x) = x^3 - 3ix^2 + (3i - 3)x - 2 + 6i; c = -1 + 3i$

MISCELLANEOUS PROBLEMS

51. Find the remainder when the first polynomial is divided by the second.
    (a) $3x^{44} + 5x^{41} + 4; x + 1$
    (b) $1988x^3 - 1989x^2 + 1990x - 1991; x - 1$
    (c) $x^9 + 512; x + 2$

52. Prove that the second polynomial is a factor of the first.
    (a) $x^n - a^n; x - a$ (n a positive integer)
    (b) $x^n + a^n; x + a$ (n an odd positive integer)
    (c) $x^5 + 32a^{10}; x + 2a^2$

53. Find all zeros of the following polynomials and give their multiplicities.
    (a) $(x^2 - 4)^3$ (b) $(x^2 - 3x + 2)^2$ (c) $(x^2 + 2x - 4)^3(x + 2)^4$

54. Show that $2x^{44} + 3x^4 + 5$ has no factor of the form $x - c$, where c is real.

55. Each of the following polynomials has $\frac{1}{2}$ as a zero. Factor each polynomial into linear factors.
   (a) $12x^3 + 4x^2 - 3x - 1$
   (b) $2x^3 - x^2 - 4x + 2$
   ☐ (c) $2x^3 - x^2 + 2x - 1$

56. Find a third degree polynomial $P(x)$ with integral coefficients that has the given numbers as zeros.
   (a) $\frac{3}{4}, 2, -\frac{2}{3}$ (b) $3, -2, -2$ (c) $3, \sqrt{2}$

57. Sketch the graph of $y = x^3 + 4x^2 - 2x$. Then find three values of x for which $y = 8$. *Hint:* One of them is an integer.

☐ 58. Refer to the graph of Problem 57 and note that there is only one real x (namely, $x = 2$) for which $y = 20$. What does this tell you about the other two solutions of $x^3 + 4x^2 - 2x = 20$? Find these other two solutions.

59. A tray is to be constructed from a piece of sheet metal 16 inches square by cutting small identical squares of length x from the corners and then folding up the flaps (Figure 1). Determine x if the volume is to be 300 cubic inches. *Hint:* There are two possible answers, one of which is an integer.

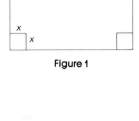

Figure 1

60. Find a polynomial $P(x) = ax^3 + bx^2 + cx + d$ that has 1, 2, and 3 as zeros and satisfies $P(0) = 36$.

61. Find $P(x) = ax^4 + bx^3 + cx^2 + dx + e$ if $P(x)$ has $\frac{1}{2}$ as a zero of multiplicity 4 and $P(0) = 1$.

62. Write $x^5 + x^4 + x^3 + x^2 + x + 1$ as a product of linear and quadratic factors with integer coefficients.

63. Show that if $P(x) = a_n x^n + a_{n-1} x^{n-1} + a_{n-2} x^{n-2} + \cdots + a_1 x + a_0$ has $n + 1$ distinct zeros, then all the coefficients must be 0.

64. Let
$$P_1(x) = x^n + a_{n-1} x^{n-1} + a_{n-2} x^{n-2} + \cdots + a_1 x + a_0$$
and
$$P_2(x) = x^n + b_{n-1} x^{n-1} + b_{n-2} x^{n-2} + \cdots + b_1 x + b_0$$

Show that if $P_1(x) = P_2(x)$ for n distinct values of x, then the two polynomials are equal for all x. *Hint:* Apply Problem 63 to $P(x) = P_1(x) - P_2(x)$.

65. Show that if c is a zero of $x^6 - 5x^5 + 3x^4 + 7x^3 + 3x^2 - 5x + 1$, then $1/c$ is also a zero.

66. Generalize Problem 65 by showing that if c is a zero of
$$a_n x^n + a_{n-1} x^{n-1} + a_{n-2} x^{n-2} + \cdots + a_1 x + a_0$$
where $a_k = a_{n-k}$ for $k = 0, 1, 2, \ldots, n$ and $a_n \neq 0$, then $1/c$ is also a zero.

67. Show that if $c_1, c_2, \ldots, c_n$ are the zeros of
$$a_n x^n + a_{n-1} x^{n-1} + a_{n-2} x^{n-2} + \cdots + a_1 x + a_0$$
with $a_n \neq 0$, then
   (a) $c_1 + c_2 + \cdots + c_n = -a_{n-1}/a_n$
   (b) $c_1 c_2 + c_1 c_3 + \cdots + c_1 c_n + c_2 c_3 + \cdots + c_2 c_n + \cdots + c_{n-1} c_n$
       $= a_{n-2}/a_n$
   (c) $c_1 c_2 \ldots c_n = (-1)^n a_0/a_n$
Hint: Use the complete factorization theorem.

68. **TEASER** Let the polynomial $x^n - kx - 1$, $n > 2$, have zeros $c_1, c_2,$ $\ldots, c_n$. Show each of the following.
 (a) If $n > 2$, then $c_1^n + c_2^n + \cdots + c_n^n = n$
 (b) If $n > 3$, then $c_1^2 + c_2^2 + \cdots + c_n^2 = 0$

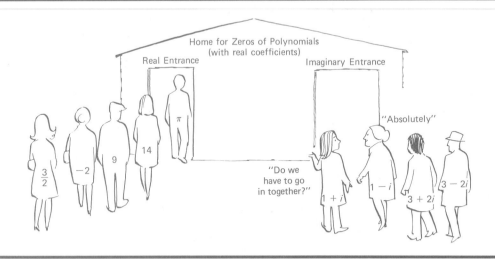

8-3   Polynomial Equations with Real Coefficients

Finding the zeros of a polynomial $P(x)$ is the same as finding the solutions of the equation $P(x) = 0$. We have solved polynomial equations before, especially in Sections 1-4 and 7-1. In particular, recall that the two solutions of a quadratic equation with real coefficients are either both real or both nonreal. For example, the equation $x^2 - 7x + 12 = 0$ has the real solutions 3 and 4. On the other hand, $x^2 - 8x + 17 = 0$ has two nonreal solutions $4 + i$ and $4 - i$.

In this section, we shall see that if a polynomial equation has real coefficients, then its nonreal solutions (if any) must occur in pairs—conjugate pairs. In the opening display, $1 + i$ and $1 - i$ must enter together or not at all; so must $3 + 2i$ and $3 - 2i$.

PROPERTIES OF CONJUGATES

We indicate the conjugate of a complex number by putting a bar over it. If $u = a + bi$, then $\bar{u} = a - bi$. For example,

$$\overline{1 + i} = 1 - i$$

$$\overline{3 - 2i} = 3 + 2i$$

$$\overline{4} = 4$$

$$\overline{2i} = -2i$$

The operation of taking the conjugate behaves nicely in both addition and multiplication. There are two pertinent properties, stated first in words.

1. The conjugate of a sum is the sum of the conjugates.
2. The conjugate of a product is the product of the conjugates.

In symbols these properties become

1. $\overline{u_1 + u_2 + \cdots + u_n} = \overline{u}_1 + \overline{u}_2 + \overline{u}_3 + \cdots + \overline{u}_n$
2. $\overline{u_1 u_2 u_3 \cdots u_n} = \overline{u}_1 \cdot \overline{u}_2 \cdot \overline{u}_3 \cdots \overline{u}_n$

A third property follows from the second property if we set all u_i's equal to u.

3. $\overline{u^n} = (\overline{u})^n$

Rather than prove these properties, we shall illustrate them.

1. $\overline{(2 + 3i) + (1 - 4i)} = \overline{2 + 3i} + \overline{1 - 4i}$
2. $\overline{(2 + 3i)(1 - 4i)} = \overline{(2 + 3i)}\,\overline{(1 - 4i)}$
3. $\overline{(2 + 3i)^3} = \overline{(2 + 3i)}^3$

Let us check that the second statement is correct. We will do it by computing both sides independently.

$$\overline{(2 + 3i)(1 - 4i)} = \overline{2 + 12 + (-8 + 3)i} = \overline{14 - 5i} = 14 + 5i$$

$$\overline{(2 + 3i)}\,\overline{(1 - 4i)} = (2 - 3i)(1 + 4i) = 2 + 12 + (8 - 3)i = 14 + 5i$$

We can use the three properties of conjugates to demonstrate some very important results. For example, we can show that if u is a solution of the equation

$$ax^3 + bx^2 + cx + d = 0$$

where a, b, c, and d are real, then $\overline{u}$ is also a solution. To do this, we must demonstrate that

$$a\overline{u}^3 + b\overline{u}^2 + c\overline{u} + d = 0$$

whenever it is given that

$$au^3 + bu^2 + cu + d = 0$$

In the latter equation, take the conjugate of both sides, using the three properties of conjugates and the fact that a real number is its own conjugate.

$$\overline{au^3 + bu^2 + cu + d} = \overline{0}$$

$$\overline{au^3} + \overline{bu^2} + \overline{cu} + \overline{d} = 0$$

$$\overline{a}\,\overline{u^3} + \overline{b}\,\overline{u^2} + \overline{c}\,\overline{u} + \overline{d} = 0$$

$$a\overline{u}^3 + b\overline{u}^2 + c\overline{u} + d = 0$$

NONREAL SOLUTIONS OCCUR IN PAIRS

We are ready to state the main theorem of this section.

CONJUGATE PAIR THEOREM

Let

$$a_n x^n + a_{n-1} x^{n-1} + \cdots + a_1 x + a_0 = 0$$

be a polynomial equation with real coefficients. If u is a solution, its conjugate $\bar{u}$ is also a solution.

We feel confident that you will be willing to accept the truth of this theorem without further argument. The formal proof would mimic the proof given above for the cubic equation.

As an illustration of one use of this theorem, suppose that we know that $3 + 4i$ is a solution of the equation

$$x^3 - 8x^2 + 37x - 50 = 0$$

Then we know that $3 - 4i$ is also a solution. We can easily find the third solution, which incidentally must be real. (Why?) Here is how we do it.

$$
\begin{array}{r|rrrr}
3 + 4i & 1 & -8 & 37 & -50 \\
& & 3 + 4i & -31 - 8i & 50 \\
\hline
3 - 4i & 1 & -5 + 4i & 6 - 8i & 0 \\
& & 3 - 4i & -6 + 8i & \\
\hline
& 1 & -2 & 0 &
\end{array}
$$

From this it follows that

$$x^3 - 8x^2 + 37x - 50 = [x - (3 + 4i)][x - (3 - 4i)][x - 2]$$

and that 2 is the third solution of our equation.

Notice something special about the product of the first two factors displayed above.

$$[x - (3 + 4i)][x - (3 - 4i)]$$

$$= x^2 - (3 + 4i)x - (3 - 4i)x + (3 + 4i)(3 - 4i)$$

$$= x^2 - 6x + 25$$

The product is a quadratic polynomial with *real* coefficients. This is not an accident. If u is any complex number, then

$$(x - u)(x - \bar{u}) = x^2 - (u + \bar{u})x + u\bar{u}$$

is a real quadratic polynomial, since both $u + \bar{u}$ and $u\bar{u}$ are real (see Problem 39). Thus the conjugate pair theorem, when combined with the complete factorization theorem of Section 8-2, has the following consequence.

REAL FACTORS THEOREM

Any polynomial with real coefficients can be factored into a product of linear and quadratic polynomials having real coefficients, where the quadratic polynomials have no real zeros.

RATIONAL SOLUTIONS

How does one get started on solving an equation of high degree? So far, all we can suggest is to guess. If you are lucky and find a solution, you can use synthetic division to reduce the degree of the equation to be solved. Eventually you may get it down to a quadratic equation, for which we have the quadratic formula.

Guessing would not be so bad if there were not so many possibilities to consider. Is there an intelligent way to guess? There is, but unfortunately it works only if the coefficients are integers, and then it only helps us find rational solutions.

Consider

$$3x^3 + 13x^2 - x - 6 = 0$$

which, as you will note, has integral coefficients. Suppose it has a rational solution c/d which is in reduced form (that is, c and d are integers without common divisors greater than 1 and $d > 0$). Then

$$3 \cdot \frac{c^3}{d^3} + 13 \cdot \frac{c^2}{d^2} - \frac{c}{d} - 6 = 0$$

or, after multiplying by d^3,

$$3c^3 + 13c^2d - cd^2 - 6d^3 = 0$$

We can rewrite this as

$$c(3c^2 + 13cd - d^2) = 6d^3$$

and also as

$$d(13c^2 - cd - 6d^2) = -3c^3$$

The first of these tells us that c divides $6d^3$ and the second that d divides $-3c^3$. But c and d have no common divisors. Therefore, c must divide 6 and d must divide 3.

The only possibilities for c are ± 1, ± 2, ± 3, and ± 6; for d, the only possibilities are 1 and 3. Thus the possible rational solutions must come from the list below.

$$\frac{c}{d}: \quad \pm 1, \pm 2, \pm 3, \pm 6, \pm \tfrac{1}{3}, \pm \tfrac{2}{3}$$

Upon checking all 12 numbers (which takes time, but a bit less time than checking *all* numbers would take!) we find that only $\tfrac{2}{3}$ works.

$$
\begin{array}{r|rrrr}
\tfrac{2}{3} & 3 & 13 & -1 & -6 \\
 & & 2 & 10 & 6 \\
\hline
 & 3 & 15 & 9 & 0
\end{array}
$$

We could prove the following theorem by using similar reasoning.

RATIONAL SOLUTION THEOREM (RATIONAL ROOT THEOREM)

Let

$$a_n x^n + a_{n-1} x^{n-1} + \cdots + a_1 x + a_0 = 0$$

have integral coefficients. If c/d is a rational solution in reduced form, then c divides a_0 and d divides a_n.

Problem Set 8-3

In Problems 1–10, write the conjugate of the number.

1. $2 + 3i$
2. $3 - 5i$
3. $4i$
4. $-6i$
5. $4 + \sqrt{6}$
6. $3 - \sqrt{5}$
7. $(2 - 3i)^8$
8. $(3 + 4i)^{12}$
9. $2(1 + 2i)^3 - 3(1 + 2i)^2 + 5$
10. $4(6 - i)^4 + 11(6 - i) - 23$

11. If $P(x)$ is a cubic polynomial with real coefficients and has -3 and $5 - i$ as zeros, what other zero does it have?

12. If $P(x)$ is a cubic polynomial with real coefficients and has 0 and $\sqrt{2} + 3i$ as zeros, what other zero does it have?

13. Suppose that $P(x)$ has real coefficients and is of the fourth degree. If it has $3 - 2i$ and $5 + 4i$ as two of its zeros, what other zeros does it have?

14. If $P(x)$ is a fourth degree polynomial with real coefficients and has $5 + 6i$ as a zero of multiplicity 2, what are its other zeros?

EXAMPLE A (Solving an Equation Given Some Solutions)   Given that -1 and $1 + 2i$ are solutions of the equation

$$2x^4 - 5x^3 + 9x^2 + x - 15 = 0$$

find the other solutions.

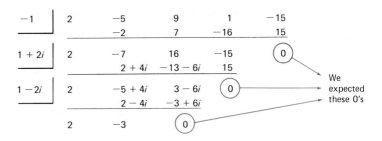

Last quotient: $2x - 3$

The 4th solution: $\dfrac{3}{2}$

Solution. Since the coefficients are real, $1 - 2i$ is a solution. The fourth solution is found by progressively using synthetic division. This division is shown at the bottom of the previous page.

In Problems 15–18, one or more solutions of the specified equations are given. Find the other solutions.

15. $2x^3 - x^2 + 2x - 1 = 0;\ i$

16. $x^3 - 3x^2 + 4x - 12 = 0;\ 2i$

17. $x^4 + x^3 + 6x^2 + 26x + 20 = 0;\ 1 + 3i$

18. $x^6 + 2x^5 + 4x^4 + 4x^3 - 4x^2 - 16x - 16 = 0;\ 2i,\ -1 + i$

EXAMPLE B (Obtaining a Polynomial Given Some of Its Zeros) Find a cubic polynomial with real coefficients that has 3 and $2 + 3i$ as zeros. Make the leading coefficient 1.

Solution. The third zero has to be $2 - 3i$. In factored form, our polynomial is

$$(x - 3)(x - 2 - 3i)(x - 2 + 3i)$$

We multiply this out in stages.

$$(x - 3)[(x - 2)^2 + 9]$$
$$(x - 3)(x^2 - 4x + 13)$$
$$x^3 - 7x^2 + 25x - 39$$

Find a polynomial with real coefficients that has the indicated degree and the given zero(s). Make the leading coefficient 1.

19. Degree 2; zero: $2 + 5i$.

20. Degree 2; zero: $\sqrt{6}i$.

21. Degree 3; zeros: $-3,\ 2i$.

22. Degree 3; zeros: $5,\ -3i$.

23. Degree 5; zeros: $2,\ 3i$ (multiplicity 2).

24. Degree 5; zeros: $1,\ 1 - i$ (multiplicity 2).

EXAMPLE C (Finding Rational Solutions) Find the rational solutions of the equation

$$3x^4 + 2x^3 + 2x^2 + 2x - 1 = 0$$

Then find the remaining solutions.

Solution. The only way that c/d (in reduced form) can be a solution is for c to be 1 or -1 and for d to be 1 or 3. This means that the possibilities for c/d are ± 1 and $\pm\frac{1}{3}$. Synthetic division shows that -1 and $\frac{1}{3}$ work.

$$\begin{array}{r|rrrrr}
-1 & 3 & 2 & 2 & 2 & -1 \\
 & & -3 & 1 & -3 & 1 \\
\hline
\frac{1}{3} & 3 & -1 & 3 & -1 & 0 \\
 & & 1 & 0 & 1 & \\
\hline
 & 3 & 0 & 3 & 0 &
\end{array}$$

Setting the final quotient, $3x^2 + 3$, equal to zero and solving, we get

$$3x^2 = -3$$

$$x^2 = -1$$

$$x = \pm i$$

The complete solution set is $\{-1, \frac{1}{3}, i, -i\}$.

In Problems 25–30, find the rational solutions of each equation. If possible, find the other solutions.

25. $x^3 - 3x^2 - x + 3 = 0$
26. $x^3 + 3x^2 - 4x - 12 = 0$
27. $2x^3 + 3x^2 - 4x + 1 = 0$
28. $5x^3 - x^2 + 5x - 1 = 0$
29. $\frac{1}{3}x^3 - \frac{1}{2}x^2 - \frac{1}{6}x + \frac{1}{6} = 0$ *Hint: Clear the equation of fractions.*
30. $\frac{2}{3}x^3 - \frac{1}{2}x^2 + \frac{2}{3}x - \frac{1}{2} = 0$

MISCELLANEOUS PROBLEMS

☐ 31. **The number $2 + i$ is one solution to $x^4 - 3x^3 + 2x^2 + x + 5 = 0$. Find the other solutions.**

☐ 32. Find the fourth degree polynomial with real coefficients and leading coefficient 1 that has 2, -4, and $2 - 3i$ as three of its zeros.

33. Find all solutions of $x^4 - 3x^3 - 20x^2 - 24x - 8 = 0$.

☐ 34. Find all solutions of $2x^4 - x^3 + x^2 - x - 1 = 0$.

35. Solve $x^5 + 6x^4 - 34x^3 + 56x^2 - 39x + 10 = 0$.

☐ 36. Solve $x^5 - 2x^4 + x - 2 = 0$. *Hint: $x^4 + 1$ is easy to factor by adding and subtracting $2x^2$.*

☐ 37. Show that a polynomial equation with real coefficients and of odd degree has at least one real solution.

38. A cubic equation with real coefficients has either 3 real solutions (not necessarily distinct), or 1 real solution and a pair of nonreal solutions. State all the possibilities for:
 (a) A fourth degree equation with real coefficients.
 (b) A fifth degree equation with real coefficients.

☐ 39. Show that if u is any complex number, then $u + \bar{u}$ and $u\bar{u}$ are real numbers.

☐ 40. Let u and v be complex numbers with $v \neq 0$. Show that $\overline{u/v} = \bar{u}/\bar{v}$. Use this to demonstrate that if $f(x)$ is a real rational function (a quotient of two polynomials with real coefficients) and if $f(a + bi) = c + di$, then $f(a - bi) = c - di$.

41. Write $x^4 + 3x^3 + 3x^2 - 3x - 4$ as a product of linear and quadratic factors with real coefficients as guaranteed by the real factors theorem.

42. Write $x^6 - 9x^5 + 38x^4 - 106x^3 + 181x^2 - 205x + 100$ as a product of linear and quadratic factors with real coefficients. *Hint:* $1 + 2i$ is a zero of multiplicity 2.

43. Write $x^8 - 1$ as a product of linear and quadratic factors with real coefficients. *Hint:* First factor as a difference of squares; then see the hint in Problem 36.

44. Let $x^n + a_{n-1}x^{n-1} + \cdots + a_1x + a_0 = 0$ have integral coefficients.
    (a) Show that all real solutions are either integral or irrational.
    (b) From (a), deduce that if m and n are positive integers and m is not a perfect nth power, then $\sqrt[n]{m}$ is irrational (in particular, $\sqrt[3]{3}$, $\sqrt[4]{17}$, and $\sqrt[5]{12}$ are irrational).

45. Find the exact value of $x = \sqrt[3]{\sqrt{5} - 2} - \sqrt[3]{\sqrt{5} + 2}$. *Hint:* Begin by showing that $x^3 + 3x + 4 = 0$.

46. **TEASER** Let (u, v, w) satisfy the following nonlinear system of equations:

$$u + v + w = 2$$

$$u^2 + v^2 + w^2 = 8$$

$$u^3 + v^3 + w^3 = 8$$

(a) Determine the cubic equation $x^3 + a_2x^2 + a_1x + a_0 = 0$ that has u, v, and w as its solutions. *Hint:* Problem 67 of Section 8-2 should be helpful.
(b) Solve this cubic equation thereby solving the system of equations.

Combining Fractions

$$\frac{5}{x} + \frac{4}{x - 1} = \frac{9x - 5}{x(x - 1)}$$

Decomposing Fractions

$$\frac{5x - 2}{(x + 2)(x - 2)} = \frac{3}{x + 2} + \frac{2}{x - 2}$$

Do and undo, the day is long enough.

Old proverb

8-4 Decomposition into Partial Fractions

The process of combining fractions is familiar to every student of algebra. We have used it often in earlier parts of this book. Less familiar, but also important—especially in calculus—is the reverse process of decomposing a complicated fraction into a sum of simpler ones. But is such a decomposition always possible; and—if so—how can it be accomplished?

Our question needs more precision. By a fraction, we here mean a real proper rational expression—that is, a quotient $N(x)/D(x)$ of two real polynomials in which the degree of $N(x)$ is less than the degree of $D(x)$. Thus we ask: Can a real proper rational expression be decomposed into a sum of simpler ones?

To begin our answer, we suppose the denominator $D(x)$ has been factored into a product of real linear and real quadratic factors (see the real factors theorem of Section 8-3). Then we break the problem into several cases.

CASE 1   (DISTINCT LINEAR FACTORS)

Consider

$$\frac{4x^2 - 3x + 2}{(x - 1)(x + 2)(x - 2)}$$

The decomposition in our opening box suggests that we look for numbers A, B, and C such that

$$\frac{4x^2 - 3x + 2}{(x - 1)(x + 2)(x - 2)} = \frac{A}{x - 1} + \frac{B}{x + 2} + \frac{C}{x - 2}$$

When we multiply both sides by $(x - 1)(x + 2)(x - 2)$ to clear of fractions, we obtain the equation

(i) $4x^2 - 3x + 2$

$$= A(x + 2)(x - 2) + B(x - 1)(x - 2) + C(x - 1)(x + 2)$$

which we wish to be an identity. After expanding and then collecting like terms, this reduces to

(ii) $4x^2 - 3x + 2 = (A + B + C)x^2 + (-3B + C)x + (-4A + 2B - 2C)$

The latter is an identity if coefficients of like terms on both sides are equal; that is, if

$$A + B + C = 4$$

$$-3B + C = -3$$

$$-4A + 2B - 2C = 2$$

We know how to solve this system and shall, in fact, solve a similar system later (see Example B). But for this particular problem, there is a better way.

Return to equation (i). If it is to be an identity (true for all x), then in particular it must be true for $x = 1$, $x = 2$, and $x = -2$. Substituting these values, in succession, in (i) gives

$$3 = -3A \qquad 12 = 4C \qquad 24 = 12B$$

(Note that our calculations were made easy by a fortunate choice of values for x.) We conclude that $A = -1$, $B = 2$, $C = 3$, and that

$$\frac{4x^2 - 3x + 2}{(x - 1)(x + 2)(x - 2)} = \frac{-1}{x - 1} + \frac{2}{x + 2} + \frac{3}{x - 2}$$

If you have any doubts about this process, try combining the fractions on the right to see that you do get the fraction on the left.

What we have done will work whenever the denominator $D(x)$ factors into a product of distinct linear factors. For each factor $dx + e$, we must have a term of the form $A/(dx + e)$ in the decomposition.

CASE 2   (REPEATED LINEAR FACTORS)

Next consider

$$\frac{5x^2 + 7x + 6}{(x - 1)(x + 2)^2}$$

in which the denominator has the repeated linear factor $(x + 2)^2$. Corresponding to this factor, we need two terms in the decomposition, namely,

$$\frac{A}{x + 2} + \frac{B}{(x + 2)^2}$$

Thus we are looking for numbers A, B, and C such that

$$\frac{5x^2 + 7x + 6}{(x + 2)^2 (x - 1)} = \frac{A}{x + 2} + \frac{B}{(x + 2)^2} + \frac{C}{x - 1}$$

After clearing fractions, this becomes

$$5x^2 + 7x + 6 = A(x + 2)(x - 1) + B(x - 1) + C(x + 2)^2$$

The two wise substitutions $x = -2$ and $x = 1$ and a third of $x = 0$ yield

$$12 = -3B \qquad 18 = 9C \qquad 6 = -2A - B + 4C$$

or $A = 3$, $B = -4$, and $C = 2$. We conclude that

$$\frac{5x^2 + 7x + 6}{(x + 2)^2 (x - 1)} = \frac{3}{x + 2} + \frac{-4}{(x + 2)^2} + \frac{2}{x - 1}$$

CASE 3   (QUADRATIC FACTORS)

Consider

$$\frac{6x^2 - 21x + 13}{(x^2 + 4)(x - 5)}$$

and note that the quadratic factor $x^2 + 4$ cannot be written as a product of real linear factors. Corresponding to it, we need (take our word for this) a term of the form $(Ax + B)/(x^2 + 4)$ in the decomposition. Thus we are looking for numbers A, B, and C such that

$$\frac{6x^2 - 21x + 13}{(x^2 + 4)(x - 5)} = \frac{Ax + B}{x^2 + 4} + \frac{C}{x - 5}$$

or, after clearing fractions,

$$6x^2 - 21x + 13 = (Ax + B)(x - 5) + C(x^2 + 4)$$

The substitutions $x = 5$, $x = 0$, and $x = 1$ yield

$$58 = 29C$$

$$13 = -5B + 4C$$

$$-2 = -4A - 4B + 5C$$

This system has the solution $C = 2$, $B = -1$, and $A = 4$. Our final result is

$$\frac{6x^2 - 21x + 13}{(x^2 + 4)(x - 5)} = \frac{4x - 1}{x^2 + 4} + \frac{2}{x - 5}$$

What if $(x^2 + 4)^2$ instead of $x^2 + 4$ had appeared in the denominator? Corresponding to it, we need a sum of two terms in the decomposition of the form

$$\frac{Ax + B}{x^2 + 4} + \frac{Cx + D}{(x^2 + 4)^2}$$

This case is illustrated in Example B of the problem set.

CASE 4 (GENERAL CASE)

Let us summarize by describing the general situation. Consider an arbitrary proper rational expression $N(x)/D(x)$, where $D(x)$ is a product of real linear or real quadratic factors or both. For each factor of the form $(dx + e)^m$, the decomposition will contain a sum of the form

$$\frac{A_1}{dx + e} + \frac{A_2}{(dx + e)^2} + \cdots + \frac{A_m}{(dx + e)^m}$$

And for each factor of the form $(ax^2 + bx + c)^n$, there is a sum of the form

$$\frac{B_1 x + C_1}{ax^2 + bx + c} + \frac{B_2 x + C_2}{(ax^2 + bx + c)^2} + \cdots + \frac{B_n x + C_n}{(ax^2 + bx + c)^n}$$

Thus, for example, the expression

$$\frac{x^4 - 2x^2 + 10}{(x + 3)(x - 1)^3(x^2 + x + 5)^2}$$

will decompose into

$$\frac{A}{x + 3} + \frac{B}{x - 1} + \frac{C}{(x - 1)^2} + \frac{D}{(x - 1)^3} + \frac{Ex + F}{x^2 + x + 5} + \frac{Gx + H}{(x^2 + x + 5)^2}$$

Clearly the method of decomposing $N(x)/D(x)$ into partial fractions rests on our ability to factor $D(x)$. This is a genuine limitation; for while the real factors theorem guarantees that it is possible, no one is likely to produce the factors of $x^{98} - 4x^{39} + 11x^2 - 17$.

Problem Set 8-4

Decompose each of the following. Make sure that the denominator is factored as much as possible before you begin.

1. $\dfrac{-x + 5}{(x + 1)(x - 1)}$

2. $\dfrac{6}{(x + 3)(x - 3)}$

3. $\dfrac{x + 7}{x^2 - 4x - 5}$

4. $\dfrac{8x - 8}{x^2 + 2x - 15}$

5. $\dfrac{x - 29}{x^2 - 3x - 4}$

6. $\dfrac{5x - 12}{x^2 - 4x}$

7. $\dfrac{4x^2 - 15x - 1}{(x - 1)(x + 2)(x - 3)}$

8. $\dfrac{x^2 + 19x + 20}{x(x + 2)(x - 5)}$

9. $\dfrac{5x^2 - 10x + 3}{x^2(x - 3)}$

10. $\dfrac{4x^2 + 26x + 48}{x(x + 4)^2}$

11. $\dfrac{6x^2 + 16x - 40}{(x^2 - 4)^2}$

12. $\dfrac{7x^3 - 48x^2 + 102x - 64}{(x^2 - 5x + 6)^2}$

13. $\dfrac{3x^2 - x + 4}{(x + 2)(x^2 + 2)}$

14. $\dfrac{2x^2 + 5x - 6}{(x - 4)(x^2 + x + 3)}$

15. $\dfrac{2x^2 + 9x - 2}{(x - 2)(x^2 + 2x + 4)}$

16. $\dfrac{2x^2 - 3x - 5}{(x + 4)(x^2 + x + 1)}$

EXAMPLE A (Improper Rational Expressions)   Find the decomposition of

$$\frac{x^3 - x^2 - 3x + 12}{x^2 + x - 2}$$

Solution.   Our discussion at the end of Section 8-1 suggests how to proceed. We begin with a long division, which results in a polynomial plus a proper rational expression. Then we use the method of this section.

$$\frac{x^3 - x^2 - 3x + 12}{x^2 + x - 2} = x - 2 + \frac{x + 8}{x^2 + x - 2}$$

$$\frac{x + 8}{x^2 + x - 2} = \frac{x + 8}{(x + 2)(x - 1)} = \frac{A}{x + 2} + \frac{B}{x - 1}$$

$$x + 8 = A(x - 1) + B(x + 2)$$

The substitutions $x = -2$ and $x = 1$ give $A = -2$ and $B = 3$. Thus

$$\frac{x^3 - x^2 - 3x + 12}{x^2 + x - 2} = x - 2 - \frac{2}{x + 2} + \frac{3}{x - 1}$$

Use the method of Example A to find the decomposition of the following.

17. $\dfrac{x^2 + 1}{x^2 - 4}$

18. $\dfrac{2x^2}{x^2 - 9}$

19. $\dfrac{2x^3 - 4x^2 + 4x + 2}{x^2 - 2x}$

20. $\dfrac{3x^3 - x^2 - 11x + 2}{x^2 - 4}$

EXAMPLE B (Repeated Quadratic Factors) Decompose

$$\frac{2x^4 - 3x^3 + 21x^2 - 13x + 50}{x(x^2 + 5)^2}$$

Solution. The expression is proper, so we proceed at once to write

$$\frac{2x^4 - 3x^3 + 21x^2 - 13x + 50}{x(x^2 + 5)^2} = \frac{A}{x} + \frac{Bx + C}{x^2 + 5} + \frac{Dx + E}{(x^2 + 5)^2}$$

As usual, we clear fractions; but this time, rather than substitute values for x, we collect terms and then equate coefficients of like terms.

$$2x^4 - 3x^3 + 21x^2 - 13x + 50$$
$$= A(x^2 + 5)^2 + (Bx + C)x(x^2 + 5) + (Dx + E)x$$
$$= (A + B)x^4 + Cx^3 + (10A + 5B + D)x^2$$
$$+ (5C + E)x + 25A$$

Thus,

$$A + B = 2$$
$$C = -3$$
$$10A + 5B + D = 21$$
$$5C + E = -13$$
$$25A = 50$$

The solution to this system is $A = 2$, $B = 0$, $C = -3$, $D = 1$, and $E = 2$. Consequently,

$$\frac{2x^4 - 3x^3 + 21x^2 - 13x + 50}{x(x^2 + 5)^2} = \frac{2}{x} - \frac{3}{x^2 + 5} + \frac{x + 2}{(x^2 + 5)^2}$$

Decompose each of the following.

21. $\dfrac{x^3}{(x^2 + 1)^2}$

22. $\dfrac{2x^3 + 4}{(x^2 + 5)^2}$

23. $\dfrac{x^4 + 4x^3 + 6x^2 + 9x + 2}{x(x^2 + x + 1)^2}$

24. $\dfrac{3x^4 - 6x^3 + 43x^2 - 46x + 170}{(x - 4)(x^2 + 7)^2}$

In Problems 25–33, obtain the partial fraction decomposition of the given expression.

25. $\dfrac{x + 23}{x^2 - x - 20}$

26. $\dfrac{3x - 18}{x^2 - 16}$

27. $\dfrac{-5x + 8}{x(x + 2)(x - 1)}$

28. $\dfrac{-16x}{x^4 + 10x^2 + 9}$

29. $\dfrac{14x^2 + 25x + 15}{(x + 2)^2(x - 5)}$

30. $\dfrac{-x^2 + 5x - 15}{(x + 3)(x^2 + 4)}$

31. $\dfrac{12x - 18}{x^3 + 27}$

32. $\dfrac{3x^4 + x^2 - 4x + 5}{(x^2 + x + 1)^2(x - 2)}$

33. $\dfrac{3x^5 - 6x^4 + 2x^3 + 4x^2 + 5x - 10}{x^4 - 2x^3}$

34. **TEASER** Demonstrate that the following is an identity.

$$\frac{1}{1 - \cos^4 2x} = \frac{1}{8}\left[\frac{1}{\sin^2 x} + \frac{1}{\cos^2 x} + \frac{4}{1 + \cos^2 2x}\right]$$

Chapter Summary

The **division law for polynomials** asserts that if $P(x)$ and $D(x)$ are any given nonconstant polynomials, then there are unique polynomials $Q(x)$ and $R(x)$ such that

$$P(x) = D(x)Q(x) + R(x)$$

where $R(x)$ is either 0 or of lower degree than $D(x)$. In fact, we can find $Q(x)$ and $R(x)$ by the **division algorithm,** which is just a fancy name for ordinary long division. When $D(x)$ has the form $x - c$, $R(x)$ will have to be a constant R, since it is of lower degree than $D(x)$. The substitution $x = c$ then gives

$$P(c) = R$$

a result known as the **remainder theorem.** An immediate consequence is the **factor theorem,** which says that c is a zero of $P(x)$ if and only if $x - c$ is a factor of $P(x)$. Division of a polynomial by $x - c$ can be greatly simplified by use of **synthetic division.**

That every nonconstant polynomial has at least one zero is guaranteed by Gauss's **fundamental theorem of algebra.** But we can say much more than that. For any nonconstant polynomial

$$P(x) = a_n x^n + a_{n-1} x^{n-1} + \cdots + a_1 x + a_0$$

there are n numbers $c_1, c_2, \ldots , c_n$ (not necessarily all different) such that

$$P(x) = a_n(x - c_1)(x - c_2) \cdots (x - c_n)$$

We call the latter result the **complete factorization theorem.**

If the polynomial equation

$$P(x) = a_n x^n + a_{n-1} x^{n-1} + \cdots + a_1 x + a_0 = 0$$

has real coefficients, then its nonreal solutions (if any) must occur in conjugate pairs $a + bi$ and $a - bi$. An immediate consequence is that $P(x)$ factors into a product of linear and quadratic factors with real coefficients, the **real factors theorem.**

If the coefficients of a polynomial are integers and if c/d is a rational zero in reduced form, then d divides the leading coefficient a_n and c divides the constant term a_0.

Any proper rational expression $N(x)/D(x)$ such that $D(x)$ has at least two distinct real factors, linear or quadratic, can be expressed as a sum of simpler rational expressions. This is called **decomposition into partial fractions.**

Chapter Review Problem Set

1. Find the quotient and remainder if the first polynomial is divided by the second.
   (a) $2x^3 - x^2 + 4x - 5;\ x^2 + 2x - 3$
   (b) $x^4 - 8x^2 + 5;\ x^2 + 3x$
   (c) $x^5 + 3;\ x + 1$

2. Use synthetic division to find the quotient and remainder if the first polynomial is divided by the second.
   (a) $x^3 - 2x^2 - 4x + 7;\ x - 2$
   (b) $2x^4 - 15x^2 + 4x - 3;\ x + 3$
   ☐ (c) $x^3 + (3 - 3i)x^2 - (9 + 15i)x - 3 - 3i;\ x - 2 - 3i$

3. Without dividing, find the remainder if $2x^4 - 6x^3 + 17$ is divided by $x - 2$; if it is divided by $x + 2$.

☐ 4. Find the zeros of the given polynomial and give their multiplicities.
   (a) $(x^2 - 1)^2(x^2 + 1)$
   (b) $x(x^2 - 2x + 4)(x + \pi)^3$
   (c) $(x^4 + 5x^2 + 6)^4$

In Problems 5 and 6, use synthetic division to show that $x - c$ is a factor of $P(x)$. Then factor $P(x)$ completely into linear factors.

5. $P(x) = 2x^3 - x^2 - 18x + 9;\ c = 3$
6. $P(x) = x^3 + 4x^2 - 7x - 28;\ c = -4$

In Problems 7 and 8, find a polynomial $P(x)$ with integral coefficients that has the given zeros. Assume each zero to be simple unless otherwise indicated.

7. $3, -2, 4$ (multiplicity 2)
☐ 8. $3 + \sqrt{7}, 3 - \sqrt{7}, 2 - i, 2 + i$
9. Show that 1 is a zero of multiplicity 2 of the polynomial $x^4 - 4x^3 - 3x^2 + 14x - 8$ and find the remaining zeros.

10. Find the polynomial $P(x) = a_3x^3 + a_2x^2 + a_1x + a_0$ which has zeros $\frac{1}{2}$, $-\frac{1}{3}$, and 4 and for which $P(2) = -42$.

11. Find the value of k so that $\sqrt{2}$ is a zero of $x^3 + 3x^2 - 2x + k$.

☐ 12. Find a cubic polynomial with real coefficients that has $4 + 3i$ and -2 as two of its zeros. Then write the polynomial as the product of a real linear and a real quadratic polynomial.

☐ 13. Let $P(x) = x^5 + a_4x^4 + a_3x^3 + a_2x^2 + a_1x + a_0$, where the a_i's are real. If $P(x)$ has $2 + 3i$ and $2 - 3i$ as zeros of multiplicity 2 and -4 as a simple zero, write $P(x)$ as the product of real linear and real quadratic polynomials.

☐ 14. Solve the equation $x^4 - 4x^3 + 24x^2 + 20x - 145 = 0$, given that $2 + 5i$ is one of its solutions.

15. The equation $2x^3 - 15x^2 + 20x - 3 = 0$ has a rational solution. Find it and then find the other solutions.

16. The equation $x^3 - x^2 - x - 7 = 0$ has at least one real solution. Why? Show that it has no rational solution.

In Problems 17–19, find the decomposition into partial fractions.

17. $\dfrac{-16x - 10}{x(x - 2)(x + 5)}$

18. $\dfrac{3x^3 - 14x^2 + 7x - 32}{(x^2 + 5)(x - 3)^2}$

19. $\dfrac{2x^5 + 3x^3 + x + 4}{x^4 + 2x^2 + 1}$

Geometry may sometimes appear to take the lead over analysis but in fact precedes it only as a servant goes before the master to clear the path and light him on his way.

James Joseph Sylvester

CHAPTER 9

Systems of Equations and Inequalities

On Algebraic
Wheels

$$\text{I} \begin{cases} x - y = 1 \\ 2x - y = 4 \end{cases} \qquad \text{II} \begin{cases} x - y = 1 \\ y = 2 \end{cases} \qquad \text{III} \begin{cases} x = 3 \\ y = 2 \end{cases}$$

On Geometric
Wheels

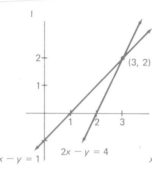 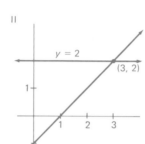 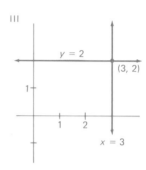

9-1 Equivalent Systems of Equations

In Section 1-4, you learned how to solve a system of two equations in two unknowns, but you probably did not think of the process as one of replacing the given system by another having the same solutions. It is this point of view that we now want to explore.

In the display above, system I is replaced by system II, which is simpler; system II is in turn replaced by system III, which is simpler yet. To go from system I to system II, we eliminated x from the second equation; we then substituted $y = 2$ in the first equation to get system III. What happened geometrically is shown in the bottom half of our display. Notice that the three pairs of lines have the same point of intersection $(3, 2)$.

Because the notion of changing from one system of equations to another having the same solutions is so important, we make a formal definition. We say that two systems of equations are **equivalent** if they have the same solutions.

OPERATIONS THAT LEAD TO EQUIVALENT SYSTEMS

Now we face a big question. What operations can we perform on a system without changing its solutions?

Operation 1 We can interchange the position of two equations.

Operation 2 We can multiply an equation by a nonzero constant, that is, we can replace an equation by a nonzero multiple of itself.

Operation 3 We can add a multiple of one equation to another, that is, we can replace an equation by the sum of that equation and a multiple of another.

Operation 3 is the workhorse of the set. We show how it is used in the example of the opening display.

$$\text{I} \quad \begin{cases} x - y = 1 \\ 2x - y = 4 \end{cases}$$

If we add -2 times the first equation to the second, we obtain

$$\text{II} \quad \begin{cases} x - y = 1 \\ y = 2 \end{cases}$$

We then add the second equation to the first. This gives

$$\text{III} \quad \begin{cases} x = 3 \\ y = 2 \end{cases}$$

This is one way to write the solution. Alternatively, we say that the solution is the ordered pair $(3, 2)$.

THE THREE POSSIBILITIES FOR A LINEAR SYSTEM

We are mainly interested in linear systems—that is, systems of linear equations—and we shall restrict our discussion to the case where there is the same number of equations as unknowns. There are three possibilities for the set of solutions: The set may be empty, it may have just one point, or it may have infinitely many points. These three cases are illustrated in Figure 1.

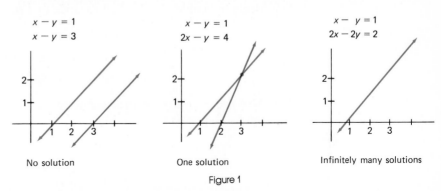

No solution   One solution   Infinitely many solutions

Figure 1

Someone is sure to object and ask if we cannot have a linear system with exactly two solutions or exactly three solutions. The answer is no. If a linear system has two solutions, it has infinitely many. This is obvious in the case of two equations in two unknowns, since two points determine a line, but it is also true for n equations in n unknowns.

LINEAR SYSTEMS IN MORE THAN TWO UNKNOWNS

When we consider large systems, it is a good idea to be very systematic about our method of attack. Our method is to reduce the system to **triangular form** and then use **back substitution.** Let us explain. Consider

$$x - 2y + z = -4$$
$$5y - 3z = 18$$
$$2z = -2$$

which is already in triangular form. (The name arises from the fact that the terms within the dotted triangle have zero coefficients.) This system is easy to solve. Solve the third equation first ($z = -1$). Then substitute that value in the second equation

$$5y - 3(-1) = 18$$

which gives $y = 3$. Finally substitute these two values in the first equation

$$x - 2(3) + (-1) = -4$$

which gives $x = 3$. Thus the solution of the system is $(3, 3, -1)$. This process, called **back substitution**, works on any linear system that is in triangular form. Just start at the bottom and work your way up.

If the system is not in triangular form initially, we try to operate on it until it is. Suppose we start with

$$2x - 4y + 2z = -8$$
$$2x + y - z = 10$$
$$3x - y + 2z = 4$$

Begin by multiplying the first equation by $\frac{1}{2}$ so that its leading coefficient is 1.

$$x - 2y + z = -4$$
$$2x + y - z = 10$$
$$3x - y + 2z = 4$$

Next add -2 times the first equation to the second equation. Also add -3 times the first equation to the third.

$$x - 2y + z = -4$$
$$5y - 3z = 18$$
$$5y - z = 16$$

Finally, add -1 times the second equation to the third.

$$x - 2y + z = -4$$
$$5y - 3z = 18$$
$$2z = -2$$

The system is now in triangular form and can be solved by back substitution. It is, in fact, the triangular system we discussed earlier. The solution is $(3, 3, -1)$.

In Examples A and B of the problem set, we show how this process works when there are infinitely many solutions and when there are no solutions.

Problem Set 9-1

Solve each of the following systems of equations.

1. $2x - 3y = 7$
   $y = -1$

2. $5x - 3y = -25$
   $y = 5$

3. $x = -2$
   $2x + 7y = 24$

4. $x = 5$
   $3x + 4y = 3$

5. $x - 3y = 7$
   $4x + y = 2$

6. $5x + 6y = 27$
   $x - y = 1$

7. $2x - y + 3z = -6$
   $2y - z = 2$
   $z = -2$

8. $x + 2y - z = -4$
   $3y + z = 2$
   $z = 5$

9. $3x - 2y + 5z = -10$
   $y - 4z = 8$
   $2y + z = 7$

10. $4x + 5y - 6z = 31$
    $y - 2z = 7$
    $5y + z = 2$

11. $x + 2y + z = 8$
    $2x - y + 3z = 15$
    $-x + 3y - 3z = -11$

12. $x + y + z = 5$
    $-4x + 2y - 3z = -9$
    $2x - 3y + 2z = 5$

13. $x - 2y + 3z = 0$
    $2x - 3y - 4z = 0$
    $x + y - 4z = 0$

14. $x + 4y - z = 0$
    $-x - 3y + 5z = 0$
    $3x + y - 2z = 0$

15. $x + y + z + w = 10$
    $y + 3z - w = 7$
    $x + y + 2z = 11$
    $x - 3y + w = -14$

16. $2x + y + z = 3$
    $y + z + w = 5$
    $4x + z + w = 0$
    $3y - z + 2w = 0$

EXAMPLE A (Infinitely Many Solutions) Solve

$$x - 2y + 3z = 10$$

$$2x - 3y - z = 8$$

$$4x - 7y + 5z = 28$$

Solution. Using Operation 3, we may eliminate x from the last two equations. First add -2 times the first equation to the second equation. Then add -4 times the first equation to the third equation. We obtain

$$x - 2y + 3z = 10$$

$$y - 7z = -12$$

$$y - 7z = -12$$

Next add -1 times the second equation to the third equation.

$$x - 2y + 3z = 10$$

$$y - 7z = -12$$

$$0 = 0$$

Finally, we solve the second equation for y in terms of z and substitute that result in the first equation.

$$y = 7z - 12$$

$$x = 2y - 3z + 10 = 2(7z - 12) - 3z + 10 = 11z - 14$$

Notice that there are infinitely many solutions; we can give any value we like to z, calculate the corresponding x and y values, and come up with a solution. Here is the format we use to list all the solutions to the system.

$$x = 11z - 14$$

$$y = 7z - 12$$

$$z \quad \text{arbitrary}$$

We could also say that the set of solutions consists of all ordered triples of the form $(11z - 14, 7z - 12, z)$. Thus if $z = 0$, we get the solution $(-14, -12, 0)$; if $z = 2$, we get $(8, 2, 2)$. Of course, it does not have to be z that is arbitrary; in our example, it could just as well be x or y. For example, if we had arranged things so y is treated as the arbitrary variable, we would have obtained

$$x = \frac{11}{7}y + \frac{34}{7}$$

$$z = \frac{1}{7}y + \frac{12}{7}$$

$$y \quad \text{arbitrary}$$

Note that the solution corresponding to $y = 2$ is $(8, 2, 2)$, which agrees with one found above.

Solve each of the following systems. Some, but not all, have infinitely many solutions.

17. $\begin{aligned} x - 4y + z &= 18 \\ 2x - 7y - 2z &= 4 \\ 3x - 11y - z &= 22 \end{aligned}$

18. $\begin{aligned} x + y - 3z &= 10 \\ 2x + 5y + z &= 18 \\ 5x + 8y - 8z &= 48 \end{aligned}$

19. $\begin{aligned} x - 2y + 3z &= -2 \\ 3x - 6y + 9z &= -6 \\ -2x + 4y - 6z &= 4 \end{aligned}$

20. $\begin{aligned} -4x + y - z &= 5 \\ 4x - y + z &= -5 \\ -24x + 6y - 6z &= 30 \end{aligned}$

21. $\begin{aligned} 2x - y + 4z &= 0 \\ 3x + 2y - z &= 0 \\ 9x - y + 11z &= 0 \end{aligned}$

22. $\begin{aligned} x + 3y - 2z &= 0 \\ 2x + y + z &= 0 \\ y - z &= 0 \end{aligned}$

EXAMPLE B (No Solution)   Solve

$$x - 2y + 3z = 10$$

$$2x - 3y - z = 8$$

$$5x - 9y + 8z = 20$$

Solution. Using Operation 3, we eliminate x from the last two equations to obtain

$$x - 2y + 3z = 10$$
$$y - 7z = -12$$
$$y - 7z = -30$$

It is already apparent that the last two equations cannot get along with each other. Let us continue anyway, putting the system in triangular form by adding -1 times the second equation to the third.

$$x - 2y + 3z = 10$$
$$y - 7z = -12$$
$$0 = -18$$

This system has no solution; we say it is **inconsistent.**

Solve the following systems or show that they are inconsistent.

23. $x - 4y + z = 18$
$2x - 7y - 2z = 4$
$3x - 11y - z = 10$

24. $x + y - 3z = 10$
$2x + 5y + z = 18$
$5x + 8y - 8z = 50$

25. $x + 3y - 2z = 10$
$2x + y + z = 4$
$5y - 5z = 16$

26. $x - 2y + 3z = -2$
$3x - 6y + 9z = -6$
$-2x + 4y - 6z = 0$

EXAMPLE C (Nonlinear Systems) Solve the following system of equations.

$$x^2 + y^2 = 25$$
$$x^2 + y^2 - 2x - 4y = 5$$

Solution. We can use the same operations on this system as we did on linear systems. Adding (-1) times the first equation to the second, we get

$$x^2 + y^2 = 25$$
$$-2x - 4y = -20$$

We solve the second equation for x in terms of y, substitute in the first equation, and then solve the resulting quadratic equation in y.

$$x = 10 - 2y$$
$$(10 - 2y)^2 + y^2 = 25$$
$$5y^2 - 40y + 75 = 0$$
$$y^2 - 8y + 15 = 0$$
$$(y - 3)(y - 5) = 0$$

From this we get $y = 3$ or 5. Substituting these values in the equation $x = 10 - 2y$ yields two solutions to the original system, $(4, 3)$ and $(0, 5)$.

Solve each of the following systems.

27. $x + 2y = 10$
    $x^2 + y^2 - 10x = 0$

28. $x + y = 10$
    $x^2 + y^2 - 10x - 10y = 0$

29. $x^2 + y^2 - 4x + 6y = 12$
    $x^2 + y^2 + 10x + 4y = 96$

30. $x^2 + y^2 - 16y = 45$
    $x^2 + y^2 + 4x - 20y = 65$

31. $y = 4x^2 - 2$
    $y = x^2 + 1$

32. $x = 3y^2 - 5$
    $x = y^2 + 3$

MISCELLANEOUS PROBLEMS

In Problems 33–40, solve the given system or show that it is inconsistent.

33. $2x - 3y = 12$
    $x + 4y = -5$

34. $2x + 5y = -2$
    $y = -\frac{2}{5}x + 3$

35. $2x - 3y = 6$
    $x + 2y = -4$

36. $x - y + 3z = 1$
    $3x - 2y + 4z = 0$
    $4x + 2y - z = 3$

37. $x + y + z = -4$
    $x + y + z = 17$
    $x + y + z = -2$

38. $.43x - .79y + 4.24z = .67$
    $3.61y - 9.74z = 2$
    $y + 1.22z = 1.67$

39. $x^2 + y^2 = 4$
    $x + 2y = 2\sqrt{5}$

40. $x - \log y = 1$
    $\log y^x = 2$

41. If the system $x + 2y = 4$ and $ax + 3y = b$ has infinitely many solutions, what are a and b?

42. Helen claims that she has \$4.40 in nickels, dimes, and quarters, that she has four times as many dimes as quarters, and that she has 40 coins in all. Is this possible? If so, determine how many coins of each kind she has.

43. A three-digit number equals 19 times the sum of its digits. If the digits are reversed, the resulting number is greater than the given number by 297. The tens digit exceeds the units digit by 3. Find the number.

44. Find the equation of the parabola $y = ax^2 + bx + c$ that goes through $(-1, 6)$, $(1, 0)$, and $(2, 3)$.

45. Find the equation and radius of the circle that goes through $(0, 0)$, $(4, 0)$, and $(\frac{72}{25}, \frac{96}{25})$. *Hint:* Writing the equation in the form $(x - h)^2 + (y - k)^2 = r^2$ is not the best way to start. Is there another way to write the equation of a circle?

46. Determine a, b, and c so that

$$\frac{-12x + 6}{(x - 1)(x + 2)(x - 3)} = \frac{a}{x - 1} + \frac{b}{x + 2} + \frac{c}{x - 3}$$

47. Determine a, b, and c so that $(x - 1)^3$ is a factor of $x^4 + ax^2 + bx + c$.

48. Find the dimensions of a rectangle whose diagonal and perimeter measure 25 and 62 meters, respectively.

49. A certain rectangle has an area of 120 square inches. Increasing the width by 4 inches and decreasing the length by 3 inches increases the area by 24 square inches. Find the dimensions of the original rectangle.

50. **TEASER** The ABC company reported the following statistics about its employees.
    (a) Average length of service for all employees: 15.9 years.

(b) Average length of service for male employees: 16.5 years.

(c) Average length of service for female employees: 14.1 years.

(d) Average hourly wage for all employees: $21.40.

(e) Average hourly wage for male employees: $22.50.

(f) Number of male employees: 300

For reasons not stated, the company did not report the number of female employees nor their average hourly wage, but you can figure them out. Do so.

Arthur Cayley, lawyer, painter, mountaineer, Cambridge professor, but most of all creative mathematician, made his biggest contributions in the field of algebra. To him we owe the idea of replacing a linear system by its matrix.

$$\begin{array}{rl} 2x + 3y - z = 1 \\ x + 4y - z = 4 \\ 3x + y + 2z = 5 \end{array} \qquad \begin{bmatrix} 2 & 3 & -1 & 1 \\ 1 & 4 & -1 & 4 \\ 3 & 1 & 2 & 5 \end{bmatrix}$$

Arthur Cayley (1821–1895)

9-2 Matrix Methods

Contrary to what many people think, mathematicians do not enjoy long, involved calculations. What they do enjoy is looking for shortcuts, for labor-saving devices, and for elegant ways of doing things. Consider the problem of solving a system of linear equations, which as you know can become very complicated. Is there any way to simplify and systematize this process? There is. It is the method of matrices (plural of matrix).

A **matrix** is just a rectangular array of numbers. One example is shown in our opening panel. It has 3 rows and 4 columns and is referred to as a 3 × 4 matrix. We follow the standard practice of enclosing a matrix in brackets.

AN EXAMPLE WITH THREE EQUATIONS

Look at our opening display again. Notice how we obtained the matrix from the system of equations. We just suppressed all the unknowns, the plus signs, and the equal signs, and supplied some 1's. We call this matrix the **matrix of the system.** We are going to solve this system, keeping track of what happens to the matrix as we move from step to step.

$$\begin{aligned} 2x + 3y - z &= 1 \\ x + 4y - z &= 4 \\ 3x + y + 2z &= 5 \end{aligned} \qquad \begin{bmatrix} 2 & 3 & -1 & 1 \\ 1 & 4 & -1 & 4 \\ 3 & 1 & 2 & 5 \end{bmatrix}$$

Interchange the first and second equations.

$$\begin{aligned} x + 4y - z &= 4 \\ 2x + 3y - z &= 1 \\ 3x + y + 2z &= 5 \end{aligned} \qquad \begin{bmatrix} 1 & 4 & -1 & 4 \\ 2 & 3 & -1 & 1 \\ 3 & 1 & 2 & 5 \end{bmatrix}$$

Add -2 times the first equation to the second; then add -3 times the first equation to the third.

$$\begin{aligned} x + 4y - z &= 4 \\ - 5y + z &= -7 \\ - 11y + 5z &= -7 \end{aligned} \qquad \begin{bmatrix} 1 & 4 & -1 & 4 \\ 0 & -5 & 1 & -7 \\ 0 & -11 & 5 & -7 \end{bmatrix}$$

Multiply the second equation by $-\frac{1}{5}$.

$$\begin{aligned} x + 4y - z &= 4 \\ y - \tfrac{1}{5}z &= \tfrac{7}{5} \\ - 11y + 5z &= -7 \end{aligned} \qquad \begin{bmatrix} 1 & 4 & -1 & 4 \\ 0 & 1 & -\frac{1}{5} & \frac{7}{5} \\ 0 & -11 & 5 & -7 \end{bmatrix}$$

Add 11 times the second equation to the third.

$$\begin{aligned} x + 4y - z &= 4 \\ y - \tfrac{1}{5}z &= \tfrac{7}{5} \\ \tfrac{14}{5}z &= \tfrac{42}{5} \end{aligned} \qquad \begin{bmatrix} 1 & 4 & -1 & 4 \\ 0 & 1 & -\frac{1}{5} & \frac{7}{5} \\ 0 & 0 & \frac{14}{5} & \frac{42}{5} \end{bmatrix}$$

Now the system is in triangular form and can be solved by backward substitution. The result is $z = 3$, $y = 2$, and $x = -1$; we say the solution is $(-1, 2, 3)$.

We make two points about what we have just done. First, the process is not unique. We happen to prefer having a leading coefficient of 1; that was the reason for our first step. One could have started by multiplying the first equation by $-\frac{1}{2}$ and adding to the second, then multiplying the first equation by $-\frac{3}{2}$ and adding to the third. Any process that ultimately puts the system in triangular form is fine.

The second and main point is this. It is unnecessary to carry along all the x's and y's. Why not work with just the numbers? Why not do all the operations on the matrix of the system? Well, why not?

EQUIVALENT MATRICES

Guided by our knowledge of systems of equations, we say that matrices **A** and **B** are **equivalent** if **B** can be obtained from **A** by applying the operations below (a finite number of times).

Operation 1 Interchanging two rows.
Operation 2 Multiplying a row by a nonzero number.
Operation 3 Replacing a row by the sum of that row and a multiple of another row.

When **A** and **B** are equivalent, we write **A** ~ **B**. If **A** ~ **B**, then **B** ~ **A**. If **A** ~ **B** and **B** ~ **C**, then **A** ~ **C**.

AN EXAMPLE WITH FOUR EQUATIONS
Consider

$$
\begin{aligned}
x + 3y + z \quad\quad &= 1 \\
2x + 7y + z - w &= -1 \\
3x - 2y \quad\quad + 4w &= 8 \\
-x + y - 3z - w &= -6
\end{aligned}
$$

To solve this system, we take its matrix and transform it to triangular form using the operations above. Here is one possible sequence of steps.

$$
\begin{bmatrix}
1 & 3 & 1 & 0 & 1 \\
2 & 7 & 1 & -1 & -1 \\
3 & -2 & 0 & 4 & 8 \\
-1 & 1 & -3 & -1 & -6
\end{bmatrix}
$$

Add -2 times the first row to the second; -3 times the first row to the third row; and 1 times the first row to the fourth row.

CAUTION

> Be sure to note that the first row of the matrix was unchanged while we performed operation 3 on the other three rows.

$$
\begin{bmatrix}
1 & 3 & 1 & 0 & 1 \\
0 & 1 & -1 & -1 & -3 \\
0 & -11 & -3 & 4 & 5 \\
0 & 4 & -2 & -1 & -5
\end{bmatrix}
$$

Add 11 times the second row to the third and -4 times the second row to the fourth.

$$\begin{bmatrix} 1 & 3 & 1 & 0 & 1 \\ 0 & 1 & -1 & -1 & -3 \\ 0 & 0 & -14 & -7 & -28 \\ 0 & 0 & 2 & 3 & 7 \end{bmatrix}$$

Multiply the third row by $-\frac{1}{14}$.

$$\begin{bmatrix} 1 & 3 & 1 & 0 & 1 \\ 0 & 1 & -1 & -1 & -3 \\ 0 & 0 & 1 & \frac{1}{2} & 2 \\ 0 & 0 & 2 & 3 & 7 \end{bmatrix}$$

Add -2 times the third row to the fourth row.

$$\begin{bmatrix} 1 & 3 & 1 & 0 & 1 \\ 0 & 1 & -1 & -1 & -3 \\ 0 & 0 & 1 & \frac{1}{2} & 2 \\ 0 & 0 & 0 & 2 & 3 \end{bmatrix}$$

This last matrix represents the system

$$\begin{aligned} x + 3y + z & = 1 \\ y - z - w & = -3 \\ z + \tfrac{1}{2}w & = 2 \\ 2w & = 3 \end{aligned}$$

If we use back substitution, we get $w = \frac{3}{2}, z = \frac{5}{4}, y = -\frac{1}{4}$, and $x = \frac{1}{2}$. The solution is $(\frac{1}{2}, -\frac{1}{4}, \frac{5}{4}, \frac{3}{2})$.

THE CASES WITH MANY SOLUTIONS AND NO SOLUTION

A system of equations need not have a unique solution; it may have infinitely many solutions or none at all. We need to be able to analyze the latter two cases by our matrix method. Fortunately, this is easy to do. Consider Example A of Section 9-1 first. Here is how we handle it using matrices.

$$\begin{bmatrix} 1 & -2 & 3 & 10 \\ 2 & -3 & -1 & 8 \\ 4 & -7 & 5 & 28 \end{bmatrix} \sim \begin{bmatrix} 1 & -2 & 3 & 10 \\ 0 & 1 & -7 & -12 \\ 0 & 1 & -7 & -12 \end{bmatrix}$$

$$\sim \begin{bmatrix} 1 & -2 & 3 & 10 \\ 0 & 1 & -7 & -12 \\ 0 & 0 & 0 & 0 \end{bmatrix}$$

The appearance of a row of zeros tells us that we have infinitely many solutions. The set of solutions is obtained by considering the equations corresponding to the first two rows.

$$x - 2y + 3z = 10$$
$$y - 7z = -12$$

When we solve for y in the second equation and substitute in the first, we obtain

$$x = 11z - 14$$
$$y = 7z - 12$$
$$z \quad \text{arbitrary}$$

Next consider the inconsistent example treated in Section 9-1 (Example B). Here is what happens when the matrix method is applied to this example.

$$\begin{bmatrix} 1 & -2 & 3 & 10 \\ 2 & -3 & -1 & 8 \\ 5 & -9 & 8 & 20 \end{bmatrix} \sim \begin{bmatrix} 1 & -2 & 3 & 10 \\ 0 & 1 & -7 & -12 \\ 0 & 1 & -7 & -30 \end{bmatrix}$$

$$\sim \begin{bmatrix} 1 & -2 & 3 & 10 \\ 0 & 1 & -7 & -12 \\ 0 & 0 & 0 & -18 \end{bmatrix}$$

We are tipped off to the inconsistency of the system by the third row of the matrix. It corresponds to the equation

$$0x + 0y + 0z = -18$$

which has no solution. Consequently, the system as a whole has no solution.

We may summarize our discussion as follows. If the process of transforming the matrix of a system of n equations in n unknowns to triangular form leads to a row in which all elements but the last one are zero, then the system is inconsistent; that is, it has no solution. If the above does not occur and we are led to a matrix with one or more rows consisting entirely of zeros, then the system has infinitely many solutions.

Problem Set 9-2

Write the matrix of each system in Problems 1–8.

1. $2x - y = 4$
   $x - 3y = -2$

2. $x + 2y = 13$
   $11x - y = 0$

3. $x - 2y + z = 3$
   $2x + y = 5$
   $x + y + 3z = -4$

4. $x + 4z = 10$
   $2y - z = 0$
   $3x - y = 20$

5. $2x = 3y - 4$
   $3x + 2 = -y$

6. $x = 4y + 3$
   $y = -2x + 5$

7. $x = 5$
   $2y + x - z = 4$
   $3x - y + 13 = 5z$

8. $z = 2$
   $2x - z = -4$
   $x + 2y + 4z = -8$

Regard each matrix in Problems 9–18 as a matrix of a linear system of equations. Tell whether the system has a unique solution, infinitely many solutions, or no solution. You need not solve any of the systems.

9. $\begin{bmatrix} 1 & -2 & 3 \\ 0 & 1 & -4 \end{bmatrix}$

10. $\begin{bmatrix} 2 & 5 & 0 \\ 0 & -3 & 5 \end{bmatrix}$

11. $\begin{bmatrix} 1 & -3 & 5 \\ 2 & -6 & -10 \end{bmatrix}$

12. $\begin{bmatrix} 2 & 1 & -4 \\ -6 & -3 & 12 \end{bmatrix}$

13. $\begin{bmatrix} 1 & -2 & 4 & -2 \\ 0 & 3 & 1 & 4 \\ 0 & 0 & 1 & -3 \end{bmatrix}$

14. $\begin{bmatrix} 5 & 4 & 0 & -11 \\ 0 & 1 & -4 & 0 \\ 0 & 0 & 2 & -4 \end{bmatrix}$

15. $\begin{bmatrix} 2 & 1 & 5 & 4 \\ 0 & 3 & -2 & 10 \\ 0 & 3 & -2 & 10 \end{bmatrix}$

16. $\begin{bmatrix} 4 & 1 & -3 & 5 \\ 0 & 0 & 1 & -4 \\ 0 & 0 & 1 & -4 \end{bmatrix}$

17. $\begin{bmatrix} 3 & 2 & -1 & 0 \\ 0 & 1 & 0 & -4 \\ 0 & 1 & 0 & 5 \end{bmatrix}$

18. $\begin{bmatrix} -1 & 5 & 6 & -3 \\ 0 & 0 & 0 & 0 \\ 0 & 0 & 0 & 4 \end{bmatrix}$

In Problems 19–30, use matrices to solve each system or to show that it has no solution.

19. $x + 2y = 5$
    $2x - 5y = -8$

20. $2x + 4y = 16$
    $3x - y = 10$

21. $3x - 2y = 1$
    $-6x + 4y = -2$

22. $x + 3y = 12$
    $5x + 15y = 12$

23. $3x - 2y + 5z = -10$
    $y - 4z = 8$
    $2y + z = 7$

24. $4x + 5y + 2z = 25$
    $y - 2z = 7$
    $5y + z = 2$

25. $x + y - 3z = 10$
    $2x + 5y + z = 18$
    $5x + 8y - 8z = 48$

26. $x - 4y + z = 18$
    $2x - 7y - 2z = 4$
    $3x - 11y - z = 22$

27. $2x + 5y + 2z = 6$
$x + 2y - z = 3$
$3x - y + 2z = 9$

28. $x - 2y + 3z = -2$
$3x - 6y + 9z = -6$
$-2x + 4y - 6z = 0$

C 29. $x + 1.2y - 2.3z = 8.1$
$1.3x + .7y + .4z = 6.2$
$.5x + 1.2y + .5z = 3.2$

30. $3x + 2y = 4$
$3x - 4y + 6z = 16$
$3x - y + z = 6$

MISCELLANEOUS PROBLEMS

Regard each matrix in Problems 31–36 as the matrix of a linear system of equations. Without solving the system, tell whether it has a unique solution, infinitely many solutions, or no solution.

31. $\begin{bmatrix} 3 & -2 & 5 \\ 0 & 1 & -3 \end{bmatrix}$

32. $\begin{bmatrix} 2 & -1 & 5 \\ -4 & 2 & 8 \end{bmatrix}$

33. $\begin{bmatrix} 2 & -1 & 4 & 6 \\ 0 & 4 & -1 & 5 \\ 0 & 0 & 2 & 1 \end{bmatrix}$

34. $\begin{bmatrix} 3 & 3 & 0 & -4 \\ 0 & 1 & -3 & 2 \\ 0 & 0 & 0 & 0 \end{bmatrix}$

35. $\begin{bmatrix} 1 & 2 & 3 & 4 & 5 \\ 0 & 3 & 2 & 1 & 0 \\ 0 & 0 & 0 & 3 & -4 \\ 0 & 0 & 0 & -9 & 15 \end{bmatrix}$

36. $\begin{bmatrix} 0 & 0 & 0 & 2 & 3 \\ 0 & 0 & 3 & 4 & 5 \\ 0 & 4 & 5 & 6 & 7 \\ 5 & 6 & 7 & 8 & 9 \end{bmatrix}$

In Problems 37–40, use matrices to solve each system or to show that it has no solution.

37. $3x - 2y + 4z = 0$
$x - y + 3z = 1$
$4x + 2y - z = 3$

38. $-4x + y - z = 5$
$4x - y + z = -5$
$-24x + 6y - 6z = 10$

39. $2x + 4y - z = 8$
$4x + 9y + 3z = 42$
$8x + 17y + z = 58$

40. $2x + y + 2z + 3w = 2$
$y - 2z + 5w = 2$
$z + 3w = 4$
$2z + 7w = 4$

Figure 2

41. Find a, b, and c so that the parabola $y = ax^2 + bx + c$ passes through the points $(-2, -32)$, $(1, 4)$, and $(3, -12)$.

42. Find the equation and radius of a circle that passes through $(3, -3)$, $(8, 2)$, and $(6, 6)$.

43. Find angles α, β, γ, and δ in Figure 2 given that $\alpha - \beta + \gamma - \delta = 110°$.

44. A chemist plans to mix three different nitric acid solutions with concentrations of 25 percent, 40 percent, and 50 percent to form 100 liters of a 32 percent solution. If she insists on using twice as much of the 25 percent solution as the 40 percent solution, how many liters of each kind should she use?

45. The local garden store stocks three brands of phosphate-potash-nitrogen fertilizer with compositions indicated in the following table.

BRAND	PHOSPHATE	POTASH	NITROGEN
A	10%	30%	60%
B	20%	40%	40%
C	20%	30%	50%

Soil analysis shows that Wanda Wiseankle needs fertilizer for her garden that is 19 percent phosphate, 34 percent potash, and 47 percent nitrogen. Can she obtain the right mixture by mixing the three brands? If so, how many pounds of each should she mix together to get 100 pounds of the desired blend?

46. TEASER   Tom, Dick, and Harry are good friends but have very different work habits. Together, they contracted to paint three identical houses. Tom and Dick painted the first house in $\frac{72}{5}$ hours; Tom and Harry painted the second house in 16 hours; Dick and Harry painted the third house in $\frac{144}{7}$ hours. How long would it have taken each boy to paint a house alone?

Cayley's Weapons

$$\begin{bmatrix} a & b \\ c & d \end{bmatrix} + \begin{bmatrix} A & B \\ C & D \end{bmatrix} = \begin{bmatrix} a + A & b + B \\ c + C & d + D \end{bmatrix}$$

$$\begin{bmatrix} a & b \\ c & d \end{bmatrix} \cdot \begin{bmatrix} A & B \\ C & D \end{bmatrix} = \begin{bmatrix} aA + bC & aB + bD \\ cA + dC & cB + dD \end{bmatrix}$$

"Cayley is forging the weapons for future generations of physicists."

P. G. Tait

9-3   The Algebra of Matrices

When Arthur Cayley introduced matrices, he had much more in mind than the application described in the previous section. There, matrices served as a device to simplify solving systems of equations. Cayley saw that these number boxes could be studied independently of equations, that they could be thought of as a new type of mathematical object. He realized that if he could give appropriate definitions of addition and multiplication, he would create a mathematical system that might stand with the real numbers and the complex numbers as a potential model for many applications. Cayley did all of this in a major paper in 1858. Some of his contemporaries saw little of significance in this new abstraction. But one of them, P. G. Tait, uttered the prophetic words quoted in the opening box. Tait was right. During the 1920's, Werner Heisenberg found that matrices were just the tool he needed to formulate his quantum mechanics. And by 1950, it was generally recognized that matrix theory provides the best model for many problems in economics and the social sciences.

To simplify our discussion, we initially consider only 2×2 matrices, that is, matrices with two rows and two columns. Examples are

$$\begin{bmatrix} -1 & 3 \\ 4 & 0 \end{bmatrix} \qquad \begin{bmatrix} \log .1 & \frac{6}{2} \\ \frac{12}{3} & \log 1 \end{bmatrix} \qquad \begin{bmatrix} a & b \\ c & d \end{bmatrix}$$

The first two of these matrices are said to be equal. In fact, two matrices are **equal** if and only if the entries in corresponding positions are equal. Be sure to distinguish the notion of equality (written $=$) from that of equivalence (written $\sim$) introduced in Section 9-2. For example,

$$\begin{bmatrix} 2 & 1 \\ -3 & 4 \end{bmatrix} \quad \text{and} \quad \begin{bmatrix} -3 & 4 \\ 2 & 1 \end{bmatrix}$$

are equivalent matrices; however, they are not equal.

ADDITION AND SUBTRACTION

Cayley's definition of addition is straightforward. To add two matrices, add the entries in corresponding positions. Thus

$$\begin{bmatrix} 1 & 3 \\ -1 & 4 \end{bmatrix} + \begin{bmatrix} 6 & -2 \\ 5 & 1 \end{bmatrix} = \begin{bmatrix} 1 + 6 & 3 + (-2) \\ -1 + 5 & 4 + 1 \end{bmatrix} = \begin{bmatrix} 7 & 1 \\ 4 & 5 \end{bmatrix}$$

and in general

$$\begin{bmatrix} a & b \\ c & d \end{bmatrix} + \begin{bmatrix} A & B \\ C & D \end{bmatrix} = \begin{bmatrix} a + A & b + B \\ c + C & d + D \end{bmatrix}$$

It is easy to check that the commutative and associative properties for addition are valid. Let $\mathbf{U}$, $\mathbf{V}$, and $\mathbf{W}$ be any three matrices.

1. (**Commutativity +**) $\mathbf{U} + \mathbf{V} = \mathbf{V} + \mathbf{U}$
2. (**Associativity +**) $\mathbf{U} + (\mathbf{V} + \mathbf{W}) = (\mathbf{U} + \mathbf{V}) + \mathbf{W}$

The matrix

$$\mathbf{O} = \begin{bmatrix} 0 & 0 \\ 0 & 0 \end{bmatrix}$$

behaves as the "zero" for matrices. And the additive inverse of the matrix

$$\mathbf{U} = \begin{bmatrix} a & b \\ c & d \end{bmatrix}$$

is given by

$$-\mathbf{U} = \begin{bmatrix} -a & -b \\ -c & -d \end{bmatrix}$$

We may summarize these statements as follows.

3. (**Additive identity +**)   There is a matrix **O** satisfying **O** + **U** = **U** + **O** = **U**.

4. (**Additive inverses**)   For each matrix **U**, there is a matrix −**U** satisfying

$$\mathbf{U} + (-\mathbf{U}) = (-\mathbf{U}) + \mathbf{U} = \mathbf{O}$$

With the existence of an additive inverse settled, we can define subtraction by **U** − **V** = **U** + (−**V**). This amounts to subtracting the entries of **V** from the corresponding entries of **U**. Thus

$$\begin{bmatrix} 1 & 3 \\ -1 & 4 \end{bmatrix} - \begin{bmatrix} 6 & -2 \\ 5 & 1 \end{bmatrix} = \begin{bmatrix} -5 & 5 \\ -6 & 3 \end{bmatrix}$$

So far, all has been straightforward and nice. But with multiplication, Cayley hit a snag.

MULTIPLICATION

Cayley's definition of multiplication may seem odd at first glance. He was led to it by consideration of a special problem that we do not have time to describe. It is enough to say that Cayley's definition is the one that proves useful in modern applications (as you will see).

Here it is in symbols.

$$\begin{bmatrix} a & b \\ c & d \end{bmatrix} \cdot \begin{bmatrix} A & B \\ C & D \end{bmatrix} = \begin{bmatrix} aA + bC & aB + bD \\ cA + dC & cB + dD \end{bmatrix}$$

Stated in words, we multiply two matrices by multiplying the rows of the left matrix by the columns of the right matrix in pairwise entry fashion, adding the results. For example, the entry in the second row and first column of the product is obtained by multiplying the entries of the second row of the left matrix by the corresponding entries of the first column of the right matrix, adding the results. Until you get used to it, it may help to use your fingers as shown in the diagram below.

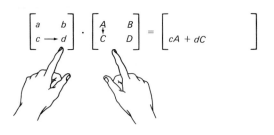

Here is an example worked out in detail.

$$\begin{bmatrix} 1 & 3 \\ -1 & 4 \end{bmatrix}\begin{bmatrix} 6 & -2 \\ 5 & 1 \end{bmatrix} = \begin{bmatrix} (1)(6) + (3)(5) & (1)(-2) + (3)(1) \\ (-1)(6) + (4)(5) & (-1)(-2) + (4)(1) \end{bmatrix}$$

$$= \begin{bmatrix} 21 & 1 \\ 14 & 6 \end{bmatrix}$$

Here is the same problem, but with the matrices multiplied in the opposite order.

$$\begin{bmatrix} 6 & -2 \\ 5 & 1 \end{bmatrix}\begin{bmatrix} 1 & 3 \\ -1 & 4 \end{bmatrix} = \begin{bmatrix} (6)(1) + (-2)(-1) & (6)(3) + (-2)(4) \\ (5)(1) + (1)(-1) & (5)(3) + (1)(4) \end{bmatrix}$$

$$= \begin{bmatrix} 8 & 10 \\ 4 & 19 \end{bmatrix}$$

Now you see the snag about which we warned you. The commutative property for multiplication fails. This is troublesome, but not fatal. We manage to get along in a world that is largely noncommutative (try removing your clothes and taking a shower in the opposite order). We just have to remember never to commute matrices under multiplication. Fortunately two other nice properties do hold.

5. **(Associativity •)** $\mathbf{U} \cdot (\mathbf{V} \cdot \mathbf{W}) = (\mathbf{U} \cdot \mathbf{V}) \cdot \mathbf{W}$
6. **(Distributivity)** $\mathbf{U} \cdot (\mathbf{V} + \mathbf{W}) = \mathbf{U} \cdot \mathbf{V} + \mathbf{U} \cdot \mathbf{W}$
$(\mathbf{V} + \mathbf{W}) \cdot \mathbf{U} = \mathbf{V} \cdot \mathbf{U} + \mathbf{W} \cdot \mathbf{U}$

We have not said anything about multiplicative inverses; that comes in the next section. We do, however, want to mention a special operation called **scalar multiplication**—that is, multiplication of a matrix by a scalar (number). To multiply a matrix by a number, multiply each entry by that number. That is,

$$k\begin{bmatrix} a & b \\ c & d \end{bmatrix} = \begin{bmatrix} ka & kb \\ kc & kd \end{bmatrix}$$

Scalar multiplication satisfies the expected properties.

7. $k(\mathbf{U} + \mathbf{V}) = k\mathbf{U} + k\mathbf{V}$
8. $(k + m)\mathbf{U} = k\mathbf{U} + m\mathbf{U}$
9. $(km)\mathbf{U} = k(m\mathbf{U})$

LARGER MATRICES AND COMPATIBILITY

So far we have considered only 2 × 2 matrices. This is an unnecessary restriction; however, to perform operations on arbitrary matrices, we must make sure they are **compatible.** For addition, this simply means that the matrices must be of the same size. Thus

$$\begin{bmatrix} 1 & -1 & 3 \\ 4 & -5 & 2 \end{bmatrix} + \begin{bmatrix} 2 & 6 & 0 \\ -3 & 2 & 4 \end{bmatrix} = \begin{bmatrix} 3 & 5 & 3 \\ 1 & -3 & 6 \end{bmatrix}$$

but

$$\begin{bmatrix} 1 & -1 & 3 \\ 4 & -5 & 2 \end{bmatrix} + \begin{bmatrix} 2 & 6 \\ -3 & 2 \end{bmatrix}$$

makes no sense.

Two matrices are compatible for multiplication if the left matrix has the same number of columns as the right matrix has rows. For example,

$$\begin{bmatrix} 1 & -1 & 3 \\ 4 & -5 & 2 \end{bmatrix} \begin{bmatrix} 2 & 1 & 5 & 0 \\ -1 & 3 & 2 & 1 \\ 1 & -2 & 0 & 2 \end{bmatrix} = \begin{bmatrix} 6 & -8 & 3 & 5 \\ 15 & -15 & 10 & -1 \end{bmatrix}$$

The left matrix is 2×3, the right one is 3×4, and the result is 2×4. In general, we can multiply an $m \times n$ matrix by an $n \times p$ matrix, the result being an $m \times p$ matrix. All of the properties mentioned earlier are valid, provided we work with compatible matrices.

A BUSINESS APPLICATION

The ABC Company sells precut lumber for two types of summer cottages, standard and deluxe. The standard model requires 30,000 board feet of lumber and 100 worker-hours of cutting; the deluxe model takes 40,000 board feet of lumber and 110 worker-hours of cutting. This year, the ABC Company buys its lumber at $.20 per board foot and pays its laborers $9.00 per hour. Next year it expects these costs to be $.25 and $10.00, respectively. This information can be displayed in matrix form as follows.

	REQUIREMENTS			UNIT COST	
	A			**B**	
	Lumber	Labor		This year	Next year
Standard	30,000	100	Lumber	$.20	$.25
Deluxe	40,000	110	Labor	$9.00	$10.00

Now we ask whether the product matrix **AB** has economic significance. It does: It gives the total dollar cost of standard and deluxe cottages both for this year and next. You can see this from the following calculation.

$$\mathbf{AB} = \begin{bmatrix} (30{,}000)(.20) + (100)(9) & (30{,}000)(.25) + (100)(10) \\ (40{,}000)(.20) + (110)(9) & (40{,}000)(.25) + (110)(10) \end{bmatrix}$$

$$= \begin{bmatrix} \$6900 & \$8500 \\ \$8990 & \$11,100 \end{bmatrix} \begin{array}{l} \text{Standard} \\ \text{Deluxe} \end{array}$$

This year   Next year

Problem Set 9-3

Calculate **A** + **B**, **A** − **B**, *and* 3**A** *in Problems 1–4.*

1. $\mathbf{A} = \begin{bmatrix} 2 & -1 \\ 3 & 7 \end{bmatrix}$, $\mathbf{B} = \begin{bmatrix} 6 & 5 \\ -2 & 3 \end{bmatrix}$

2. $\mathbf{A} = \begin{bmatrix} -1 & 0 \\ 5 & 4 \end{bmatrix}$, $\mathbf{B} = \begin{bmatrix} 2 & -2 \\ 3 & 7 \end{bmatrix}$

3. $\mathbf{A} = \begin{bmatrix} 3 & -2 & 5 \\ 4 & 0 & -3 \end{bmatrix}$, $\mathbf{B} = \begin{bmatrix} 2 & 6 & -1 \\ 4 & 3 & -3 \end{bmatrix}$

4. $\mathbf{A} = \begin{bmatrix} 1 & 2 & 3 \\ 4 & 5 & 6 \\ 7 & 8 & 9 \end{bmatrix}$, $\mathbf{B} = \begin{bmatrix} -1 & -2 & -2 \\ -4 & -5 & -6 \\ -7 & -8 & -9 \end{bmatrix}$

Calculate **AB** *and* **BA** *if possible in Problems 5–12.*

5. $\mathbf{A} = \begin{bmatrix} 2 & -1 \\ 3 & 7 \end{bmatrix}$, $\mathbf{B} = \begin{bmatrix} 6 & 5 \\ -2 & 3 \end{bmatrix}$

6. $\mathbf{A} = \begin{bmatrix} -1 & 0 \\ 5 & 4 \end{bmatrix}$, $\mathbf{B} = \begin{bmatrix} 2 & -2 \\ 3 & 7 \end{bmatrix}$

7. $\mathbf{A} = \begin{bmatrix} 1 & -1 & 2 \\ 3 & 4 & -4 \\ 2 & 1 & 3 \end{bmatrix}$, $\mathbf{B} = \begin{bmatrix} 0 & 2 & -3 \\ 1 & 2 & 3 \\ -1 & -2 & 4 \end{bmatrix}$

8. $\mathbf{A} = \begin{bmatrix} -2 & 5 & 1 \\ 0 & -2 & 3 \\ 1 & 2 & -1 \end{bmatrix}$, $\mathbf{B} = \begin{bmatrix} -3 & 4 & 1 \\ 2 & 5 & 1 \\ 1 & 2 & 3 \end{bmatrix}$

9. $\mathbf{A} = \begin{bmatrix} 1 & -2 & 3 & 4 \\ 3 & 2 & -5 & 1 \end{bmatrix}$, $\mathbf{B} = \begin{bmatrix} 1 & 2 \\ 3 & 4 \end{bmatrix}$

10. $\mathbf{A} = \begin{bmatrix} -1 & 3 \\ 4 & 2 \\ 1 & 5 \end{bmatrix}$, $\mathbf{B} = \begin{bmatrix} -1 & 2 & 3 & 4 \\ 0 & -3 & 2 & 1 \end{bmatrix}$

11. $\mathbf{A} = \begin{bmatrix} 3 & 1 & -1 \\ 2 & 4 & 2 \\ -3 & 2 & -1 \end{bmatrix}$, $\mathbf{B} = \begin{bmatrix} 1 \\ 2 \\ 3 \end{bmatrix}$

12. $\mathbf{A} = \begin{bmatrix} 1 & 2 & -1 \end{bmatrix}$, $\mathbf{B} = \begin{bmatrix} 4 & 3 \\ 0 & 2 \\ -1 & 4 \end{bmatrix}$

13. Calculate $\mathbf{AB}$ and $\mathbf{BA}$ for

$$\mathbf{A} = \begin{bmatrix} 0 & 0 \\ 0 & 0 \end{bmatrix} \qquad \mathbf{B} = \begin{bmatrix} 2 & -1 \\ 3 & 4 \end{bmatrix}$$

14. State the general property illustrated by Problem 13.

15. Find $\mathbf{X}$ if

$$\begin{bmatrix} 2 & 1 & -3 \\ 1 & 5 & 0 \end{bmatrix} + \mathbf{X} = 2\begin{bmatrix} -1 & 4 & 3 \\ -2 & 0 & 4 \end{bmatrix}$$

16. Solve for $\mathbf{X}$.

$$-3\mathbf{X} + 2\begin{bmatrix} 1 & -2 \\ 5 & 6 \end{bmatrix} = -\begin{bmatrix} 5 & -14 \\ 8 & 15 \end{bmatrix}$$

17. Calculate $\mathbf{A}(\mathbf{B} + \mathbf{C})$ and $\mathbf{AB} + \mathbf{AC}$ for

$$\mathbf{A} = \begin{bmatrix} 2 & -1 \\ 3 & 4 \end{bmatrix} \qquad \mathbf{B} = \begin{bmatrix} 2 & 4 \\ 6 & 1 \end{bmatrix} \qquad \mathbf{C} = \begin{bmatrix} -1 & -2 \\ 3 & 6 \end{bmatrix}$$

What property does this illustrate?

18. Calculate $(\mathbf{A} + \mathbf{B})(\mathbf{A} - \mathbf{B})$ and $\mathbf{A}^2 - \mathbf{B}^2$ for

$$\mathbf{A} = \begin{bmatrix} 3 & -2 \\ 1 & 4 \end{bmatrix} \qquad \mathbf{B} = \begin{bmatrix} 6 & -3 \\ 2 & 5 \end{bmatrix}$$

Why are your answers different?

□c 19. Find the entry in the third row and second column of the product

$$\begin{bmatrix} 1.39 & 4.13 & -2.78 \\ 4.72 & -3.69 & 5.41 \\ 8.09 & -6.73 & 5.03 \end{bmatrix}\begin{bmatrix} 5.45 & 6.31 \\ 7.24 & -5.32 \\ 6.06 & 1.34 \end{bmatrix}$$

□c 20. Find the entry in the second row and first column of the product in Problem 19.

MISCELLANEOUS PROBLEMS

21. Compute $\mathbf{A} - 2\mathbf{B}$, $\mathbf{AB}$, and $\mathbf{A}^2$ for

$$\mathbf{A} = \begin{bmatrix} 4 & -1 & 3 \\ 2 & 5 & 3 \\ 6 & 2 & 1 \end{bmatrix} \qquad \mathbf{B} = \begin{bmatrix} 1 & -3 & 2 \\ 5 & 0 & 3 \\ -5 & 2 & 1 \end{bmatrix}$$

22. Let $\mathbf{A}$ and $\mathbf{B}$ be 3×4 matrices, $\mathbf{C}$ a 4×3 matrix, and $\mathbf{D}$ a 3×3 matrix. Which of the following do not make sense, that is, do not satisfy the compatibility conditions?

(a) $\mathbf{AB}$ (b) $\mathbf{AC}$ (c) $\mathbf{AC} - \mathbf{D}$

(d) $(\mathbf{A} - \mathbf{B})\mathbf{C}$ (e) $(\mathbf{AC})\mathbf{D}$ (f) $\mathbf{A}(\mathbf{CD})$
(g) $\mathbf{A}^2$ (h) $(\mathbf{CA})^2$ (i) $\mathbf{C}(\mathbf{A} + 2\mathbf{B})$

23. Calculate $\mathbf{AB}$ and $\mathbf{BA}$ for

$$\mathbf{A} = [1 \quad 2 \quad 3 \quad 4] \qquad \mathbf{B} = \begin{bmatrix} 2 \\ 1 \\ -1 \\ -2 \end{bmatrix}$$

24. Show that if $\mathbf{AB}$ and $\mathbf{BA}$ both make sense, then $\mathbf{AB}$ and $\mathbf{BA}$ are both square matrices.

25. If $(\mathbf{A} + \mathbf{B})^2 = \mathbf{A}^2 + 2\mathbf{AB} + \mathbf{B}^2$, what conclusions can you draw about $\mathbf{A}$ and $\mathbf{B}$?

26. If the ith row of $\mathbf{A}$ consists of all zeros, what is true about the ith row of $\mathbf{AB}$ (assuming $\mathbf{AB}$ makes sense)?

27. Let

$$\mathbf{A} = \begin{bmatrix} 0 & a \\ 0 & 0 \end{bmatrix} \qquad \mathbf{B} = \begin{bmatrix} 0 & a & b \\ 0 & 0 & c \\ 0 & 0 & 0 \end{bmatrix}$$

Calculate $\mathbf{A}^2$ and $\mathbf{B}^3$ and then make a conjecture.

28. A matrix of the form

$$\mathbf{A} = \begin{bmatrix} 1 & a & b \\ 0 & 1 & c \\ 0 & 0 & 1 \end{bmatrix}$$

where a, b, and c are any real numbers is called a Heisenberg matrix. What is true about the product of two such matrices?

29. Let

$$\mathbf{A} = \begin{bmatrix} 3 & 0 & 0 \\ 0 & -4 & 0 \\ 0 & 0 & 5 \end{bmatrix}$$

If $\mathbf{B}$ is any 3×3 matrix, what does multiplication on the left by $\mathbf{A}$ do to $\mathbf{B}$? Multiplication on the right by $\mathbf{A}$?

30. Calculate $\mathbf{A}^2$ and $\mathbf{A}^3$ for the matrix $\mathbf{A}$ of Problem 29. State a general result about raising a diagonal matrix to a positive integral power.

31. Art, Bob, and Curt work for a company that makes Flukes, Gizmos, and Horks. They are paid for their labor on a piecework basis, receiving $1 for each Fluke, $2 for each Gizmo, and $3 for each Hork. Below are matrices $\mathbf{U}$ and $\mathbf{V}$ representing their outputs on Monday and Tuesday. Matrix $\mathbf{X}$ is the wage/unit matrix.

	MONDAY'S OUTPUT			TUESDAY'S OUTPUT			WAGE/UNIT
	U			**V**			**X**
	F	G	H	F	G	H	
Art	4	3	2	3	6	1	F 1
Bob	5	1	2	4	2	2	G 2
Curt	3	4	1	5	1	3	H 3

Compute the following matrices and decide what they represent.
(a) $\mathbf{UX}$ (b) $\mathbf{VX}$ (c) $\mathbf{U} + \mathbf{V}$ (d) $(\mathbf{U} + \mathbf{V})\mathbf{X}$

32. Four friends, A, B, C, and D, have unlisted telephone numbers. Whether or not one person knows another's number is indicated by the matrix $\mathbf{U}$ below, where 1 indicates knowing and 0 indicates not knowing. For example, the 1 in row 3 and column 1 means that C knows A's number.

$$\mathbf{U} = \begin{array}{c} \\ A \\ B \\ C \\ D \end{array} \begin{array}{c} \begin{array}{cccc} A & B & C & D \end{array} \\ \left[\begin{array}{cccc} 1 & 0 & 1 & 0 \\ 0 & 1 & 1 & 0 \\ 1 & 0 & 1 & 1 \\ 0 & 1 & 0 & 1 \end{array} \right] \end{array}$$

(a) Calculate $\mathbf{U}^2$.
(b) Interpret $\mathbf{U}^2$ in terms of the possibility of each person being able to get a telephone message to another.
(c) Can D get a message to A via one other person?
(d) Interpret $\mathbf{U}^3$.

33. Consider the set $\mathbf{C}$ of all 2×2 matrices of the form

$$\begin{bmatrix} a & b \\ -b & a \end{bmatrix}$$

where a and b are real numbers.
(a) Let

$$\mathbf{U} = \begin{bmatrix} u_1 & u_2 \\ -u_2 & u_1 \end{bmatrix} \quad \text{and} \quad \mathbf{V} = \begin{bmatrix} v_1 & v_2 \\ -v_2 & v_1 \end{bmatrix}$$

be two such matrices.
Calculate $\mathbf{U} + \mathbf{V}$ and $\mathbf{UV}$. Note that both $\mathbf{U} + \mathbf{V}$ and $\mathbf{UV}$ are in $\mathbf{C}$.
(b) Let $\mathbf{I} = \begin{bmatrix} 1 & 0 \\ 0 & 1 \end{bmatrix}$ and $\mathbf{J} = \begin{bmatrix} 0 & 1 \\ -1 & 0 \end{bmatrix}$. Calculate $\mathbf{I}^2$ and $\mathbf{J}^2$.
(c) Note that $\mathbf{U} = u_1\mathbf{I} + u_2\mathbf{J}$ and $\mathbf{V} = v_1\mathbf{I} + v_2\mathbf{J}$. Write $\mathbf{U} + \mathbf{V}$ and $\mathbf{UV}$ in terms of $\mathbf{I}$ and $\mathbf{J}$.
[i] (d) What does all this have to do with the complex numbers?

34. **TEASER** Find the four square roots of the matrix

$$\begin{bmatrix} 7 & 10 \\ 15 & 22 \end{bmatrix}$$

"What was that?" inquired Alice. "Reeling and Writing, of course, to begin with," the mock turtle replied, "and then the different branches of Arithmetic—Ambition, Distraction, Uglification, and Derision."

from *Alice's Adventures in Wonderland*

by Lewis Carroll

$$\frac{\begin{bmatrix} 2 & 3 \\ -4 & 1 \end{bmatrix}}{\begin{bmatrix} 6 & 7 \\ 1 & 2 \end{bmatrix}} = \begin{bmatrix} ? & ? \\ ? & ? \end{bmatrix}$$

9-4 Multiplicative Inverses

Even for ordinary numbers, the notion of division seems more difficult than that of addition, subtraction, and multiplication. Certainly this is true for division of matrices. Look at the example displayed above. It could tempt more than a mock turtle to derision. However, Arthur Cayley saw no need to sneer. He noted that in the case of numbers,

$$\frac{U}{V} = U \cdot \frac{1}{V} = U \cdot V^{-1}$$

What is needed is a concept of "one" for matrices; then we need the concept of multiplicative inverse. The first is easy.

THE MULTIPLICATIVE IDENTITY FOR MATRICES

Let

$$\mathbf{I} = \begin{bmatrix} 1 & 0 \\ 0 & 1 \end{bmatrix}$$

Then for any 2×2 matrix **U**,

$$\mathbf{UI} = \mathbf{U} = \mathbf{IU}$$

This can be checked by noting that

$$\begin{bmatrix} a & b \\ c & d \end{bmatrix}\begin{bmatrix} 1 & 0 \\ 0 & 1 \end{bmatrix} = \begin{bmatrix} a & b \\ c & d \end{bmatrix} = \begin{bmatrix} 1 & 0 \\ 0 & 1 \end{bmatrix}\begin{bmatrix} a & b \\ c & d \end{bmatrix}$$

The symbol **I** is chosen because it is often called the **multiplicative identity.** For 3×3 matrices, the multiplicative identity has the form

$$\begin{bmatrix} 1 & 0 & 0 \\ 0 & 1 & 0 \\ 0 & 0 & 1 \end{bmatrix}$$

You should be able to guess its form for 4×4 and higher order matrices.

INVERSES OF 2 × 2 MATRICES

Suppose we want to find the multiplicative inverse of

$$\mathbf{V} = \begin{bmatrix} 6 & 7 \\ 1 & 2 \end{bmatrix}$$

We are looking for a matrix

$$\mathbf{W} = \begin{bmatrix} a & b \\ c & d \end{bmatrix}$$

that satisfies $\mathbf{VW} = \mathbf{I}$ and $\mathbf{WV} = \mathbf{I}$. Taking $\mathbf{VW} = \mathbf{I}$ first, we want

$$\begin{bmatrix} 6 & 7 \\ 1 & 2 \end{bmatrix} \begin{bmatrix} a & b \\ c & d \end{bmatrix} = \begin{bmatrix} 1 & 0 \\ 0 & 1 \end{bmatrix}$$

which means

$$\begin{bmatrix} 6a + 7c & 6b + 7d \\ a + 2c & b + 2d \end{bmatrix} = \begin{bmatrix} 1 & 0 \\ 0 & 1 \end{bmatrix}$$

or

$$6a + 7c = 1 \qquad 6b + 7d = 0$$
$$a + 2c = 0 \qquad b + 2d = 1$$

When these four equations are solved for a, b, c, d, we have

$$\mathbf{W} = \begin{bmatrix} \frac{2}{5} & -\frac{7}{5} \\ -\frac{1}{5} & \frac{6}{5} \end{bmatrix}$$

as a tentative solution to our problem. We say tentative, because so far we know only that $\mathbf{VW} = \mathbf{I}$. Happily, $\mathbf{W}$ works on the other side of $\mathbf{V}$ too, as we can check. (In this exceptional case, we do have commutativity.)

$$\mathbf{WV} = \begin{bmatrix} \frac{2}{5} & -\frac{7}{5} \\ -\frac{1}{5} & \frac{6}{5} \end{bmatrix} \begin{bmatrix} 6 & 7 \\ 1 & 2 \end{bmatrix} = \begin{bmatrix} 1 & 0 \\ 0 & 1 \end{bmatrix}$$

Success! $\mathbf{W}$ is the inverse of $\mathbf{V}$; we denote it by the symbol $\mathbf{V}^{-1}$.

The process just described can be carried out for any specific 2 × 2 matrix, or better, it can be carried out for a general 2 × 2 matrix. But before we give the result, we make an important comment. There is no reason to think that every 2 × 2 matrix has a multiplicative inverse. Remember that the number 0 does not have such an inverse; neither does the matrix **O**. But here is a mild surprise. Many other 2 × 2 matrices do not have inverses. The following theorem identifies in a very precise way those that do, and then gives a formula for their inverses.

THEOREM (MULTIPLICATIVE INVERSES)

The matrix

$$\mathbf{V} = \begin{bmatrix} a & b \\ c & d \end{bmatrix}$$

has a multiplicative inverse if and only if $D = ad - bc$ is nonzero. If $D \neq 0$, then

$$\mathbf{V}^{-1} = \begin{bmatrix} \dfrac{d}{D} & -\dfrac{b}{D} \\[2ex] -\dfrac{c}{D} & \dfrac{a}{D} \end{bmatrix}$$

Thus the number D determines whether a matrix has an inverse. This number, which we shall call a *determinant,* will be studied in detail in the next section. Each 2×2 matrix has such a number associated with it. Let us look at two examples.

$$\mathbf{X} = \begin{bmatrix} 2 & -3 \\ -4 & 6 \end{bmatrix} \qquad\qquad \mathbf{Y} = \begin{bmatrix} 5 & -3 \\ -4 & 3 \end{bmatrix}$$

$$D = (2)(6) - (-3)(-4) = 0 \qquad D = (5)(3) - (-3)(-4) = 3$$

$$\mathbf{X}^{-1} \text{ does not exist} \qquad \mathbf{Y}^{-1} = \begin{bmatrix} \frac{3}{3} & \frac{3}{3} \\ \frac{4}{3} & \frac{5}{3} \end{bmatrix}$$

INVERSES FOR HIGHER-ORDER MATRICES

There is a theorem like the one above for square matrices of any size, which Cayley found in 1858. It is complicated and, rather than try to state it, we are going to illustrate a process which yields the inverse of a matrix whenever it exists. Briefly described, it is this. Take any square matrix $\mathbf{V}$ and write the corresponding identity matrix $\mathbf{I}$ next to it on the right. By using the three row operations of Section 9-2, attempt to reduce $\mathbf{V}$ to the identity matrix while simultaneously performing the same operations on $\mathbf{I}$. If you can reduce $\mathbf{V}$ to $\mathbf{I}$, you will simultaneously turn $\mathbf{I}$ into $\mathbf{V}^{-1}$. If you cannot reduce $\mathbf{V}$ to $\mathbf{I}$, $\mathbf{V}$ has no inverse.

Here is an illustration for the 2×2 matrix $\mathbf{V}$ that we used earlier.

$$\left[\begin{array}{cc|cc} 6 & 7 & 1 & 0 \\ 1 & 2 & 0 & 1 \end{array} \right]$$

$$\sim \left[\begin{array}{cc|cc} 1 & 2 & 0 & 1 \\ 6 & 7 & 1 & 0 \end{array} \right] \qquad \text{(interchange rows)}$$

$$\sim \left[\begin{array}{cc|cc} 1 & 2 & 0 & 1 \\ 0 & -5 & 1 & -6 \end{array} \right] \qquad \text{(add } -6 \text{ times row 1 to row 2)}$$

$$\sim \begin{bmatrix} 1 & 2 & | & 0 & 1 \\ 0 & 1 & | & -\frac{1}{5} & \frac{6}{5} \end{bmatrix} \qquad \text{(divide row 2 by } -5\text{)}$$

$$\sim \begin{bmatrix} 1 & 0 & | & \frac{2}{5} & -\frac{7}{5} \\ 0 & 1 & | & -\frac{1}{5} & \frac{6}{5} \end{bmatrix} \qquad \text{(add } -2 \text{ times row 2 to row 1)}$$

Notice that the matrix $\mathbf{V}^{-1}$ that we obtained earlier appears on the right. We illustrate the same process for a 3×3 matrix in Example A of the problem set.

AN APPLICATION

Consider the system of equations

$$2x + 6y + 6z = 8$$
$$2x + 7y + 6z = 10$$
$$2x + 7y + 7z = 9$$

If we introduce matrices

$$\mathbf{A} = \begin{bmatrix} 2 & 6 & 6 \\ 2 & 7 & 6 \\ 2 & 7 & 7 \end{bmatrix} \qquad \mathbf{X} = \begin{bmatrix} x \\ y \\ z \end{bmatrix} \qquad \mathbf{B} = \begin{bmatrix} 8 \\ 10 \\ 9 \end{bmatrix}$$

this system can be written in the form

$$\mathbf{AX} = \mathbf{B}$$

Now divide both sides by $\mathbf{A}$, by which we mean, of course, multiply both sides by $\mathbf{A}^{-1}$. We must be more precise. Multiply both sides on the left by $\mathbf{A}^{-1}$ (do not forget the lack of commutativity).

$$\mathbf{A}^{-1}\mathbf{AX} = \mathbf{A}^{-1}\mathbf{B}$$

$$\mathbf{IX} = \mathbf{A}^{-1}\mathbf{B}$$

$$\mathbf{X} = \mathbf{A}^{-1}\mathbf{B}$$

In Example A on page 337–338, $\mathbf{A}^{-1}$ is found to be

$$\mathbf{A}^{-1} = \begin{bmatrix} \frac{7}{2} & 0 & -3 \\ -1 & 1 & 0 \\ 0 & -1 & 1 \end{bmatrix}$$

Thus

$$\mathbf{X} = \begin{bmatrix} \frac{7}{2} & 0 & -3 \\ -1 & 1 & 0 \\ 0 & -1 & 1 \end{bmatrix} \begin{bmatrix} 8 \\ 10 \\ 9 \end{bmatrix} = \begin{bmatrix} 1 \\ 2 \\ -1 \end{bmatrix}$$

and therefore $(1, 2, -1)$ is the solution to our system.

This method of solution is particularly useful when many systems with the same coefficient matrix **A** are under consideration. Once we have $\mathbf{A}^{-1}$, we can obtain any solution simply by doing an easy matrix multiplication. If only one system is being studied, the method of Section 9-2 is best.

Problem Set 9-4

Find the multiplicative inverse of each matrix. Use matrix multiplication as a check.

1. $\begin{bmatrix} 2 & 3 \\ -1 & -1 \end{bmatrix}$

2. $\begin{bmatrix} 4 & 3 \\ 1 & 2 \end{bmatrix}$

3. $\begin{bmatrix} 6 & -14 \\ 0 & 2 \end{bmatrix}$

4. $\begin{bmatrix} 0 & 3 \\ 2 & 4 \end{bmatrix}$

5. $\begin{bmatrix} 1 & 0 \\ 0 & 1 \end{bmatrix}$

6. $\begin{bmatrix} 4 & 0 \\ 0 & 5 \end{bmatrix}$

7. $\begin{bmatrix} a & 0 \\ 0 & b \end{bmatrix}$

8. $\begin{bmatrix} 3 & 0 & 0 \\ 0 & 4 & 0 \\ 0 & 0 & 5 \end{bmatrix}$

EXAMPLE A (Inverses of Large Matrices)   Find the multiplicative inverse of

$$\begin{bmatrix} 2 & 6 & 6 \\ 2 & 7 & 6 \\ 2 & 7 & 7 \end{bmatrix}$$

Solution.   We use the reduction method described in the text.

$$\begin{bmatrix} 2 & 6 & 6 & | & 1 & 0 & 0 \\ 2 & 7 & 6 & | & 0 & 1 & 0 \\ 2 & 7 & 7 & | & 0 & 0 & 1 \end{bmatrix}$$

$$\sim \begin{bmatrix} 1 & 3 & 3 & | & \frac{1}{2} & 0 & 0 \\ 2 & 7 & 6 & | & 0 & 1 & 0 \\ 2 & 7 & 7 & | & 0 & 0 & 1 \end{bmatrix} \qquad \text{(divide row 1 by 2)}$$

$$\sim \begin{bmatrix} 1 & 3 & 3 & | & \frac{1}{2} & 0 & 0 \\ 0 & 1 & 0 & | & -1 & 1 & 0 \\ 0 & 1 & 1 & | & -1 & 0 & 1 \end{bmatrix} \qquad \begin{array}{l} \text{(add } -2 \text{ times row 1 to} \\ \text{row 2 and to row 3)} \end{array}$$

$$\sim \begin{bmatrix} 1 & 3 & 3 & | & \frac{1}{2} & 0 & 0 \\ 0 & 1 & 0 & | & -1 & 1 & 0 \\ 0 & 0 & 1 & | & 0 & -1 & 1 \end{bmatrix} \qquad \begin{array}{l} \text{(add } -1 \text{ times row 2 to} \\ \text{row 3)} \end{array}$$

$$\sim \begin{bmatrix} 1 & 0 & 3 & | & \frac{7}{2} & -3 & 0 \\ 0 & 1 & 0 & | & -1 & 1 & 0 \\ 0 & 0 & 1 & | & 0 & -1 & 1 \end{bmatrix} \qquad \begin{array}{l} \text{(add } -3 \text{ times row 2 to} \\ \text{row 1)} \end{array}$$

$$\sim \begin{bmatrix} 1 & 0 & 0 & | & \frac{7}{2} & 0 & -3 \\ 0 & 1 & 0 & | & -1 & 1 & 0 \\ 0 & 0 & 1 & | & 0 & -1 & 1 \end{bmatrix} \qquad \begin{array}{l} \text{(add } -3 \text{ times row 3 to} \\ \text{row 1)} \end{array}$$

Thus the desired inverse is

$$\begin{bmatrix} \frac{7}{2} & 0 & -3 \\ -1 & 1 & 0 \\ 0 & -1 & 1 \end{bmatrix}$$

Use the method illustrated above to find the multiplicative inverse of each of the following.

9. $\begin{bmatrix} 1 & 3 \\ 2 & 4 \end{bmatrix}$

10. $\begin{bmatrix} 2 & 6 \\ 3 & 1 \end{bmatrix}$

11. $\begin{bmatrix} 1 & 1 & 1 \\ 1 & -1 & 2 \\ 3 & 2 & 0 \end{bmatrix}$

12. $\begin{bmatrix} 2 & 1 & 1 \\ 1 & 3 & 1 \\ -1 & 4 & 0 \end{bmatrix}$

13. $\begin{bmatrix} 3 & 1 & 2 \\ 4 & 1 & -6 \\ 1 & 0 & 1 \end{bmatrix}$

14. $\begin{bmatrix} 2 & 4 & 6 \\ 3 & 2 & -5 \\ 2 & 3 & 1 \end{bmatrix}$

15. $\begin{bmatrix} 1 & 2 & 1 & 1 \\ 0 & 2 & 3 & 2 \\ 0 & 0 & 1 & 3 \\ 0 & 0 & 0 & 4 \end{bmatrix}$

16. $\begin{bmatrix} 1 & 1 & 1 & 1 \\ 1 & 1 & 1 & -1 \\ 1 & 1 & -1 & 1 \\ 1 & -1 & 1 & 1 \end{bmatrix}$

*Solve the following systems by making use of the inverses you found in Problems 11–14. Begin by writing the system in the matrix form **AX = B**.*

17. $x + y + z = 2$
    $x - y + 2z = -1$
    $3x + 2y = 5$

18. $2x + y + z = 4$
    $x + 3y + z = 5$
    $-x + 4y = 0$

19. $3x + y + 2z = 3$
    $4x + y - 6z = 2$
    $x + z = 6$

20. $2x + 4y + 6z = 9$
    $3x + 2y - 5z = 2$
    $2x + 3y + z = 4$

EXAMPLE B (Matrices Without Inverses) Try to find the multiplicative inverse of

$$U = \begin{bmatrix} 1 & 4 & 2 \\ 0 & 2 & 4 \\ 0 & -3 & -6 \end{bmatrix}$$

Solution.

$$\left[\begin{array}{ccc|ccc} 1 & 4 & 2 & 1 & 0 & 0 \\ 0 & 2 & 4 & 0 & 1 & 0 \\ 0 & -3 & -6 & 0 & 0 & 1 \end{array} \right] \sim \left[\begin{array}{ccc|ccc} 1 & 4 & 2 & 1 & 0 & 0 \\ 0 & 1 & 2 & 0 & \frac{1}{2} & 0 \\ 0 & -3 & -6 & 0 & 0 & 1 \end{array} \right]$$

$$\sim \left[\begin{array}{ccc|ccc} 1 & 4 & 2 & 1 & 0 & 0 \\ 0 & 1 & 2 & 0 & \frac{1}{2} & 0 \\ 0 & 0 & 0 & 0 & \frac{3}{2} & 1 \end{array} \right]$$

Since we got a row of zeros in the left half above, we know we can never reduce it to the identity matrix $\mathbf{I}$. The matrix $\mathbf{U}$ does not have an inverse.

Show that neither of the following matrices has a multiplicative inverse.

21. $\begin{bmatrix} 1 & 3 & 4 \\ 2 & 1 & -1 \\ 4 & 7 & 7 \end{bmatrix}$

22. $\begin{bmatrix} 2 & -2 & 4 \\ 5 & 3 & 2 \\ 3 & 5 & -2 \end{bmatrix}$

MISCELLANEOUS PROBLEMS

In Problems 23–26, find the multiplicative inverse or indicate that it does not exist.

23. $\begin{bmatrix} 4 & -3 \\ 5 & -\frac{15}{4} \end{bmatrix}$

24. $\begin{bmatrix} 3 & -1 \\ 4 & 2 \end{bmatrix}$

25. $\begin{bmatrix} 1 & -2 & 1 \\ 3 & 0 & 2 \\ 1 & 2 & \frac{1}{2} \end{bmatrix}$

26. $\begin{bmatrix} -2 & 4 & 2 \\ 3 & 5 & 6 \\ 1 & 9 & 8 \end{bmatrix}$

27. Find the multiplicative inverse of

$$\begin{bmatrix} 2 & 0 & 0 \\ 0 & 3 & 0 \\ 0 & 0 & -4 \end{bmatrix}$$

28. Give a formula for $\mathbf{U}^{-1}$ if

$$\mathbf{U} = \begin{bmatrix} a & 0 & 0 \\ 0 & b & 0 \\ 0 & 0 & c \end{bmatrix}$$

When does the matrix $\mathbf{U}$ fail to have an inverse?

29. Use your result from Problem 25 to solve the system

$$x - 2y + z = a$$
$$3x \quad\quad + 2z = b$$
$$x + 2y + \tfrac{1}{2}z = c$$

30. Let $\mathbf{A}$ and $\mathbf{B}$ be 3×3 matrices with inverses $\mathbf{A}^{-1}$ and $\mathbf{B}^{-1}$. Show that $\mathbf{AB}$ has an inverse given by $\mathbf{B}^{-1}\mathbf{A}^{-1}$. *Hint:* The product in either order must be $\mathbf{I}$.

31. Show that

$$\begin{bmatrix} 1 & -1 \\ 3 & -3 \end{bmatrix} \begin{bmatrix} 2 & -4 \\ 2 & -4 \end{bmatrix} = \begin{bmatrix} 0 & 0 \\ 0 & 0 \end{bmatrix}$$

Thus $\mathbf{AB} = \mathbf{O}$ but neither $\mathbf{A}$ nor $\mathbf{B}$ is $\mathbf{O}$. This is another way in which matrices differ from ordinary numbers.

32. Suppose $\mathbf{AB} = \mathbf{O}$ and $\mathbf{A}$ has a multiplicative inverse. Show that $\mathbf{B} = \mathbf{O}$. See Problem 31.

33. Find the inverse of the matrix

$$\begin{bmatrix} 1 & 1 & 1 & 1 \\ 1 & 2 & 2 & 2 \\ 1 & 2 & 1 & 1 \\ 1 & 2 & 1 & 2 \end{bmatrix}$$

34. Consider the matrices

$$\mathbf{A} = \begin{bmatrix} 0 & 1 & 0 \\ 0 & 0 & 1 \\ 1 & 0 & 0 \end{bmatrix} \quad \text{and} \quad \mathbf{B} = \begin{bmatrix} 0 & 1 & 0 & 0 \\ 0 & 0 & 1 & 0 \\ 0 & 0 & 0 & 1 \\ 1 & 0 & 0 & 0 \end{bmatrix}$$

(a) Show that $\mathbf{A}^3 = \mathbf{I}$ and $\mathbf{B}^4 = \mathbf{I}$.
(b) What does (a) allow you to conclude about $\mathbf{A}^{-1}$ and $\mathbf{B}^{-1}$?
(c) Conjecture a generalization of (a).

35. Show that the inverse of a Heisenberg matrix (see Problem 28 of Section 9-3) is a Heisenberg matrix.

36. **TEASER** The matrices

$$\mathbf{A} = \begin{bmatrix} 1 & \frac{1}{2} \\ \frac{1}{2} & \frac{1}{3} \end{bmatrix} \quad \text{and} \quad \mathbf{B} = \begin{bmatrix} 1 & \frac{1}{2} & \frac{1}{3} \\ \frac{1}{2} & \frac{1}{3} & \frac{1}{4} \\ \frac{1}{3} & \frac{1}{4} & \frac{1}{5} \end{bmatrix}$$

and their $n \times n$ generalizations are called Hilbert matrices; they play an important role in numerical analysis.

(a) Find $\mathbf{A}^{-1}$ and $\mathbf{B}^{-1}$.
(b) Let $\mathbf{C}$ be the column matrix with entries $(\frac{11}{6}, \frac{13}{12}, \frac{47}{60})$. Solve the equation $\mathbf{BX} = \mathbf{C}$ for $\mathbf{X}$.

Matrix	Determinant	Value of Determinant
$\begin{bmatrix} a & b \\ c & d \end{bmatrix}$	$\begin{vmatrix} a & b \\ c & d \end{vmatrix}$	$ad - bc$
$\begin{bmatrix} a_1 & b_1 & c_1 \\ a_2 & b_2 & c_2 \\ a_3 & b_3 & c_3 \end{bmatrix}$	$\begin{vmatrix} a_1 & b_1 & c_1 \\ a_2 & b_2 & c_2 \\ a_3 & b_3 & c_3 \end{vmatrix}$	$a_1 b_2 c_3 + a_2 b_3 c_1 + a_3 b_1 c_2$ $- a_1 b_3 c_2 - a_2 b_1 c_3 - a_3 b_2 c_1$

9-5 Second- and Third-Order Determinants

The notion of a determinant is usually attributed to the German mathematician Gottfried Wilhelm Leibniz (1646–1716), but it seems that Seki Kōwa of Japan had the idea somewhat earlier. It grew out of the study of systems of equations.

SECOND-ORDER DETERMINANTS

Consider the general system of two equations in two unknowns

$$ax + by = r$$
$$cx + dy = s$$

If we multiply the second equation by a and then add $-c$ times the first equation to it, we obtain the equivalent triangular system.

$$ax + \qquad by = r$$
$$(ad - bc)y = as - cr$$

If $ad - bc \neq 0$, we can solve this system by backward substitution.

$$x = \frac{rd - bs}{ad - bc}$$

$$y = \frac{as - rc}{ad - bc}$$

These formulas are hard to remember unless we associate special symbols with the numbers $ad - bc$, $rd - bs$, and $as - rc$. For the first of these, we propose

$$\begin{vmatrix} a & b \\ c & d \end{vmatrix} = ad - bc$$

The symbol on the left is called a **second-order determinant,** and we say that $ad - bc$ is its value. Thus

$$\begin{vmatrix} -2 & -1 \\ 5 & 6 \end{vmatrix} = (-2)(6) - (-1)(5) = -7$$

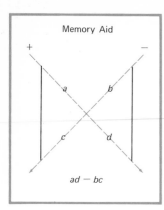

Memory Aid

$ad - bc$

Figure 3

Figure 3 may help you remember how to make the evaluation. With this new symbol, we can write the solution to

$$ax + by = r$$
$$cx + dy = s$$

as

$$x = \frac{rd - bs}{ad - bc} = \frac{\begin{vmatrix} r & b \\ s & d \end{vmatrix}}{\begin{vmatrix} a & b \\ c & d \end{vmatrix}}$$

$$y = \frac{as - rc}{ad - bc} = \frac{\begin{vmatrix} a & r \\ c & s \end{vmatrix}}{\begin{vmatrix} a & b \\ c & d \end{vmatrix}}$$

These results are easy to remember when we notice that the denominator is the determinant of the coefficient matrix, and that the numerator is the same except that the coefficients of the unknown we are seeking are replaced by the constants from the right side of the system.

Here is an example.

$$3x - 2y = 7$$
$$4x + 5y = 2$$

$$x = \frac{\begin{vmatrix} 7 & -2 \\ 2 & 5 \end{vmatrix}}{\begin{vmatrix} 3 & -2 \\ 4 & 5 \end{vmatrix}} = \frac{(7)(5) - (-2)(2)}{(3)(5) - (-2)(4)} = \frac{39}{23}$$

$$y = \frac{\begin{vmatrix} 3 & 7 \\ 4 & 2 \end{vmatrix}}{\begin{vmatrix} 3 & -2 \\ 4 & 5 \end{vmatrix}} = \frac{(3)(2) - (7)(4)}{23} = -\frac{22}{23}$$

The choice of the name *determinant* is appropriate, for the determinants of a system completely *determine* its character.

1. If $ad - bc \neq 0$, the system has a unique solution, the one given at the top of this page.
2. If $ad - bc = 0$, $as - rc = 0$, and $rd - bs = 0$, then a, b, and r are proportional to c, d, and s and the system has infinitely many solutions.

Here is an example.

$$3x - 2y = 7 \qquad \frac{3}{6} = \frac{-2}{-4} = \frac{7}{14}$$
$$6x - 4y = 14$$

3. If $ad - bc = 0$ and $as - rc \neq 0$ or $rd - bs \neq 0$, then a and b are proportional to c and d, but this proportionality does not extend to r and s; the system has no solution. This is illustrated by the following.

$$3x - 2y = 7 \qquad \frac{3}{6} = \frac{-2}{-4} \neq \frac{7}{10}$$
$$6x - 4y = 10$$

THIRD-ORDER DETERMINANTS

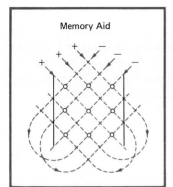

Memory Aid

Figure 4

When we consider the general system of three equations in three unknowns

$$a_1 x + b_1 y + c_1 z = d_1$$
$$a_2 x + b_2 y + c_2 z = d_2$$
$$a_3 x + b_3 y + c_3 z = d_3$$

things get more complicated, but the results are similar. The appropriate determinant symbol and its corresponding value are

$$\begin{vmatrix} a_1 & b_1 & c_1 \\ a_2 & b_2 & c_2 \\ a_3 & b_3 & c_3 \end{vmatrix} = a_1 b_2 c_3 + b_1 c_2 a_3 + c_1 b_3 a_2 - c_1 b_2 a_3 - b_1 a_2 c_3 - a_1 b_3 c_2$$

There are six terms in the sum on the right, three with a plus sign and three with a minus sign. The diagrams in Figures 4 and 5 will help you remember the products that enter each term. Just follow the arrows. Here is an example.

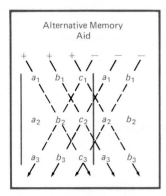

Alternative Memory Aid

Figure 5

$$\begin{vmatrix} 3 & 2 & 4 \\ 4 & -2 & 6 \\ 8 & 3 & 5 \end{vmatrix} = (3)(-2)(5) + (2)(6)(8) + (4)(3)(4)$$
$$-(4)(-2)(8) - (2)(4)(5) - (3)(3)(6)$$
$$= 84$$

CRAMER'S RULE

We saw that the solutions for x and y in a second-order system could be written as the quotients of two determinants. That fact generalizes to the third-order case. We present it without proof. Consider

$$a_1 x + b_1 y + c_1 z = d_1$$
$$a_2 x + b_2 y + c_2 z = d_2$$
$$a_3 x + b_3 y + c_3 z = d_3$$

If

$$D = \begin{vmatrix} a_1 & b_1 & c_1 \\ a_2 & b_2 & c_2 \\ a_3 & b_3 & c_3 \end{vmatrix} \neq 0$$

then the system above has a unique solution given by

$$x = \frac{1}{D} \begin{vmatrix} d_1 & b_1 & c_1 \\ d_2 & b_2 & c_2 \\ d_3 & b_3 & c_3 \end{vmatrix} \qquad y = \frac{1}{D} \begin{vmatrix} a_1 & d_1 & c_1 \\ a_2 & d_2 & c_2 \\ a_3 & d_3 & c_3 \end{vmatrix} \qquad z = \frac{1}{D} \begin{vmatrix} a_1 & b_1 & d_1 \\ a_2 & b_2 & d_2 \\ a_3 & b_3 & d_3 \end{vmatrix}$$

The pattern is the same as in the second-order situation. The denominator D is the determinant of the coefficient matrix. The numerator in each case is obtained from D by replacing the coefficients of the unknown by the constants from the right side of the system.

This method of solving a system of equations is named after one of its discoverers, Gabriel Cramer (1704–1752). Historically, it has been a popular method. However, note that even for a system of three equations in three unknowns, it requires the evaluation of four determinants. The method of matrices (Section 9-2) is considerably more efficient, both for hand and computer calculation. Consequently, Cramer's rule is now primarily of theoretical rather than practical interest.

PROPERTIES OF DETERMINANTS

We are interested in how the matrix operations considered in Section 9-2 affect the values of the corresponding determinants.

1. *Interchanging two rows changes the sign of the determinant; for example,*

$$\begin{vmatrix} a & b \\ c & d \end{vmatrix} = - \begin{vmatrix} c & d \\ a & b \end{vmatrix}$$

CAUTION

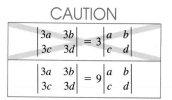

$$\begin{vmatrix} 3a & 3b \\ 3c & 3d \end{vmatrix} = 3 \begin{vmatrix} a & b \\ c & d \end{vmatrix}$$

$$\begin{vmatrix} 3a & 3b \\ 3c & 3d \end{vmatrix} = 9 \begin{vmatrix} a & b \\ c & d \end{vmatrix}$$

2. *Multiplying a row by a constant k multiplies the value of the determinant by k; for example,*

$$\begin{vmatrix} ka & kb \\ c & d \end{vmatrix} = k \begin{vmatrix} a & b \\ c & d \end{vmatrix}$$

3. *Adding a multiple of one row to another does not affect the value of the determinant; for example,*

$$\begin{vmatrix} a & b \\ c + ka & d + kb \end{vmatrix} = \begin{vmatrix} a & b \\ c & d \end{vmatrix}$$

We mention also the effect of a new operation.

4. *Interchanging the rows and columns (pairwise) does not affect the value of the determinant; for example,*

$$\begin{vmatrix} a & c \\ b & d \end{vmatrix} = \begin{vmatrix} a & b \\ c & d \end{vmatrix}$$

Though they are harder to prove in the third-order case, we emphasize that all four properties hold for both second- and third-order determinants. We offer only one proof, a proof of Property 3 in the second-order case.

$$\begin{vmatrix} a & b \\ c + ka & d + kb \end{vmatrix} = a(d + kb) - b(c + ka)$$

$$= ad + akb - bc - bka$$

$$= ad - bc$$

$$= \begin{vmatrix} a & b \\ c & d \end{vmatrix}$$

Property 3 can be a great aid in evaluating a determinant. Using it, we can transform a matrix to triangular form without changing the value of its determinant. But *the determinant of a triangular matrix is just the product of the elements on the main diagonal,* since this is the only nonzero term in the determinant formula. Here is an example.

$$\begin{vmatrix} 1 & 3 & 4 \\ -1 & -2 & 3 \\ 2 & -6 & 11 \end{vmatrix} = \begin{vmatrix} 1 & 3 & 4 \\ 0 & 1 & 7 \\ 0 & -12 & 3 \end{vmatrix} = \begin{vmatrix} 1 & 3 & 4 \\ 0 & 1 & 7 \\ 0 & 0 & 87 \end{vmatrix} = 87$$

Problem Set 9-5

Evaluate each of the determinants in Problems 1–8 by inspection.

1. $\begin{vmatrix} 4 & 0 \\ 0 & -2 \end{vmatrix}$

2. $\begin{vmatrix} 8 & 0 \\ 5 & 0 \end{vmatrix}$

3. $\begin{vmatrix} 11 & 4 \\ 0 & 2 \end{vmatrix}$

4. $\begin{vmatrix} 2 & -1 & 5 \\ 0 & 4 & 2 \\ 0 & 0 & -1 \end{vmatrix}$

5. $\begin{vmatrix} -1 & -7 & 9 \\ 0 & 5 & 4 \\ 0 & 0 & 10 \end{vmatrix}$

6. $\begin{vmatrix} 3 & -2 & 1 \\ 0 & 0 & 0 \\ 1 & 5 & -8 \end{vmatrix}$

7. $\begin{vmatrix} 3 & 0 & 8 \\ 10 & 0 & 2 \\ -1 & 0 & -9 \end{vmatrix}$

8. $\begin{vmatrix} 9 & 0 & 0 \\ 0 & 0 & -2 \\ 0 & 4 & 0 \end{vmatrix}$

9. If

$$\begin{vmatrix} a_1 & b_1 & c_1 \\ a_2 & b_2 & c_2 \\ a_3 & b_3 & c_3 \end{vmatrix} = 12$$

find the value of each of the following determinants.

(a) $\begin{vmatrix} a_1 & a_2 & a_3 \\ b_1 & b_2 & b_3 \\ c_1 & c_2 & c_3 \end{vmatrix}$
(b) $\begin{vmatrix} a_3 & b_3 & c_3 \\ a_2 & b_2 & c_2 \\ a_1 & b_1 & c_1 \end{vmatrix}$

(c) $\begin{vmatrix} a_1 & b_1 & c_1 \\ a_2 & b_2 & c_2 \\ 3a_3 & 3b_3 & 3c_3 \end{vmatrix}$
(d) $\begin{vmatrix} a_1 + 3a_3 & b_1 + 3b_3 & c_1 + 3c_3 \\ a_2 & b_2 & c_2 \\ a_3 & b_3 & c_3 \end{vmatrix}$

Evaluate each of the determinants in Problems 10–17.

10. $\begin{vmatrix} 3 & 2 \\ 5 & 6 \end{vmatrix}$
11. $\begin{vmatrix} 5 & 3 \\ 5 & -3 \end{vmatrix}$

12. $\begin{vmatrix} 3 & 0 & 0 \\ -2 & 5 & 4 \\ 1 & 2 & -9 \end{vmatrix}$
13. $\begin{vmatrix} 4 & 8 & -2 \\ 1 & -2 & 0 \\ 2 & 4 & 0 \end{vmatrix}$

14. $\begin{vmatrix} 3 & 2 & -4 \\ 1 & 0 & 5 \\ 4 & -2 & 3 \end{vmatrix}$
15. $\begin{vmatrix} 2 & 4 & 1 \\ 1 & 3 & 6 \\ 2 & 3 & -1 \end{vmatrix}$

c 16. $\begin{vmatrix} 5.1 & -3.2 & 2.6 \\ 1.3 & 4.5 & 2.3 \\ 3.4 & -2.2 & 1.9 \end{vmatrix}$
c 17. $\begin{vmatrix} 2.03 & 5.41 & -3.14 \\ 0 & 6.22 & 0 \\ -1.93 & 7.13 & 6.34 \end{vmatrix}$

Use Cramer's rule to solve the system of equations in Problems 18–21.

18. $2x - 3y = -11$
    $x + 2y = -2$

19. $5x + y = 7$
    $3x - 4y = 18$

20. $2x + 4y + z = 15$
    $x + 3y + 6z = 15$
    $2x + 3y - z = 11$

21. $5x - 3y + 2z = 18$
    $x + 4y + 2z = -4$
    $3x - 2y + z = 11$

MISCELLANEOUS PROBLEMS

22. Establish Property 2 for a general second-order determinant.
23. Evaluate each of the following determinants (the easy way).

(a) $\begin{vmatrix} 1 & 2 & 3 \\ 0 & 0 & 0 \\ 1.9 & 2.9 & 3.9 \end{vmatrix}$
(b) $\begin{vmatrix} 1 & 2 & 3 \\ 1.1 & 2.2 & 3.3 \\ 1.9 & 2.9 & 3.9 \end{vmatrix}$
(c) $\begin{vmatrix} 1.1 & 2.2 & 3.3 \\ 4.4 & 5.5 & 6.6 \\ 5.5 & 7.7 & 9.9 \end{vmatrix}$

(d) $\begin{vmatrix} 1 & 2 & 3 \\ 0 & 2 & 3 \\ 0 & 0 & 3 \end{vmatrix}$
(e) $\begin{vmatrix} 1 & 1 & 1 \\ 1 & 2 & 3 \\ 1 & 3 & 6 \end{vmatrix}$
(f) $\begin{vmatrix} 1 & 1 & 1 \\ 1 & 2 & 4 \\ 1 & 3 & 9 \end{vmatrix}$

24. Show that the value of a third-order determinant is 0 if either of the following are true.
    (a) Two rows are proportional.
    (b) The sum of two rows is the third row.

25. Consider the system of equations

$$kx + y = k^2$$
$$x + ky = 1$$

For what values of k does this system have (a) unique solution; (b) infinitely many solutions; (c) no solution?

26. If $\mathbf{C} = \mathbf{AB}$, where $\mathbf{A}$ and $\mathbf{B}$ are square matrices of the same size, then their determinants satisfy $|\mathbf{C}| = |\mathbf{A}||\mathbf{B}|$. Prove this fact for 2×2 matrices.

27. Suppose that $\mathbf{A}$ and $\mathbf{B}$ are 3×3 matrices with $|\mathbf{A}| = -2$. Use Problem 26 to evaluate (a) $|\mathbf{A}^5|$; (b) $|\mathbf{A}^{-1}|$; (c) $|\mathbf{B}^{-1}\mathbf{AB}|$; (d) $|3\mathbf{A}^3|$.

28. Show that

$$\begin{vmatrix} 1 & a & a^2 \\ 1 & b & b^2 \\ 1 & c & c^2 \end{vmatrix} = (a - b)(b - c)(c - a)$$

29. Let $(0, 0)$, (a, b), (c, d), and $(a + c, b + d)$ be the vertices of a parallelogram with a, b, c, and d being positive numbers. Show that the area of the parallelogram is the determinant

$$\begin{vmatrix} a & b \\ c & d \end{vmatrix}$$

30. Consider the following determinant equation.

$$\begin{vmatrix} x & y & 1 \\ a & b & 1 \\ c & d & 1 \end{vmatrix} = 0 \qquad \text{where } (a, b) \neq (c, d)$$

(a) Show that the above equation is the equation of a line in the xy-plane.
(b) How can you tell immediately that the points (a, b) and (c, d) are on this line?
(c) Write a determinant equation for the line that passes through the points $(5, -1)$ and $(4, 11)$.

31. Determine the polynomial $P(a, b, c)$ for which

$$\begin{vmatrix} a & b & c \\ b & c & a \\ c & a & b \end{vmatrix} = (a + b + c)P(a, b, c)$$

32. **TEASER** If $\mathbf{A}$ is a square matrix, then $P(x) = |\mathbf{A} - x\mathbf{I}|$ is called the *characteristic polynomial* of the matrix. Let

$$\mathbf{A} = \begin{vmatrix} 1 & 1 & -1 \\ 0 & 0 & -1 \\ 1 & 1 & 4 \end{vmatrix}$$

(a) Write the polynomial $P(x)$ in the form $ax^3 + bx^2 + cx + d$.
(b) Solve the equation $P(x) = 0$. The solutions are called the *characteristic values* of $\mathbf{A}$.
(c) Show that $P(\mathbf{A}) = 0$, that is, $\mathbf{A}$ satisfies its own characteristic equation.

James Joseph Sylvester
1814–1897

A Mathematician and a Poet

One of the most consistent workers on the theory of determinants over a period of 50 years was the Englishman, James Joseph Sylvester. Known as a poet, a wit, and a mathematician, he taught in England and in America. During his stay at Johns Hopkins University in Baltimore (1877-1883), he helped establish one of the first graduate programs in mathematics in America. Under his tutelage, mathematics began to flourish in the United States.

9-6 Higher-Order Determinants

Having defined determinants for 2×2 and 3×3 matrices, we expect to do it for 4×4 matrices, 5×5 matrices, and so on. Our problem is to do it in such a way that Cramer's rule and the determinant properties of Section 9-5 still hold. This will take some work.

MINORS

We begin by introducing the standard notation for a general $n \times n$ matrix.

$$\begin{bmatrix} a_{11} & a_{12} & a_{13} & \cdots & a_{1n} \\ a_{21} & a_{22} & a_{23} & \cdots & a_{2n} \\ a_{31} & a_{32} & a_{33} & \cdots & a_{3n} \\ \cdot & \cdot & \cdot & & \cdot \\ \cdot & \cdot & \cdot & & \cdot \\ \cdot & \cdot & \cdot & & \cdot \\ a_{n1} & a_{n2} & a_{n3} & \cdots & a_{nn} \end{bmatrix}$$

Note the use of the double subscript on each entry: the first subscript gives the row in which a_{ij} is and the second gives the column. For example, a_{32} is the entry in the third row and second column.

Associated with each entry a_{ij} in an $n \times n$ matrix is a determinant M_{ij} of order $n - 1$ called the **minor** of a_{ij}. It is obtained by taking the determinant of the submatrix that results when we blot out the row and column in which a_{ij} stands. For example, the minor M_{13} of a_{13} in the 4×4 matrix

$$\begin{bmatrix} a_{11} & a_{12} & a_{13} & a_{14} \\ a_{21} & a_{22} & a_{23} & a_{24} \\ a_{31} & a_{32} & a_{33} & a_{34} \\ a_{41} & a_{42} & a_{43} & a_{44} \end{bmatrix}$$

is the third-order determinant.

$$\begin{vmatrix} a_{21} & a_{22} & a_{24} \\ a_{31} & a_{32} & a_{34} \\ a_{41} & a_{42} & a_{44} \end{vmatrix}$$

THE GENERAL nth-ORDER DETERMINANT

Here is the definition to which we have been leading.

$$\begin{vmatrix} a_{11} & a_{12} & \cdots & a_{1n} \\ a_{21} & a_{22} & \cdots & a_{2n} \\ \vdots & \vdots & & \vdots \\ a_{n1} & a_{n2} & \cdots & a_{nn} \end{vmatrix} = a_{11}M_{11} - a_{12}M_{12} + a_{13}M_{13} \cdots + (-1)^{n+1}a_{1n}M_{1n}$$

There are three important questions to answer regarding this definition.

Does this definition really define? Only if the minors M_{ij} can be evaluated. They are themselves determinants, but here is the key point: They are of order $n - 1$, one less than the order of the determinant we started with. They can, in turn, be expressed in terms of determinants of order $n - 2$, and so on, using the same definition. Thus, for example, a fifth-order determinant can be expressed in terms of fourth-order determinants, and these fourth-order determinants can be expressed in terms of third-order determinants. But we know how to evaluate third-order determinants from Section 9-5.

Is this definition consistent with the earlier definition when applied to third-order determinants? Yes, for if we apply it to a general third-order determinant, we get

$$\begin{vmatrix} a_1 & b_1 & c_1 \\ a_2 & b_2 & c_2 \\ a_3 & b_3 & c_3 \end{vmatrix} = a_1\begin{vmatrix} b_2 & c_2 \\ b_3 & c_3 \end{vmatrix} - b_1\begin{vmatrix} a_2 & c_2 \\ a_3 & c_3 \end{vmatrix} + c_1\begin{vmatrix} a_2 & b_2 \\ a_3 & b_3 \end{vmatrix}$$

$$= a_1b_2c_3 - a_1c_2b_3 - b_1a_2c_3 + b_1c_2a_3 + c_1a_2b_3 - c_1b_2a_3$$

This is the same value we gave in Section 9-5.

Does this definition preserve Cramer's rule and the properties of Section 9-5? Yes, it does. We shall not prove this because the proofs are lengthy and difficult.

EXPANSION ACCORDING TO ANY ROW OR COLUMN

Our definition expressed the value of a determinant in terms of the entries and minors of the first row; we call it an expansion according to the first row. It is a remarkable fact that we can expand a determinant according to any row or column (and always get the same answer).

Before we can show what we mean, we must explain a sign convention. We associate a plus or minus sign with every position in a matrix. To the ij-position, we assign a plus sign if $i + j$ is even and a minus sign otherwise. Thus for a 4×4 matrix, we have this pattern of signs.

$$\begin{bmatrix} + & - & + & - \\ - & + & - & + \\ + & - & + & - \\ - & + & - & + \end{bmatrix}$$

There is always a $+$ in the upper left position and then the signs alternate.

With this understanding about signs, we may expand according to any row or column. For example, to evaluate a fourth-order determinant, we can expand according to the second column if we wish. We multiply each entry in that column by its minor, prefixing each product with a plus or minus sign according to the pattern above. Then we add the results.

$$\begin{vmatrix} a_{11} & a_{12} & a_{13} & a_{14} \\ a_{21} & a_{22} & a_{23} & a_{24} \\ a_{31} & a_{32} & a_{33} & a_{34} \\ a_{41} & a_{42} & a_{43} & a_{44} \end{vmatrix} = -a_{12}M_{12} + a_{22}M_{22} - a_{32}M_{32} + a_{42}M_{42}$$

EXAMPLE

To evaluate

$$\begin{vmatrix} 6 & 0 & 4 & -1 \\ 2 & 0 & -1 & 4 \\ -2 & 4 & -2 & 3 \\ 4 & 0 & 5 & -4 \end{vmatrix}$$

it is obviously best to expand according to the second column, since three of the four resulting terms are zero. The single nonzero term is just $(-1)(4)$ times the minor M_{32}—that is,

$$-4\begin{vmatrix} 6 & 4 & -1 \\ 2 & -1 & 4 \\ 4 & 5 & -4 \end{vmatrix}$$

We could now evaluate this third-order determinant as in Section 9-5. But having seen the usefulness of zeros, let us take a different tack. It is easy to get two zeros in the first column. Simply add -3 times the second row to the first row and -2 times the second row to the third. We get

$$-4 \begin{vmatrix} 0 & 7 & -13 \\ 2 & -1 & 4 \\ 0 & 7 & -12 \end{vmatrix}$$

Finally, expand according to the first column.

$$(-4)(-1)(2)\begin{vmatrix} 7 & -13 \\ 7 & -12 \end{vmatrix} = 8(-84 + 91) = 56$$

The reason for the factor of -1 is that the entry 2 is in a minus position in the 3×3 pattern of signs.

Problem Set 9-6

Evaluate each of the determinants in Problems 1–6 according to a row or column of your choice. Make a good choice or suffer the consequences!

1. $\begin{vmatrix} 3 & -2 & 4 \\ 1 & 5 & 0 \\ 3 & 10 & 0 \end{vmatrix}$

2. $\begin{vmatrix} 4 & 0 & -6 \\ -2 & 3 & 5 \\ 1 & 0 & 8 \end{vmatrix}$

3. $\begin{vmatrix} 1 & 2 & 3 \\ 0 & 2 & 3 \\ 1 & 3 & 4 \end{vmatrix}$

4. $\begin{vmatrix} 2 & -1 & -1 \\ 3 & 4 & 2 \\ 0 & -1 & -1 \end{vmatrix}$

5. $\begin{vmatrix} 3 & 0 & 0 & 0 \\ -1 & 1 & 4 & 2 \\ 2 & 0 & 2 & -3 \\ -4 & 0 & 1 & 5 \end{vmatrix}$

6. $\begin{vmatrix} 0 & 5 & 0 & 0 \\ 1 & -3 & 0 & 2 \\ 4 & 1 & 2 & 8 \\ -3 & 2 & 0 & 5 \end{vmatrix}$

Evaluate each of the determinants in Problems 7–10 by first getting some zeros in a row or column and then expanding according to that row or column.

7. $\begin{vmatrix} 3 & 5 & -10 \\ 2 & 4 & 6 \\ -3 & -5 & 12 \end{vmatrix}$

8. $\begin{vmatrix} 2 & -1 & 2 \\ 4 & 3 & 4 \\ 7 & -5 & 10 \end{vmatrix}$

9. $\begin{vmatrix} 1 & -2 & 1 & 4 \\ -2 & 5 & -3 & 1 \\ 0 & 7 & -4 & 2 \\ 3 & -2 & 2 & 6 \end{vmatrix}$

10. $\begin{vmatrix} 1 & -2 & 0 & -4 \\ 3 & -4 & 3 & -10 \\ 2 & 1 & -2 & 1 \\ 4 & -5 & 1 & 4 \end{vmatrix}$

11. Solve the following system for x only.

$$x - 2y + z + 4w = 1$$
$$-2x + 5y - 3z + w = -2$$
$$7y - 4z + 2w = 3$$
$$3x - 2y + 2z + 6w = 6$$

(Make use of your answer to Problem 9.)

12. Solve the following system for z only.

$$x - 2y - 4w = -14$$
$$3x - 4y + 3z - 10w = -28$$
$$2x + y - 2z + w = 0$$
$$4x - 5y + z + 4w = 9$$

(Make use of your answer to Problem 10.)

Evaluate the determinants in Problems 13–18.

13. $\begin{vmatrix} 2 & -3 & 2 \\ 1 & 0 & -4 \\ -1 & 0 & 6 \end{vmatrix}$

14. $\begin{vmatrix} 3 & 1 & -5 \\ 2 & -2 & 7 \\ 1 & 0 & -1 \end{vmatrix}$

15. $\begin{vmatrix} 2 & -3 & 4 & 5 \\ 2 & -3 & 4 & 7 \\ 1 & 6 & 4 & 5 \\ 2 & 6 & 4 & -8 \end{vmatrix}$

16. $\begin{vmatrix} 2 & 2 & 3 & 7 \\ 1 & 2 & 3 & -2 \\ 4 & -3 & 9 & 6 \\ 1 & 2 & 3 & -1 \end{vmatrix}$

17. $\begin{vmatrix} 1 & 2 & -3 & 1 & 2 \\ -1 & 0 & 2 & 5 & -3 \\ 5 & 0 & 0 & -2 & 4 \\ 0 & 0 & 0 & 6 & 3 \\ 0 & 0 & 0 & 2 & -7 \end{vmatrix}$

18. $\begin{vmatrix} 1 & 2 & 3 & 4 & 5 \\ 2 & 1 & 1 & 1 & 1 \\ 3 & 1 & 1 & 1 & 1 \\ 4 & 1 & 1 & 1 & 1 \\ 5 & 1 & 1 & 1 & 1 \end{vmatrix}$

Hint: Subtract row 2 from row 3.

19. Evaluate the following determinant.

$$\begin{vmatrix} a & b & c & d \\ 0 & e & f & g \\ 0 & 0 & h & i \\ 0 & 0 & 0 & j \end{vmatrix}$$

Conjecture a general result about the determinant of a triangular matrix.

[c] 20. Use the result of Problem 19 to evaluate

$$\begin{vmatrix} 2.12 & 3.14 & -1.61 & 1.72 \\ 0 & -2.36 & 5.91 & 7.82 \\ 0 & 0 & 1.46 & 3.34 \\ 0 & 0 & 0 & 3.31 \end{vmatrix}$$

21. Evaluate by reducing to triangular form and using Problem 19.

$$\begin{vmatrix} 1 & 2 & 2.6 & 1.5 \\ 2.3 & 5.6 & -1.3 & 9.8 \\ 2.7 & 1.3 & 4.2 & -1.9 \\ 5.5 & 6.2 & 3.0 & 1.4 \end{vmatrix}$$

22. Show that

$$\begin{vmatrix} a_1 + d_1 & b_1 & c_1 \\ a_2 + d_2 & b_2 & c_2 \\ a_3 + d_3 & b_3 & c_3 \end{vmatrix} = \begin{vmatrix} a_1 & b_1 & c_1 \\ a_2 & b_2 & c_2 \\ a_3 & b_3 & c_3 \end{vmatrix} + \begin{vmatrix} d_1 & b_1 & c_1 \\ d_2 & b_2 & c_2 \\ d_3 & b_3 & c_3 \end{vmatrix}$$

MISCELLANEOUS PROBLEMS

23. Evaluate each of the following determinants.

(a) $\begin{vmatrix} 1 & 0 & 0 & 2 \\ 2.7 & 5 & 0 & 8.9 \\ 3.4 & 0 & 6 & 9.1 \\ 3 & 0 & 0 & 4 \end{vmatrix}$ (b) $\begin{vmatrix} a & 0 & 0 & b \\ ? & e & 0 & ? \\ ? & 0 & f & ? \\ c & 0 & 0 & d \end{vmatrix}$

24. Solve for x.

$$\begin{vmatrix} 2-x & 0 & 0 & 5 \\ 2.7 & 3-x & 0 & 8.9 \\ 3.4 & 0 & 4-x & 9.1 \\ 5 & 0 & 0 & 2-x \end{vmatrix} = 0$$

25. Evaluate the determinant below. *Hint:* Subtract the first row from each of the second and third rows.

$$\begin{vmatrix} n+1 & n+2 & n+3 \\ n+4 & n+5 & n+6 \\ n+7 & n+8 & n+9 \end{vmatrix}$$

26. Show that if the entries in a determinant are all integers, then the value of the determinant is an integer. From this and Cramer's rule, draw a conclusion about the nature of the solution to a system of n linear equations in n unknowns if all the constants in the system are integers and the determinant of coefficients is nonzero.

27. Evaluate the given determinants, in which the entries come from Pascal's triangle (see Section 10-4).

$$D_2 = \begin{vmatrix} 1 & 1 \\ 1 & 2 \end{vmatrix}$$

$$D_3 = \begin{vmatrix} 1 & 1 & 1 \\ 1 & 2 & 3 \\ 1 & 3 & 6 \end{vmatrix}$$

$$D_4 = \begin{vmatrix} 1 & 1 & 1 & 1 \\ 1 & 2 & 3 & 4 \\ 1 & 3 & 6 & 10 \\ 1 & 4 & 10 & 20 \end{vmatrix}$$

28. Based on the results of Problem 27, make a conjecture about the value of D_n, the nth-order determinant obtained from Pascal's triangle. Then support your conjecture by describing a systematic way of evaluating D_n.

29. Solve the given system. It should be easy after Problem 27.

$$\begin{aligned} x + y + z + w &= 0 \\ x + 2y + 3z + 4w &= 0 \\ x + 3y + 6z + 10w &= 0 \\ x + 4y + 10z + 20w &= 1 \end{aligned}$$

30. Evaluate the determinants.

$$E_1 = \begin{vmatrix} 2 & 1 \\ 1 & 2 \end{vmatrix} \qquad E_2 = \begin{vmatrix} 2 & 1 & 0 \\ 1 & 2 & 1 \\ 0 & 1 & 2 \end{vmatrix} \qquad E_3 = \begin{vmatrix} 2 & 1 & 0 & 0 \\ 1 & 2 & 1 & 0 \\ 0 & 1 & 2 & 1 \\ 0 & 0 & 1 & 2 \end{vmatrix}$$

Now generalize by conjecturing the value of the nth-order determinant E_n that has 2's on the main diagonal, 1's adjacent to this diagonal on either side, and 0's elsewhere. Then prove your conjecture. *Hint:* Expand according to the first row to show that $E_n = 2E_{n-1} - E_{n-2}$ and from this argue that your conjecture must be correct.

31. Let $a_1, a_2, \ldots, a_n$ and $b_1, b_2, \ldots, b_n$ be two sequences of numbers.
    (a) Let C_n be the $n \times n$ matrix with entries $c_{ij} = a_i b_j$. Evaluate $|C_n|$ for $n = 2$, 3, and 4 and make a conjecture.
    (b) Do the same if $c_{ij} = a_i - b_j$.

32. **TEASER** Generalize Problem 28 of Section 9-5 by evaluating the following determinant and writing your answer as the product of six linear factors.

$$\begin{vmatrix} 1 & a & a^2 & a^3 \\ 1 & b & b^2 & b^3 \\ 1 & c & c^2 & c^3 \\ 1 & d & d^2 & d^3 \end{vmatrix}$$

Corn

Profit : $40 per acre

Labor : 2 hours per acre

Maximizing Profit

Farmer Brown has 480 acres of land on which he can grow either corn or wheat. He figures that he has 800 hours of labor available during the crucial summer season. Given the profit margins and labor requirements shown at the right, how many acres of each should he plant to maximize his profit? What is this maximum profit?

Wheat

Profit : $30 per acre

Labor : 1 hour per acre

9-7   Systems of Inequalities

At first glance you might think that Farmer Brown should put all of his land into corn. Unfortunately, however, that requires 960 hours of labor and he has only 800 available. Well, maybe he should plant 400 acres of corn, using his allocated 800 hours of labor on them, and let the remaining 80 acres lie idle. Or would it be wise to at least plant enough wheat so all his land is in use? This problem is complicated enough so that no one is likely to find the best solution without a lot of work. And would not a method be better than blind experimenting? That is our subject—a method for handling Farmer Brown's problem and others of the same type.

Like all individuals and businesses, Farmer Brown must operate within certain limitations; we call them **constraints.** Suppose he plants x acres of corn and y acres of wheat. His constraints can be translated into inequalities.

$$\text{Land constraint:} \qquad x + y \le 480$$

$$\text{Labor constraint:} \qquad 2x + y \le 800$$

$$\text{Nonnegativity constraints:} \quad x \ge 0 \qquad y \ge 0$$

His task is to maximize the profit $P = 40x + 30y$ subject to these constraints. Before we can solve his problem, we will need to know more about inequalities.

THE GRAPH OF A LINEAR INEQUALITY

The best way to visualize an inequality is by means of its graph. Consider, for example,

$$2x + y \le 6$$

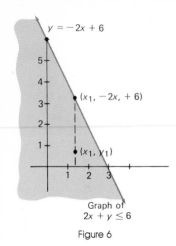

$y = -2x + 6$

$(x_1, -2x, + 6)$

(x_1, y_1)

Graph of
$2x + y \leq 6$

Figure 6

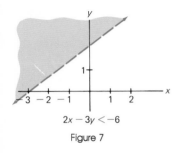

$2x - 3y < -6$

Figure 7

which can be rewritten as

$$y \leq -2x + 6$$

The complete graph consists of those points which satisfy $y = -2x + 6$ (a line), together with those that satisfy $y < -2x + 6$ (the points below the line). To see that this description is correct, note that for any abscissa x_1, the point $(x_1, -2x_1 + 6)$ is on the line $y = -2x + 6$. The point (x_1, y_1) is directly below that point if and only if $y_1 < -2x_1 + 6$ (Figure 6). Thus the graph of $y \leq -2x + 6$ is the **closed half-plane** that we have shaded on the diagram. We refer to it as *closed* because the edge $y = -2x + 6$ is included. Correspondingly, the graph of $y < -2x + 6$ is called an **open half-plane.**

The graph of any linear inequality in x and y is a half-plane, open or closed. To sketch the graph, first draw the corresponding edge. Then determine the correct half-plane by taking a sample point, not on the edge, and checking to see if it satisfies the inequality.

To illustrate this procedure, consider

$$2x - 3y < -6$$

Its graph does not include the line $2x - 3y = -6$, although that line is crucial in determining the graph. We therefore show it as a dotted line. Since the sample point $(0, 0)$ does not satisfy the inequality, we choose the half-plane on the opposite side of the line from it. The complete graph is the shaded open half-plane shown in Figure 7.

GRAPHING A SYSTEM OF LINEAR INEQUALITIES

The graph of a system of inequalities like Farmer Brown's constraints

$$x + y \leq 480$$

$$2x + y \leq 800$$

$$x \geq 0 \qquad y \geq 0$$

is simply the intersection of the graphs of the individual inequalities. We can construct the graph in stages as we do in Figure 8, though we are confident that you will quickly learn to do it in one operation.

The diagram on the right of Figure 8 is the one we want. All the points in the shaded region F satisfy the four inequalities simultaneously. The points $(0, 0)$, $(400, 0)$, $(320, 160)$, and $(0, 480)$ are called the **vertices** (or corner points) of F. Incidentally, the point $(320, 160)$ was obtained by solving the two equations $2x + y = 800$ and $x + y = 480$ simultaneously.

The region F has three important properties (Figure 9).

1. It is polygonal (its boundary consists of line segments).
2. It is convex (if points P and Q are in the region, then the line segment PQ lies entirely within the region).
3. It is bounded (it can be enclosed in a circle).

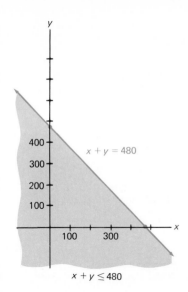

$x + y \leq 480$

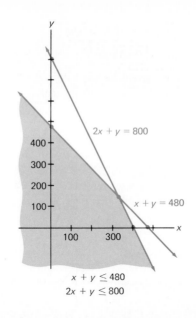

$x + y \leq 480$
$2x + y \leq 800$

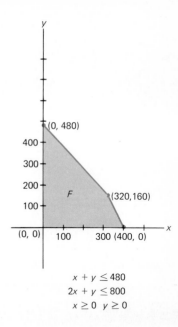

$x + y \leq 480$
$2x + y \leq 800$
$x \geq 0 \quad y \geq 0$

Figure 8

As a matter of fact, every region that arises as the solution set of a system of linear inequalities is polygonal and convex, though it need not be bounded. The shaded region in Figure 10 could not be the solution set for a system of linear inequalities because it is not convex.

LINEAR PROGRAMMING PROBLEMS

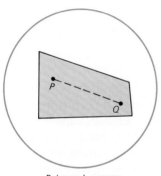

Polygonal, convex, and bounded

Figure 9

It is time that we solved Farmer Brown's problem.
Maximize

$$P = 40x + 30y$$

subject to

$$\begin{cases} x + y \leq 480 \\ 2x + y \leq 800 \\ x \geq 0 \quad y \geq 0 \end{cases}$$

Any problem that asks us to find the maximum (or minimum) of a linear function subject to linear inequality constraints is called a **linear programming problem.** Here is a method for solving such problems.

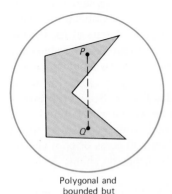

Polygonal and bounded but not convex

Figure 10

1. Graph the solution set corresponding to the inequality constraints.
2. Find the coordinates of the vertices of the solution set.
3. Evaluate the linear function that you want to maximize (or minimize) at each of these vertices. The largest of these gives the maximum, while the smallest gives the minimum.

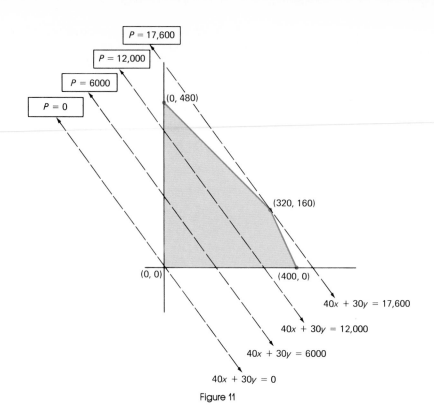

$P = 17,600$

$P = 12,000$

$P = 6000$

$P = 0$

(0, 480)

(320, 160)

(0, 0)   (400, 0)

$40x + 30y = 17,600$

$40x + 30y = 12,000$

$40x + 30y = 6000$

$40x + 30y = 0$

Figure 11

Vertex	$P = 40x + 30y$
(0, 0)	0
(0, 480)	14,400
(320, 160)	17,600
(400, 0)	16,000

Figure 12

Vertex	$P = 80x + 30y$
(0, 0)	0
(0, 480)	14,400
(320, 160)	30,400
(400, 0)	32,000

Figure 13

To see why this method works, consider the diagram for Farmer Brown's problem (Figure 11). The dotted lines are profit lines, each with slope $-\frac{4}{3}$; they are the graphs of $40x + 30y = P$ for various values of P. All the points on a dotted line give the same total profit. Imagine a profit line moving from left to right across the shaded region with the slope constant, as indicated in Figure 11 . During this motion, the profit is zero at (0, 0) and increases to its maximum of 17,600 at (320,160). It should be clear that such a moving line (no matter what its slope) will always enter the shaded set at a vertex and leave it at another vertex. The particular vertex depends upon the slope of the profit lines. In Farmer Brown's problem, the minimum profit of \$0 occurs at (0, 0); the maximum profit of \$17,600 occurs at (320, 160). In Figure 12, we show the total profit for each of the four vertices. Clearly Farmer Brown should plant 320 acres of corn and 160 acres of wheat.

Now suppose the price of corn goes up so that Farmer Brown can expect a profit of \$80 per acre on corn but still only \$30 per acre on wheat. Would this change his strategy? In Figure 13, we show his total profit $P = 80x + 30y$ at each of the four vertices. Evidently he should plant 400 acres of corn and no wheat to achieve maximum profit. Note that this means he should leave 80 acres of his land idle.

Finally, suppose that the profit per acre is \$40 both for wheat and for corn. The table for this case (Figure 14) shows the same total profit at the vertices (0, 480) and (320, 160). This means that the moving profit line leaves

Vertex	$P = 40x + 40y$
(0, 0)	0
(0, 480)	19,200
(320, 160)	19,200
(400, 0)	16,000

Figure 14

the shaded region along the side determined by those two vertices, so that every point on that side gives a maximum profit. It is still true, however, that the maximum profit occurs at a vertex.

The situation with an unbounded constraint set is slightly more complicated. It is discussed in Example A. Here we simply point out that in the unbounded case, there may not be a maximum (or minimum), but if there is one, it will still occur at a vertex.

Problem Set 9-7

In Problems 1–6, graph the solution set of each inequality in the xy-plane.

1. $4x + y \leq 8$ 2. $2x + 5y \leq 20$ 3. $x \leq 3$
4. $y \leq -2$ 5. $4x - y \geq 8$ 6. $2x - 5y \geq -20$

In Problems 7–10, graph the solution set of the given system. On the graph, label the coordinates of the vertices.

7. $4x + y \leq 8$
   $2x + 3y \leq 14$
   $x \geq 0 \quad y \geq 0$

8. $2x + 5y \leq 20$
   $4x + y \leq 22$
   $x \geq 0 \quad y \geq 0$

9. $4x + y \leq 8$
   $x - y \leq -2$
   $x \geq 0$

10. $2x + 5y \leq 20$
    $x - 2y \geq 1$
    $y \geq 0$

In Problems 11–14, find the maximum and minimum values of the given linear function P subject to the given inequalities.

11. $P = 2x + y$; the inequalities of Problem 7.
12. $P = 3x + 2y$; the inequalities of Problem 8.
13. $P = 2x - y$; the inequalities of Problem 9.
14. $P = 3x - 2y$; the inequalities of Problem 10.

EXAMPLE A (Unbounded Region)   Find the maximum and minimum values of the function $3x + 4y$ subject to the constraints

$$\begin{cases} 3x + 2y \geq 13 \\ x + y \geq 5 \\ x \geq 1 \\ y \geq 0 \end{cases}$$

Solution.   We proceed to graph the solution set of our system, noting that the region must lie above the lines $3x + 2y = 13$ and $x + y = 5$ and to the right of the line $x = 1$. It is shown in Figure 15. Notice that the region is unbounded and has (5, 0), (3, 2), and (1, 5) as its vertices.

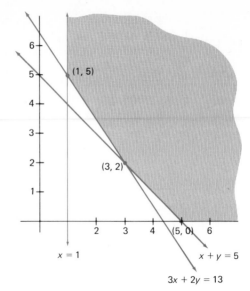

$x = 1$ $x + y = 5$
$3x + 2y = 13$
Figure 15

The point $(1, 4)$ at which the lines $x + y = 5$ and $x = 1$ intersect is not a vertex. It should be clear right away that $3x + 4y$ does not assume a maximum value in our region; its values can be made as large as we please by increasing x and y. To find the minimum value, we calculate $3x + 4y$ at the three vertices.

$$(5, 0): \quad 3 \cdot 5 + 4 \cdot 0 = 15$$

$$(3, 2): \quad 3 \cdot 3 + 4 \cdot 2 = 17$$

$$(1, 5): \quad 3 \cdot 1 + 4 \cdot 5 = 23$$

The minimum value of $3x + 4y$ is 15.

Solve each of the following problems.

15. Minimize $5x + 2y$ subject to

$$\begin{cases} x + y \geq 4 \\ \phantom{x +{}} x \geq 2 \\ \phantom{x +{}} y \geq 0 \end{cases}$$

16. Minimize $2x - y$ subject to

$$\begin{cases} x - 2y \geq 2 \\ \phantom{x -{}} y \geq 2 \end{cases}$$

17. Minimize $2x + y$ subject to

$$\begin{cases} 4x + y \geq 7 \\ 2x + 3y \geq 6 \\ \phantom{2x +{}} x \geq 1 \\ \phantom{2x +{}} y \geq 0 \end{cases}$$

18. Minimize $3x + 2y$ subject to

$$\begin{cases} x - 2y \leq 2 \\ x - 2y \geq -2 \\ 3x - 2y \geq 10 \end{cases}$$

EXAMPLE B (Systems with Nonlinear Inequalities)   Graph the solution set of the following system of inequalities.

$$\begin{cases} y \geq 2x^2 \\ y \leq 2x + 4 \end{cases}$$

Solution. It helps to find the points at which the parabola $y = 2x^2$ intersects the line $y = 2x + 4$. Eliminating y between the two equations and then solving for x, we get

$$2x^2 = 2x + 4$$

$$x^2 - x - 2 = 0$$

$$(x - 2)(x + 1) = 0$$

$$x = 2 \qquad x = -1$$

The corresponding values of y are 8 and 2, respectively; so the points of intersection are $(-1, 2)$ and $(2, 8)$. Making use of these points, we draw the parabola and the line. Since the point $(0, 2)$ satisfies the inequality $y \geq 2x^2$, the desired region is above and including the parabola. The graph of the linear inequality $y \leq 2x + 4$ is to the right of and including the line. The shaded region in Figure 16 is the graph we want.

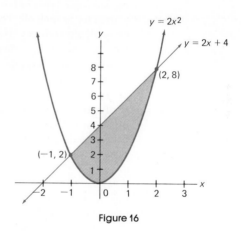

Figure 16

In Problems 19–22, graph the solution set of each system of inequalities.

19. $y \leq 4x - x^2$
$y \leq x$
$x \geq 0 \qquad y \geq 0$

20. $y \geq 2^x$
$y \leq 8$
$x \geq 0$

21. $y \leq \log_2 x$
$x \leq 8$
$y \geq 0$

22. $x^2 + y^2 \geq 9$
$0 \leq x \leq 3$
$0 \leq y \leq 3$

MISCELLANEOUS PROBLEMS

In Problems 23 and 24, graph the solution set and label the coordinates of the vertices.

23. $4x + y \leq 8$
$x - y \geq -2$
$x \geq 0 \qquad y \geq 0$

24. $2x + 5y \leq 20$
$x - 2y \leq 1$
$x \geq 0$

25. Find the maximum and minimum values of $P = 2x - y$ subject to the inequalities of Problem 23.

26. Find the maximum and minimum values of $P = 3x - 2y$ subject to the inequalities of Problem 24.

27. Find the maximum and minimum values of $x + y$ (if they exist) subject to the following inequalities.

$$x - y \geq -1$$
$$x - 2y \leq 5$$
$$3x + y \leq 10$$
$$x \geq 0 \qquad y \geq 0$$

28. Find the maximum and minimum values of $x - 2y$ (if they exist) subject to the inequalities of Problem 27.

29. A company makes a single product on two production lines, A and B. A labor force of 900 hours per week is available, and weekly running costs shall not exceed $1500. It takes 4 hours to produce one item on production line A and 3 hours on production line B. The cost per item is $5 on line A and $6 on line B. Find the largest number of items that can be produced in one week.

30. An oil refinery has a maximum production of 2000 barrels of oil per day. It produces two types of oil; type A, which is used for gasoline and type B, which is used for heating oil. There is a requirement that at least 300 barrels of type B be produced each day. If the profit is $3 a barrel for type A and $2 a barrel for type B, find the maximum profit per day.

31. A manufacturer of trailers wishes to determine how many camper units and how many house trailers she should produce in order to make the best possible use of her resources. She has 42 units of wood, 56 worker-weeks of labor and 16 units of aluminum. (Assume that all other needed resources are available and have no effect on her decision.) The amount of each resource needed to produce each camper and each trailer is given below.

	Wood	Worker-weeks	Aluminum
Per camper	3	7	3
Per trailer	6	7	1

If the manufacturer realizes a profit of $600 on a camper and $800 on a trailer, what should be her production in order to maximize her profit?

32. A shoemaker has a supply of 100 square feet of type A leather which is used for soles and 600 square feet of type B leather used for the rest of the shoe. The average shoe uses $\frac{1}{4}$ square feet of type A leather and 1 square foot of type B leather. The average boot uses $\frac{1}{4}$ square feet and 3 square feet of types A and B leather, respectively. If shoes sell at $40 a pair and boots at $60 a pair, find the maximum income.

33. Suppose that the minimum monthly requirements for one person are 60 units of carbohydrates, 40 units of protein, and 35 units of fat. Two foods A and B contain the following numbers of units of the three diet components per pound.

	Carbohydrates	Protein	Fat
A	5	3	5
B	2	2	1

If food A costs $3.00 a pound and food B costs $1.40 a pound, how many pounds of each should a person purchase per month to minimize the cost?

34. A grain farmer has 100 acres available for sowing oats and wheat. The seed oats costs $5 per acre and the seed wheat costs $8 per acre. The labor costs are $20 per acre for oats and $12 per acre for wheat. The farmer expects an income from oats of $220 per acre and from wheat of $250 per acre. How many acres of each crop should he sow to maximize his profit, if he does not wish to spend more than $620 for seed and $1800 for labor?

35. Sketch the polygon with vertices $(0, 3)$, $(4, 7)$, $(3, 0)$, and $(2, 4)$ taken in cyclic order. Then, find the maximum value of $|y - 2x| + y + x$ on this polygon.

36. **TEASER** If $P = (a, b)$ is a point in the plane, then $tP = (ta, tb)$. Thus, if $0 \le t \le 1$, the set of points of the form $tP + (1 - t)Q$ is just the line segment PQ (see point-of-division formula, page 142). It follows that a set A is convex if whenever P and Q are in A, all points of the form $tp + (1 - t)Q, 0 \le t \le 1$ are also in A.

   (a) Let P, Q, and R be three fixed points in the plane and consider the set H of all points of the form $t_1 P + t_2 Q + t_3 R$, where the t's are nonnegative and $t_1 + t_2 + t_3 = 1$. Show that H is convex.

   (b) Let P, Q, R, and S be four fixed points in the plane and consider the set K of all points of the form $t_1 P + t_2 Q + t_3 R + t_4 S$ where the t's are non-negative and $t_1 + t_2 + t_3 + t_4 = 1$. Show that K is convex.

   (c) Describe the sets H and K geometrically.

Chapter Summary

Two systems of equations are **equivalent** if they have the same solutions. Three elementary operations (multiplying an equation by a nonzero constant, interchanging two equations, and adding a multiple of one equation to another) lead to equivalent systems. Use of these operations allows us to transform a system of linear equations to **triangular form** and then to solve the system by **back substitution** or to show it is **inconsistent.**

A **matrix** is a rectangular array of numbers. In solving a system of linear equations, it is efficient to work with just the **matrix of the system.** We solve the system by transforming its matrix to triangular form using the three operations mentioned above.

Addition and multiplication are defined for matrices with resulting algebraic rules. Matrices behave much like numbers, with the exception that the commutative law for multiplication fails. Even the notion of **multiplicative inverse** has meaning, though the process for finding such an inverse is lengthy.

Associated with every square matrix is a symbol called its **determinant.** For 2 × 2 and 3 × 3 matrices, the value of the determinant can be found by using certain arrow diagrams. For higher order cases, the value of a determinant is found by expanding it in terms of the elements of a row (or column) and their **minors** (determinants whose order is lower than the given determinant by 1). In doing this, it is helpful to know what happens to the determinant of a matrix when any of the three elementary operations are applied to it. **Cramer's rule** provides a direct way of solving a system of n equations in n unknowns using determinants.

The graph of a **linear inequality** in x and y is a **half-plane** (closed or open according as the inequality sign does or does not include the equal sign). The graph of a **system of linear inequalities** is the intersection of the half-planes corresponding to the separate inequalities. Such a graph is always **polygonal** and **convex** but may be **bounded** or **unbounded**. A **linear programming problem** asks us to find the maximum (or minimum) of a linear function (such as $2x + 5y$) subject to a system of linear inequalities called **constraints.** The maximum (or minimum) always occurs at a **vertex,** that is, at a corner point of the graph of the inequality constraints.

Chapter Review Problem Set

In Problems 1–5, solve each system or show that there is no solution.

1. $3x + y = 12$
   $2x - y = -2$

2. $x - 3y + 2z = 5$
   $2y - z = 1$
   $z = 3$

3. $2x - y + 3z = 10$
   $y - 2z = 4$
   $-2y + 4z = 8$

4. $x - 3y + 2z = -5$
   $4x + y - z = 4$
   $5x + 11y - 8z = 20$

5. $x + 2y + 3z + 4w = 20$
   $y - 4z + w = 6$
   $z + 2w = 4$
   $2z + 4w = 8$

6. Evaluate each expression for

$$A = \begin{bmatrix} 1 & 1 & 1 \\ 3 & 1 & -1 \\ 2 & 2 & -1 \end{bmatrix} \quad \text{and} \quad B = \begin{bmatrix} -3 & 6 & -4 \\ 2 & -3 & 5 \\ 1 & 9 & 2 \end{bmatrix}$$

(a) $2A + B$ (b) $A - 2B$ (c) AB (d) BA

7. Find A^{-1} for the matrix A of Problem 6.

8. Consider the system

$$x + y + z = 4$$
$$3x + y - z = -4$$
$$2x + 2y - z = -1$$

Write this system in matrix form and then use the result of Problem 7 to solve it.

Evaluate the determinants in Problems 9–13.

9. $\begin{vmatrix} -2 & 5 \\ 2 & -6 \end{vmatrix}$

10. $\begin{vmatrix} 2 & 3 \\ -4 & -6 \end{vmatrix}$

11. $\begin{vmatrix} -2 & 1 & 4 \\ 0 & 5 & -1 \\ 4 & 0 & 3 \end{vmatrix}$
12. $\begin{vmatrix} 1 & -2 & 3 \\ 4 & 1 & 5 \\ 7 & -5 & 14 \end{vmatrix}$

13. $\begin{vmatrix} 6 & 0 & 0 & 0 \\ -1 & 3 & 1 & 0 \\ 3 & 2 & 3 & 2 \\ 4 & 5 & 1 & -4 \end{vmatrix}$

14. Use Cramer's rule to solve the following system.

$$2x + y - z = -4$$
$$x - 3y - 2z = -1$$
$$3x + 2y + 3z = 11$$

In Problems 15 and 16, graph the solution set of the given system. On the graph, label the coordinates of the vertices.

15. $x + y \leq 7$
$3x + y \leq 15$
$x \geq 0 \quad y \geq 0$

16. $x - 2y + 4 \geq 0$
$x + y - 11 \geq 0$
$x \geq 0 \quad y \geq 0$

17. Find the maximum value of the function $P = x + 2y$ subject to the inequalities of Problem 15.

18. Find the minimum value of the function $P = x + 2y$ subject to the inequalities of Problem 16.

19. A certain company has 100 employees, some of whom get $4 an hour, others $5, and the rest $8. Half as many make $8 an hour as $5 an hour. If the total paid out in hourly wages is $544, find the number of employees who make $8 an hour.

20. A tailor has 110 yards of cotton material and 160 yards of woolen material. It takes $1\frac{1}{2}$ yards of cotton and 1 yard of wool to make a suit, while a dress requires 1 yard of cotton and 2 yards of wool. If a suit sells for $100 and a dress for $80, how many of each should the tailor make to maximize the total income?

*Method consists entirely in
properly ordering and
arranging things to which
we should pay attention.*
                    René Descartes

CHAPTER 10

Sequences
and Mathematical
Induction

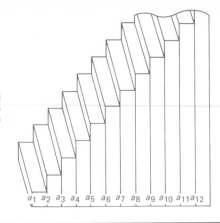

Jacob's Stairway

In the skyscraper that Jacob owns, there is a stairway going from ground level to the very top. The first step is 8 inches high. After that each step is 9 inches above the previous one. How high above the ground is the 800th step?

10-1 Arithmetic Sequences

The word *sequence* is commonly used in ordinary language. For example, your history teacher may talk about a sequence of events that led to World War II (for instance, the Versailles Treaty, world depression, Hitler's ascendancy, Munich Agreement). What characterizes this sequence is the notion of one event following another in a definite order. There is a first event, a second event, a third event, and so on. We might even give them labels.

E_1: Versailles Treaty
E_2: World depression
E_3: Hitler's ascendancy
E_4: Munich Agreement

We use a similar notation for what are called number sequences.

NUMBER SEQUENCES

A **number sequence** is an arrangement of numbers in which there is a first number, a second number, a third number, and so on. A typical example is the sequence of step heights (above the ground) in our opening panel. We may write this sequence as

$$a_1, a_2, a_3, a_4, \ldots$$

Then

$$a_1 = 8$$

$$a_2 = 8 + 1(9) = 17$$

$$a_3 = 8 + 2(9) = 26$$

$$a_4 = 8 + 3(9) = 35$$
$$\vdots$$

If you study the pattern carefully, you will see that

$$a_{800} = 8 + 799(9) = 7199$$

The 800th step of Jacob's stairway is 7199 inches (almost 600 feet) above the ground.

Note that a_3 stands for the third term, a_4 for the fourth term, and so on. The subscript indicates the position of the term in the sequence. For the general term—that is, the nth term—we use the symbol a_n. Three dots indicate that the sequence continues according to the pattern established.

There is another way to describe a number sequence. A **number sequence** is a function whose domain is the set of positive integers. This means it is a rule that associates with each positive integer n a definite number a_n. In conformity with Chapter 2, we could use the notation $a(n)$, but tradition dictates that we use a_n instead. We usually specify functions by giving formulas; this is true of sequences also.

Consider the sequence

$$b_1, b_2, b_3, b_4, \ldots$$

where b_n is given by the formula

$$b_n = \frac{1}{2n - 1}$$

Then

$$b_1 = \frac{1}{2 \cdot 1 - 1} = 1$$

$$b_2 = \frac{1}{2 \cdot 2 - 1} = \frac{1}{3}$$

$$b_3 = \frac{1}{2 \cdot 3 - 1} = \frac{1}{5}$$

$$b_4 = \frac{1}{2 \cdot 4 - 1} = \frac{1}{7}$$
$$\vdots$$

Of the two number sequences mentioned so far, the first interests us most. It is an example of an arithmetic sequence.

ARITHMETIC SEQUENCES

Consider the following number sequences. When you see a pattern, fill in the boxes.

(a) 5, 9, 13, 17, □, □, . . .

(b) 2, 2.5, 3, 3.5, □, □, . . .

(c) 8, 5, 2, −1, □, □, . . .

What is it that these three sequences have in common? Simply this: In each case, you can get a term by adding a fixed number to the preceding term. In (a), you add 4 each time, in (b) you add 0.5, and in (c), you add −3. Such sequences are called **arithmetic sequences.** If we denote such a sequence by $a_1, a_2, a_3, \ldots$, it satisfies the recursion formula

$$a_n = a_{n-1} + d$$

where d is a fixed number called the **common difference.**

We have used the phrase recursion formula. A **recursion formula** relates the nth term of a sequence to the term just before it, namely, the $(n - 1)$st term. If we denote the nth term of our three sequences by a_n, b_n, and c_n, respectively, then

$$a_n = a_{n-1} + 4$$

$$b_n = b_{n-1} + .5$$

$$c_n = c_{n-1} - 3$$

Of more use than a recursion formula is an explicit formula. An **explicit formula** relates the nth term directly to the subscript n. Figure 1 shows us how to find such a formula for an arithmetic sequence. Notice that the number of d's to be added to a_1 is one less than the subscript n (you would add 13d's to get a_{14}). This means that

$$a_n = a_1 + (n - 1)d$$

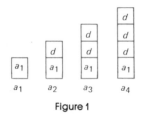

Figure 1

Now we can give explicit formulas for each of the sequences (a), (b), and (c) above.

$$a_n = 5 + (n - 1)4 = 1 + 4n$$

$$b_n = 2 + (n - 1)(.5) = 1.5 + .5n$$

$$c_n = 8 + (n - 1)(-3) = 11 - 3n$$

ARITHMETIC SEQUENCES AND LINEAR FUNCTIONS

We have said that a sequence is a function whose domain is the set of positive integers. Functions are best visualized by drawing their graphs. Consider the sequence b_n discussed above; its explicit formula is

$$b_n = 1.5 + .5n$$

Its graph is shown in Figure 2.

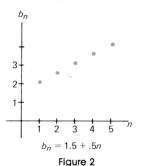

$b_n = 1.5 + .5n$

Figure 2

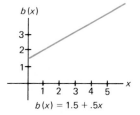

$b(x) = 1.5 + .5x$

Figure 3

Even a cursory look at this graph suggests that the points lie along a straight line. Consider now the function

$$b(x) = 1.5 + 0.5x = 0.5x + 1.5$$

where x is allowed to be any real number. This is a linear function, being of the form $mx + b$ (see Section 2-2). Its graph is a straight line (Figure 3), and its values at $x = 1, 2, 3, \ldots$ are equal to $b_1, b_2, b_3, \ldots$.

The relationship illustrated above between an arithmetic sequence and a linear function holds in general. An arithmetic sequence is just a linear function whose domain has been restricted to the positive integers.

SUMS OF ARITHMETIC SEQUENCES

There is an old story about Carl Gauss that aptly illustrates the next idea. We are not sure if the story is true, but if not, it should be.

When Gauss was about 10 years old, he was admitted to an arithmetic class. To keep the class busy, the teacher often assigned long addition problems. One day he asked his students to add the numbers from 1 to 100. Hardly had he made the assignment when young Gauss laid his slate on the teacher's desk with the answer 5050 written on it.

Here is how Gauss probably thought about the problem.

$$1 + 2 + \ldots + 49 + 50 + 51 + 52 + \ldots + 99 + 100$$

Each of the indicated pairs has 101 as its sum and there are 50 such pairs. Thus the answer is $50(101) = 5050$. For a 10-year-old boy, that is good thinking.

Gauss's trick works perfectly well on any arithmetic sequence where we want to add an even number of terms. And there is a slight modification that works whether the number of terms to be added is even or odd.

Suppose $a_1, a_2, a_3, \ldots$ is an arithmetic sequence and let

$$A_n = a_1 + a_2 + a_3 + \cdots + a_{n-1} + a_n$$

Write this sum twice, once forwards and once backwards, and then add.

$$
\begin{aligned}
A_n &= a_1 + a_2 + \cdots + a_{n-1} + a_n \\
\underline{A_n} &= \underline{a_n + a_{n-1} + \cdots + a_2 + a_1} \\
2A_n &= (a_1 + a_n) + (a_2 + a_{n-1}) + \cdots + (a_2 + a_{n-1}) + (a_1 + a_n)
\end{aligned}
$$

Each group on the right has the same sum, namely, $a_1 + a_n$. For example,

$$a_2 + a_{n-1} = a_1 + d + a_n - d = a_1 + a_n$$

There are n such groups and so

$$2A_n = n(a_1 + a_n)$$

$$A_n = \frac{n}{2}(a_1 + a_n)$$

We call this the **sum formula** for an arithmetic sequence. You can remember this sum as being n times the average term $(a_1 + a_n)/2$.

Here is how we apply this formula to Gauss's problem. We want the sum of 100 terms of the sequence $1, 2, 3, \ldots$, that is, we want A_{100}. Here $n = 100$, $a_1 = 1$, and $a_n = 100$. Therefore

$$A_{100} = \frac{100}{2}(1 + 100) = 50(101) = 5050$$

For a second example, suppose we want to add the first 350 odd numbers, that is, the first 350 terms of the sequence $1, 3, 5, \ldots$. We can calculate the 350th odd number from the formula $a_n = a_1 + (n - 1)d$.

$$a_{350} = 1 + (349)2 = 699$$

Then we use the sum formula with $n = 350$.

$$A_{350} = 1 + 3 + 5 + \cdots + 699$$

$$= \frac{350}{2}(1 + 699) = 122{,}500$$

SIGMA NOTATION

There is a convenient shorthand that is frequently employed in connection with sums. The first letter of the word *sum* is *s*; the Greek letter for *S* is Σ(sigma). We use Σ in mathematics to stand for the operation of summation. In particular,

$$\sum_{i=1}^{n} a_i = a_1 + a_2 + a_3 + \cdots + a_n$$

The symbol $i = 1$ underneath the sigma tells where to start adding the terms a_i and the n at the top tells where to stop. Thus

$$\sum_{i=1}^{4} a_i = a_1 + a_2 + a_3 + a_4$$

$$\sum_{i=3}^{7} b_i = b_3 + b_4 + b_5 + b_6 + b_7$$

$$\sum_{i=1}^{5} i^2 = 1^2 + 2^2 + 3^2 + 4^2 + 5^2$$

$$\sum_{i=1}^{30} 3i = 3 + 6 + 9 + \cdots + 90$$

If $a_1, a_2, a_3, \ldots$ is an *arithmetic sequence,* then the sum formula previously derived may be written

$$\sum_{i=1}^{n} a_i = \frac{n}{2}(a_1 + a_n)$$

Problem Set 10-1

1. Look for a pattern in each of the following sequences and use it to fill in the boxes. Which of these sequences is arithmetic?
   (a) 2, 7, 12, 17, 22, □, □, . . .
   (b) 2, 6, 11, 17, 24, □, □, . . .
   (c) 5, 3, 1, −1, −3, □, □, . . .
   (d) 1, 2, 4, 8, 16, □, □, . . .
   (e) 1, 4, 9, 16, 25, □, □, . . .

2. Follow the instructions of Problem 1 for the following sequences.
   (a) 5, 5.4, 5.8, 6.2, 6.6, □, □, . . .
   (b) 7, 5, 3, 1, −1, □, □, . . .
   (c) $0, \frac{1}{2}, \frac{2}{3}, \frac{3}{4}, \frac{4}{5},$ □, □, . . .
   (d) 2, 3, 5, 7, 11, □, □, . . .
   (e) 1, −2, 4, −8, 16, □, □, . . .

EXAMPLE A (Using Recursion and Explicit Formulas) Find a_6 if a_n is given by (a) the recursion formula $a_n = 2a_{n-1} + 1$ with $a_1 = \frac{1}{2}$; (b) the explicit formula $a_n = n^2 - 3n + 1$.

Solution. (a) The great disadvantage in using a recursion formula is that we must first calculate all terms preceding the one we want.

$$a_1 = \frac{1}{2}$$
$$a_2 = 2a_1 + 1 = 2(\tfrac{1}{2}) + 1 = 2$$
$$a_3 = 2a_2 + 1 = 2(2) + 1 = 5$$
$$a_4 = 2a_3 + 1 = 2(5) + 1 = 11$$
$$a_5 = 2a_4 + 1 = 2(11) + 1 = 23$$
$$a_6 = 2a_5 + 1 = 2(23) + 1 = 47$$

(b) An explicit formula allows us to calculate any term directly.

$$a_6 = 6^2 - 3 \cdot 6 + 1 = 19$$

Calculate a_6 in each of the following.

3. $a_n = \dfrac{n}{n^2 + 1}$

4. $a_n = (a_{n-1})^2, \; a_1 = 2$

5. $a_n = 3a_{n-1} - 6, \; a_1 = 2$

6. $a_n = n^2 + 4n - 16$

7. $a_n = n^3 - 6n^2 + 5n$

8. $a_n = (-2)^n - 3n$

EXAMPLE B (Arithmetic Sequences). The sequence

$$4, 7, 10, 13, 16, \ldots$$

is arithmetic. Find an explicit formula for the nth term a_n and use it to calculate a_{81}.

Solution. First we note that the common difference is $d = 3$. Thus

$$a_n = a_1 + (n - 1)d = 4 + (n - 1)3$$

and

$$a_{81} = 4 + (80)3 = 244$$

Find an explicit formula for a_n for each of the following arithmetic sequences. Then calculate a_{51}.

9. $-1, 5, 11, 17, \ldots$
10. $4, 6, 8, 10, \ldots$
11. $2, 2.3, 2.6, 2.9, \ldots$
12. $4, 4.2, 4.4, 4.6, \ldots$
13. $28, 24, 20, 16, \ldots$
14. $4, 3.8, 3.6, 3.4, \ldots$
15. $a_1 = 5$ and $a_{40} = 24.5$
16. $a_1 = 6$ and $a_{30} = -52$

EXAMPLE C (Sums of Arithmetic Sequences) Find the sum of the first 81 terms of the arithmetic sequence

$$4, 7, 10, 13, 16, \ldots$$

Solution. Recall that if $A_n = a_1 + a_2 + \cdots + a_n$, then $A_n = (n/2)(a_1 + a_n)$. Thus $A_{81} = \frac{81}{2}(a_1 + a_{81})$. Now a_{81} was calculated in Example B to be 244. Therefore,

$$A_{81} = \frac{81}{2}(4 + 244) = 10{,}044$$

17. Calculate the sum of the first 51 terms of the sequence in Problem 9.
18. Find the sum of the first 51 terms of the sequence in Problem 10.
19. Calculate each sum.
    (a) $2 + 4 + 6 + \cdots + 200$
    (b) $1 + 3 + 5 + \cdots + 199$
    (c) $3 + 6 + 9 + \cdots + 198$
    Hint: Before using the sum formula, you have to determine n. In part (a), n is 100 since we are adding the doubles of the integers from 1 to 100.
20. Calculate each sum.
    (a) $4 + 8 + 12 + \cdots + 100$
    (b) $10 + 15 + 20 + \cdots + 200$
    (c) $6 + 9 + 12 + \cdots + 72$
21. The bottom rung of a tapered ladder is 30 centimeters long and the top rung is

Figure 4

15 centimeters long (Figure 4). If there are 17 rungs, how many centimeters of rung material are needed to make the ladder, assuming no waste?

22. A clock strikes once at 1:00, twice at 2:00, and so on. How many times does it strike between 10:30 A.M. on Monday and 10:30 P.M. on Tuesday?

23. How many multiples of 9 are there between 200 and 300? Find their sum.

24. If Ronnie is paid $10 on January 1, $20 on January 2, $30 on January 3, and so on, how much does he earn during January?

25. Calculate each sum.

(a) $\displaystyle\sum_{i=2}^{6} i^2$ (b) $\displaystyle\sum_{i=1}^{4} \frac{2}{i}$

(c) $\displaystyle\sum_{i=1}^{100} (3i + 2)$ (d) $\displaystyle\sum_{i=2}^{100} (2i - 3)$

26. Calculate each sum.

(a) $\displaystyle\sum_{i=1}^{6} 2^i$ (b) $\displaystyle\sum_{i=1}^{5} (i^2 - 2i)$

(c) $\displaystyle\sum_{i=1}^{101} (2i - 6)$ (d) $\displaystyle\sum_{i=3}^{102} (3i + 5)$

27. Write in sigma notation.
(a) $b_3 + b_4 + \cdots + b_{20}$
(b) $1^2 + 2^2 + \cdots + 19^2$
(c) $1 + \dfrac{1}{2} + \dfrac{1}{3} + \cdots + \dfrac{1}{n}$

28. Write in sigma notation.
(a) $a_6 + a_7 + a_8 + \cdots + a_{70}$
(b) $2^3 + 3^3 + 4^3 + \cdots + 100^3$
(c) $1 + 3 + 5 + 7 + \cdots + 99$

MISCELLANEOUS PROBLEMS

Calculate a_5 in Problems 29–32.

29. $a_n = \dfrac{2n + 1}{n^3}$ 30. $a_n = n^2 + 4n - 6$

31. $a_n = n - (a_{n-1})^2,\ a_1 = 2$ 32. $a_n = \frac{3}{2}a_{n-1} + 4,\ a_1 = \frac{8}{3}$

33. Find each of the following for the arithmetic sequence 20, 19.25, 18.5, 17.75, . . .
(a) The common difference d.
(b) The 51st term.
(c) The sum of the first 51 terms.

34. Let a_n be an arithmetic sequence with $a_{19} = 42$ and $a_{39} = 54$. Find (a) d, (b) a_1, (c) a_{14}, and (d) k, given that $a_k = 64.2$.

35. Calculate the sum $6 + 6.8 + 7.6 + \cdots + 37.2 + 38$.

36. If 15, a, b, c, d, 24, . . . is an arithmetic sequence, find a, b, c, and d.

37. Let a_n be an arithmetic sequence and, as usual, let $A_n = a_1 + a_2 + \cdots + a_n$. If $A_5 = 50$ and $A_{20} = 650$, find A_{15}.

38. Find the sum of all multiples of 7 between 300 and 450.
39. Calculate each sum.

(a) $\displaystyle\sum_{i=1}^{100} (2i + 1)$

(b) $\displaystyle\sum_{i=1}^{100} \left(\frac{i}{i+1} - \frac{i-1}{i} \right)$

(c) $\displaystyle\sum_{i=1}^{100} \left(\frac{1}{i(i+1)} - \frac{1}{(i+1)(i+2)} \right)$

Hint: Write out the first three or four terms to see a pattern.

40. Write in sigma notation.
    (a) $c_3 + c_4 + c_5 + \cdots + c_{112}$
    (b) $4^2 + 5^2 + 6^2 + \cdots + 104^2$
    (c) $35 + 40 + 45 + \cdots + 185$

41. At a club meeting with 300 people present, everyone shook hands with every other person exactly once. How many handshakes were there? *Hint:* Person A shook hands with how many people, person B shook hands with how many people not already counted, and so on.

42. Mary learned 20 new French words on January 1, 24 new French words on January 2, 28 new French words on January 3, and so on, through January 31. By how much did she increase her French vocabulary during January?

43. A pile of logs has 70 logs in the bottom layer, 69 logs in the second layer, and so on to the top layer with 10 logs. How many logs are in the pile?

44. Calculate the following sum.
$$-1^2 + 2^2 - 3^2 + 4^2 - 5^2 + 6^2 - \cdots - 99^2 + 100^2$$

Hint: Group in a clever way.

45. Calculate the following sum.

$$\tfrac{1}{2} + (\tfrac{1}{3} + \tfrac{2}{3}) + (\tfrac{1}{4} + \tfrac{2}{4} + \tfrac{3}{4}) + \cdots + (\tfrac{1}{100} + \tfrac{2}{100} + \tfrac{3}{100} + \cdots + \tfrac{99}{100})$$

46. Show that the sum of n consecutive integers plus n^2 is equal to the sum of the next n consecutive integers.

47. Approximately how long is the playing groove in a $33\frac{1}{3}$-rpm record that takes 18 minutes to play if the groove starts 6 inches from the center and ends 3 inches from the center? To approximate, assume that each revolution produces a groove that is circular.

48. **TEASER** Find the sum of all the digits in the integers from 1 to 999,999.

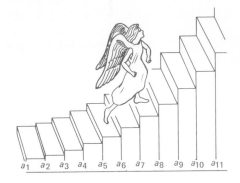

Jacob's Golden Staircase

In his dreams, Jacob saw a golden staircase with angels walking up and down. The first step was 8 inches high but after that each step was 5/4 as high above the ground as the previous one. How high above the ground was the 800th step?

a_1 a_2 a_3 a_4 a_5 a_6 a_7 a_8 a_9 a_{10} a_{11}

10-2 Geometric Sequences

The staircase of Jacob's dream is most certainly one for angels, not for people. The 800th step actually stands 3.4×10^{73} miles high. By way of comparison, it is 9.3×10^6 miles to the sun and 2.5×10^{13} miles to Alpha Centauri, our nearest star beyond the sun. You might say the golden staircase reaches to heaven.

To see how to calculate the height of the 800th step, notice the pattern of heights for the first few steps and then generalize.

$$a_1 = 8$$

$$a_2 = 8\left(\frac{5}{4}\right)$$

$$a_3 = 8\left(\frac{5}{4}\right)^2$$

$$a_4 = 8\left(\frac{5}{4}\right)^3$$

$$\vdots$$

$$a_{800} = 8\left(\frac{5}{4}\right)^{799}$$

With a pocket calculator, it is easy to calculate $8(\frac{5}{4})^{799}$ and then change this number of inches to miles; the result is the figure given above.

FORMULAS

In the sequence above, each term was $\frac{5}{4}$ times the preceding one. You should be able to find a similar pattern in each of the following sequences. When you do, fill in the boxes.

(a) 3, 6, 12, 24, □, □, . . .

(b) 12, 4, $\frac{4}{3}$, $\frac{4}{9}$, □, □, . . .

(c) .6, 6, 60, 600, □, □, . . .

The common feature of these three sequences is that in each case, you can get a term by multiplying the preceding term by a fixed number. In sequence (a), you multiply by 2; in (b), by $\frac{1}{3}$; and in (c), by 10. We call such sequences **geometric sequences.** Thus a geometric sequence $a_1, a_2, a_3, \ldots$ satisfies the recursion formula

$$a_n = ra_{n-1}$$

where r is a fixed number called the **common ratio.**

To obtain the corresponding explicit formula, note that

$$a_2 = ra_1$$
$$a_3 = ra_2 = r(ra_1) = r^2a_1$$
$$a_4 = ra_3 = r(r^2a_1) = r^3a_1$$

In each case, the exponent on r is one less than the subscript on a. Thus

$$a_n = a_1r^{n-1}$$

From this, we can get explicit formulas for each of the sequences (a), (b), and (c) on page 419.

$$a_n = 3 \cdot 2^{n-1}$$

$$b_n = 12\left(\frac{1}{3}\right)^{n-1}$$

$$c_n = (.6)(10)^{n-1}$$

GEOMETRIC SEQUENCES AND EXPONENTIAL FUNCTIONS

Let us consider sequence (b) once more; its explicit formula is

$$b_n = 12\left(\frac{1}{3}\right)^{n-1} = 36\left(\frac{1}{3}\right)^n$$

We have graphed this sequence (Figure 5) and also the exponential function

$$b(x) = 36\left(\frac{1}{3}\right)^x$$

in Figure 6. (See Section 3-2 for a discussion of exponential functions.) It should be clear that the sequence b_n is the function $b(x)$ with its domain restricted to the positive integers.

What we have observed in this example is true in general. A geometric sequence is simply an exponential function with its domain restricted to the positive integers.

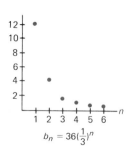

$b_n = 36(\frac{1}{3})^n$

Figure 5

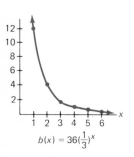

$b(x) = 36(\frac{1}{3})^x$

Figure 6

SUMS OF GEOMETRIC SEQUENCES

There is an old legend about geometric sequences and chessboards. When the king of Persia learned to play chess, he was so enchanted with the game that he determined to reward the inventor, a man named Sessa. Calling Sessa to the palace, the king promised to fulfill any request he might make. With an air of modesty, wily Sessa asked for one grain of wheat for the first square of the chessboard, two for the second, four for the third, and so on (Figure 7). The king was amused at such an odd request; nevertheless, he called a servant, told him to get a bag of wheat, and start counting. To the king's surprise, it soon became apparent that Sessa's request could never be fulfilled. The world's total production of wheat for a whole century would not be sufficient.

Sessa was really asking for

$$1 + 2 + 2^2 + 2^3 + \cdots + 2^{63}$$

grains of wheat, the sum of the first 64 terms of the geometric sequence 1, 2, 4, 8, ... We are going to develop a formula for this sum and for all others that arise from adding the terms of a geometric sequence.

Let $a_1, a_2, a_3, \ldots$ be a geometric sequence with ratio $r \neq 1$. As usual, let

$$A_n = a_1 + a_2 + a_3 + \cdots + a_n$$

which can be written

$$A_n = a_1 + a_1r + a_1r^2 + \cdots + a_1r^{n-1}$$

Now multiply A_n by r, subtract the result from A_n, and use a little algebra to solve for A_n. We obtain

$$A_n = a_1 + a_1r + a_1r^2 + \cdots + a_1r^{n-1}$$
$$\underline{rA_n = \qquad a_1r + a_1r^2 + \cdots + a_1r^{n-1} + a_1r^n}$$
$$A_n - rA_n = a_1 + 0 + 0 + \cdots + 0 - a_1r^n$$
$$A_n(1 - r) = a_1(1 - r^n)$$

$$\boxed{A_n = \frac{a_1(1 - r^n)}{1 - r} \qquad r \neq 1}$$

In sigma notation, this is

$$\sum_{i=1}^{n} a_i = \frac{a_1(1 - r^n)}{1 - r} \qquad r \neq 1$$

In the case where $r = 1$,

$$\sum_{i=1}^{n} a_i = a_1 + a_1 + \cdots + a_1 = na_1$$

Applying the sum formula to Sessa's problem (using $n = 64$, $a_1 = 1$, and $r = 2$), we get

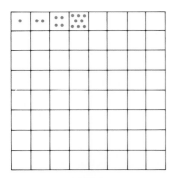

Figure 7

$$A_{64} = \frac{1(1 - 2^{64})}{1 - 2} = 2^{64} - 1$$

Ignoring the -1 and using the approximation $2^{10} \approx 1000$ gives

$$A_{64} \approx 2^{64} \approx 2^4(1000)^6 = 1.6 \times 10^{19}$$

If you do this problem on a calculator, you will get $A_{64} \approx 1.845 \times 10^{19}$. Thus if a bushel of wheat has one million grains, A_{64} grains would amount to more than 1.8×10^{13}, or 18 trillion, bushels. That exceeds the world's total production of wheat in 1 century.

THE SUM OF THE WHOLE SEQUENCE

Is it possible to add infinitely many numbers? Do the following sums make sense?

$$\frac{1}{2} + \frac{1}{4} + \frac{1}{8} + \frac{1}{16} + \cdots$$

$$1 + 3 + 9 + 27 + \cdots$$

Questions like this have intrigued great thinkers since Zeno first introduced his famous paradoxes of the infinite over 2000 years ago. We now show that we can make sense out of the first of these two sums but not the second.

Consider a string of length 1 kilometer. We may imagine cutting it into infinitely many pieces, as indicated in Figure 8.

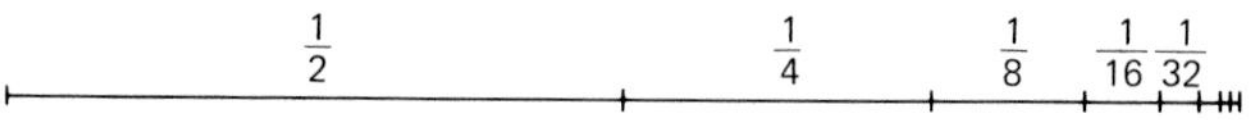

Figure 8

Since these pieces together make a string of length 1, it seems natural to say

$$\frac{1}{2} + \frac{1}{4} + \frac{1}{8} + \frac{1}{16} + \cdots = 1$$

Let us look at it another way. The sum of the first n terms of the geometric sequence $\frac{1}{2}, \frac{1}{4}, \frac{1}{8}, \frac{1}{16}, \ldots$ is given by

$$A_n = \frac{\frac{1}{2}[1 - (\frac{1}{2})^n]}{1 - \frac{1}{2}} = 1 - \left(\frac{1}{2}\right)^n$$

As n gets larger and larger (tends to infinity), $(\frac{1}{2})^n$ gets smaller and smaller (approaches 0). Thus A_n tends to 1 as n tends to infinity. We therefore say that 1 is the sum of *all* the terms of this sequence.

Now consider any geometric sequence with ratio r satisfying $|r| < 1$. We claim that when n gets large, r^n approaches 0. (As evidence, try calculating $(.99)^{100}$, $(.99)^{1000}$, and $(.99)^{10,000}$ on your calculator.) Thus, as n gets large,

$$A_n = \frac{a_1(1 - r^n)}{1 - r}$$

approaches $a_1/(1 - r)$. We write

$$\sum_{i=1}^{\infty} a_i = \frac{a_1}{1 - r} \qquad |r| < 1$$

For an important use of this formula in a familiar context, see the example in the problem set.

We emphasize that what we have just done is valid if $|r| < 1$. There is no way to make sense out of adding all the terms of a geometric sequence if $|r| \geq 1$.

Problem Set 10-2

1. Fill in the boxes.
   (a) $\frac{1}{2}$, 1, 2, 4, □, □, . . .
   (b) 8, 4, 2, 1, □, □, . . .
   (c) .3, .03, .003, .0003, □, □, . . .

2. Fill in the boxes.
   (a) 1, 3, 9, 27, □, □, . . .
   (b) 27, 9, 3, 1, □, □, . . .
   (c) .2, .02, .002, .0002, □, □, . . .

3. Determine r for each of the sequences in Problem 1 and write an explicit formula for the nth term.

4. Write a formula for the nth term of each sequence in Problem 2.

5. Evaluate the 30th term of each sequence in Problem 1.

6. Evaluate the 20th term of each sequence in Problem 2.

7. Use the sum formula to find the sum of the first five terms of each sequence in Problem 1.

8. Use the sum formula to find the sum of the first five terms of each sequence in Problem 2.

9. Find the sum of the first 30 terms of each sequence in Problem 1.

10. Find the sum of the first 20 terms of each sequence in Problem 2.

11. A certain culture of bacteria doubles every week. If there are 100 bacteria now, how many will there be after 10 full weeks?

12. A water lily grows so rapidly that each day it covers twice the area it covered the day before. At the end of 20 days, it completely covers a pond. If we start with two lilies, how long will it take to cover the same pond?

13. Johnny is paid $1 on January 1, $2 on January 2, $4 on January 3, and so on. Approximately how much will he earn during January?

14. If you were offered 1¢ today, 2¢ tomorrow, 4¢ the third day, and so on for 20 days or a lump sum of $10,000, which would you choose? Show why.

15. Calculate:

   (a) $\sum_{i=1}^{\infty} \left(\frac{1}{3}\right)^i$ (b) $\sum_{i=2}^{\infty} \left(\frac{2}{5}\right)^i$

16. Calculate:

(a) $\displaystyle\sum_{i=1}^{\infty} \left(\frac{2}{3}\right)^i$ (b) $\displaystyle\sum_{i=3}^{\infty} \left(\frac{1}{6}\right)^i$

17. A ball is dropped from a height of 10 feet. At each bounce, it rises to a height of one-half the previous height. How far will it travel altogether (up and down) by the time it comes to rest? *Hint:* Think of the total distance as being the sum of the "down" distances $(10 + 5 + \frac{5}{2} + \cdots)$ and the "up" distances $(5 + \frac{5}{2} + \frac{5}{4} + \cdots)$.

18. Do Problem 17 assuming the ball rises to $\frac{2}{3}$ its previous height at each bounce.

EXAMPLE (Repeating Decimals) Show that $.333\overline{3} \ldots$ and $.2323\overline{23} \ldots$ are rational numbers by using the methods of this section.

Solution.

$$.333\overline{3} = \frac{3}{10} + \frac{3}{100} + \frac{3}{1000} + \cdots$$

Thus we must add all the terms of an infinite geometric sequence with ratio $\frac{1}{10}$. Using the formula $a_1/(1 - r)$, we get

$$.333\overline{3} = \frac{\frac{3}{10}}{1 - \frac{1}{10}} = \frac{\frac{3}{10}}{\frac{9}{10}} = \frac{1}{3}$$

Similarly,

$$.2323\overline{23} = \frac{23}{100} + \frac{23}{10,000} + \frac{23}{1,000,000} + \cdots$$

$$= \frac{\frac{23}{100}}{1 - \frac{1}{100}} = \frac{\frac{23}{100}}{\frac{99}{100}} = \frac{23}{99}$$

Use this method to express each of the following as the ratio of two integers.

19. $.11\overline{1}$ 20. $.77\overline{7}$ 21. $.2525\overline{25}$

22. $.99\overline{9}$ 23. $1.2343\overline{34}$ 24. $.341\overline{41}$

MISCELLANEOUS PROBLEMS

25. If $a_n = 625(0.2)^{n-1}$, find a_1, a_2, a_3, a_4, and a_5.

26. Which of the following sequences are geometric, which are arithmetic, and which are neither?
    (a) 130, 65, 32.5, 16.25, . . .
    (b) $1, \frac{1}{2}, \frac{1}{3}, \frac{1}{4}, \ldots$
    (c) 100(1.05), 100(1.07), 100(1.09), 100(1.11), . . .
    (d) 100(1.05), 100(1.05)², 100(1.05)³, 100(1.05)⁴, . . .
    (e) 1, 3, 6, 10, . . .
    (f) 3, −6, 12, −24, . . .

27. Write an explicit formula for each geometric or arithmetic sequence in Problem 26.

28. Use the formula for the sum of an infinite geometric sequence to express each of the following repeating decimals as a ratio of two integers.
    (a) $.499999\ldots = .4\overline{9}$ (b) $.1234234234\ldots = .1\overline{234}$

☐ *Recall from Section 3-3 that if a sum of P dollars is invested today at a compound rate of i per conversion period, then the accumulated value after n periods is given by* $P(1 + i)^n$. *The sequence of accumulated values*

$$P(1 + i),\ P(1 + i)^2,\ P(1 + i)^3,\ P(1 + i)^4,\ \ldots$$

is geometric with ratio $1 + i$. *In problems 29–33, write a formula for the answer and then use a calculator to evaluate it.*

29. If $1 is put in the bank at 8 percent interest compounded annually, it will be worth $(1.08)^n$ dollars after n years. How much will $100 be worth after 10 years? When will the amount first exceed $250?

30. If $1 is put in the bank at 8 percent interest compounded quarterly, it will be worth $(1.02)^n$ dollars after n quarters. How much will $100 be worth after 10 years (40 quarters)? When will the amount first exceed $250?

31. Suppose Karen puts $100 in the bank today and $100 at the beginning of each of the following 9 years. If this money earns interest at 8 percent compounded annually, what will it be worth at the end of 10 years?

32. José makes 40 deposits of $25 each in a bank at intervals of three months, making the first deposit today. If money earns interest at 8 percent compounded quarterly, what will it all be worth at the end of 10 years (40 quarters)?

33. Suppose the government pumps an extra billion dollars into the economy. Assume that each business and individual saves 25 percent of its income and spends the rest, so that of the initial one billion dollars, 75 percent is re-spent by individuals and businesses. Of that amount, 75 percent is spent, and so forth. What is the total increase in spending due to the government action? (This is called the *multiplier effect* in economics.)

34. Given an arbitrary triangle of perimeter 10, a second triangle is formed by joining the midpoints of the first, a third triangle is formed by joining the midpoints of the second, and so on forever. Find the total length of all line segments in the resulting configuration.

35. Find the area of the painted region in Figure 9, which consists of an infinite sequence of 30°-60°-90° triangles.

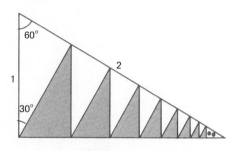

Figure 9

36. If the pattern in Figure 10 is continued indefinitely, what fraction of the area of the original square will be painted?

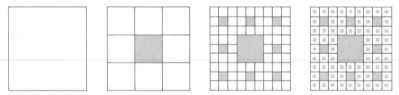

Figure 10

37. Expand in powers of 3; then evaluate P.

$$P = (1 + 3)(1 + 3^2)(1 + 3^4)(1 + 3^8)(1 + 3^{16})$$

38. By considering $S - rS$, find S.
    (a) $S = r + 2r^2 + 3r^3 + 4r^4 + \cdots$, $|r| < 1$.
    (b) $S = \frac{1}{3} + 2(\frac{1}{3})^2 + 3(\frac{1}{3})^3 + 4(\frac{1}{3})^4 + \cdots$

39. In a geometric sequence, the sum of the first two terms is 5 and the sum of the first six terms is 65. What is the sum of the first four terms? *Hint:* Call the first term a and the ratio r. Then $a + ar = 5$, so $a = 5/(1 + r)$. This allows you to write 65 in terms of r alone. Solve for r.

40. Imagine a huge maze with infinitely many adjoining cells, each having a square base 1 meter by 1 meter. The first cell has walls 1 meter high, the second cell has walls $\frac{1}{2}$ meter high, the third cell has walls $\frac{1}{4}$ meter high, and so on.
    (a) How much paint would it take to fill the maze with paint?
    (b) How much paint would it take to paint the floors of all the cells?
    (c) Explain this apparent contradiction.

41. Starting 100 miles apart, Tom and Joel ride toward each other on their bicycles, Tom going at 8 miles per hour and Joel at 12 miles per hour. Tom's dog, Corky, starts with Tom running toward Joel at 25 miles per hour. When Corky meets Joel, he immediately turns tail and heads back to Tom. Reaching Tom, Corky again turns tail and heads toward Joel, and so on. How far did Corky run by the time Tom and Joel met? This can be answered using geometric sequences, but, if you are clever, you will find a better way.

42. **TEASER** Sally walked 4 miles north, then 2 miles east, 1 mile south, $\frac{1}{2}$ mile west, $\frac{1}{4}$ mile north, and so on. If she continued this pattern indefinitely, how far from her initial point did she end?

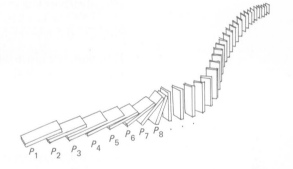

The Principle of Mathematical Induction

Let $P_1, P_2, P_3, \ldots$ be a sequence of statements with the following two properties:

1. P_1 is true.
2. The truth of P_k implies the truth of P_{k+1} $(P_k \Rightarrow P_{k+1})$.

Then the statement P_n is true for every positive integer n.

10-3 Mathematical Induction

The principle of mathematical induction deals with a sequence of statements. A **statement** is a sentence which is either true or false. In a sequence of statements, there is a statement corresponding to each positive integer. Here are four examples.

$$P_n: \quad \frac{1}{1 \cdot 2} + \frac{1}{2 \cdot 3} + \frac{1}{3 \cdot 4} + \cdots + \frac{1}{n(n+1)} = \frac{n}{n+1}$$

$Q_n: \quad n^2 - n + 41$ is a prime number.

$R_n: \quad (a+b)^n = a^n + b^n$

$$S_n: \quad 1 + 2 + 3 + \cdots + n = \frac{n^2 + n - 6}{2}$$

To be sure we understand the notation, let us write each of these statements for the case $n = 3$.

$$P_3: \quad \frac{1}{1 \cdot 2} + \frac{1}{2 \cdot 3} + \frac{1}{3 \cdot 4} = \frac{3}{4}$$

$Q_3: \quad 3^2 - 3 + 41$ is a prime number.

$R_3: \quad (a+b)^3 = a^3 + b^3$

$$S_3: \quad 1 + 2 + 3 = \frac{3^2 + 3 - 6}{2}$$

n	$n^2 - n + 41$
1	41
2	43
3	47
4	53
5	61
6	71
7	83
.	.
.	.
.	.
40	1601
41	$1681 = 41^2$

Figure 11

Of these, P_3 and Q_3 are true, while R_3 and S_3 are false; you should verify this fact. A careful study of these four sequences will indicate the wide range of behavior that sequences of statements can display.

While it certainly is not obvious, we claim that P_n is true for every positive integer n; we are going to prove it soon. Q_n is a well-known sequence. It was thought by some to be true for all n and, in fact, it is true for $n = 1, 2, 3, \ldots,$ 40 (see Figure 11). However, it fails for $n = 41$, a fact that allows us to make an important point. Establishing the truth of Q_n for a finite number of cases, no matter how many, does not prove its truth for *all n*. Sequences R_n and S_n are rather hopeless cases, since R_n is true only for $n = 1$ and S_n is never true.

PROOF BY MATHEMATICAL INDUCTION

How does one prove that something is true for all n? The tool uniquely designed for this purpose is the **principle of mathematical induction;** it was stated in our opening display. Let us use mathematical induction to show that

$$P_n: \quad \frac{1}{1 \cdot 2} + \frac{1}{2 \cdot 3} + \frac{1}{3 \cdot 4} + \cdots + \frac{1}{(n-1)n} + \frac{1}{n(n+1)} = \frac{n}{n+1}$$

is true for every positive integer n. There are two steps to the proof. We must show that

1. P_1 is true;
2. $P_k \Rightarrow P_{k+1}$; that is, the truth of P_k implies the truth of P_{k+1}.

The first step is easy. P_1 is just the statement

$$\frac{1}{1 \cdot 2} = \frac{1}{1+1}$$

which is clearly true.

To handle the second step ($P_k \Rightarrow P_{k+1}$), it is a good idea to write down the statements corresponding to P_k and P_{k+1} (at least on scratch paper). We get them by substituting k and $k + 1$ for n in the statement for P_n.

$$P_k: \quad \frac{1}{1 \cdot 2} + \frac{1}{2 \cdot 3} + \cdots + \frac{1}{(k-1)k} + \frac{1}{k(k+1)} = \frac{k}{k+1}$$

$$P_{k+1}: \quad \frac{1}{1 \cdot 2} + \frac{1}{2 \cdot 3} + \cdots + \frac{1}{k(k+1)} + \frac{1}{(k+1)(k+2)} = \frac{k+1}{k+2}$$

Notice that the left side of P_{k+1} is the same as that of P_k except for the addition of one more term, $1/(k+1)(k+2)$.

Suppose for the moment that P_k is true and consider how this assumption allows us to simplify the left side of P_{k+1}.

$$\left[\frac{1}{1 \cdot 2} + \frac{1}{2 \cdot 3} + \cdots + \frac{1}{k(k+1)} \right] + \frac{1}{(k+1)(k+2)} = \frac{k}{k+1} + \frac{1}{(k+1)(k+2)}$$

$$= \frac{k(k+2) + 1}{(k+1)(k+2)}$$

$$= \frac{(k+1)^2}{(k+1)(k+2)}$$

$$= \frac{k+1}{k+2}$$

If you read this chain of equalities from top to bottom, you will see that we have established the truth of P_{k+1} but under the *assumption that P_k is true*. That is, we have established that the truth of P_k implies the truth of P_{k+1}.

SOME COMMENTS ABOUT MATHEMATICAL INDUCTION

Students never have any trouble with the verification step (showing that P_1 is true). The inductive step (showing that $P_k \Rightarrow P_{k+1}$) is harder and more subtle. In that step, we do *not* prove that P_k or P_{k+1} is true, but rather that the truth of P_k implies the truth of P_{k+1}. For a vivid illustration of the difference, we point out that in the fourth example of our opening paragraph, the truth of S_k does imply the truth of S_{k+1} ($S_k \Rightarrow S_{k+1}$) and yet not a single statement in that sequence is true (see Problem 30). To put it another way, what $S_k \Rightarrow S_{k+1}$ means is that *if* S_k were true, then S_{k+1} would be true also. It is like saying that if spinach were ice cream, then kids would want two helpings at every meal.

Perhaps the dominoes in the opening display can help illuminate the idea. For all the dominoes to fall it is sufficient that

1. the first domino is pushed over;
2. if any domino falls (say the kth one), it pushes over the next one (the $(k + 1)$st one).

Figure 12 illustrates what happens to the dominoes in the four examples of our opening paragraph. Study them carefully.

Why They Fall and Why They Don't

$P_n : \dfrac{1}{1 \cdot 2} + \dfrac{1}{2 \cdot 3} + \cdots + \dfrac{1}{n(n + 1)} = \dfrac{n}{n + 1}$ P_1 is true $P_k \Rightarrow P_{k + 1}$	 $P_1 P_2 P_3 P_4 P_5 P_6 \cdots$ First domino is pushed over. Each falling domino pushes over the next one.
$Q_n : n^2 - n + 41$ is prime. $Q_1, Q_2, \ldots, Q_{40}$ are true. $Q_k \not\Rightarrow Q_{k + 1}$	 $Q_{35}\, Q_{36}\, Q_{37}\, Q_{38}\, Q_{39}\, Q_{40} \quad Q_{41} \quad Q_{42}$ First 40 dominoes are pushed over. 41st domino remains standing.
$R_n : (a + b)^n = a^n + b^n$ R_1 is true. $R_k \not\Rightarrow R_{k + 1}$	 $R_1 \quad R_2 \quad R_3 \quad R_4 \quad R_5 \quad R_6 \quad R_7$ First domino is pushed over but dominoes are spaced too far apart to push each other over.
$S_n : 1 + 2 + 3 + \ldots + n = \dfrac{n^2 + n - 6}{2}$ S_1 is false. $S_k \Rightarrow S_{k + 1}$	 $S_1 S_2 S_3 S_4 S_5$ Spacing is just right but no one can push over the first domino.

Figure 12

ANOTHER EXAMPLE

Consider the statement

$$P_n: \quad 1^2 + 2^2 + 3^2 + \cdots + n^2 = \frac{n(n + 1)(2n + 1)}{6}$$

We are going to prove that P_n is true for all n by mathematical induction. For $n = 1$, k, and $k + 1$, the statements P_n are:

$$P_1: \quad 1^2 = \frac{1(2)(3)}{6}$$

$$P_k: \quad 1^2 + 2^2 + 3^2 + \cdots + k^2 = \frac{k(k + 1)(2k + 1)}{6}$$

$$P_{k+1}: \quad 1^2 + 2^2 + 3^2 + \cdots + k^2 + (k + 1)^2 = \frac{(k + 1)(k + 2)(2k + 3)}{6}$$

Clearly P_1 is true.

Assuming that P_k is true, we can write the left side of P_{k+1} as shown below.

$$
\begin{aligned}
1^2 + 2^2 + 3^2 + \cdots + k^2 + (k + 1)^2 &= \frac{k(k + 1)(2k + 1)}{6} + (k + 1)^2 \\
&= \frac{(k + 1)[k(2k + 1) + 6(k + 1)]}{6} \\
&= \frac{(k + 1)(2k^2 + 7k + 6)}{6} \\
&= \frac{(k + 1)(k + 2)(2k + 3)}{6}
\end{aligned}
$$

Thus the truth of P_k does imply the truth of P_{k+1}. We conclude by mathematical induction that P_n is true for every positive integer n. Incidentally, the result just proved will be used in calculus.

Problem Set 10-3

In Problems 1–8, prove by mathematical induction that P_n is true for every positive integer n.

1. $P_n: \quad 1 + 2 + 3 + \cdots + n = \dfrac{n(n + 1)}{2}$

2. $P_n: \quad 1 + 3 + 5 + \cdots + (2n - 1) = n^2$

3. $P_n: \quad 3 + 7 + 11 + \cdots + (4n - 1) = n(2n + 1)$

4. $P_n: \quad 2 + 9 + 16 + \cdots + (7n - 5) = \dfrac{n(7n - 3)}{2}$

5. P_n: $1 \cdot 2 + 2 \cdot 3 + 3 \cdot 4 + \cdots + n(n + 1) = \frac{1}{3}n(n + 1)(n + 2)$

6. P_n: $\dfrac{1}{1 \cdot 3} + \dfrac{1}{3 \cdot 5} + \dfrac{1}{5 \cdot 7} + \cdots + \dfrac{1}{(2n - 1)(2n + 1)} = \dfrac{n}{2n + 1}$

7. P_n: $2 + 2^2 + 2^3 + \cdots + 2^n = 2(2^n - 1)$

8. P_n: $1^2 + 3^2 + 5^2 + \cdots + (2n - 1)^2 = \dfrac{n(2n - 1)(2n + 1)}{3}$

In Problems 9–18, tell what you can conclude from the information given about the sequence of statements. For example, if you are given that P_4 is true and that $P_k \Rightarrow P_{k+1}$ for any k, then you can conclude that P_n is true for every integer $n \geq 4$.

9. P_8 is true and $P_k \Rightarrow P_{k+1}$.

10. P_8 is not true and $P_k \Rightarrow P_{k+1}$.

11. P_1 is true but P_k does not imply P_{k+1}.

12. $P_1, P_2, \ldots, P_{1000}$ are all true.

13. P_1 is true and $P_k \Rightarrow P_{k+2}$.

14. P_{40} is true and $P_k \Rightarrow P_{k-1}$.

15. P_1 and P_2 are true; P_k and P_{k+1} together imply P_{k+2}.

16. P_1 and P_2 are true and $P_k \Rightarrow P_{k+2}$.

17. P_1 is true and $P_k \Rightarrow P_{4k}$.

18. P_1 is true, $P_k \Rightarrow P_{4k}$, and $P_k \Rightarrow P_{k-1}$.

EXAMPLE A (Mathematical Induction Applied to Inequalities)   Show that the following statement is true for every integer $n \geq 4$.

$$3^n > 2^n + 20$$

Solution.   Let P_n represent the given statement. You might check that P_1, P_2, and P_3 are false. However, that does not matter to us. What we need to do is to show that P_4 is true and that $P_k \Rightarrow P_{k+1}$ for any $k \geq 4$.

$$P_4: \quad 3^4 > 2^4 + 20$$
$$P_k: \quad 3^k > 2^k + 20$$
$$P_{k+1}: \quad 3^{k+1} > 2^{k+1} + 20$$

Clearly P_4 is true (81 is greater than 36). Next we assume P_k to be true (where $k \geq 4$) and seek to show that this would force P_{k+1} to be true. Working with the left side of P_{k+1} and using the assumption that $3^k > 2^k + 20$, we get

$$3^{k+1} = 3 \cdot 3^k > 3(2^k + 20) > 2(2^k + 20) = 2^{k+1} + 40 > 2^{k+1} + 20$$

Therefore P_{k+1} is true, provided P_k is true. We conclude that P_n is true for every integer $n \geq 4$.

In Problems 19–24, find the smallest positive integer n for which the given statement is true. Then prove that the statement is true for all integers greater than that smallest value.

19. $n + 5 < 2^n$

20. $3n \leq 3^n$

21. $\log_{10} n < n$ *Hint:* $k + 1 < 10k$.

22. $n^2 \leq 2^n$ *Hint:* $k^2 + 2k + 1 = k(k + 2 + 1/k) < k(k + k)$.

23. $(1 + x)^n \geq 1 + nx$, where $x \geq -1$

24. $|\sin nx| \leq |\sin x| \cdot n$ for all x

EXAMPLE B (Mathematical Induction and Divisibility)   Prove that the statement

$$P_n: \quad x - y \text{ is a factor of } x^n - y^n$$

is true for every positive integer n.

Solution.   Trivially, $x - y$ is a factor of $x - y$ since $x - y = 1(x - y)$; so P_1 is true. Now suppose that P_k is true, that is, that

$$x - y \text{ is a factor of } x^k - y^k$$

This means that there is a polynomial $Q(x, y)$ such that

$$x^k - y^k = Q(x, y)(x - y)$$

Using this assumption, we may write

$$x^{k+1} - y^{k+1} = x^{k+1} - x^k y + x^k y - y^{k+1}$$
$$= x^k(x - y) + y(x^k - y^k)$$
$$= x^k(x - y) + yQ(x, y)(x - y)$$
$$= [x^k + yQ(x, y)](x - y)$$

Thus $x - y$ is a factor of $x^{k+1} - y^{k+1}$. We have shown that $P_k \Rightarrow P_{k+1}$ and that P_1 is true; we therefore conclude that P_n is true for all n.

Use mathematical induction to prove that each of the following is true for every positive integer n.

25. $x + y$ is a factor of $x^{2n} - y^{2n}$. *Hint:* $x^{2k+2} - y^{2k+2} = x^{2k+2} - x^{2k}y^2 + x^{2k}y^2 - y^{2k+2}$.

26. $x + y$ is a factor of $x^{2n-1} + y^{2n-1}$.

27. $n^2 - n$ is even (that is, has 2 as a factor).

28. $n^3 - n$ is divisible by 6.

MISCELLANEOUS PROBLEMS

29. The following four formulas can all be proved by mathematical induction. We proved (b) in the text; you prove the others.
    (a) $1 + 2 + 3 + \cdots + n = \frac{1}{2}n(n + 1)$
    (b) $1^2 + 2^2 + 3^2 + \cdots + n^2 = \frac{1}{6}n(n + 1)(2n + 1)$
    (c) $1^3 + 2^3 + 3^3 + \cdots + n^3 = \frac{1}{4}n^2(n + 1)^2$
    (d) $1^4 + 2^4 + 3^4 + \cdots + n^4 = \frac{1}{30}n(n + 1)(6n^3 + 9n^2 + n - 1)$
    From (a) and (c), another interesting formula follows, namely,

$$1^3 + 2^3 + 3^3 + \cdots + n^3 = (1 + 2 + 3 + \cdots + n)^2$$

30. Consider the statement

$$S_n: \quad 1 + 2 + 3 + \cdots + n = \frac{n^2 + n - 6}{2}$$

Show that:

(a) $S_k \Rightarrow S_{k+1}$ for $k \geq 1$;

(b) S_n is not true for any positive integer n.

31. Use the results of Problem 29 to evaluate each of the following.

(a) $\displaystyle\sum_{k=1}^{100} (3k + 1)$ (b) $\displaystyle\sum_{k=1}^{10} (k^2 - 3k)$

(c) $\displaystyle\sum_{k=1}^{10} (k^3 + 3k^2 + 3k + 1)$ (d) $\displaystyle\sum_{k=1}^{n} (6k^2 + 2k)$

32. In a popular song titled *The Twelve Days of Christmas*, my true love gave me 1 gift on the first day, $(2 + 1)$ gifts on the second day, $(3 + 2 + 1)$ gifts on the third day, and so on.

(a) How many gifts did I get all together during the 12 days?

(b) How many gifts would I get all together in a Christmas that had n days?

33. Prove that for $n \geq 2$,

$$\left(1 - \frac{1}{4}\right)\left(1 - \frac{1}{9}\right)\left(1 - \frac{1}{16}\right)\cdots\left(1 - \frac{1}{n^2}\right) = \frac{n + 1}{2n}$$

34. Prove that for $n \geq 1$.

$$\frac{1}{\sqrt{1}} + \frac{1}{\sqrt{2}} + \frac{1}{\sqrt{3}} + \cdots + \frac{1}{\sqrt{n}} < 2\sqrt{n}..$$

35. Prove that for $n \geq 3$.

$$\frac{1}{n + 1} + \frac{1}{n + 2} + \frac{1}{n + 3} + \cdots + \frac{1}{2n} > \frac{3}{5}$$

$n = 4$

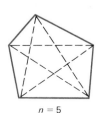

$n = 5$

Figure 13

36. Prove that the number of diagonals in an n-sided convex polygon is $n(n - 3)/2$ for $n \geq 3$. The diagrams in Figure 13 show the situation for $n = 4$ and $n = 5$.

37. Prove that the sum of the measures of the interior angles in an n-sided polygon (without holes or self-intersections) is $(n - 2)180°$. What is the sum of the measures of the exterior angles of such a polygon?

38. Let $f_1 = 1$, $f_2 = 1$, and $f_{n+2} = f_{n+1} + f_n$ for $n \geq 1$ determine the **Fibonacci sequence** and let $F_n = f_1 + f_2 + f_3 + \cdots + f_n$. Prove by mathematical induction that $F_n = f_{n+2} - 1$ for all n.

39. For the Fibonacci sequence of Problem 38, prove that for $n \geq 1$,

$$f_1^2 + f_2^2 + f_3^2 + \cdots + f_n^2 = f_n f_{n+1}$$

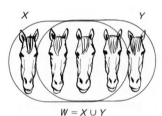

$W = X \cup Y$

Figure 14

40. What is wrong with the following argument?

Theorem. All horses in the world have the same color.

Proof. Let P_n be the statement: All the horses in any set of n horses are identically colored. Certainly P_1 is true. Suppose that P_k is true, that is, that all the horses in any set of k horses are identically colored. Let W be any set of $k + 1$ horses. Now we may think of W as the union of two overlapping sets X and Y, each with k horses. (The situation for $k = 4$ is shown in Figure 14). By assumption, the horses in X are identically colored and the horses in Y are identically colored. Since X and Y overlap, all the horses in $X \cup Y$ must be identically colored. We conclude that P_n is true for all n. Thus the set of all horses in the world (some finite number) have the same color.

41. Let $a_0 = 0$, $a_1 = 1$, and $a_{n+2} = (a_{n+1} + a_n)/2$ for $n \geq 0$. Prove that for $n \geq 0$.

$$a_n = \tfrac{2}{3}[1 - (-\tfrac{1}{2})^n]$$

Hint: In the inductive step, show that P_k and P_{k+1} together imply P_{k+2}.

42. **TEASER** Let f_n be the Fibonacci sequence of Problem 38. Use mathematical induction (as in the hint to Problem 41) to prove that

$$f_n = \frac{1}{\sqrt{5}}\left[\left(\frac{1 + \sqrt{5}}{2}\right)^n - \left(\frac{1 - \sqrt{5}}{2}\right)^n\right]$$

Pascal's Triangle

```
                    1
                1       1
            1       2       1
        1       3       3       1
    1       4       6       4       1
 1      5      10      10      5       1
1     6     15      20     15     6      1
```

Blaise Pascal (1623–1662)

10-4   The Binomial Formula

The triangular array of numbers in our opening display has intrigued professional mathematicians and amateurs for centuries. It is named for the gifted mathematician-philosopher Blaise Pascal, who wrote an early treatise on the subject. Notice that the array has 1's down the sides and that any interior term can be obtained by adding the two neighbors immediately above. For example, $15 = 10 + 5$, as the dotted triangle suggests.

Of the hundreds of interesting facts about this array of numbers that are known, it is its intimate relation to binomial powers that is most important. Consider the following results.

$$(x + y)^0 = \qquad\qquad 1$$

$$(x + y)^1 = \qquad\qquad x + y$$

$$(x + y)^2 = \qquad\qquad x^2 + 2xy + y^2$$

$$(x + y)^3 = \qquad\qquad x^3 + 3x^2y + 3xy^2 + y^3$$

$$(x + y)^4 = \qquad x^4 + 4x^3y + 6x^2y^2 + 4xy^3 + y^4$$

$$(x + y)^5 = x^5 + 5x^4y + 10x^3y^2 + 10x^2y^3 + 5xy^4 + y^5$$

The coefficients are just the numbers in Pascal's triangle.

Suppose that we wanted to find the coefficient of $x^{18}y^{32}$ in $(x + y)^{50}$. One way to proceed would be to generate Pascal's triangle, one row at a time, until we got to row 50. It would be much better to have an explicit formula for each coefficient, a matter to which we now turn.

THE BINOMIAL COEFFICIENT $\binom{n}{r}$

We define the symbol $\binom{n}{r}$, to be called a **binomial coefficient,** by

$$\binom{n}{r} = \frac{n(n - 1)(n - 2) \cdots (n - r + 2)(n - r + 1)}{r(r - 1)(r - 2) \cdots 3 \cdot 2 \cdot 1}$$

Here n and r are positive integers with $1 \le r \le n$. For example,

$$\binom{5}{3} = \frac{5 \cdot 4 \cdot 3}{3 \cdot 2 \cdot 1} = 10 \qquad \binom{5}{4} = \frac{5 \cdot 4 \cdot 3 \cdot 2}{4 \cdot 3 \cdot 2 \cdot 1} = 5 \qquad \binom{5}{5} = \frac{5 \cdot 4 \cdot 3 \cdot 2 \cdot 1}{5 \cdot 4 \cdot 3 \cdot 2 \cdot 1} = 1$$

the last three coefficients in $(x + y)^5$. You can remember the definition of $\binom{n}{r}$ by noting that both numerator and denominator are products of r integers starting at n and r, respectively, with the factors steadily decreasing by 1. Thus

$$\binom{12}{9} = \frac{12 \cdot 11 \cdot 10 \cdot 9 \cdot 8 \cdot 7 \cdot 6 \cdot 5 \cdot 4}{9 \cdot 8 \cdot 7 \cdot 6 \cdot 5 \cdot 4 \cdot 3 \cdot 2 \cdot 1} = \frac{\overset{2}{\cancel{12}} \cdot 11 \cdot 10}{\cancel{3} \cdot \cancel{2} \cdot 1} = 220$$

In evaluating one of these symbols, do not fail to cancel all common factors in the denominator and numerator first. This is always possible and the result is always an integer.

The product in the denominator of $\binom{n}{r}$ occurs often enough in mathematics to be given a special name, **r factorial,** and a special symbol, $r!$

$$r! = r(r - 1)(r - 2) \cdots 3 \cdot 2 \cdot 1$$

For example, $3! = 3 \cdot 2 \cdot 1 = 6$, $4! = 4 \cdot 3 \cdot 2 \cdot 1 = 24$, and $50!$ is so large that you need a calculator to evaluate it (some calculators have a special factorial key).

Consider again the definition of $\binom{n}{r}$. If we multiply both numerator and denominator by $(n - r)!$, we obtain

$$\binom{n}{r} = \frac{n!}{r!(n - r)!}$$

a result that we shall need shortly. Also we define $0! = 1$. Then in order that the last boxed formula hold true for $r = 0$, we must have $\binom{n}{0} = 1$. Finally, that formula implies that

$$\binom{n}{r} = \binom{n}{n - r}$$

For example, $\binom{6}{4} = \binom{6}{2}$ and $\binom{50}{45} = \binom{50}{5}$. This corresponds to the fact that Pascal's triangle is symmetric about its vertical median.

THE BINOMIAL FORMULA

Here is the result toward which we have been aiming.

$$(x + y)^n = \binom{n}{0}x^n y^0 + \binom{n}{1}x^{n-1}y^1 + \cdots + \binom{n}{n-1}x^1 y^{n-1} + \binom{n}{n}x^0 y^n$$

It is called the **binomial formula** and plays a significant role in many parts of mathematics, including calculus.

Suppose we wish to expand $(x + y)^6$. The coefficients are

$$\binom{6}{0} \quad \binom{6}{1} \quad \binom{6}{2} \quad \binom{6}{3} \quad \binom{6}{4} \quad \binom{6}{5} \quad \binom{6}{6}$$

Now, without calculating, we know $\binom{6}{0} = \binom{6}{6} = 1$; it is almost as trivial that $\binom{6}{1} = \binom{6}{5} = 6$. Further

$$\binom{6}{2} = \binom{6}{4} = \frac{6 \cdot 5}{2 \cdot 1} = 15 \qquad \binom{6}{3} = \frac{6 \cdot 5 \cdot 4}{3 \cdot 2 \cdot 1} = 20$$

Thus

$$(x + y)^6 = \binom{6}{0}x^6 + \binom{6}{1}x^5 y^1 + \binom{6}{2}x^4 y^2$$

$$+ \binom{6}{3}x^3y^3 + \binom{6}{4}x^2y^4 + \binom{6}{5}x^1y^5 + \binom{6}{0}y^6$$

$$= x^6 + 6x^5y + 15x^4y^2 + 20x^3y^3 + 15x^2y^4 + 6xy^5 + y^6$$

This same result applies to the expansion of $(2a - b^2)^6$. We simply think of $2a$ as x and $-b^2$ as y. Thus

$$[2a + (-b^2)]^6 = (2a)^6 + 6(2a)^5(-b^2) + 15(2a)^4(-b^2)^2$$

$$+ 20(2a)^3(-b^2)^3 + 15(2a)^2(-b^2)^4 + 6(2a)(-b^2)^5 + (-b^2)^6$$

$$= 64a^6 - 192a^5b^2 + 240a^4b^4 - 160a^3b^6$$

$$+ 60a^2b^8 - 12ab^{10} + b^{12}$$

As another illustration, consider finding the first four terms in the expansion of $(x + y)^{100}$. We will need to know that

$$\binom{100}{0} = 1 \qquad \binom{100}{1} = 100 \qquad \binom{100}{2} = \frac{100 \cdot 99}{2 \cdot 1} = 4950$$

$$\binom{100}{3} = \frac{100 \cdot 99 \cdot 98}{3 \cdot 2 \cdot 1} = 161{,}700$$

Thus

$$(x + y)^{100} = x^{100} + 100x^{99}y + 4950x^{98}y^2 + 161{,}700x^{97}y^3 + \cdots$$

PROOF OF THE BINOMIAL FORMULA

It does not take much faith to believe that the binomial formula is valid for $(x + y)^{10}$, since the result is easily checked. Would it be valid for $(x + y)^{10{,}000}$? A rigorous proof should erase any lingering doubts.

The binomial formula is a sequence of statements, one for each positive integer n. This calls for proof by mathematical induction. We note first that

$$(x + y)^1 = \binom{1}{0}x + \binom{1}{1}y = x + y$$

so the first statement is true. Next, suppose the kth statement to be true; that is, suppose

$$(x + y)^k = \binom{k}{0}x^k + \binom{k}{1}x^{k-1}y + \binom{k}{2}x^{k-2}y^2 + \cdots + \binom{k}{k}y^k$$

We must show that this supposition implies the $(k + 1)$st statement, namely, $(x + y)^{k+1}$

$$= \binom{k+1}{0}x^{k+1} + \binom{k+1}{1}x^k y + \binom{k+1}{2}x^{k-1}y^2 + \cdots + \binom{k+1}{k+1}y^{k+1}$$

To do this, we multiply both sides of the equality for $(x + y)^k$ by $x + y$. On the right side, we accomplish this by multiplying first by x, then by y, and adding the results as shown below. Then $(x + y)^{k+1}$ is equal to

$$\binom{k}{0}x^{k+1} + \boxed{\binom{k}{1}}x^k y + \boxed{\binom{k}{2}}x^{k-1}y^2 + \cdots + \boxed{\binom{k}{k}}xy^k$$

$$+ \boxed{\binom{k}{0}}x^k y + \boxed{\binom{k}{1}}x^{k-1}y^2 + \cdots + \boxed{\binom{k}{k-1}}xy^k + \binom{k}{k}y^{k+1}$$

The first and last coefficients have the right values, namely, 1. It remains to show that the coefficients in the rectangles have the required sum. For example, we want

$$\binom{k}{1} + \binom{k}{0} = \binom{k+1}{1} \qquad \binom{k}{2} + \binom{k}{1} = \binom{k+1}{2}$$

and, more generally, we want

$$\binom{k}{r} + \binom{k}{r-1} = \binom{k+1}{r}$$

If we can establish the boxed result, we will be done. Incidently, this result corresponds to the fact in Pascal's triangle that we get an interior term by adding the adjacent neighbors above it.

Here is the required demonstration.

$$\binom{k}{r} + \binom{k}{r-1} = \frac{k!}{r!(k-r)!} + \frac{k!}{(r-1)!(k-r+1)!}$$

$$= \frac{k!}{(r-1)!(k-r)!}\left[\frac{1}{r} + \frac{1}{k-r+1}\right]$$

$$= \frac{k!}{(r-1)!(k-r)!}\left[\frac{k+1}{r(k-r+1)}\right]$$

$$= \frac{(k+1)!}{r!(k-r+1)!} = \binom{k+1}{r}$$

Problem Set 10-4

Calculate each of the following.

1. $\binom{7}{2}$ 2. $\binom{8}{6}$ 3. $\binom{7}{4}$ 4. $\binom{10}{3}$

5. $\binom{20}{3}$ 6. $\binom{10}{8}$ ⚀ 7. $\binom{100}{5}$ ⚀ 8. $\binom{40}{10}$

9. $7!$ 10. $\dfrac{10!}{6!}$ ⚀ 11. $15!$ ⚀ 12. $20!$

In Problems 13–20, expand and simplify.

13. $(x + y)^3$ 14. $(x - y)^3$ 15. $(x - 2y)^3$ 16. $(3x + b)^3$

17. $(c^2 - 3d^3)^4$ 18. $(xy - 2z^2)^4$ 19. $(a + b)^7$ 20. $(2x - y)^7$

Write the first three terms of each expansion in Problems 21–26 in simplified form.

21. $(x + y)^{20}$ 22. $(x + y)^{30}$ 23. $\left(x + \dfrac{1}{x^5}\right)^{20}$

24. $\left(xy^2 + \dfrac{1}{y}\right)^{14}$ 25. $(a - b)^{50}$ 26. $(2a + 3b)^{40}$

EXAMPLE A (Finding a Specific Term of a Binomial Expansion)   Find the term in the expansion of $(2x + y^2)^{10}$ that involves y^{12}.

Solution.   This term will arise from raising y^2 to the 6th power. It is therefore

$$\binom{10}{6}(2x)^4(y^2)^6 = 210 \cdot 16x^4y^{12} = 3360x^4y^{12}$$

27. Find the term in the expansion of $(y^2 - z^3)^{10}$ that involves z^9.
28. Find the term in the expansion of $(3x - y^3)^{10}$ that involves y^{24}.
29. Find the term in the expansion of $(2a - b)^{12}$ that involves a^3.
30. Find the term in the expansion of $(x^2 - 2/x)^5$ that involves x^4.

EXAMPLE B (An Application to Compound Interest)   If \$100 is invested at 12 percent compounded monthly, it will accumulate to $100(1.01)^{12}$ dollars by the end of one year. Use the binomial formula to approximate this amount.

Solution.
$$100(1.01)^{12} = 100(1 + .01)^{12}$$

$$= 100\left[1 + 12(.01) + \frac{12 \cdot 11}{2}(.01)^2 + \frac{12 \cdot 11 \cdot 10}{6}(.01)^3 + \cdots\right]$$

$$= 100[1 + .12 + .0066 + .00022 + \cdots]$$

$$\approx 100(1.12682) \approx 112.86$$

This answer of \$112.68 is accurate to the nearest penny since the last nine terms of the expansion do not add up to as much as a penny.

In Problems 31–34 use the first three terms of a binomial expansion to find an approximate value of the given expression.

31. $20(1.02)^8$ 32. $100(1.002)^{20}$ 33. $500(1.005)^{20}$ 34. $200(1.04)^{10}$

35. Bacteria multiply in a certain medium so that by the end of k hours their number N is $N = 100(1.02)^k$. Approximate the number of bacteria after 20 hours.
36. Do Problem 35 assuming $N = 1000(1.01)^k$.

MISCELLANEOUS PROBLEMS

37. Expand and simplify.

    (a) $\left(x^2 - \dfrac{3}{x}\right)^4$

    (b) $\dfrac{(x + h)^5 - x^5}{h}$

38. Find and simplify the first three terms of $(a + 3b^2)^{10}$.

39. Find and simplify the term in the expansion of $(x - 2y^2)^8$ that involves y^6.

40. Find the term in the expanded and simplified form that does not involve h (a procedure important in calculus).

    (a) $\dfrac{(x + h)^{20} - x^{20}}{h}$

    (b) $\dfrac{3(x + h)^5 + 2(x + h)^3 - 3x^5 - 2x^3}{h}$

41. Given that i is the imaginary unit, calculate

    (a) $(2 + i)^5$;

    (b) $(1 - 2i)^4$.

42. Find the constant term in the expansion of $\left(x^3 - \dfrac{1}{2x^2}\right)^{10}$.

43. Without using a calculator, show that $(1.002)^{20} < 1.0408$.

44. Without using a calculator, find $(.998)^{12}$ correct to 5 decimal places.

45. By substituting certain values for x and y in the binomial formula, simplify

$$\binom{n}{0} + \binom{n}{1} + \binom{n}{2} + \cdots + \binom{n}{n-1} + \binom{n}{n}$$

46. Simplify

$$\binom{n}{0} - \binom{n}{1} + \binom{n}{2} - \binom{n}{3} + \cdots + (-1)^n\binom{n}{n}$$

47. In the expansion of the trinomial $(x + y + z)^n$, the coefficient of $x^r y^s z^t$ where $r + s + t = n$ is $\dfrac{n!}{r!s!t!}$.

    (a) Expand $(x + y + z)^3$.

    (b) Find the coefficient of the term $x^2 y^4 z$ in the expansion of $(2x + y + z)^7$.

48. Find the sum of all the coefficients in the expansion of the trinomial $(x + y + z)^n$.

49. Find a simple formula for $\displaystyle\sum_{k=0}^{n} \binom{n}{k} 2^k$.

50. Find the sum of the coefficients in $(4x^3 - x)^6$ after it is expanded and simplified. *Hint:* This is a simple problem when looked at the right way.

51. Let $P(x)$ be the nth degree polynomial defined by

$$P(x) = 1 + x + \frac{x(x-1)}{2!} + \frac{x(x-1)(x-2)}{3!} + \cdots + \frac{x(x-1)(x-2)\cdots(x-n+1)}{n!}$$

    Find a simple formula for each of the following.

    (a) $P(k)$, $k = 0, 1, 2, \ldots, n$.

    (b) $P(n + 1)$

    (c) $P(n + 2)$

52. **TEASER** Define $\dbinom{n}{k} = 0$ in the case $k > n$. Use the binomial expansion of $(1 + i)^n$ to find simple formulas for each of the following.

(a) $\dbinom{n}{1} - \dbinom{n}{3} + \dbinom{n}{5} - \dbinom{n}{7} + \cdots$

(b) $\dbinom{n}{0} - \dbinom{n}{2} + \dbinom{n}{4} - \dbinom{n}{6} + \cdots$

Chapter Summary

A **number sequence** $a_1, a_2, a_3, \ldots$ is a function that associates with each positive integer n a number a_n. Such a sequence may be described by an **explicit formula** (for instance, $a_n = 2n + 1$), by a **recursion formula** (for instance, $a_n = 3a_{n-1}$), or by giving enough terms so a pattern is evident (for instance, 1, 11, 21, 31, 41, . . .).

If any term in a sequence can be obtained by adding a fixed number d to the preceding term, we call it an **arithmetic sequence.** There are three key formulas associated with this type of sequence.

$$\text{Recursion formula:} \quad a_n = a_{n-1} + d$$

$$\text{Explicit formula:} \quad a_n = a_1 + (n - 1)d$$

$$\text{Sum formula:} \quad A_n = \frac{n}{2}(a_1 + a_n)$$

In the last formula, A_n represents

$$A_n = a_1 + a_2 + \cdots + a_n = \sum_{i=1}^{n} a_i$$

A **geometric sequence** is one in which any term results from multiplying the previous term by a fixed number r. The corresponding key formulas are

$$\text{Recursion formula:} \quad a_n = ra_{n-1}$$

$$\text{Explicit formula:} \quad a_n = a_1 r^{n-1}$$

$$\text{Sum formula:} \quad A_n = \frac{a_1(1 - r^n)}{1 - r}, \, r \neq 1$$

In the last formula, we may ask what happens as n grows larger and larger. If $|r| < 1$, the value of A_n gets closer and closer to $a_1/(1 - r)$, which we regard as the sum of *all* the terms of the sequence.

Often in mathematics, we wish to demonstrate that a whole **sequence of statements** P_n is true. For this, a powerful tool is the **principle of mathematical induction,** which asserts that if P_1 is true and if the truth of P_k implies the truth of P_{k+1}, then all the statements of the sequence are true.

After defining the binomial coefficient $\dbinom{n}{r}$ by

$$\dbinom{n}{r} = \frac{n(n - 1)(n - 2) \cdots (n - r + 1)}{r(r - 1)(r - 2) \cdots 3 \cdot 2 \cdot 1}$$

we obtain the **binomial formula**

$$(x + y)^n = x^n + \binom{n}{1}x^{n-1}y + \binom{n}{2}x^{n-2}y^2 + \cdots + \binom{n}{n-1}xy^{n-1} + y^n$$

which is proved by mathematical induction.

Chapter Review Problem Set

Problems 1–6 refer to the sequences below.

(a) 2, 5, 8, 11, 14, . . .

(b) 2, 6, 18, 54, . . .

(c) 2, 1.5, 1, 0.5, 0, . . .

(d) 2, 4, 6, 10, 16, . . .

(e) $2, \frac{2}{3}, \frac{2}{9}, \frac{2}{27}, \frac{2}{81}, \ldots$

1. Which of these sequences are arithmetic? Which are geometric?
2. Give a recursion formula for each of the sequences (a)–(c).
3. Give an explicit formula for sequences (a) and (b).
4. Find the sum of the first 67 terms of sequence (a).
5. Write a formula for the sum of the first 100 terms of sequence (b).
6. Find the sum of *all* the terms of sequence (e).
7. If $a_n = 3a_{n-1} - a_{n-2}$, $a_1 = 1$, and $a_2 = 2$, find a_6.
8. If $a_n = n^2 - n$, find $A_5 = \sum_{n=1}^{5} a_n$.
9. Calculate $2 + 4 + 6 + 8 + \cdots + 1000$.
10. Write $.55\overline{55}$ as a ratio of two integers.
11. If $100 is put in the bank today earning 8 percent interest compounded quarterly, write a formula for its value at the end of 12 years.
12. Show by mathematical induction that
    (a) $5 + 9 + \cdots + (4n + 1) = 2n^2 + 3n$;
    (b) $n! > 2^n$ when $n \geq 4$.
13. Suppose P_3 is true and $P_k \Rightarrow P_{k+3}$. What can we conclude about the sequence P_n?
14. Evaluate
    (a) $8!$ (b) $\binom{9}{6}$ (c) $\binom{40}{38}$
15. Find the first 4 terms in simplified form in the expansion of $(x + 2y)^{10}$.
16. Find the term involving a^3b^6 in the expansion of $(a - b^2)^6$.
17. Find $(1.002)^{20}$ accurate to 4 decimal places by using the binomial formula.

Appollonius' metric treatment of the conic sections—ellipses, hyperbolas, and parabolas—was one of the great mathematical achievements of antiquity. The importance of conic sections for pure and applied mathematics (for example, the orbits of the planets and of electrons in the hydrogen atom are conic sections) can hardly be overestimated.

Richard Courant
and Herbert Robbins

CHAPTER 11

Analytic Geometry

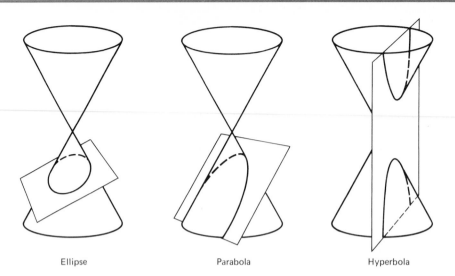

| Ellipse | Parabola | Hyperbola |

11-1 Parabolas

Analytic geometry could well be called algebraic geometry for it is the study of geometric concepts such as curves and surfaces by means of algebra. Among the most important curves are the conic sections, curves that are obtained by intersecting a cone of two nappes with a plane (see our opening display). We are especially interested in the three general cases of an ellipse, parabola, and hyperbola though we shall also consider certain limiting forms like a circle, two intersecting lines, and so on. We begin with the parabola, a curve already discussed in Section 2-3. Here we give a very general treatment based on the geometric definition given to us by the Greeks.

THE GEOMETRIC DEFINITION OF A PARABOLA

A parabola is the set of points P that are equidistant from a fixed line l (the directrix) and a fixed point F (the focus). In other words, a parabola is the set of points P in Figure 1 satisfying

$$d(P, L) = d(P, F)$$

Here, L is the point of l closest to P and, as usual, $d(A, B)$ denotes the distance between the points A and B.

A little thought convinces us that a parabola is a two-armed curve opening ever wider and symmetric with respect to the line through the focus perpendicular to the directrix. This line is called the **axis of symmetry** and the point

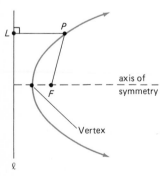

Figure 1

where this line intersects the parabola is called the **vertex**. Note that the vertex is the point of the parabola closest to the directrix.

THE EQUATION OF A PARABOLA

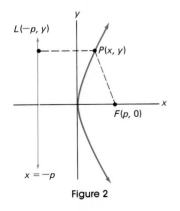

Figure 2

Place the parabola in the coordinate system so that its axis is the x-axis and its vertex is the origin (Figure 2). Let the focus be to the right of the origin, for example, at $(p, 0)$; then the directrix is the line $x = -p$. If $P(x, y)$ is any point on the curve, it must satisfy

$$d(P, F) = d(P, L)$$

which, because of the distance formula, assumes the form

$$\sqrt{(x - p)^2 + (y - 0)^2} = \sqrt{(x + p)^2 + (y - y)^2}$$

Since both sides are positive, this is equivalent to the result when both sides are squared.

$$x^2 - 2px + p^2 + y^2 = x^2 + 2px + p^2$$

This, in turn, simplifies to

$$y^2 = 4px$$

The final equation is called the **standard equation of the parabola.** It is easy to write, simple to graph, and has a form which is straightforward to interpret. For example, we may replace y by $-y$ without affecting the equation, which means that the graph is symmetric with respect to the x-axis. It crosses the x-axis at the origin which is the vertex. The positive number p measures the distance from the focus to the vertex.

The equation just derived has three other variants. If we interchange the roles of x and y (giving $x^2 = 4py$), we have the equation of a parabola that opens upward with the y-axis as its axis. The corresponding parabolas which open to the left and down have equations $y^2 = -4px$ and $x^2 = -4py$, respectively. All of this is summarized in Figure 3 on the next page.

What would be the equation of the parabola with vertex at the origin and focus at the point $(0, -3)$? This is a parabola of the fourth type in Figure 3; it turns down with $p = 3$. We conclude that its equation is $x^2 = -4(3)y = -12y$.

Conversely, suppose that we want to find the focus of the parabola with equation $y^2 = -2x$. Write the equation as $y^2 = -4 \cdot \frac{1}{2} \cdot x$, which is of the third type in Figure 3. We conclude that the parabola opens left, that $p = \frac{1}{2}$, and that the focus is at $(-\frac{1}{2}, 0)$.

In Section 2-3, we claimed that the graph of an equation of the form $y = ax^2$ was a parabola. Thus, for example, the graph of $y = \frac{1}{3}x^2$ should be a parabola. To confirm this, write the equation in the form $x^2 = 3y = 4 \cdot \frac{3}{4}y$. This is the equation of a parabola with vertex at the origin and opening up. Since $p = \frac{3}{4}$, its focus is at $(0, \frac{3}{4})$.

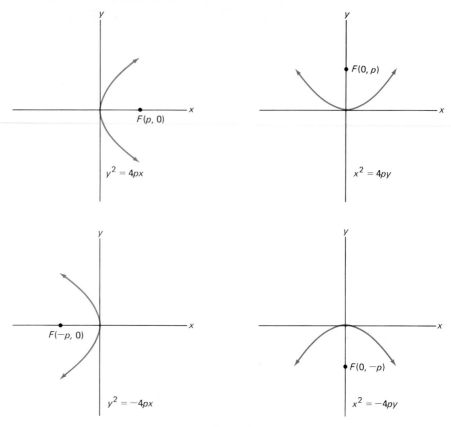

$y^2 = 4px$

$F(p, 0)$

$x^2 = 4py$

$F(0, p)$

$y^2 = -4px$

$F(-p, 0)$

$x^2 = -4py$

$F(0, -p)$

Figure 3

APPLICATIONS OF THE PARABOLA

Perhaps the most important property of the parabola is its *optical property*. Consider a cup-shaped mirror with a parabolic cross section (Figure 4). If a light source is placed at the focus, the resulting rays of light are reflected from the mirror in a beam in which all the rays are parallel to the axis (Figure 5). This fact is used in designing search lights. Conversely, if parallel light rays (as from a star) hit a parabolic mirror, they will be "focused" at the focus. This is the basis for the design of one type of reflecting telescope. The optical property of the parabola is usually demonstrated by means of calculus, but

Cross section of a parabolic mirror
with light source at focus

Figure 4

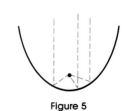

Figure 5

there is a way to do it that involves only geometry and algebra (see Problems 34 and 35).

In calculus, it is shown that the path of a projectile is a parabola and that the cables of a suspension bridge have a parabolic shape. We touch on these and other applications in the problem set.

Problem Set 11-1

Each equation below determines a parabola with vertex at the origin. In what direction does the parabola open?

1. $x^2 = 8y$ 2. $y^2 = -2x$ 3. $y^2 = 6x$
4. $x^2 = -3y$ 5. $3y^2 = -5x$ 6. $y = -2x^2$

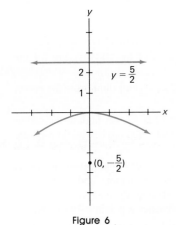

$y = \frac{5}{2}$

$(0, -\frac{5}{2})$

Figure 6

EXAMPLE A (Finding the Focus and Directrix)   Determine the focus and directrix of the parabola with equation $y = -\frac{1}{10}x^2$. Then sketch its graph.

Solution.   We first write the equation in standard form by solving for the quadratic term and then factoring out 4.

$$x^2 = -10y = -4(\tfrac{5}{2})y$$

This is the equation of a parabola with $p = \frac{5}{2}$; it opens downward. The focus is $(0, -\frac{5}{2})$ and the directrix is the line $y = \frac{5}{2}$. All this and the graph are shown in Figure 6.

In Problems 7–14, find p and then sketch the graph, showing the focus and directrix.

7. $x^2 = -8y$ 8. $y^2 = 3x$ 9. $y^2 = \frac{1}{2}x$
10. $y = -3x^2$ 11. $y = \frac{1}{2}x^2$ 12. $6x = -4y^2$
13. $9x = 4y^2$ 14. $x = .125y^2$
15. Determine the coordinates of two points on the parabola $y = 4x^2$ with y-coordinate 1.
16. Determine two points on the parabola $y^2 = 8x$ with the same x-coordinate as the focus.

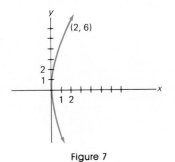

$(2, 6)$

Figure 7

EXAMPLE B (Parabolas with Side Conditions)   Find the equation of the parabola that has vertex at the origin, opens to the right, and passes through $(2, 6)$. See Figure 7.

Solution.   The equation has the form $y^2 = 4px$. Since $(2, 6)$ lies on the parabola, $6^2 = 4p(2)$, or $p = \frac{36}{8} = \frac{9}{2}$. Thus $y^2 = (4)(\frac{9}{2})x$, which simplifies to $y^2 = 18x$.

Find the equation of the parabola with vertex at $(0, 0)$ that satisfies the given conditions.

17. Opens up; goes through (2, 6).
18. Opens down; goes through (−2, −4).
19. Directrix is $y = 3$.
20. Focus is (−4, 0).
21. Goes through (1, 2) and (1, −2).
22. Goes through (1, 4) and (2, 16).

MISCELLANEOUS PROBLEMS

23. Find the focus and directrix of the parabola with equation $4x = -5y^2$.
24. Find the equation of the parabola with vertex at (0, 0) which in addition satisfies the following condition.
    (a) Its focus is at $(\frac{5}{2}, 0)$
    (b) It opens down and passes through (3, −10)
25. The chord of a parabola through the focus and perpendicular to the axis is called the **latus rectum** of the parabola. Find the length of the latus rectum for the parabola $4px = y^2$.
26. The chord of a parabola that is perpendicular to the axis and 1 unit from the vertex has length 3 units. How long is its latus rectum?
27. A door in the shape of a parabolic arch (Figure 8) is 12 feet high at the center and 5 feet wide at the base. A rectangular box 9 feet tall is to be slid through the door. What is the widest the box can be?
28. The path of a projectile fired from ground level is a parabola opening down. If the greatest height reached by the projectile is 100 meters and if its range (horizontal reach) is 800 meters, what is the horizontal distance from the point of firing to the point where the projectile first reaches a height of 64 meters?
29. The cables for the central span of a suspension bridge take the shape of a parabola as shown in Figure 9. If the towers are 800 meters apart and the cables are attached to them at points 400 meters above the floor of the bridge, how long is the vertical strut that is 100 meters from the tower? Assume the vertex of the parabola is on the floor of the bridge.
30. In Figure 10, AP is parallel to the x-axis and Q is the midpoint of AP. Find the equation of the path of $Q(x_1, y_1)$ as $P(x, y)$ moves along the path $x^2 = 4py$.

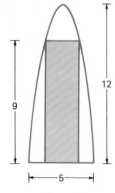

Figure 8

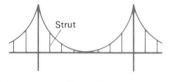

Figure 9

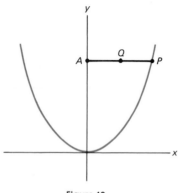

Figure 10

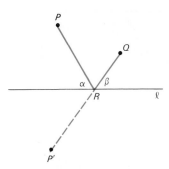

Figure 11

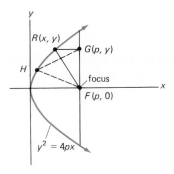

Figure 12

31. Suppose that a submarine has been ordered to follow a path that keeps it equidistant from a circular island of radius r and a straight shoreline that is $2p$ units from the edge of the island. Derive the equation of the submarine's path, assuming that the shoreline has equation $x = -p$ and that the center of the island is on the x-axis.

32. Show that there is no point $P(x, y)$ on the parabola $x^2 = 8y$ for which OP is perpendicular to PF, F being the focus of the parabola. *Hint:* Two lines with slopes m_1 and m_2 are perpendicular if and only if $m_1 m_2 = -1$.

33. An equilateral triangle is inscribed in the parabola $y^2 = 4px$ with one vertex at the origin. Find the length of a side of the triangle.

34. Consider a line l, two fixed points P and Q on the same side of l, and a (variable) point R on l. Use Figure 11 to show that the distance $\overline{PR} + \overline{RQ}$ is minimized precisely when $\alpha = \beta$. *Note:* Since a light ray is known to be reflected from a mirror l so that the angle of incidence equals the angle of reflection, we see that a light ray from P to l to Q picks the shortest path.

35. (Optical property of the parabola) Imagine the parabola $y^2 = 4px$ of Figure 12 to be a mirror with points F, R, G, and H as indicated and with RG parallel to the x-axis.
    (a) Show that $\overline{FR} + \overline{RG} = 2p$.
    (b) Show that $\overline{FH} + \overline{HG} > 2p$.

    Conclude from Problem 34 that a light ray from the focus to a parabolic mirror is reflected parallel to the axis of the parabola.

36. **TEASER** Consider the parabola $y = x^2$ (Figure 13). Let T_1 be the triangle with vertices on this parabola at a, c, and b with c midway between a and b. Let T_2 be the union of the two triangles with vertices on the parabola at a, d, c and c, e, b, respectively, with d midway between a and c and e midway between c and b. In a similar manner, let T_3 be the union of four triangles with vertices on the parabola, and so on.
    (a) Show that the area of T_1 is given by $A(T_1) = (b - a)^3/8$.
    (b) Show that $A(T_2) = A(T_1)/4$.
    (c) Find the area of the curved parabolic segment below the line PQ.

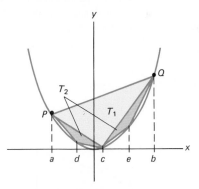

Figure 13

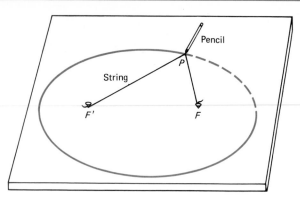

This shows the drawing of an ellipse. A string is tacked down at its ends by thumbtacks. A pencil pulls the string taut.

11-2   Ellipses

Our opening display suggests the geometric definition of the ellipse. An **ellipse** is the set of points P for which the sum of the distances from two fixed points F' and F is a constant. In other words, an ellipse is the set of points P in Figure 14 satisfying

$$d(P, F') + d(P, F) = 2a$$

for some positive constant a.

The two fixed points F' and F are called **foci** (plural of focus) and the point midway between the foci is the **center** of the ellipse. We call the line through the two foci the **major axis** of the ellipse; the line through the center and perpendicular to the major axis is its **minor axis**. Note that the ellipse is symmetric with respect to both its major and minor axes. Finally, the intersection of the ellipse with the major axis determines the two points A' and A, which are called **vertices**. All this is shown in Figure 14.

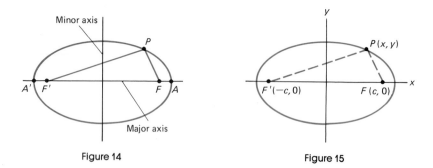

Figure 14

Figure 15

THE EQUATION OF AN ELLIPSE

Place the ellipse in the coordinate system so that its center is at the origin with the major axis along the x-axis. We may suppose the two foci F' and F to be located at $(-c, 0)$ and $(c, 0)$, where c is a positive constant (Figure 15). Then $d(P, F') + d(P, F) = 2a$ combined with the distance formula yields

$$\sqrt{(x + c)^2 + (y - 0)^2} + \sqrt{(x - c)^2 + (y - 0)^2} = 2a$$

or, equivalently,

$$\sqrt{(x + c)^2 + y^2} = 2a - \sqrt{(x - c)^2 + y^2}$$

After squaring both sides, we obtain

$$(x + c)^2 + y^2 = 4a^2 - 4a\sqrt{(x - c)^2 + y^2} + (x - c)^2 + y^2$$

and this in turn simplifies to

$$4cx - 4a^2 = -4a\sqrt{(x - c)^2 + y^2}$$

If we now divide both sides by 4 and square again, we get

$$(cx - a^2)^2 = a^2[(x - c)^2 + y^2]$$

$$c^2x^2 - 2a^2cx + a^4 = a^2[x^2 - 2cx + c^2 + y^2]$$

$$a^4 - a^2c^2 = a^2x^2 - c^2x^2 + a^2y^2$$

$$a^2(a^2 - c^2) = (a^2 - c^2)x^2 + a^2y^2$$

Finally, divide both sides by $a^2(a^2 - c^2)$ and interchange the two sides of the equation to obtain

$$\frac{x^2}{a^2} + \frac{y^2}{a^2 - c^2} = 1$$

It is clear from Figure 15 that $a > c$, so we may let $b^2 = a^2 - c^2$. This results in what we shall call the **standard equation of the ellipse**, namely,

$$\frac{x^2}{a^2} + \frac{y^2}{b^2} = 1$$

INTERPRETING a, b, AND c

We have used c to denote the distance from the center to a focus. If we apply the defining condition for the ellipse to the vertex A, we obtain

$$d(A, F') + d(A, F) = 2a$$

which implies (see Figure 14) that the distance between the two vertices is $2a$. For this reason, we refer to the number $2a$ as the **major diameter** of the ellipse. Since $b^2 + c^2 = a^2$, b and c are the legs of a right triangle with hypotenuse a and it follows that $2b$ is the **minor diameter** of the ellipse. All this is

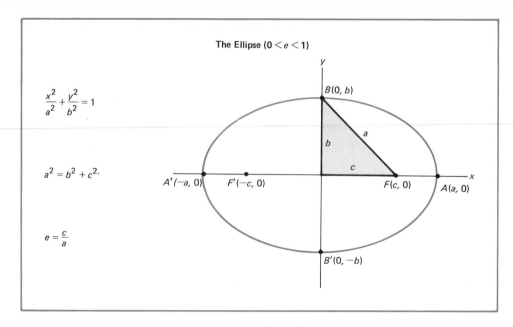

The Ellipse ($0 < e < 1$)

$$\frac{x^2}{a^2} + \frac{y^2}{b^2} = 1$$

$a^2 = b^2 + c^2,$

$e = \frac{c}{a}$

Figure 16

summarized in Figure 16, where you should note especially the significance of the triangle with sides a, b, and c. We call it the fundamental triangle for the ellipse.

The number $e = c/a$, which varies between 0 and 1, measures the **eccentricity** of the ellipse. If e is near 1, the ellipse is very eccentric (long and narrow); if e is near 0, the ellipse is almost circular (Figure 17). Sometimes, a circle is referred to as an ellipse of eccentricity 0; for in this case $c = 0$ and $a = b$ and the standard equation takes the form

$$\frac{x^2}{a^2} + \frac{y^2}{a^2} = 1$$

which is equivalent to the familiar circle equation $x^2 + y^2 = a^2$.

As an example of the equation of an ellipse, consider

$$\frac{x^2}{36} + \frac{y^2}{4} = 1$$

Note that $a = 6$ and $b = 2$, so this is the equation of an ellipse with center at the origin, major diameter $2a = 12$ and minor diameter $2b = 4$. Since $c = \sqrt{a^2 - b^2} = \sqrt{32} = 4\sqrt{2}$, the foci are at $(\pm 4\sqrt{2}, 0)$ and the eccentricity is $4\sqrt{2}/6 \approx .94$ (see Figure 18).

One other observation should be made. If we interchange the roles of x and y, then the standard equation takes the form

$$\frac{x^2}{b^2} + \frac{y^2}{a^2} = 1$$

e near 1

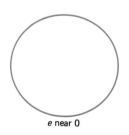

e near 0

Figure 17

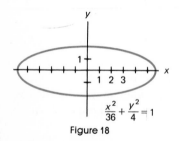

$$\frac{x^2}{36} + \frac{y^2}{4} = 1$$

Figure 18

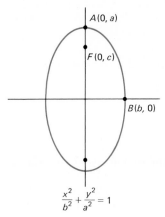

$$\frac{x^2}{b^2} + \frac{y^2}{a^2} = 1$$

Figure 19

The major axis is now the y-axis; the vertices and foci lie on it (Figure 19). For example, the equation

$$\frac{x^2}{16} + \frac{y^2}{25} = 1$$

represents a *vertical* ellipse (the major axis is vertical). Its vertices are at $(0, \pm 5)$, its foci are at $(0, \pm 3)$, its x-intercepts are at $(\pm 4, 0)$, and its eccentricity is $3/5 = .6$. One can always tell from the equation whether the corresponding ellipse is horizontal or vertical by noting whether the larger square in the denominator is in the x-term or the y-term.

APPLICATIONS

Like the parabola, the ellipse has an important optical property. If we imagine the ellipse to represent a mirror, then a light ray emanating from one focus will be reflected from the ellipse back through the other focus (see Problem 33 for a demonstration of this fact). This is the basis for the whispering gallery effect resulting from the elliptical shaped domes of St. Paul's Cathedral in London and the National Statuary Hall in the United States Capitol.

A much more significant application is the observation by Kepler that the planets move around the sun in elliptical orbits. Later, Newton established that this is a consequence of the fact that the gravitational force of attraction between two bodies is inversely proportional to the square of the distance between them. For the same reason, the electrons in the Bohr model of the hydrogen atom travel in elliptical orbits.

Problem Set 11-2

In Problems 1-8, decide whether the ellipse with the given equation is horizontal or vertical and then determine the major and minor diameters. In Problems 5–8, you will first have to rewrite the equation in standard form.

1. $\dfrac{x^2}{7} + \dfrac{y^2}{16} = 1$

2. $\dfrac{x^2}{9} + \dfrac{y^2}{8} = 1$

3. $\dfrac{x^2}{36} + \dfrac{y^2}{20} = 1$

4. $\dfrac{x^2}{12} + \dfrac{y^2}{25} = 1$

5. $4x^2 + 9y^2 = 4$

6. $9x^2 + 8y^2 = 18$

7. $4k^2x^2 + k^2y^2 = 1$

8. $k^2x^2 + (k^2 + 1)y^2 = k^2$

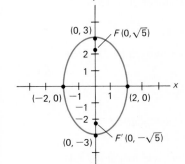

Figure 20

EXAMPLE A (Graphing the Equation of an Ellipse) Graph the equation $x^2/4 + y^2/9 = 1$, showing all the key features.

Solution. We identify this as the equation of a vertical ellipse (because the larger denominator is in the y term.) Also $a = 3$, $b = 2$, and $c = \sqrt{9 - 4} = \sqrt{5}$, from which we determine the four intercepts and the foci as shown in Figure 20.

In Problems 9–12, decide whether the corresponding ellipse is horizontal or vertical, determine a, b, and c, and sketch the graph.

9. $\dfrac{x^2}{25} + \dfrac{y^2}{9} = 1$

10. $\dfrac{x^2}{16} + \dfrac{y^2}{9} = 1$

11. $\dfrac{x^2}{1} + \dfrac{y^2}{4} = 1$

12. $\dfrac{x^2}{25} + \dfrac{y^2}{169} = 1$

EXAMPLE B (Finding Ellipses with Given Properties)   Write the equations of the three ellipses with vertices at $(0, \pm 8)$ and foci at (a) $(0, \pm 7)$; (b) $(0, \pm 4)$; (c) $(0, \pm 1)$. Determine the eccentricity e in each case. Sketch the graphs.

Solution.   In each case, the ellipse is vertical. From the formulas $b = \sqrt{a^2 - c^2}$ and $e = c/a$, we determine the following.

(a) $a = 8$, $c = 7$, $b = \sqrt{15}$, $e = \frac{7}{8}$

(b) $a = 8$, $c = 4$, $b = \sqrt{48}$, $e = \frac{1}{2}$

(c) $a = 8$, $c = 1$, $b = \sqrt{63}$, $e = \frac{1}{8}$

The three graphs and the corresponding equations are shown in Figure 21. Note that the smaller e is, the more circular the ellipse.

(a) $\dfrac{x^2}{15} + \dfrac{y^2}{64} = 1$

(b) $\dfrac{x^2}{48} + \dfrac{y^2}{64} = 1$

(c) $\dfrac{x^2}{63} + \dfrac{y^2}{64} = 1$

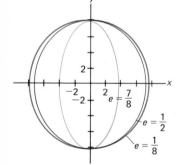

Figure 21

In Problems 13–20, write the equation of the ellipse that satisfies the given conditions, determine its eccentricity, and sketch its graph.

13. Vertices at $(0, \pm 5)$, foci at $(0, \pm 3)$.
14. Vertices at $(0, \pm 10)$, foci at $(0, \pm 8)$.
15. Center at $(0, 0)$, a vertex at $(-7, 0)$, a focus at $(3, 0)$.
16. Center at $(0, 0)$, a focus at $(6, 0)$, major diameter 20.
17. Horizontal, center at $(0, 0)$, major diameter 14, minor diameter 4.
18. Foci at $(\pm 10, 0)$, minor diameter 10.
19. Vertices at $(\pm 9, 0)$, curve passes through $(3, \sqrt{8})$.
20. Ends of minor diameter at $(\pm 4, 0)$, curve passes through $(\sqrt{2}, 4\sqrt{3})$.

MISCELLANEOUS PROBLEMS

21. Determine a, b, c, and e for the ellipse $4x^2 + 25y^2 = 100$.
22. Find the eccentricity of the ellipse $8x^2 + 2y^2 = 8$.
23. Find the equation of the ellipse with eccentricity $\frac{1}{3}$ and foci at $(0, \pm 4)$.

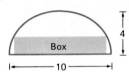

Figure 22

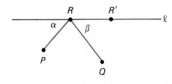

Figure 23

24. Find the equation of the ellipse that goes through $(\frac{1}{4}, \sqrt{3}/2)$ and has vertices at $(0, \pm1)$.

25. A door has the shape of an elliptical arch (a half-ellipse) that is 10 feet wide and 4 feet high at the center (Figure 22). A box 2 feet high is to be pushed through the door. How wide can it be?

26. How long is the **latus rectum** (chord through the focus perpendicular to the major axis) for the ellipse $x^2/a^2 + y^2/b^2 = 1$?

27. Assume that the center of the earth (a sphere of radius 4000 miles) is at one focus of the elliptical path of a satellite. If the satellite's nearest approach to the surface of the earth is 2000 miles and its farthest distance away is 10,000 miles, what are the major and minor diameters of the elliptical path?

28. ABC is a right triangle with the right angle at B. A and B are the foci of an ellipse and C is on the ellipse. Determine the major and minor diameters of the ellipse given that $\overline{AB} = 8$ and $\overline{BC} = 6$.

29. The area of the ellipse $x^2/a^2 + y^2/b^2 = 1$ is πab. Find the area of the ellipse $11x^2 + 7y^2 = 77$.

30. A square with sides parallel to the coordinate axes is inscribed in the ellipse $b^2x^2 + a^2y^2 = a^2b^2$. Determine the area of the square.

31. A dog's collar is attached by a ring to a loop of rope 32 feet long. The loop of rope is thrown over two stakes 12 feet apart.
    (a) How much area can the dog cover?
    (b) If the dog should manage to nudge the rope over the top of one of the stakes, how much would this increase the area it can cover?

32. Let P be a point on a 16-foot ladder 7 feet from the top end. As the ladder slides with its top end against a wall (the y-axis) and its bottom end along the ground (the x-axis), P traces a curve. Find the equation of this curve.

33. (Optical property of the ellipse)   In Figure 23, let P and Q be the foci of an ellipse, R be a point on the ellipse, l be the tangent line at R, and R' be any other point of l.
    (a) Show that $\overline{PR'} + \overline{R'Q} > \overline{PR} + \overline{RQ}$.
    (b) Show that $\alpha = \beta$. (See Problem 34 of Section 11-1.)

From this we conclude that a light ray from one focus P of an elliptic mirror is reflected back through the other focus.

34. **TEASER**   Two ellipses with the same eccentricity e are such that the major diameter of the smaller ellipse coincides with the minor diameter of the larger ellipse and the area of the smaller ellipse is 19 percent of the area of the larger one. Use this information to determine e.

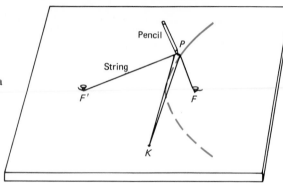

This shows how to draw a hyperbola. Take a string with a knot at K and tack its ends down with thumbtacks. Insert pencil as shown and pull string taut. Pull on knot at K.

11-3   Hyperbolas

Our opening display hints at the geometric definition of a hyperbola. A **hyperbola** is the set of points P for which the difference of the distances from two fixed points F' and F is a constant. More precisely, a hyperbola is the set of points P in Figure 24 satisfying

$$\left| d(P, F') - d(P, F) \right| = 2a$$

for some constant a.

As with the ellipse, the two fixed points F' and F are called **foci** and the point midway between the foci is the **center** of the hyperbola. The line through the foci is the **major axis** (or transverse axis) of the hyperbola and the line through the center and perpendicular to the major axis is the **minor axis** (or conjugate axis). Also, the points A' and A where the hyperbola intersects the major axis are the **vertices** of the hyperbola. By applying the defining condition with $P = A$, we see that $2a$ is the distance between the vertices. Also, if $2c$ denotes the distance between the foci, then $2c$ must be greater than $2a$, that is $c > a$.

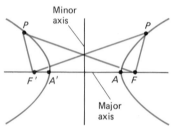

Figure 24

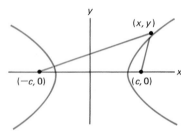

Figure 25

THE EQUATION OF A HYPERBOLA

Place the hyperbola in the coordinate system so that its center is at the origin with the major axis along the x-axis and the foci at $(-c, 0)$ and $(c, 0)$ as in Figure 25. Then the condition $|d(P, F') - d(P, F)| = 2a$ combined with the distance formula yields

$$\left|\sqrt{(x + c)^2 + (y - 0)^2} - \sqrt{(x - c)^2 + (y - 0)^2}\right| = 2a$$

If we now employ the same kind of procedure used in obtaining the equation of the ellipse (square both sides, simplify, square again, and simplify), we get

$$(c^2 - a^2)x^2 - a^2y^2 = a^2(c^2 - a^2)$$

or, equivalently, after dividing both sides by $a^2(c^2 - a^2)$,

$$\frac{x^2}{a^2} - \frac{y^2}{c^2 - a^2} = 1$$

Finally, we let $b^2 = c^2 - a^2$ to obtain what is called the **standard equation of the hyperbola**, namely,

$$\frac{x^2}{a^2} - \frac{y^2}{b^2} = 1$$

Note that both x and y occur to the second power, which corresponds to the fact that the graph of this equation is symmetric with respect to both the x- and y-axes as well as the origin.

INTERPRETING a, b, AND c

We have already noted that $2c$ is the distance between the foci and $2a$ is the distance between the vertices of the hyperbola. To interpret b, observe that if we solve the standard equation for y in terms of x, we get

$$y = \pm\frac{b}{a}\sqrt{x^2 - a^2}$$

For large x, $\sqrt{x^2 - a^2}$ behaves much like x; in fact, as $|x|$ gets larger and larger, $x - \sqrt{x^2 - a^2}$ approaches zero (Problem 26) and hence so does b/a times this quantity. Thus the two branches of the hyperbola $y = \pm\frac{b}{a}\sqrt{x^2 - a^2}$ approach the two lines $y = \pm\frac{b}{a}x$. We say that the hyperbola has these lines as (oblique) asymptotes (see Section 2-5).

Since $c^2 = a^2 + b^2$, the numbers a, b, and c determine a right triangle, which we call the fundamental triangle for the hyperbola. Its role in determining the asymptotes of the hyperbola is clear from Figure 26 on the next page. This figure summarizes all the key facts for a hyperbola.

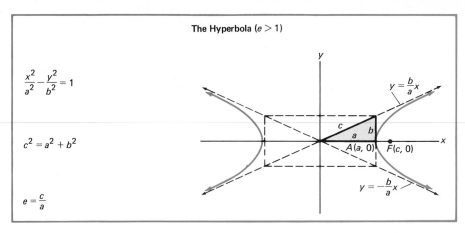

The Hyperbola ($e > 1$)

$$\frac{x^2}{a^2} - \frac{y^2}{b^2} = 1$$

$$c^2 = a^2 + b^2$$

$$e = \frac{c}{a}$$

Figure 26

The number $e = c/a$, which in this case is greater than 1, is called the **eccentricity** of the hyperbola. If e is near 1, then b is small relative to a and the hyperbola is very thin; if e is large the hyperbola is fat.

As a first example, consider the equation

$$\frac{x^2}{9} - \frac{y^2}{16} = 1$$

In this case, $a = 3$, $b = 4$, and $c = \sqrt{a^2 + b^2} = 5$. Thus the vertices are at $(\pm 3, 0)$, the foci are at $(\pm 5, 0)$, and the asymptotes have equations $y = \pm \frac{4}{3} x$. All this is shown in Figure 27.

Again we should consider what happens if we interchange the roles of x and y. The standard equation then takes the form

$$\frac{y^2}{a^2} - \frac{x^2}{b^2} = 1$$

This equation represents a hyperbola with major axis along the y-axis (we call it a *vertical hyperbola*). Its vertices are at $(0, \pm a)$ and its foci are at $(0, \pm c)$. As an example, consider the equation

$$\frac{y^2}{9} - \frac{x^2}{16} = 1$$

which again has $a = 3$ and $b = 4$. Its graph has vertices at $(0, \pm 3)$ and foci at $(0, \pm 5)$. Note how the fundamental triangle determines the asymptotes and helps us draw the graph (Figure 28). The hyperbolas of Figures 27 and 28 have the same eccentricity, namely, $e = c/a = 5/3$.

We make one final important observation. It is not the relative sizes of the denominators in the x- and y-terms that determine whether the hyperbola is vertical or horizontal (as it was with the ellipse). Rather, this is determined by whether in the standard form the minus is associated with the x- or the y-term.

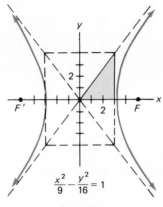

$$\frac{x^2}{9} - \frac{y^2}{16} = 1$$

Figure 27

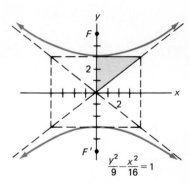

$$\frac{y^2}{9} - \frac{x^2}{16} = 1$$

Figure 28

APPLICATIONS

The hyperbola, too, has an optical property, as is illustrated in Figure 29. If we imagine one branch of the hyperbola to be a mirror, then a light ray from the opposite focus upon hitting the mirror will be reflected away along the line which passes through the nearby focus. The optical properties of the parabola and the hyperbola are combined in one design for a reflecting telescope (Figure 30). Other applications are treated in the problem set.

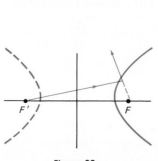

Figure 29

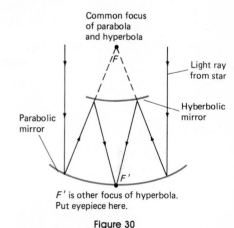

F' is other focus of hyperbola.
Put eyepiece here.

Figure 30

Problem Set 11-3

In Problems 1–8, decide whether the given equation determines a horizontal or vertical hyperbola. Also find a (the distance from the center to a vertex), b, and c. Be sure the equation is in standard form before you try to give the answers.

1. $\dfrac{x^2}{16} - \dfrac{y^2}{36} = 1$ 2. $-\dfrac{x^2}{1} + \dfrac{y^2}{8} = 1$ 3. $\dfrac{x^2}{16} - \dfrac{y^2}{9} = -1$

4. $\dfrac{x^2}{4} - \dfrac{y^2}{9} = 1$ 5. $4x^2 - 16y^2 = 1$ 6. $25y^2 - 9x^2 = 1$

7. $4x^2 - y^2 = 16$ 8. $k^2y^2 - 4k^2x^2 = 1$

EXAMPLE A (Graphing the Equation of a Hyperbola)   Sketch the graph of $x^2/4 - y^2/9 = 1$, showing all the important features.

Solution.   The graph is a hyperbola and, since the minus sign is associated with the y-term, the major axis is horizontal. We conclude that $a = 2$, $b = 3$, and $c = \sqrt{4 + 9} = \sqrt{13}$. The asymptotes are the lines $y = \pm\frac{3}{2}x$. With this information, we may sketch the graph shown in Figure 31.

In Problems 9–12, decide whether the corresponding hyperbola is horizontal or vertical, give the values of a, b, and c, and sketch the graph. Be sure to show the asymptotes.

9. $\dfrac{x^2}{25} - \dfrac{y^2}{9} = 1$ 10. $\dfrac{x^2}{16} - \dfrac{y^2}{9} = 1$

11. $\dfrac{y^2}{64} - \dfrac{x^2}{36} = 1$ 12. $\dfrac{x^2}{4} - \dfrac{y^2}{4} = 1$

EXAMPLE B   (Finding Hyperbolas Satisfying Given Conditions)   Find the equation of the hyperbola with vertices at $(\pm 2, 0)$ that passes through $(2\sqrt{2}, 4)$. Sketch its graph.

Solution.   Since the vertices are on the x-axis, the hyperbola is horizontal with $a = 2$. The equation has the form

$$\frac{x^2}{4} - \frac{y^2}{b^2} = 1$$

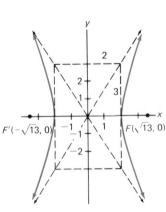

Figure 31

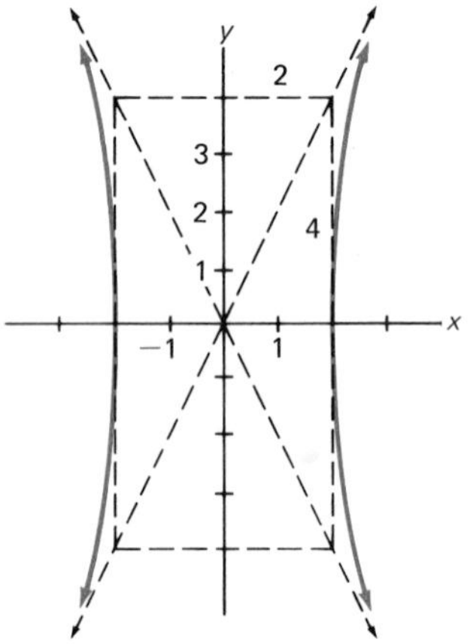

Figure 32

Since the point $(2\sqrt{2}, 4)$ is on the graph,

$$\frac{(2\sqrt{2})^2}{4} - \frac{4^2}{b^2} = 1$$

which gives $b = 4$. Thus the equation is

$$\frac{x^2}{4} - \frac{y^2}{16} = 1$$

The graph is shown in Figure 32.

In Problems 13–18, find the equation of the hyperbola satisfying the given conditions and sketch its graph, displaying the asymptotes.

13. Vertices at $(0, \pm 3)$ and going through $(2, 5)$.
14. Vertices at $(\pm 3, 0)$ and going through $(2\sqrt{3}, 9)$.
15. Foci at $(\pm 4, 0)$, vertices at $(\pm 1, 0)$.
16. Vertices at $(\pm 5, 0)$, equations of asymptotes $y = \pm x$.
17. Vertices at $(\pm 3, 0)$, equations of asymptotes $y = \pm 2x$.
18. Vertices at $(0, \pm 3)$, eccentricity $e = \frac{4}{3}$.

MISCELLANEOUS PROBLEMS

19. Find the equation of the hyperbola centered at the origin with a focus at $(0, 8)$ and a vertex at $(0, -6)$.
20. Determine the eccentricity of the hyperbola with equation $16x^2 - 20y^2 = 320$.
21. A conic has eccentricity 3 and foci at $(\pm 12, 0)$. Find its equation.
22. Find the equations of the asymptotes of the vertical hyperbola with eccentricity 2 and center at the origin.
23. How long is the **focal chord** (chord through a focus perpendicular to the major axis) of the hyperbola $x^2/9 - y^2/16 = 1$.
24. Generalize Problem 23 by finding the length of the focal chord for the hyperbola $x^2/a^2 - y^2/b^2 = 1$.
25. Find the eccentricity of the hyperbola with asymptotes $y = \pm x$.
26. Show that $x - \sqrt{x^2 - a^2}$ approaches 0 as $|x|$ gets larger and larger. *Hint:* Multiply and divide by $x + \sqrt{x^2 - a^2}$.
27. A ball shot from $(-5, 0)$ hit the right branch of the hyperbolic bangboard $x^2/16 - y^2/9 = 1$ at the point $(8, 3\sqrt{3})$. What was the ball's y-coordinate when its x-coordinate was 10?
28. The rectangle $PQRS$ with sides parallel to the coordinate axes is inscribed in the hyperbola $x^2/4 - y^2/9 = 1$ as shown in Figure 33. Find the coordinates of P if the area of the rectangle is $6\sqrt{5}$.
29. Andrew, located at $(0, -2200)$, fired a rifle. The sound echoed off a cliff at $(0, 2200)$ to Brian, located at the point (x, y). Brian heard this echo 6 seconds after he heard the original shot. Find the xy-equation of the curve on which Brian is located. Assume that distances are in feet and that sound travels 1100 feet per second.

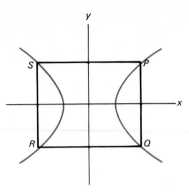

Figure 33

30. **TEASER** Amy, Betty, and Cindy, located at $(-8, 0)$, $(8, 0)$, and $(8, 10)$, respectively, recorded the exact times when they heard an explosion. On comparing notes, they discovered that Betty and Cindy heard the explosion at the same time but that Amy heard it 12 seconds later. Assuming that distances are in kilometers and that sound travels $\frac{1}{3}$ kilometer per second, determine the point of the explosion.

Conic Sections		Limiting Forms	
		4. Parallel lines: $y^2 = 3$	
1. Parabola: $y^2 = 3x$		5. Single line: $y^2 = 0$	
		6. Circle: $x^2 + y^2 = 4$	
2. Ellipse: $\dfrac{x^2}{9} + \dfrac{y^2}{4} = 1$		7. Point: $2x^2 + y^2 = 0$	•
		8. Empty set: $3x^2 + y^2 = -1$	
3. Hyperbola: $\dfrac{x^2}{9} - \dfrac{y^2}{4} = 1$		9. Intersecting lines: $2x^2 - y^2 = 0$	

11-4 Translation of Axes

An astute—or perhaps even a casual—observer will note that the standard equations of the three conic sections involve the second power of x or y. This observation suggests a question. Suppose we graph a polynomial equation that is of second degree in x and y. Will it always be a conic section? The answer is no; that is, it is no unless we admit the six limiting forms of the conic sections illustrated in the right half of our opening display. But if we do admit them, the answer is yes. In particular, we claim that the graph of any equation of the form

$$Ax^2 + Cy^2 + Dx + Ey + F = 0 \qquad (A, C \text{ not both } 0)$$

is a conic section or one of its limiting forms. We will show you why by moving the coordinate axes in just the right way.

TRANSLATIONS

Consider the equation

$$x^2 + y^2 - 4x - 6y - 12 = 0$$

If we let

$$x = u + 2 \qquad y = v + 3$$

this equation becomes

$$(u + 2)^2 + (v + 3)^2 - 4(u + 2) - 6(v + 3) - 12 = 0$$

or

$$u^2 + 4u + 4 + v^2 + 6v + 9 - 4u - 8 - 6v - 18 - 12 = 0$$

This simplifies to

$$u^2 + v^2 = 25$$

which we recognize as the equation of a circle of radius 5.

　　To understand what we have just done, introduce new coordinate axes in the plane (the u- and v-axes) parallel to the old x- and y-axes, but with the new origin at $x = 2$ and $y = 3$ (see Figure 34). Each point now has two sets of coordinates: (x, y) and (u, v). They are related by the equations $x = u + 2$ and $y = v + 3$. In the new coordinate system, which is called a **translation** of the old one, the uv-equation represents a circle of radius 5 centered at the origin. In the old coordinate system, the xy-equation must also have represented a circle of radius 5, but centered at (2, 3).

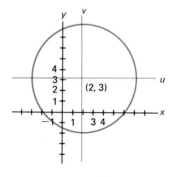

$$x^2 + y^2 - 4x - 6y - 12 = 0$$
$$\downarrow$$
$$\begin{cases} x = u + 2 \\ y = v + 3 \end{cases}$$
$$\downarrow$$
$$u^2 + v^2 = 25$$

Figure 34

　　Let us see what happens in general to coordinates of points under a translation of axes. If u- and v-axes are introduced with the same directions as the old x- and y-axes so that the new origin has coordinates (h, k) relative to the old axes (Figure 35), then the two sets of coordinates for a point P are connected by

$$u = x - h \qquad v = y - k$$

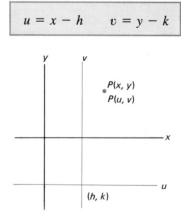

Figure 35

or, equivalently, by

$$x = u + h \qquad y = v + k$$

The shape of a curve is not changed by a translation of axes since it is the axes, not the curve, that are moved. But as we saw in the example above, the resulting change in the equation may enable us to recognize the curve.

The concept of a translation was discussed from a slightly different perspective in Sections 2-3 and 2-6. There we translated the graph; here we are translating the axes.

COMPLETING THE SQUARE

Given an equation, how do we know what translation to make? Here an old algebraic friend, completing the square, comes to our aid. As a typical example, consider

$$x^2 + y^2 - 6x + 8y + 10 = 0$$

We first rewrite the equation and then complete each square.

$$(x^2 - 6x \quad) + (y^2 + 8y \quad) = -10$$
$$(x^2 - 6x + 9) + (y^2 + 8y + 16) = -10 + 9 + 16$$
$$(x - 3)^2 + (y + 4)^2 \quad = 15$$

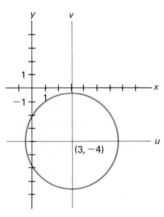

Figure 36

We can recognize this as the equation of a circle of radius $\sqrt{15}$, centered at $(3, -4)$. In terms of translations (Figure 36), we note that the substitutions $u = x - 3$ and $v = y + 4$ transform the equation into

$$u^2 + v^2 = 15$$

As a second example, consider

$$4x^2 + 9y^2 - 24x + 18y + 9 = 0$$

We may rewrite this equation successively as

$$4(x^2 - 6x \quad) + 9(y^2 + 2y \quad) = -9$$
$$4(x^2 - 6x + 9) + 9(y^2 + 2y + 1) = -9 + 36 + 9$$
$$4(x - 3)^2 + 9(y + 1)^2 \quad = 36$$
$$\frac{(x - 3)^2}{9} + \frac{(y + 1)^2}{4} \quad = 1$$

The translation $u = x - 3$ and $v = y + 1$ transforms this to

$$\frac{u^2}{9} + \frac{v^2}{4} = 1$$

which we recognize as an ellipse with $a = 3$ and $b = 2$ (Figure 37).

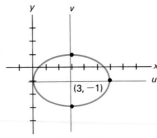

$$4x^2 + 9y^2 - 24x + 18y + 9 = 0$$

$$\downarrow$$

$$\begin{cases} x = u + 3 \\ y = v - 1 \end{cases}$$

$$\downarrow$$

$$\frac{u^2}{9} + \frac{v^2}{4} = 1$$

Figure 37

THE GENERAL CASE

Consider the general equation

$$Ax^2 + Cy^2 + Dx + Ey + F = 0$$

For the moment, we assume that at least one of the coefficients A or C is different from zero. When we apply the process of completing of the square, we transform this equation into one of several forms, the most typical being the following.

1. $(y - k)^2 = 4p(x - h)$

2. $\dfrac{(x - h)^2}{a^2} + \dfrac{(y - k)^2}{b^2} = 1$

3. $\dfrac{(x - h)^2}{a^2} - \dfrac{(y - k)^2}{b^2} = 1$

Perhaps these are already recognizable as the equations of a parabola with vertex at (h, k), an ellipse with center at (h, k), and a hyperbola with center at (h, k). But to remove any doubt, we may translate the axes by the substitutions $u = x - h$ and $v = y - k$, thereby obtaining

1. $v^2 = 4pu$

2. $\dfrac{u^2}{a^2} + \dfrac{v^2}{b^2} = 1$

3. $\dfrac{u^2}{a^2} - \dfrac{v^2}{b^2} = 1$

Our work may also yield these equations with u and v interchanged or we may get one of the six limiting forms illustrated in our opening display. There are no other possibilities.

Problem Set 11-4

In Problems 1–6, make the indicated change of variables (a translation of axes) and then name the conic section or limiting form represented by the equation.

1. $x^2 + 2y^2 - 4y = 0$; $x = u$, $y = v + 1$
2. $x^2 - 4y^2 - 4x - 5 = 0$; $x = u + 2$, $y = v$
3. $x^2 + y^2 - 4x + 2y = -4$; $x = u + 2$, $y = v - 1$
4. $x^2 + y^2 - 4x + 2y = -5$; $x = u + 2$, $y = v - 1$
5. $x^2 - 6x - 4y + 13 = 0$; $x = u + 3$, $y = v + 1$
6. $x^2 - 4x + 1 = 0$; $x = u + 2$, $y = v$

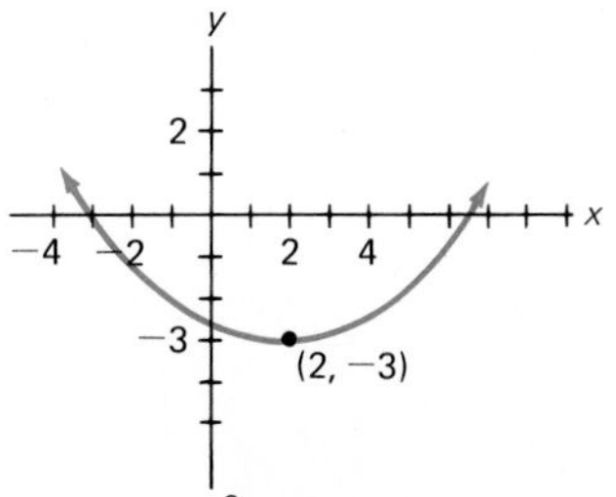

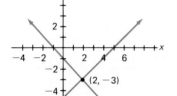

$(x - 2)^2 = 8(y + 3)$

Figure 38

EXAMPLE A (Graphing Conics) In the xy-plane, sketch the graphs of (a) $(x - 2)^2 = 8(y + 3)$; (b) $(x - 2)^2 - (y + 3)^2 = 0$.

Solution. (a) We could formally make the translation of axes corresponding to $u = x - 2$ and $v = y + 3$, which would yield $u^2 = 8v$. But perhaps we can do this mentally and thereby recognize that $(x - 2)^2 = 8(y + 3)$ is the equation of a vertical parabola with vertex at $(2, -3)$ and $p = 2$. With that information, we can make the sketch shown in Figure 38.

(b) The mental substitutions $u = x - 2$ and $v = y + 3$ transform the second equation into $u^2 - v^2 = 0$, which is equivalent to $(u - v) \times (u + v) = 0$. We recognize this as the equation of two intersecting lines, $v = u$ and $v = -u$. In terms of x and y, this gives $y + 3 = x - 2$ and $y + 3 = -x + 2$ (see Figure 39).

Sketch the graphs of each of the following equations.

7. $\dfrac{(x + 3)^2}{4} + \dfrac{(y + 2)^2}{16} = 1$

8. $(x + 3)^2 + (y - 4)^2 = 25$

9. $\dfrac{(x + 3)^2}{4} - \dfrac{(y + 2)^2}{16} = 1$

10. $4(x + 3) = (y + 2)^2$

11. $(x + 2)^2 = 8(y - 1)$

12. $(x + 2)^2 = 4$

13. $(y - 1)^2 = 16$

14. $\dfrac{(x + 3)^2}{4} + \dfrac{(y - 2)^2}{8} = 0$

Figure 39

$(x - 2)^2 - (y + 3)^2 = 0$

EXAMPLE B (Identifying Conics by Completing Squares) Identify the conic whose equation is

$$4x^2 - 8x - 2y^2 + 16y = 0$$

Solution. We complete the squares.

$$4(x^2 - 2x + \quad) - 2(y^2 - 8y + \quad) = 0$$

$$4(x^2 - 2x + 1) - 2(y^2 - 8y + 16) = 4 - 32$$

$$4(x - 1)^2 - 2(y - 4)^2 = -28$$

$$-\frac{(x - 1)^2}{7} + \frac{(y - 4)^2}{14} = 1$$

We recognize this as the equation of a vertical hyperbola with center at $(1, 4)$.

Identify the conics determined by the equations in Problems 15–24.

15. $4x^2 + 16x + 4y^2 - 8y = 0$
16. $x^2 + 2x + 4y^2 - 8y = 0$
17. $4x^2 - 16x + y^2 - 8y = -6$
18. $4x^2 - 16x - y^2 - 8y = 2$
19. $4x^2 - 16x + y^2 - 8y = -32$
20. $4x^2 - 16x + y^2 - 8y = -40$
21. $4x^2 - 16x + y - 8 = 0$
22. $4x^2 - 16x + 12 = 0$
23. $4x^2 - 16x - 9y^2 + 18y + 7 = 0$
24. $4x^2 - 16x - 9y^2 + 18y + 8 = 0$
25. Sketch the graph of $9x^2 - 18x + 4y^2 + 16y = 11$.
26. Sketch the graph of $4x^2 + 16x - 16y + 32 = 0$.
27. Determine the distance between the vertices of the graph of $-9x^2 + 18x + 4y^2 + 24y = 9$.
28. Find the focus and the directrix of the parabola with equation $x^2 - 4x + 8y = 0$.
29. Find the focus and directrix of the parabola with equation $2y^2 - 4y - 10x = 0$.
30. Find the foci of the ellipse with equation $16(x - 1)^2 + 25(y + 2)^2 = 400$.

MISCELLANEOUS PROBLEMS

31. Sketch the graph of each of the following.

   (a) $\dfrac{(x + 5)^2}{16} + \dfrac{(y - 3)^2}{9} = 1$ (b) $\dfrac{(x + 5)^2}{16} - \dfrac{(y - 3)^2}{9} = 1$

32. Identify the curve with the given equation.
   (a) $4x^2 + 9y^2 - 16x + 54y + 61 = 0$
   (b) $x^2 + 8x + 8y = 0$
   (c) $x^2 - 4y^2 + 6x + 16y = 16$
   (d) $4x^2 + 9y^2 + 16x - 18y + 25 = 0$

33. Name the conic with equation $y^2 + ax^2 = x$ for the various values of a.

34. Find the equation of the parabola with the line $y = 4$ as directrix and the point $(2, -1)$ as focus.

35. Write the equation of the parabola with vertex $(4, 5)$ and focus $(3, 5)$.

36. Write the equation of the ellipse with vertices at $(2, -2)$ and $(2, 10)$ as ends of the major diameter and $(2, 6)$ as a focus.

37. Write the equation of the ellipse with foci $(\pm 2, 2)$ that goes through the origin.

38. Write the equation of the hyperbola with foci $(0, 0)$ and $(4, 0)$ that passes through $(9, 12)$

39. Find the equation of the hyperbola with the lines $y = 2x - 10$ and $y = -2x + 2$ as asymptotes and one focus at $(3, 2)$.

40. Transform the equation $xy - 2x + 3y = 18$ by translation of axes so that the new equation has no first degree terms. Sketch the graph showing both sets of axes.

41. A curve C goes through the three points $(-1, 2)$, $(0, 0)$, and $(3, 6)$. Write the equation for C if C is;

(a) A vertical parabola;

(b) A horizontal parabola;

(c) A circle.

42. Find the equation of the hyperbola with eccentricity 2 that has the y-axis as one directrix and the corresponding focus at $(6, 0)$.

43. Find the equation of the circle that goes through the two foci and the upper y-intercept of the ellipse $36x^2 + 100y^2 = 3600$.

44. **TEASER** Let C be an arbitrary horizontal ellipse that intersects the parabola $y = x^2$ in four points (x_1, y_1), (x_2, y_2), (x_3, y_3), and (x_4, y_4). Prove that $x_1 + x_2 + x_3 + x_4 = 0$.

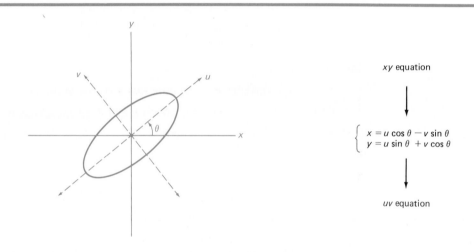

11-5   Rotation of Axes

We want you to observe two facts about what we have done in the first four sections of this chapter. First, all the conic sections we have considered so far were oriented in the coordinate system with their major axis parallel to either the x-axis or the y-axis. Second, none of the equations of these conics had an xy-term. We would not mention these facts unless there were a connection between them. To see the connection, we need to discuss rotation of axes.

ROTATIONS

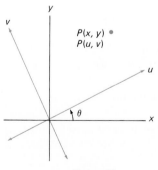

Figure 40

Introduce a new pair of axes, called the u- and v-axes, into the xy-plane. These axes have the same origin as the old x- and y-axes, but they are rotated so that the positive u-axis makes an angle θ with the positive x-axis (see the diagram in the opening panel and also the one in Figure 40). A point P then has two sets of coordinates: (x, y) and (u, v). How are they related?

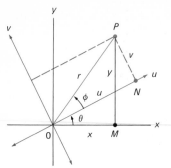

Figure 41

Draw a line segment from the origin O to P, let r denote the length of OP, and let φ denote the angle from the u-axis to OP. Then x, y, u, and v will have the geometric interpretations indicated in Figure 41.

Looking at the right triangle OPM, we see that

$$\cos(\varphi + \theta) = \frac{x}{r}$$

so

$$x = r\cos(\varphi + \theta) = r(\cos\varphi\cos\theta - \sin\varphi\sin\theta)$$
$$= (r\cos\varphi)\cos\theta - (r\sin\varphi)\sin\theta$$

Next, the right triangle OPN tells us that $u = r\cos\varphi$ and $v = r\sin\varphi$. Thus

$$\boxed{x = u\cos\theta - v\sin\theta}$$

Similarly

$$y = r\sin(\varphi + \theta) = r(\sin\varphi\cos\theta + \cos\varphi\sin\theta)$$
$$= (r\sin\varphi)\cos\theta + (r\cos\varphi)\sin\theta$$

so

$$\boxed{y = u\sin\theta + v\cos\theta}$$

We call the boxed results **rotation formulas.**

A SIMPLE EXAMPLE

Consider the equation

$$xy = 1$$

Let us make a rotation of axes through $45°$ to see what happens to this equation. The required substitutions are

$$x = u\cos 45° - v\sin 45° = \frac{\sqrt{2}}{2}(u - v)$$

$$y = u\sin 45° + v\cos 45° = \frac{\sqrt{2}}{2}(u + v)$$

When we make these substitutions in $xy = 1$, we obtain

$$\frac{\sqrt{2}}{2}(u - v)\frac{\sqrt{2}}{2}(u + v) = 1$$

which simplifies to

$$\frac{u^2}{2} - \frac{v^2}{2} = 1$$

We recognize this as the equation of a hyperbola with $a = b = \sqrt{2}$. Note how the cross-product term xy disappeared as a result of the rotation. The choice of a 45° angle was just right to make this happen (see Figure 42).

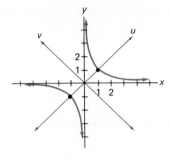

$$xy = 1$$
$$\downarrow$$
$$\begin{cases} x = u \cos 45° - v \sin 45° \\ y = u \sin 45° + v \cos 45° \end{cases}$$
$$\downarrow$$
$$\frac{u^2}{2} - \frac{v^2}{2} = 1$$

Figure 42

THE GENERAL SECOND DEGREE EQUATION

How do we know what rotation to make? Consider the most general second degree equation in x and y:

$$Ax^2 + Bxy + Cy^2 + Dx + Ey + F = 0$$

If we make the substitutions

$$x = u \cos \theta - v \sin \theta$$
$$y = u \sin \theta + v \cos \theta$$

this equation takes the form

$$au^2 + buv + cv^2 + du + ev + f = 0$$

where a, b, c, d, e and f are numbers which depend upon θ. We could find values for all of them, but we really care only about b. When we do the necessary algebra, we find

$$b = B(\cos^2 \theta - \sin^2 \theta) - 2(A - C) \sin \theta \cos \theta$$
$$= B \cos 2\theta - (A - C) \sin 2\theta$$

We would like to have $b = 0$; that is,

$$B \cos 2\theta = (A - C) \sin 2\theta$$

This will occur if

$$\cot 2\theta = \frac{A - C}{B}$$

This formula is the answer to our question: to eliminate the cross-product (xy) term, choose θ so it satisfies this formula. In the example $xy = 1$, we have $A = 0$, $B = 1$, and $C = 0$ so we choose θ to satisfy

$$\cot 2\theta = \frac{0 - 0}{1} = 0$$

One angle that works is $\theta = 45°$. We could also use $\theta = 135°$ or $\theta = -225°$, but it is customary to choose a first quadrant angle.

ANOTHER EXAMPLE

Consider the equation

$$4x^2 + 2\sqrt{3}\,xy + 2y^2 + 10\sqrt{3}x + 10y = 5$$

To remove the cross-product term, we rotate the axes through an angle θ satisfying

$$\cot 2\theta = \frac{A - C}{B} = \frac{4 - 2}{2\sqrt{3}} = \frac{1}{\sqrt{3}}$$

this means that $2\theta = 60°$ and so $\theta = 30°$. When we use this value of θ in the rotation formulas, we obtain

$$x = u \cdot \frac{\sqrt{3}}{2} - v \cdot \frac{1}{2} = \frac{\sqrt{3}\,u - v}{2}$$

$$y = u \cdot \frac{1}{2} + v \cdot \frac{\sqrt{3}}{2} = \frac{u + \sqrt{3}\,v}{2}$$

Substituting these in the original equation gives

$$4\frac{(\sqrt{3}\,u - v)^2}{4} + 2\sqrt{3}\frac{(\sqrt{3}\,u - v)(u + \sqrt{3}v)}{4}$$

$$+ 2\frac{(u + \sqrt{3}\,v)^2}{4} + 10\sqrt{3}\frac{\sqrt{3}\,u - v}{2} + 10\frac{u + \sqrt{3}\,v}{2} = 5$$

After collecting terms and simplifying, we have

$$5u^2 + v^2 + 20u = 5$$

Next we complete the squares.

$$5(u^2 + 4u + 4) + v^2 = 5 + 20$$

$$\frac{(u + 2)^2}{5} + \frac{v^2}{25} = 1$$

As a final step, we make the translation determined by $r = u + 2$ and $s = v$, which gives

$$\frac{r^2}{5} + \frac{s^2}{25} = 1$$

This is the equation of a vertical ellipse in the rs-coordinate system. It has major diameter of length 10 and minor diameter of length $2\sqrt{5}$. All of this is shown in Figure 43.

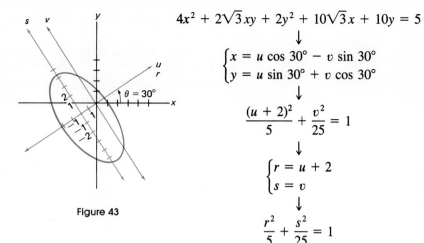

$$4x^2 + 2\sqrt{3}xy + 2y^2 + 10\sqrt{3}x + 10y = 5$$

$$\downarrow$$

$$\begin{cases} x = u \cos 30° - v \sin 30° \\ y = u \sin 30° + v \cos 30° \end{cases}$$

$$\downarrow$$

$$\frac{(u + 2)^2}{5} + \frac{v^2}{25} = 1$$

$$\downarrow$$

$$\begin{cases} r = u + 2 \\ s = v \end{cases}$$

$$\downarrow$$

$$\frac{r^2}{5} + \frac{s^2}{25} = 1$$

Figure 43

Problem Set 11-5

In the text, we derived the following rotation formulas.

$$x = u \cos \theta - v \sin \theta$$

$$y = u \sin \theta + v \cos \theta$$

In Problems 1–10, transform the given xy-equation to a uv-equation by a rotation through the specified angle θ.

1. $y = \sqrt{3}x$; $\theta = 60°$
2. $y = x$; $\theta = 45°$
3. $x^2 + 4y^2 = 16$; $\theta = 90°$
4. $4y^2 - x^2 = 4$; $\theta = 90°$
5. $y^2 = 4\sqrt{2}x$; $\theta = 45°$
6. $x^2 = -\sqrt{2}y + 3$; $\theta = 45°$
7. $x^2 - xy + y^2 = 4$; $\theta = 45°$
8. $x^2 + 3xy + y^2 = 10$; $\theta = 45°$
9. $6x^2 - 24xy - y^2 = 30$; $\theta = \cos^{-1}(\frac{3}{5})$
10. $3x^2 - \sqrt{3}xy + 2y^2 = 39$; $\theta = 60°$

EXAMPLE A (Eliminating the xy-term) By rotation of axes, eliminate the xy-term from

$$x^2 + 24xy + 8y^2 = 136$$

and then draw its graph.

Solution. We review the example in the text, noting that we must choose θ to satisfy

$$\cot 2\theta = \frac{A - C}{B} = \frac{1 - 8}{24} = -\frac{7}{24}$$

Here our problem is complicated by the fact that 2θ is not a special angle.

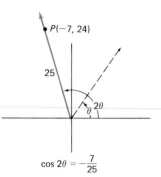

P(−7, 24)

25

2θ

$\cos 2\theta = -\dfrac{7}{25}$

Figure 44

How shall we find the values of $\sin \theta$ and $\cos \theta$ needed for the rotation formulas?

First, we place 2θ in standard position (see Figure 44), noting that $P(-7, 24)$ is on its terminal side. Since P is a distance $r = \sqrt{(-7)^2 + (24)^2} = 25$ from the origin, $\cos 2\theta = -\dfrac{7}{25}$.

Second, we recall the half-angle formulas.

$$\sin \theta = \pm\sqrt{\frac{1 - \cos 2\theta}{2}} \qquad \cos \theta = \pm\sqrt{\frac{1 + \cos 2\theta}{2}}$$

Since our θ is in the first quadrant, we use the plus sign in both cases. We obtain

$$\sin \theta = \sqrt{\frac{1 + \frac{7}{25}}{2}} = \frac{4}{5} \qquad \cos \theta = \sqrt{\frac{1 - \frac{7}{25}}{2}} = \frac{3}{5}$$

These, in turn, give the rotation formulas

$$x = \frac{3u - 4v}{5} \qquad y = \frac{4u + 3v}{5}$$

All this was preliminary; our main task is to substitute these expressions for x and y in the original equation and simplify.

$$\left(\frac{3u - 4v}{5}\right)^2 + 24\left(\frac{3u - 4v}{5}\right)\left(\frac{4u + 3v}{5}\right) + 8\left(\frac{4u + 3v}{5}\right)^2 = 136$$

After multiplying by 25 and collecting terms, we have

$$425u^2 - 200v^2 = 136 \cdot 25$$

or

$$\frac{u^2}{8} - \frac{v^2}{17} = 1$$

We summarize the process below and show the graph in Figure 45.

$$x^2 + 24xy + 8y^2 = 136$$
$$\downarrow$$
$$\begin{cases} x = \frac{3}{5}u - \frac{4}{5}v \\ y = \frac{4}{5}u + \frac{3}{5}v \end{cases}$$
$$\downarrow$$
$$\frac{u^2}{8} - \frac{v^2}{17} = 1$$

In Problems 11–20, eliminate the xy-term by a suitable rotation of axes and then, if necessary, translate axes (complete the squares) to put the equation in standard form. Finally, graph the equation showing all axes used. (Note: Some problems involve special angles, but several do not.)

11. $3x^2 + 10xy + 3y^2 + 8 = 0$
12. $2x^2 + xy + 2y^2 = 90$
13. $4x^2 - 3xy = 18$
14. $4xy - 3y^2 = 64$

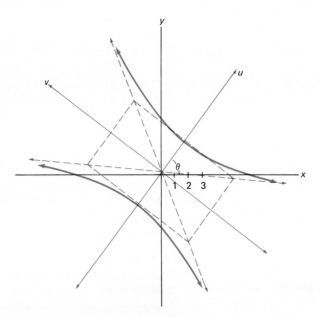

Figure 45

15. $x^2 - 2\sqrt{3}\,xy + 3y^2 - 12\sqrt{3}x - 12y = 0$

16. $x^2 + 2\sqrt{3}\,xy + 3y^2 + 8\sqrt{3}x - 8y = 0$

17. $13x^2 + 6\sqrt{3}\,xy + 7y^2 - 32 = 0$

18. $17x^2 + 12xy + 8y^2 + 17 = 0$

19. $9x^2 - 24xy + 16y^2 - 60x + 80y + 75 = 0$

20. $16x^2 + 24xy + 9y^2 - 20x - 15y - 150 = 0$

EXAMPLE B (The Inverse Rotation Formulas)   For a rotation of axes through angle θ, obtain the formulas that express u and v in terms of x and y. Then use the result to obtain the uv-coordinates of the point that has xy-coordinates $(4, 2\sqrt{3})$ if the angle θ is 30°.

Solution.   Consider the rotation formulas at the beginning of this problem set. Multiply the first one by cos θ and the second by sin θ, then add.

$$x \cos \theta = u \cos^2 \theta - v \sin \theta \cos \theta$$

$$y \sin \theta = u \sin^2 \theta + v \sin \theta \cos \theta$$

$$\overline{x \cos \theta + y \sin \theta = u(\cos^2 \theta + \sin^2 \theta)} \qquad = u$$

Similarly, multiply the first formula by $-\sin \theta$ and the second by cos θ, and add. The two resulting formulas are

$$u = x \cos \theta + y \sin \theta$$

$$v = -x \sin \theta + y \cos \theta$$

To find the values of u and v, simply substitute $x = 4$, $y = 2\sqrt{3}$, and $\theta = 30°$ in the above formulas. This gives

$$u = 4 \cdot \frac{\sqrt{3}}{2} + 2\sqrt{3} \cdot \frac{1}{2} = 3\sqrt{3}$$

$$v = -4 \cdot \frac{1}{2} + 2\sqrt{3} \cdot \frac{\sqrt{3}}{2} = 1$$

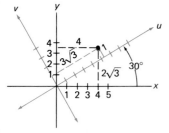

Figure 46

The geometric interpretation of these numbers is shown in Figure 46.

In Problems 21–26, find u and v for the given values of x, y, and θ. Then make a diagram to check that your answers make sense.

21. $(5, -3)$; $60°$
22. $(-2, 5)$; $60°$
23. $(3\sqrt{2}, \sqrt{2})$; $45°$
24. $(5/\sqrt{2}, -5/\sqrt{2})$; $45°$
25. $(3, 4)$; $\tan^{-1}(\frac{4}{3})$
26. $(5, -12)$; $\arctan(\frac{5}{12})$
27. Find the xy-equation that simplifies to $u^2 = 4v$ when the axes are rotated through an angle of $60°$.
28. Find the xy-equation that simplifies to $u^2 - 4v^2 = 4$ when the axes are rotated through an angle of $30°$.

MISCELLANEOUS PROBLEMS

29. Transform the equation $2x^2 + \sqrt{3}xy + y^2 = 5$ to a uv-equation by rotating the axes through $30°$. Use the result to identify the corresponding curve.
30. Without any algebra, determine the uv-equation corresponding to the equation $x^2/16 + y^2/9 = 1$ when the axes are rotated through $90°$. Then do the algebra to corroborate your answer.
31. Without any algebra, determine the uv-equation corresponding to $(x - 2\sqrt{2})^2 + (y - 2\sqrt{2})^2 = 16$ when the axes are rotated through $45°$.
32. Find the angle θ through which the axes must be rotated so that the circle $(x - 4)^2 + (y - 3)^2 = 4$ lies above the u-axis and is tangent to it. Write the uv-equation of this circle.
33. By a rotation of axes, remove the cross-product term from $13x^2 + 24xy + 3y^2 = 105$ and identify the corresponding conic section.
34. Transform the equation $(y^2 - x^2)(y + x) = 8\sqrt{2}$ to a uv-equation by rotating the axes through $45°$. Sketch the graph showing both sets of axes.
35. The graph of $x \cos \alpha + y \sin \alpha = d$ is a line. Show that the perpendicular distance from the origin to this line is $|d|$ by making a rotation of axes through the angle α.
36. Use Problem 35 to show that the perpendicular distance from the origin to the line $ax + by = c$ is $|c|/\sqrt{a^2 + b^2}$.
37. Use the result of Problem 36 to find the perpendicular distance from the origin to the line $5x + 12y = 39$.
38. When $Ax^2 + Bxy + Cy^2 = K$ is transformed to $au^2 + buv + cv^2 = K$ by a rotation of axes, it turns out that $A + C = a + c$ and $B^2 - 4AC = b^2 - 4ac$. (Ambitious students will find showing this to be a straightforward but somewhat

lengthy algebraic exercise.) Use these results to transform $x^2 - 8xy + 7y^2 = 9$ to $au^2 + cv^2 = 9$ without actually carrying out the rotation.

39. Recall that the area of an ellipse with major diameter $2a$ and minor diameter $2b$ is πab. Use the first sentence of Problem 38 to show that if $A + C$ and $4AC - B^2$ are both positive, then $Ax^2 + Bxy + Cy^2 = 1$ is the equation of an ellipse with area $2\pi/\sqrt{4AC - B^2}$.

40. **TEASER** The graph of $x^2 - 2xy + 3y^2 = 32$ is an ellipse and therefore can be circumscribed by a rectangle with sides parallel to the x- and y-axes. Find the vertices of this rectangle.

Which Curve Has the Simpler Equation?

Parabola

Four-leaved Rose

11-6 The Polar Coordinate System

The question we have asked above makes no sense unless coordinate axes are present. Most people would then choose the parabola as having the simpler equation; but the question is still more subtle than one might think. You already know that the complexity of the equation of a curve depends on the placement of the coordinate axes. Placed just right, the equation of the parabola might be as simple as $y = x^2$. Placed less wisely, the equation might be as complicated as $x - 3 = -(y + 7)^2$, or even worse. However, the four- leaved rose has a very messy equation no matter where the x- and y-axes are placed.

But there is another aspect to the question, one that Fermat and Descartes did not think about when they gave us Cartesian coordinates. There are many different kinds of coordinate systems, that is, different ways of specifying the position of a point. One of these systems, when placed the best possible way, gives the four-leaved rose a delightfully simple equation (see Example B). This system is called the **polar coordinate system;** it simplifies many problems that arise in calculus.

POLAR COORDINATES

In place of two perpendicular axes as in Cartesian coordinates, we introduce in the plane a single horizontal ray, called the **polar axis,** emanating from a fixed point O, called the **pole.** On the polar axis, we mark off the positive half of a number scale with zero at the pole. Any point P other than the pole is the intersection of a unique circle with center O and a unique ray emanating from O, (Figure 47). If r is the radius of the circle and θ is the angle the ray makes with the polar axis, then (r, θ) are the polar coordinates of P.

Points specified by polar coordinates are easiest to plot if we use polar graph paper. The grid on this paper consists of concentric circles and rays emanating from their common center. We have reproduced such a grid in Figure 48 and plotted a few points.

Of course, we can measure the angle θ in degrees as well as radians. More significantly, notice that while a pair of coordinates (r, θ) determines a unique point $P(r, \theta)$, each point has many different pairs of polar coordinates. For example,

$$\left(2, \frac{3\pi}{2}\right) \qquad \left(2, -\frac{\pi}{2}\right) \qquad \left(2, \frac{7\pi}{2}\right)$$

are all coordinates for the same point.

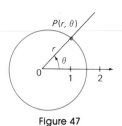

Figure 47

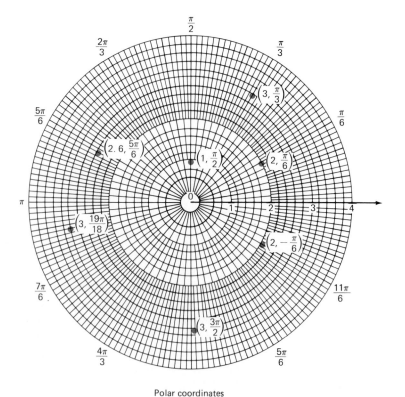

Polar coordinates

Figure 48

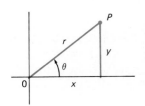

Figure 49

RELATION TO CARTESIAN COORDINATES

Let the positive x-axis of the Cartesian coordinate system serve also as the polar axis of a polar coordinate system, the origin coinciding with the pole (Figure 49). The Cartesian coordinates and polar coordinates are related by two pairs of simple equations.

$$x = r \cos \theta \qquad r^2 = x^2 + y^2$$

$$y = r \sin \theta \qquad \tan \theta = \frac{y}{x}$$

For example, if $(4, \pi/6)$ are the polar coordinates of a point, then its Cartesian coordinates are

$$x = 4 \cos \frac{\pi}{6} = 4 \cdot \frac{\sqrt{3}}{2} = 2\sqrt{3}$$

$$y = 4 \sin \frac{\pi}{6} = 4 \cdot \frac{1}{2} = 2$$

On the other hand, if $(-3, \sqrt{3})$ are the Cartesian coordinates of a point (Figure 50), then

$$r = \sqrt{(-3)^2 + (\sqrt{3})^2} = \sqrt{12} = 2\sqrt{3}$$

$$\tan \theta = \frac{\sqrt{3}}{-3}$$

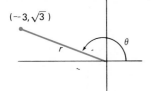

$(-3, \sqrt{3})$

Figure 50

Since the point is in the second quadrant, we choose $5\pi/6$ as an appropriate value of θ. Thus one choice of polar coordinates for the point in question is $(2\sqrt{3}, 5\pi/6)$.

POLAR GRAPHS

The simplest polar equations are $r = k$ and $\theta = k$, where k is a constant. The graph of the first is a circle; the graph of the second is a ray emanating from the origin. Examples are shown in Figure 51. Equations like

$$r = 4 \sin^2\theta \quad \text{and} \quad r = 1 + \cos 2\theta$$

are more complicated. To graph such equations, we suggest making a table of values, plotting the corresponding points, and then connecting those points with a smooth curve. As an example, consider the equation

$$r = \frac{1}{1 - \cos \theta}$$

In Figure 52, we have constructed a table of values and drawn the corresponding graph. It looks suspiciously like a parabola, and in the next section, we will verify that this suspicion is correct.

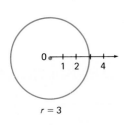

$r = 3$

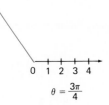

$\theta = \frac{3\pi}{4}$

Figure 51

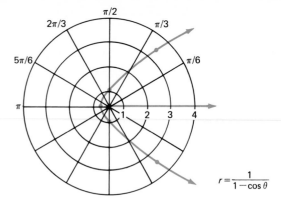

r	θ
—	0
3.4	$\pi/4$
1	$\pi/2$
.6	$3\pi/4$
.5	π
.6	$5\pi/4$
1	$3\pi/2$
3.4	$7\pi/4$
—	2π

$$r = \frac{1}{1 - \cos\theta}$$

Figure 52

LIMAÇONS

We consider next equations of the form

$$r = a \pm b \cos\theta \quad \text{and} \quad r = a \pm b \sin\theta$$

with a and b positive. Their graphs are called **limaçons** with the special case $a = b$ giving a curve called a **cardioid** (a heartlike curve). We assert that these graphs have the shapes shown in Figure 53.

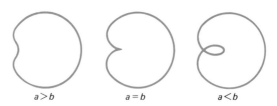

Figure 53

(In the case $a < b$, we allow r to be negative, a matter discussed in Example B in the problem set.)

Consider as an example the equation

$$r = 2(1 + \cos\theta)$$

Its graph (together with a table of values) is the cardioid shown in Figure 54.

Problem Set 11-6

Graph each of the following points given in polar coordinates. Polar graph paper will simplify the graphing process.

1. $\left(3, \dfrac{\pi}{4}\right)$ 2. $\left(2, \dfrac{\pi}{3}\right)$ 3. $\left(\dfrac{3}{2}, \dfrac{5\pi}{6}\right)$ 4. $\left(1, \dfrac{5\pi}{3}\right)$

5. $(3, \pi)$ 6. $\left(2, \dfrac{\pi}{2}\right)$ 7. $(3, -\pi)$ 8. $\left(2, -\dfrac{3\pi}{2}\right)$

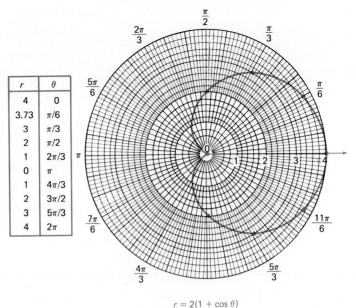

r	θ
4	0
3.73	$\pi/6$
3	$\pi/3$
2	$\pi/2$
1	$2\pi/3$
0	π
1	$4\pi/3$
2	$3\pi/2$
3	$5\pi/3$
4	2π

$$r = 2(1 + \cos \theta)$$

Figure 54

9. $(4, 70°)$ 10. $(3, 190°)$ 11. $\left(\dfrac{5}{2}, \dfrac{7\pi}{3}\right)$ 12. $\left(\dfrac{7}{2}, \dfrac{11\pi}{4}\right)$

Find the Cartesian coordinates of the point having the given polar coordinates.

13. $\left(4, \dfrac{\pi}{4}\right)$ 14. $\left(6, \dfrac{\pi}{6}\right)$ 15. $(3, \pi)$ 16. $\left(2, \dfrac{3\pi}{2}\right)$

17. $\left(10, \dfrac{4\pi}{3}\right)$ 18. $\left(8, \dfrac{11\pi}{6}\right)$ 19. $\left(2, -\dfrac{\pi}{4}\right)$ 20. $\left(3, -\dfrac{2\pi}{3}\right)$

Find polar coordintes for the point with the given Cartesiancoordinates.

21. $(4, 0)$ 22. $(0, 3)$ 23. $(-2, 0)$ 24. $(0, -5)$
25. $(2, 2)$ 26. $(2, -2)$ 27. $(-2, 2)$ 28. $(-2, -2)$
29. $(1, -\sqrt{3})$ 30. $(-2\sqrt{3}, 2)$ 31. $(3, -\sqrt{3})$ 32. $(-\sqrt{3}, -3)$

Graph each of the following equations. Use polar graph paper if it is available.

33. $r = 2$ 34. $r = 5$
35. $\theta = \pi/3$ 36. $\theta = -2\pi/3$
37. $r = |\theta|$ (with θ in radians) 38. $r = \theta^2$
39. $r = 2(1 - \cos \theta)$ 40. $r = 3(1 + \sin \theta)$
41. $r = 2 + \cos \theta$ 42. $r = 2 - \sin \theta$

EXAMPLE A (Transforming Equations) (a) Change the Cartesian equation $(x^2 + y^2)^2 = x^2 - y^2$ to a polar equation. (b) Change $r = 2 \sin 2\theta$ to a Cartesian equation.

Solution. (a) Replacing $x^2 + y^2$ by r^2, x by $r \cos \theta$, and y by $r \sin \theta$, we get

$$(r^2)^2 = r^2 \cos^2 \theta - r^2 \sin^2 \theta$$
$$r^4 = r^2(\cos^2 \theta - \sin^2 \theta)$$
$$r^2 = \cos 2\theta$$

Dividing by r^2 at the last step did no harm since the graph of the last equation passes through the pole $r = 0$.

(b)
$$r = 2 \sin \theta$$
$$r = 2 \cdot 2 \sin \theta \cos \theta$$

Multiplying both sides by r^2 gives

$$r^3 = 4(r \sin \theta)(r \cos \theta)$$
$$(x^2 + y^2)^{3/2} = 4yx$$

Transform to a polar equation.

43. $x^2 + y^2 = 4$ 44. $\sqrt{x^2 + y^2} = 6$

45. $y = x^2$ 46. $x^2 + (y - 1)^2 = 1$

Transform to a Cartesian equation.

47. $\tan \theta = 2$ 48. $r = 3 \cos \theta$ 49. $r = \cos 2\theta$ 50. $r^2 = \cos \theta$

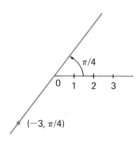

Figure 55

EXAMPLE B (Allowing Negative Values for r)   It is sometimes useful to allow r to be negative. By the point $(-3, \pi/4)$, we shall mean the point 3 units from the pole on the ray in the opposite direction from the ray for $\theta = \pi/4$ (see Figure 55). Allowing r to be negative, graph

$$r = 2 \sin 2\theta$$

Solution.   We begin with a table of values (Figure 56), plot the corresponding points, and then sketch the graph (Figure 57).

θ	0	$\dfrac{\pi}{12}$	$\dfrac{\pi}{6}$	$\dfrac{\pi}{4}$	$\dfrac{\pi}{3}$	$\dfrac{5\pi}{12}$	$\dfrac{\pi}{2}$	$\dfrac{7\pi}{12}$	$\dfrac{3\pi}{4}$	$\dfrac{11\pi}{12}$	π	$\dfrac{5\pi}{4}$	$\dfrac{3\pi}{2}$	$\dfrac{7\pi}{4}$	2π
2θ	0	$\dfrac{\pi}{6}$	$\dfrac{\pi}{3}$	$\dfrac{\pi}{2}$	$\dfrac{2\pi}{3}$	$\dfrac{5\pi}{6}$	π	$\dfrac{7\pi}{6}$	$\dfrac{3\pi}{2}$	$\dfrac{11\pi}{6}$	2π	$\dfrac{5\pi}{2}$	3π	$\dfrac{7\pi}{2}$	4π
r	0	1	$\sqrt{3}$	2	$\sqrt{3}$	1	0	-1	-2	-1	0	2	0	-2	0

         a b c d

Figure 56

Note: The four leaves correspond to the four parts (a), (b), (c), and (d) of the table of values. For example, leaf (b) results from values of θ between $\pi/2$ and π where r is negative. This graph is the four-leaved rose of our opening display. Its Cartesian equation was obtained in (b) of Example A.

Graph each of the following, allowing r to be negative.

51. $r = 3 \cos 2\theta$ 52. $r = \cos 3\theta$ 53. $r = \sin 3\theta$

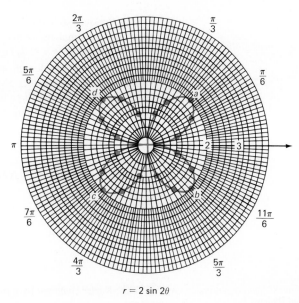

$r = 2 \sin 2\theta$

Figure 57

54. $r = 4 \cos \theta$ 55. $r = \sin 4\theta$ 56. $r = \cos 4\theta$

MISCELLANEOUS PROBLEMS

57. Transform to a Cartesian equation and identify the corresponding curve.
    (a) $r = 5/(3 \sin \theta - 2 \cos \theta)$ (b) $r = 4 \cos \theta - 6 \sin \theta$

58. Transform to a polar equation.
    (a) $x^2 = 4y$ (b) $(x - 5)^2 + (y + 2)^2 = 29$

Graph each of the polar equations in Problems 59–64.

59. $r = 4$ 60. $\theta = -\pi/3$

61. $r = 2(1 - \sin \theta)$ 62. $r = 1/\theta,\ \theta > 0$

63. $r^2 = \sin 2\theta$ *Caution:* Avoid values of θ that make r^2 negative.

64. $r = 2^\theta$ *Note:* Use both negative and positive values for θ.

65. Sketch the graphs of each pair of equations and find their points of intersection.
    (a) $r = 4 \cos \theta,\ r \cos \theta = 1$ (b) $r = 2\sqrt{3} \sin \theta,\ r = 2(1 + \cos \theta)$

66. Find the polar coordinates of the midpoint of the line segment joining the points with polar coordinates $(4, 2\pi/3)$ and $(8, \pi/6)$.

67. Show the distance d between the points with polar coordinates (r_1, θ_1) and (r_2, θ_2) is given by

$$d = \sqrt{r_1^2 + r_2^2 - 2r_1 r_2 \cos(\theta_2 - \theta_1)}$$

and use this result to find the distance between $(4, 2\pi/3)$ and $(8, \pi/6)$.

68. Show that a circle of radius a and center (a, α) has polar equation $r = 2a \cos(\theta - \alpha)$. *Hint:* Law of cosines.

69. Find a formula for the area of the polar rectangle $0 < a < r < b$, $\alpha \le \theta \le \beta,\ \beta - \alpha < \pi$.

70. A point P moves so that its distance from the pole is always equal to its distance from the horizontal line $r \sin \theta = 4$. Show that the equation of the resulting curve (a parabola) is $r = 4/(1 + \sin \theta)$.

71. A line segment L of length 4 has its two endpoints on the x- and y-axes, respectively. The point P is on L and is such that the line OP from the pole to P is perpendicular to L. Show that the set of points P satisfying this condition is a four-leaved rose by finding its polar equation.

72. **TEASER** Let F and F' be fixed points with polar coordinates $(a, 0)$ and $(-a, 0)$, respectively. A point P moves so that the product of its distances from F and F' is equal to the constant a^2 (that is, $\overline{PF} \cdot \overline{PF'} = a^2$). Find a simple polar equation (of the form $r^2 = f(\theta)$) for the resulting curve and sketch its graph.

If the Greeks had not cultivated conic sections, Kepler could not have superseded Ptolemy.

William Whewell

The planets move around the sun in ellipses; the sun is at one focus of these ellipses.

Kepler's First Law

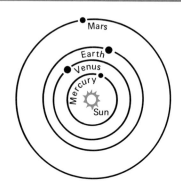

11-7   Polar Equations of Conics

Historians have suggested that the Greeks studied the conic sections simply to satisfy their intellectual cravings after the ideal and that they hardly dreamed that these curves would have important physical applications. Today we know that they describe the motions of the moon, of the planets, of comets, of space probes, and even of tiny electrons as they orbit the nucleus of an atom. In the study of such motions, it is not the Cartesian equations of the conics that prove most useful. It is rather their polar equations that play a central role.

A NEW APPROACH TO THE CONICS

The definitions we have given for the three conics (parabola, ellipse, and hyperbola) are quite dissimilar in form. However, there is another approach to the three curves that treats them in a uniform way; it is this approach that will eventually lead us to the polar equations for these curves.

In the plane, consider a fixed line l (the **directrix**) and a fixed point F (the **focus**). Let a point P move so that its distance $d(P, F)$ from the focus is a constant e times its distance from the directrix $d(P, L)$, that is, so that

$$d(P, F) = ed(P, L)$$

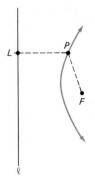

as suggested by Figure 58. Here the constant e (called the **eccentricity**) may be any positive number. Of course, if $e = 1$, we are on familiar ground; the corresponding curve is a parabola. But what if $e \neq 1$?

To get a feeling for this new situation, try graphing the path of P for the two cases $e = \frac{1}{2}$ and $e = 2$. If you do it carefully, you should get curves that look like an ellipse and a hyperbola, respectively. In fact, we claim that the equation $d(P, F) = ed(P, L)$ can serve as the definition of an ellipse when $0 < e < 1$ and of a hyperbola when $e > 1$. We demonstrate this now.

Figure 58

THE NEW DEFINITIONS ARE EQUIVALENT TO THE OLD ONES

Suppose that $e \neq 1$ and consider the curve determined by the defining equation $d(P, F) = ed(P, L)$. It is fairly easy to see that this curve must be symmetric with respect to the line through the focus and perpendicular to the directrix and that the curve must cross this line twice say at A' and A. Place the curve in the coordinate system so that A' and A have coordinates $(-a, 0)$ and $(a, 0)$, respectively. Let the directrix be the line $x = k$ and the focus be the point $(c, 0)$ with a, k, and c all positive. There are two possible arrangements (Figure 59) depending on whether $0 < e < 1$ or $e > 1$. (Do not let the appearance of the curves in the figure lead you to the conclusion that we have proved anything yet.)

Apply the equation $d(P, F) = ed(P, L)$ first with $P = A'$ and then with $P = A$ to obtain the pair of equations

$$a - c = e(k - a) = ek - ea$$
$$a + c = e(k + a) = ek + ea$$

Solve this pair of equations for a and c to get

$$c = ea \quad \text{and} \quad k = \frac{a}{e}$$

and note for later reference that this implies $e = c/a$. Now let $P(x, y)$ be any

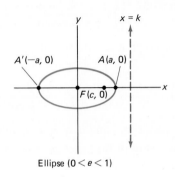

Ellipse $(0 < e < 1)$

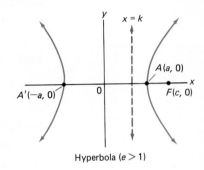

Hyperbola $(e > 1)$

Figure 59

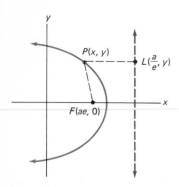

Figure 60

point on the curve. Then $L(a/e, y)$ is its projection on the directrix (see Figure 60 for the case $0 < e < 1$) and the condition $d(P, F) = ed(P, L)$ translates to

$$\sqrt{(x - ae)^2 + y^2} = e\sqrt{\left(x - \frac{a}{e}\right)^2}$$

Squaring both sides and collecting like terms yields

$$x^2 - 2aex + a^2e^2 + y^2 = e^2\left(x^2 - 2\frac{a}{e}x + \frac{a^2}{e^2}\right)$$

or

$$(1 - e^2)x^2 + y^2 = a^2(1 - e^2)$$

or

$$\frac{x^2}{a^2} + \frac{y^2}{(1 - e^2)a^2} = 1$$

If $0 < e < 1$, this is the standard equation of an ellipse; if $e > 1$, it is the standard equation of a hyperbola. Morever since $e = c/a$, our use of e in this section is consistent with our usage in Sections 11-2 and 11-3.

POLAR EQUATIONS OF THE CONICS

To simplify matters, we will place the conic in the polar coordinate system so that the focus is at the pole (origin) and the directrix is d units away as in Figure 61. The defining equation $d(P, F) = ed(P, L)$ takes the form

$$r = e(d - r\cos(\theta - \theta_0))$$

which is equivalent to

$$\boxed{r = \frac{ed}{1 + e\cos(\theta - \theta_0)}}$$

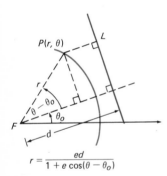

$$r = \frac{ed}{1 + e\cos(\theta - \theta_0)}$$

Figure 61

As an example, consider a case where $\theta_0 = 0$, namely,

$$r = \frac{2}{1 + \frac{1}{2}\cos\theta} = \frac{\frac{1}{2}\cdot 4}{1 + \frac{1}{2}\cos\theta}$$

Since $e = \frac{1}{2}$, the graph is an ellipse, the one shown in Figure 62. On the other hand,

$$r = \frac{12}{3 + 4\cos\theta} = \frac{4}{1 + \frac{4}{3}\cos\theta} = \frac{\frac{4}{3}\cdot 3}{1 + \frac{4}{3}\cos\theta}$$

is a hyperbola with $e = \frac{4}{3}$ and $d = 3$. Examples A–C in the problem set give a complete discussion of the cases $\theta_0 = 0$, $\pi/2$, π, and $3\pi/2$.

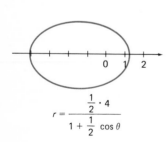

$$r = \frac{\frac{1}{2}\cdot 4}{1 + \frac{1}{2}\cos\theta}$$

Figure 62

POLAR EQUATIONS OF A LINE

If the line passes through the pole, it has the exceedingly simple equation $\theta = \theta_0$, with θ_0 a constant. If the line does not go through the pole, it is some distance d from it. Let θ_0 be the angle from the polar axis to the perpendicular drawn from the pole to the given line (Figure 63). Then if $P(r, \theta)$ is a point on the line,

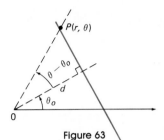

Figure 63

$$\cos(\theta - \theta_0) = \frac{d}{r}$$

or

$$r = \frac{d}{\cos(\theta - \theta_0)}$$

For example,

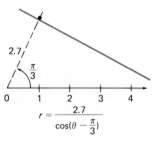

$$r = \frac{2.7}{\cos(\theta - \frac{\pi}{3})}$$

Figure 64

$$r = \frac{2.7}{\cos(\theta - \pi/3)}$$

is the equation of the line shown in Figure 64.

THE POLAR EQUATION OF A CIRCLE

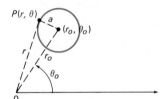

Figure 65

If the circle is centered at the pole, its polar equation is simply $r = a$, where a is the radius of the circle. If the center is at (r_0, θ_0), then by the law of cosines (see Figure 65)

$$a^2 = r^2 + r_0^2 - 2rr_0 \cos(\theta - \theta_0)$$

which is too complicated to be of much use. However, if the circle passes through the pole so $r_0 = a$, the equation simplifies to

$$r^2 = 2ra \cos(\theta - \theta_0)$$

or, after dividing by r,

$$r = 2a \cos(\theta - \theta_0)$$

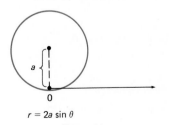

$r = 2a \cos \theta$

$r = 2a \sin \theta$

Figure 66

The cases $\theta_0 = 0$ and $\theta_0 = \pi/2$ are particularly nice. The first gives $r = 2a \cos \theta$; the second gives $r = 2a \cos(\theta - \pi/2)$, which is equivalent to $r = 2a \sin \theta$. The graphs for these two cases are shown in Figure 66.

In Figure 67, we have summarized most of the results of this section.

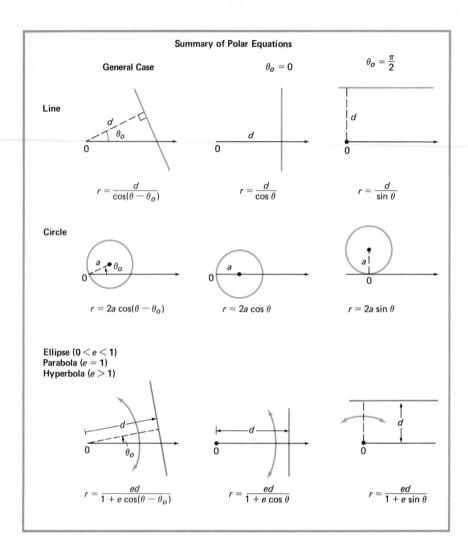

Summary of Polar Equations

Figure 67

Problem Set 11-7

In doing the following problems, it will be helpful to keep three things in mind.

1. *The summary of polar equations in Figure 67.*
2. *The addition formulas* $\cos(\theta \mp \theta_0) = \cos\theta\cos\theta_0 \pm \sin\theta\sin\theta_0$.
3. *The relations* $x = r\cos\theta$ *and* $y = r\sin\theta$.

In Problems 1–4, write a simple polar equation for each equation.

1. $x = 4$ 2. $y = 3$ 3. $x = -3$ 4. $y = -2$

Each of the polar equations in Problems 5–10 represents a line. Sketch its graph and then write its xy-equation in the form Ax + By = C.

5. $r = \dfrac{6}{\cos \theta}$

6. $r = \dfrac{3}{\sin \theta}$

7. $r = \dfrac{4}{\cos(\theta - \pi/3)}$

8. $r = \dfrac{10}{\cos(\theta - \pi/4)}$

9. $r = \dfrac{5}{\cos(\theta + \pi/4)}$

10. $r = \dfrac{4}{\cos(\theta + \pi/3)}$

In Problems 11–14, write the polar equation of the circle which passes through the pole and whose center has the given polar coordinates. Then find the corresponding xy-equations.

11. (4, 0)   12. (3, 90°)   13. (5, π/3)   14. (8, 150°)

EXAMPLE A (Conics with Directrix Perpendicular to the Polar Axis) Refer to Figure 68. If the directrix l is perpendicular to the polar axis, then $\theta_0 = 0$ or $\theta_0 = \pi$, and the polar equation of the conic takes one of the two forms.

$$\theta_0 = 0 \qquad\qquad \theta_0 = \pi$$

$$r = \frac{ed}{1 + e \cos \theta} \qquad r = \frac{ed}{1 - e \cos \theta}$$

In the first case, the directrix is to the right of the focus; in the second, it is to the left.

Identify each of the following conics by name, find its eccentricity, and write the xy-equation of its directrix.

(a) $r = \dfrac{4}{3 + 3 \cos \theta}$

(b) $r = \dfrac{5}{2 - 3 \cos \theta}$

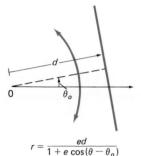

$$r = \frac{ed}{1 + e \cos(\theta - \theta_o)}$$

Figure 68

Solution.

(a) Divide numerator and denominator by 3, obtaining

$$r = \frac{\frac{4}{3}}{1 + \cos \theta}$$

The conic is a parabola, since the eccentricity $e = 1$. The equation of the directrix is $x = \frac{4}{3}$.

(b) The equation can be rewritten as

$$r = \frac{\frac{5}{2}}{1 - \frac{3}{2} \cos \theta} = \frac{\frac{3}{2} \cdot \frac{5}{3}}{1 - \frac{3}{2} \cos \theta}$$

The conic is a hyperbola with $e = \frac{3}{2}$ and directrix $x = -\frac{5}{3}$.

In Problems 15–22, identify the conic by name, give its eccentricity, and write the equation of its directrix (in xy-coordinates).

15. $r = \dfrac{4}{1 + \frac{2}{3} \cos \theta}$

16. $r = \dfrac{\frac{9}{2}}{1 + \frac{3}{4} \cos \theta}$

17. $r = \dfrac{5}{2 + 4 \cos \theta}$

18. $r = \dfrac{3}{1 + \cos \theta}$ 19. $r = \dfrac{7}{1 - \cos \theta}$ 20. $r = \dfrac{\frac{1}{2}}{1 - \frac{3}{2} \cos \theta}$

21. $r = \dfrac{\frac{1}{2}}{\frac{3}{2} - \cos \theta}$ 22. $r = \dfrac{3}{6 - 6 \cos \theta}$

EXAMPLE B (Conics with Directrix Parallel to the Polar Axis)   Refer to the diagram of Example A. If the directrix is parallel to the polar axis and above it, then $\theta_0 = \pi/2$; if it is below the polar axis, then $\theta_0 = 3\pi/2$. The corresponding equations can be simplified to

$$\theta_0 = \frac{\pi}{2} \qquad\qquad\qquad \theta_0 = \frac{3\pi}{2}$$

$$r = \frac{ed}{1 + e \sin \theta} \qquad\qquad r = \frac{ed}{1 - e \sin \theta}$$

(a) Derive the first of these equations.

(b) Identify the conic $r = 5/(2 - \sin \theta)$ by name, give its eccentricity, and write the xy-equation of its directrix.

Solution.

(a) The equation of the conic with $\theta_0 = \pi/2$ is

$$r = \frac{ed}{1 + e \cos (\theta - \frac{\pi}{2})}$$

Since

$$\cos(\theta - \tfrac{\pi}{2}) = \cos \theta \cos \tfrac{\pi}{2} + \sin \theta \sin \tfrac{\pi}{2} = \sin \theta$$

we get

$$r = \frac{ed}{1 + e \sin \theta}$$

(b) Dividing numerator and denominator by 2, we obtain

$$r = \frac{\frac{5}{2}}{1 - \frac{1}{2} \sin \theta} = \frac{\frac{1}{2} \cdot 5}{1 - \frac{1}{2} \sin \theta}$$

The conic is an ellipse with eccentricity $\frac{1}{2}$. The directrix is below the polar axis and has xy-equation $y = -5$.

In Problems 23–28, identify the conic by name, give its eccentricity, and write the xy-equation of its directrix.

23. $r = \dfrac{5}{1 + \sin \theta}$ 24. $r = \dfrac{2}{1 + \frac{2}{3} \sin \theta}$ 25. $r = \dfrac{6}{2 - \sin \theta}$

26. $r = \dfrac{5}{2 - 4 \sin \theta}$ 27. $r = \dfrac{4}{2 + \frac{5}{2} \sin \theta}$ 28. $r = \dfrac{5}{4 - 3 \sin \theta}$

EXAMPLE C (Graphing Conics in Polar Coordinates)   Graph the conic whose polar equation is

$$r = \frac{6}{1 + 2 \cos \theta}$$

Solution.   We recognize this as the polar equation of a hyperbola ($e = 2$) with major axis along the polar axis. Next we make the small table of values shown in Figure 69 and plot the corresponding points (marked with dots). The points marked with a cross are obtained by symmetry $(\cos(-\theta) = \cos \theta)$.

Graph each of the following conics.

29. $r = \dfrac{4}{1 + \frac{2}{3} \cos \theta}$ 30. $r = \dfrac{6}{1 + \frac{3}{4} \sin \theta}$ 31. $r = \dfrac{5}{1 + \sin \theta}$

32. $r = \dfrac{4}{2 + 2 \cos \theta}$ 33. $r = \dfrac{18}{2 + 3 \cos \theta}$ 34. $r = \dfrac{3}{1 - 2 \cos \theta}$

r	θ
2	0°
3	60°
6	90°
−8.2	150°
−6	180°

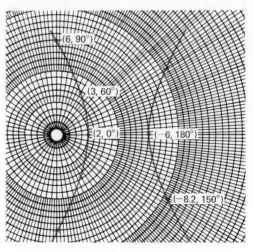

Figure 69

MISCELLANEOUS PROBLEMS

35. Write the polar equation $r = f(\theta)$ that corresponds to each of the following Cartesian equations.
    (a) $y = 3$ 　　　　　　　　　　　　　(b) $x^2 + y^2 = 9$
    (c) $(x + 9)^2 + y^2 = 81$ 　　　　　(d) $(x - 3)^2 + (y - 3)^2 = 18$
36. Graph each of the following polar equations.
    (a) $r = -4/\cos \theta$ 　　　　　　　(b) $r = -8 \sin \theta$
    (c) $r = 3$ 　　　　　　　　　　　　　(d) $r = 6 \cos(\theta - \pi/3)$
    (e) $r = 6/(1 - \cos \theta)$ 　　　　(f) $r = 6/(2 + \cos \theta)$
37. Determine the polar equations for the parabolas with Cartesian equations given by (a) $4(x + 1) = y^2$ and (b) $-8(y - 2) = x^2$. *Hint:* In each case, the focus is at the origin.

38. Find the polar equation $r = f(\theta)$ corresponding to the parabola with polar equation $y^2 = 16x$. Why is this polar equation unlike any discussed in this section?

39. Write the Cartesian equation of the curve whose polar equation is

$$\cos^2\theta + \sin\theta\cos\theta - 6\sin^2\theta = 0$$

40. Find the points of intersection (in polar coordinates) of the circles with polar equations $r = 3\sin\theta$ and $r = \sqrt{3}\cos\theta$.

41. Find e and d for the ellipse with polar equation $r = 8/(2 + \cos\theta)$. Also find the major and minor diameters of this ellipse.

42. Show that the ellipse $r = ed/(1 + e\cos\theta)$ has major diameter $2ed/(1 - e^2)$ and minor diameter $2ed/\sqrt{1 - e^2}$

43. Find the length of the latus rectum (chord through the focus perpendicular to the major axis) for the ellipse $r = 8/(2 + \cos\theta)$.

44. Express the length of the latus rectum of the conic $r = ed/(1 + e\cos\theta)$ in terms of e and d.

45. The graph of $r = 2a + a\cos\theta$ for $a > 0$ is an example of a curve called a *limaçon* (Section 11-6).
    (a) Sketch this graph for the case $a = 3$.
    (b) Show that every chord through the pole has length $4a$ (a nice property it shares with the circle $r = 2a$).

46. **TEASER** Sketch the graph of the polar equation $r = 1/\theta$ for $\theta \geq \pi/2$ and show that this curve has infinite length.

Every physicist knows that a good way to specify the path of a particle is to give its coordinates as functions of the elapsed time t. Suppose a point $P(x, y)$ moves in the plane so that $x = 3 \cos t$ and $y = 2 \sin t$. What is the shape of its path?

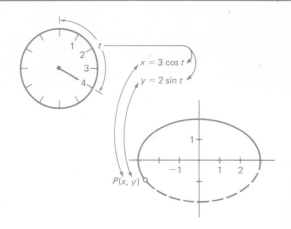

11-8 Parametric Equations

So far, we have described the conic sections by their Cartesian equations and by their polar equations. There is still another way to describe the conic sections, as well as many other curves. It arises naturally in the study of motion in physics, but its use goes far beyond that particular application.

Imagine that the xy-coordinates of a point on a curve are specified not by giving a relationship between x and y, but rather by telling how x and y are related to a third variable. For example, it may be that as time t advances from $t = a$ to $t = b$, a point $P(x, y)$ traces out a curve in the xy-plane. Then both x and y are functions of t. That is,

$$x = f(t) \quad \text{and} \quad y = g(t) \qquad a \le t \le b$$

We call the boxed equations the **parametric equations** of a curve with t as parameter. A **parameter** is simply an auxiliary variable on which other variables depend.

AN EXAMPLE

Let

$$x = 2t \quad \text{and} \quad y = t^2 - 3 \qquad -1 \le t \le 3$$

These equations can be used to make a table of values and then to draw a graph. We illustrate in Figure 70 on the next page.

The curve just drawn looks suspiciously like part of a parabola. We can demonstrate that this is true by **eliminating the parameter** t. Solve the first equation for t, giving $t = x/2$. Then substitute this value of t in the second equation. We obtain

t	x	y
-1	-2	-2
$-1/2$	-1	$-11/4$
0	0	-3
$1/2$	1	$-11/4$
1	2	-2
2	4	1
3	6	6

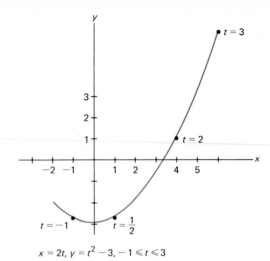

$$x = 2t, \; y = t^2 - 3, \; -1 \leqslant t \leqslant 3$$

Figure 70

$$y = \left(\frac{x}{2}\right)^2 - 3 = \frac{1}{4}x^2 - 3$$

which we do recognize as the equation of a parabola.

PARAMETRIC EQUATIONS OF A LINE

Consider first a line l which passes through the points $(0, 0)$ and (a, b). We claim that

$$x = at \quad \text{and} \quad y = bt$$

are a pair of parametric equations for this line. To see why, note that $t = 0$ and $t = 1$ yield the given points. Moreover, if we eliminate t, we get $y = (b/a)x$, which is the equation of a line. This example shows, incidentally, that a parametric representation is not unique since there are many choices for (a, b). (See Problem 29 for more evidence on this point.)

Next we translate the above line by replacing x by $x - x_1$ and y by $y - y_1$. This gives

$$\boxed{x = x_1 + at \quad \text{and} \quad y = y_1 + bt}$$

These are the parametric equations of a line through (x_1, y_1) parallel to the line with which we started (Figure 71).

Finally, consider the line through (x_1, y_1) with slope m. This line is parallel to the line through the points $(0, 0)$ and $(1, m)$ and so has parametric equations

$$x = x_1 + t \quad \text{and} \quad y = y_1 + mt$$

Figure 71

Note that if we eliminate t by solving the first equation for t and substituting, it in the second, we get $y - y_1 = m(x - x_1)$, the point-slope form for the equation of a line.

THE CIRCLE AND THE ELLIPSE

If you think about the definitions of sine and cosine for a moment, you already know one set of parametric equations for a circle of radius a centered at the origin.

$$x = a \cos t \quad \text{and} \quad y = a \sin t \qquad 0 \leq t \leq 2\pi$$

For other possibilities, see Problem 29.

Consider next an ellipse centered at $(0, 0)$ and passing through $(a, 0)$ and $(0, b)$. We claim that

$$x = a \cos t \quad \text{and} \quad y = b \sin t \qquad 0 \leq t \leq 2\pi$$

are parametric equations for it. To see that this is correct, we shall eliminate the parameter t. One way to do this is to solve for $\cos t$ and $\sin t$, respectively, square the results, and add. This gives

$$\frac{x^2}{a^2} + \frac{y^2}{b^2} = \cos^2 t + \sin^2 t = 1$$

which we recognize as the xy-equation of an ellipse centered at $(0, 0)$ and passing through $(a, 0)$ and $(0, b)$.

Now we can answer the question asked in the opening display of this section. The parametric equations

$$x = 3 \cos t \quad \text{and} \quad y = 2 \sin t$$

determine an ellipse with center at $(0, 0)$ and passing through $(3, 0)$ and $(0, 2)$.

THE HYPERBOLA

By analogy with the situation for the ellipse, we are led to

$$x = a \sec t \quad \text{and} \quad y = b \tan t \qquad 0 \leq t \leq 2\pi, t \neq \frac{\pi}{2}, t \neq \frac{3\pi}{2}$$

as parametric equations for a hyperbola. Note that when we solve for $\sec t$ and $\tan t$ in the two equations, square the results, and subtract, we obtain

$$\frac{x^2}{a^2} - \frac{y^2}{b^2} = \sec^2 t - \tan^2 t = 1$$

There is another set of parametric equations for the hyperbola. It involves an important pair of functions called the *hyperbolic sine* and the *hyperbolic cosine*. These functions are discussed in Problem 36.

THE CYCLOID

So far, our discussion of parametric equations has concentrated on the conic sections. Actually, there are other important curves where parametric representation is almost essential. The cycloid provides an example where the xy-equation is so complicated it is rarely used.

Consider a wheel of radius a which is free to roll along the x-axis. As the wheel turns, a point P on the rim traces out a curve called the **cycloid** (Figure 72).

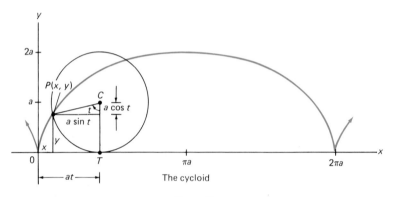

The cycloid

Figure 72

Assume P is initially at the origin and let C and T be as indicated in the diagram, with t denoting the radian measure of angle TCP. Then the arc PT and the segment OT have the same length and so the center C of the rolling circle is at (at, a). Using a little trigonometry, we conclude that

$$x = at - a \sin t = a(t - \sin t)$$

and

$$y = a - a \cos t = a(1 - \cos t)$$

Problem Set 11-8

In Problems 1–4, write parametric equations for the line that passes through the two given points.

1. $(0, 0)$, $(2, -3)$
2. $(0, 0)$, $(-3, 6)$
3. $(1, 2)$, $(4 -5)$
4. $(-2, 3)$, $(3, 7)$
5. Find the slope and y-intercept of the line with parametric equations $x = 3 + 2t$ and $y = -5 - 4t$. *Hint:* Eliminate the parameter t.

6. Write the equation of the line with parametric equations $x = -2 - 3t$ and $y = 4 + 9t$ in the form $Ax + By + C = 0$.

In Problems 7–16, eliminate the parameter to determine the corresponding xy-equation.

7. $x = 3s + 1, y = -2s + 5$ 8. $x = 3s + 1, y = s^2$

9. $x = 2t - 1, y = 2t^2 + t$ 10. $x = 3t, y = t^2 - 3t + 1$

11. $x = 2 \cos t, y = 2 \sin t$ 12. $x = 3 \sin t, y = 3 \cos t$

13. $x = 2 \cos t, y = 3 \sin t$ 14. $x = 6 \sin t, y = \cos t$

15. $x = 3t + 1, y = t^3$ 16. $x = 2 \sec t, y = 3 \tan t$

EXAMPLE A (More Graphing)   Sketch the graph of the curve with parametric equations $x = 8 \cos^3 t$ and $y = 8 \sin^3 t$, $0 \le t \le 2\pi$.

Solution.   A table of values and the graph are shown in Figure 73. This curve is called a *hypocycloid* (see Problem 39).

Sketch the graph of each of the following for the indicated interval.

17. $x = 2t - 1, y = t^2 + 2; -2 \le t \le 2$

18. $x = 2t^2, y = 3 - 2t; -2 \le t \le 2$

19. $x = t^3, y = t^2; -2 \le t \le 2$

20. $x = 2^t, y = 3t; -3 \le t \le 2$

21. $x = \dfrac{1 - t^2}{1 + t^2}, y = \dfrac{2t}{1 + t^2};$ all t

22. $x = \dfrac{t^2}{1 + t^2}, y = \dfrac{t^3}{1 + t^2};$ all t

$\boxed{\text{c}}$ 23. $x = 8t - 4 \sin t, y = 8 - 4 \cos t; 0 \le t \le 4\pi$
      Note: This curve is called a *curtate cycloid* (see Problem 37).

t	x	y
0	8	0
$\pi/6$	5.2	1
$\pi/4$	2.8	2.8
$\pi/3$	1	5.2
$\pi/2$	0	8
$3\pi/4$	-2.8	2.8
π	-8	0
$5\pi/4$	-2.8	-2.8
$3\pi/2$	0	-8
$7\pi/4$	2.8	-2.8

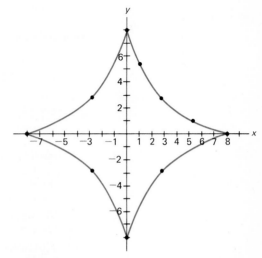

Figure 73

© 24. $x = 6t - 8 \sin t$, $y = 6 - 8 \cos t$; $0 \le t \le 4\pi$

    Note: This curve is called a *prolate cycloid* (see Problem 38).

EXAMPLE B (A Projectile Problem)   The path of a projectile fired at 64 feet per second from ground level at an angle of 60° with the horizontal is given by the parametric equations

$$x = 32t \quad \text{and} \quad y = -16t^2 + 32\sqrt{3}\,t$$

where the origin is at the point of firing and the *x*-axis is along the (horizontal) ground in the plane of the projectile's flight (Figure 74). Find
(a) the *xy*-equation of the path;
(b) the total time of flight;
(c) the range, that is, the value of *x* where the projectile strikes the ground;
(d) the greatest height reached.

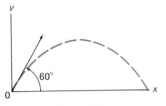

Figure 74

Solution.
(a) Since $t = x/32$,

$$y = -16\left(\frac{x}{32}\right)^2 + 32\sqrt{3}\left(\frac{x}{32}\right)$$

$$= -\tfrac{1}{64}x^2 + \sqrt{3}\,x$$

which is the equation of a parabola.
(b) We find the values of t for which $y = 0$.

$$y = -16t^2 + 32\sqrt{3}\,t = -16t(t - 2\sqrt{3})$$

Thus $y = 0$ when $t = 0$ (time of firing) and when $t = 2\sqrt{3}$ (time of landing). The time of flight is $2\sqrt{3}$ seconds.
(c) The range is $32(2\sqrt{3}) = 64\sqrt{3}$ feet.
(d) The maximum height occurs when $t = \tfrac{1}{2}(2\sqrt{3}) = \sqrt{3}$. At this time,

$$y = -16(\sqrt{3})^2 + 32\sqrt{3}(\sqrt{3}) = 48 \text{ feet}$$

Answer the four questions of Example B for the data below.

25. $x = 64\sqrt{3}\,t$, $y = -16t^2 + 64t$
26. $x = 48\sqrt{2}\,t$, $y = -16t^2 + 48\sqrt{2}\,t$

MISCELLANEOUS PROBLEMS

27. *Determine the Cartesian equation corresponding to each given pair of parametric equations.*
   (a) $x = 3 - 2t$, $y = 4 + 3t$
   (b) $x = 3t$, $y = 4 \cos t$
   (c) $x = 2 \sec t$, $y = 3 \tan t$
   (d) $x = 1 - t^3$, $y = 2t - 1$
   (e) $x = t^2 + 2t$, $y = \sqrt[3]{t} - t^2 - 2t$

28. Write parametric equations for each of the following.
   (a) The line that passes through the points $(2, -1)$ and $(4, 3)$.

(b) The line through $(4, -2)$ and parallel to the line with parametric equations $x = 3 + 2t, y = -2 + t$.

(c) The ellipse with Cartesian equation $x^2/9 + y^2/25 = 1$.

(d) The circle $(x - 4)^2 + (y + 2)^2 = 25$

29. Show that all of the following parametrizations represent the same curve (one quarter of a circle).

(a) $x = 2 \cos t, y = 2 \sin t; 0 \le t \le \pi/2$

(b) $x = \sqrt{t}, y = \sqrt{4 - t}; 0 \le t \le 4$

(c) $x = t + 1, y = \sqrt{3 - 2t - t^2}; -1 \le t \le 1$

(d) $x = (2 - 2t)/(1 + t), y = 4\sqrt{t}/(1 + t); 0 \le t \le 1$

30. Show that the parametric equations $x = \sqrt{2t + 1}, y = \sqrt{8t}, t \ge 0$, represent part of a hyperbola. Sketch that part.

31. Sketch the graph of $x = 2 + 3 \cos t, y = 1 + 4 \sin t, 0 \le t \le 2\pi$, and determine the corresponding Cartesian equation.

32. Sketch the graph of $x = \sin t, y = \tan t, -\pi/2 < t < \pi/2$, and determine the corresponding Cartesian equation.

33. The path of a projectile fired from level ground with a speed of v_0 feet per second at an angle α with the ground, is given by the parametric equations

$$x = (v_0 \cos \alpha)t \qquad y = -16t^2 + (v_0 \sin \alpha)t$$

(a) Show that the path is a parabola.

(b) Find the time of flight.

(c) Show that the range is $(v_0^2/32) \sin 2\alpha$.

(d) For a given v_0, what value of α gives the largest possible range?

34. Show that the parametric equations $x = t^{-1/2} \cos t, y = t^{-1/2} \sin t, t > 0$, represent the same curve as the spiral whose polar equation is $r^2\theta = 1, r > 0, \theta > 0$.

35. Show that the graph of $x = 5 \sin^2 t - 4 \cos^2 t, y = 4 \cos^2 t + 5 \sin^2 t$, $0 \le t \le \pi/2$, is a line segment and find its endpoints. What curve do you get when t varies from $\pi/2$ to π?

36. Define two (important) functions called the **hyperbolic sine** and **hyperbolic cosine** by

$$\sinh t = \frac{e^t - e^{-t}}{2} \qquad \cosh t = \frac{e^t + e^{-t}}{2}$$

(a) Show that $\cosh^2 t - \sinh^2 t = 1$

(b) Show that $x = a \cosh t, y = a \sinh t$ give a parameterization for one branch of a hyperbola.

37. Modify the text discussion of the cycloid (and its accompanying diagram) to handle the case where the point P is $b < a$ units from the center of the wheel. You should obtain the parametric equations

$$x = at - b \sin t \qquad y = a - b \cos t$$

The graph of these equations is called a *curtate cycloid* (see Problem 23).

38. Follow the instructions of Problem 37 for the case $b > a$ (a flanged wheel, as on a train) showing that you get the same parametric equations. The graph of these equations is now called a *prolate cycloid* (see Problem 24).

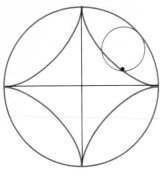

Figure 75

39. Suppose that a wheel of radius b rolls inside a circle of radius $a = 4b$ (Figure 75). Show that the parametric equations for a point P on the rim of the rolling wheel are

$$x = a \cos^3 \theta \qquad y = a \sin^3 \theta$$

provided the fixed wheel is centered at $(0, 0)$ and P is initially at $(a, 0)$. The resulting curve is called a *hypocycloid* (see Example A).

40. **TEASER**   A wheel of radius 3 and centered at the origin is rotating counterclockwise at 2 radians per second so that the point P on its rim is at $(3, 0)$ when time $t = 0$. Another wheel of radius 1 and centered at $(8, 0)$ is also rotating but clockwise at 2 radians per second so that the point Q on its rim is at $(9, 0)$ at $t = 0$. A knot K is tied at the middle of an elastic string and then one end of the string is attached to P and the other to Q. The string is short enough to remain taut at all times.
    (a) Obtain the parametric equations for the path of the knot K, using time t as the parameter.
    (b) Write the Cartesian equation for the path of K and use it to describe the path in geometric terms.

Chapter Summary

A **parabola** is the set of points that are equidistant from a fixed line (the **directrix**) and a fixed point (the **focus**). An **ellipse** is the set of points for which the sum of the distances from two fixed points (the **foci**) is a constant. A **hyperbola** is the set of points for which the difference of the distances from two fixed points (the **foci**) is a constant. These three curves (called **conics**) together with their limiting forms (circle, point, empty set, line, two intersecting lines, two parallel lines) are the chief objects of study in this chapter.

When these curves are placed in the Cartesian coordinate plane in an advantageous way, their xy-equations take the **standard forms** below.

$$\text{Parabola:} \qquad y^2 = 4px$$

$$\text{Ellipse:} \qquad \frac{x^2}{a^2} + \frac{y^2}{b^2} = 1$$

$$\text{Hyperbola:} \qquad \frac{x^2}{a^2} - \frac{y^2}{b^2} = 1$$

The curves can, of course, be placed in the plane in other ways; in these cases, their equations are more complicated. However, they can be brought to standard form by a **translation** or a **rotation** of axes. Even the complicated equations are always of the form

$$Ax^2 + Bxy + Cy^2 + Dx + Ey + F = 0$$

Conversely, the graph of any equation of this type is one of the conics or the six limiting forms.

Cartesian coordinates (x, y) are not the only way to specify the position of a point. **Polar coordinates** (r, θ) also determine points, and **polar equations** determine curves. In fact, some very beautiful curves (**limaçons**, **cardioids**, **roses**) have simple polar equations but complicated Cartesian equations. Moreover, for purposes of astronomy, the conics are best described by polar equations, since their equations all take the form

$$r = \frac{ed}{1 + e\cos(\theta - \theta_0)}$$

This equation determines an ellipse if e (called the **eccentricity**) satisfies $0 < e < 1$, a parabola if $e = 1$, and a hyperbola if $e > 1$.

Finally, many curves (including the conics) can be described by giving **parametric equations** $x = f(t)$ and $y = g(t)$ in which both x and y are specified in terms of a **parameter** t.

Chapter Review Problem Set

1. Determine the number of the description that best fits the given equation.

   _____ (a) $(x - 1)^2 + y^2 + 3 = 0$
   _____ (b) $x = y^2 + 2y + 3$
   _____ (c) $x^2 - 4y^2 = 0$
   _____ (d) $4(x - 1)^2 + y^2 = 3$
   _____ (e) $4(x - 1)^2 + 4y^2 = 3$
   _____ (f) $4(x - 1)^2 - 4y^2 = -3$
   _____ (g) $4(x - 1)^2 - 4y^2 = 3$
   _____ (h) $4(x - 1)^2 = 0$
   _____ (i) $4x^2 - 4 = 0$
   _____ (j) $(x - 1)^2 + 4y^2 = 0$
   _____ (k) $4x - 4y = 3$

   (i) horizontal ellipse
   (ii) vertical ellipse
   (iii) horizontal parabola
   (iv) vertical parabola
   (v) horizontal hyperbola
   (vi) vertical hyperbola
   (vii) circle
   (viii) intersecting lines
   (ix) parallel lines
   (x) single line
   (xi) single point
   (xii) empty set

2. Find the focus and vertex for the parabola with equation

   $$y = x^2 + 4x + 1$$

3. Find the xy-equation of the parabola that has vertex at the origin, opens up, and passes through $(2, 5)$.

4. Write the xy-equation of a vertical ellipse centered at $(2, 1)$ with major diameter of length 10 and minor diameter of length 6.

5. Sketch the graph of the ellipse with equation $x^2 + 6x + 4y^2 = 7$. Determine its eccentricity.

6. Find the xy-equation of a hyperbola with vertices $(0, \pm 3)$ that passes through $(4, 5)$.

7. Find the vertices of the hyperbola with equation $x^2 - 6x - 2y^2 - 4y = 27$.

8. Eliminate the xy-term from

   $$3x^2 + \sqrt{3}\,xy + 2y^2 = 28$$

by making a rotation of axes. Find the distance between the vertices of this conic section.

9. Plot the points with the following polar coordinates.
   (a) $(3, 2\pi/3)$ (b) $(2, -3\pi/2)$ (c) $(6, 210°)$

10. Find the Cartesian coordinates of the points in Problem 9.

11. Find the polar coordinates of the point having the given Cartesian coordinates.
   (a) $(5, 0)$ (b) $(2\sqrt{2}, -2\sqrt{2}$ (c) $(-2\sqrt{3}, 2)$

12. Graph each of the following polar equations.
   (a) $r = 4$ (b) $r = 4 \sin \theta$ (c) $r = 4 \cos 3\theta$

13. Transform $xy = 4$ to a polar equation. Transform $r = \sin 2\theta$ to a Cartesian equation.

14. Determine the number of the description that best fits the given polar equation.

   _____ (a) $r = 6 \sin \theta$ (i) a vertical line
   _____ (b) $r = 6/(1 + 2 \cos \theta)$ (ii) a nonvertical line
   _____ (c) $r = 3/\cos(\theta - \pi/4)$ (iii) a circle
   _____ (d) $r = 12/(2 + \cos \theta)$ (iv) a parabola
   _____ (e) $r = 6$ (v) an ellipse with $e = \frac{1}{2}$
   _____ (f) $r = 6/\sin \theta$ (vi) an ellipse with $e = \frac{1}{4}$
   _____ (g) $r = 6/(2 + 2 \cos \theta)$ (vii) a hyperbola
   _____ (h) $r = 12/(4 + \cos \theta)$ (viii) none of the above

15. Write the equation of a circle of radius 5 centered at the origin (pole) in
   (a) Cartesian coordinates;
   (b) polar coordinates;
   (c) parametric form using the polar angle θ as parameter.

16. Sketch the graph of the Witch of Agnesi, which has parametric equations

$$x = 4 \cot t \qquad y = 4 \sin^2 t$$

Appendix

Table A. Natural logarithms: See Section 3-5.

Table B. Common logarithms: See Sections 3-6 and 3-7.

Table C. Trigonometric functions (degrees): See Sections 4-1 and 4-5.

Table D. Trigonometric functions (radians): See Sections 4-2 and 4-5.

To find values between those given in any of these tables, we suggest a process called **linear interpolation.** If we know $f(a)$ and $f(b)$ and want $f(c)$, where c is between a and b, we may write

$$f(c) \approx f(a) + d$$

where d is obtained by pretending that the graph of $y = f(x)$ is a straight line on the interval $a \leq x \leq b$. A complete description is given in Section 3-6; the following examples illustrate this process.

EXAMPLE A (Natural Logarithms)   Find ln 2.133.

Solution.

$$.010 \left[.003 \begin{bmatrix} \ln 2.130 = .7561 \\ \ln 2.133 = \quad ? \\ \ln 2.240 = .7608 \end{bmatrix} d \right] .0047$$

$$\frac{d}{.0047} = \frac{.003}{.010} = .3$$

$$d = .3(.0047) \approx .0014$$

$$\ln 2.133 \approx \ln 2.130 + d = .7561 + .0014 = .7575$$

EXAMPLE B (Common Logarithms)   Find log 63.26.

Solution.

$$.10 \left[.06 \begin{bmatrix} \log 63.20 = 1.8007 \\ \log 63.26 = \quad ? \\ \log 63.30 = 1.8014 \end{bmatrix} d \right] .0007$$

$$\frac{d}{.0007} = \frac{.06}{.10} = .6$$

$$d = .6(.0007) \approx .0004$$

$$\log 63.26 \approx \log 63.20 + d = 1.8007 + .0004 = 1.8011$$

EXAMPLE C (Trigonometric Functions (Degrees))   Find sin 57.44°.

Solution.   Note that we must use the degree column on the right and the bottom caption in Table C.

$$.10 \left[.04 \begin{bmatrix} \sin 57.40° = .8425 \\ \sin 57.44° = \quad ? \\ \sin 57.50° = .8434 \end{bmatrix} d \right] .0009$$

$$\frac{d}{.0009} = \frac{.04}{.10} = .4$$

$$d = .4(.0009) \approx .0004$$

$$\sin 57.44° \approx \sin 57.40° + d = .8425 + .0004 = .8429$$

EXAMPLE D (Trigonometric Functions (Radians)) Find $\cos(1.436)$.

Solution.   The cosine is a decreasing function on the interval $0 \le t \le \pi/2$. This causes d to be negative.

$$.010\left[.006\left[\begin{array}{l}\cos 1.430 = .14033 \\ \cos 1.436 = \quad ? \\ \cos 1.440 = .13042\end{array}\right]d\right] - .00991$$

$$\frac{d}{-.00991} = \frac{.006}{.010} = .6$$

$$d = .6(-.00991) \approx -.00595$$

$$\cos(1.436) \approx \cos(1.430) + d = .14033 - .00595 = .13438$$

EXAMPLE E (Angle, Given a Trigonometric Function) Find θ if $\tan \theta = .43600$. Give the answer in radians.

Solution.   We use Table D and find that θ is between .41 and .42.

$$.01\left[d\left[\begin{array}{l}\tan .41 = .43463 \\ \tan \quad \theta = .43600 \\ \tan .42 = .44657\end{array}\right].00137\right].01194$$

$$\frac{d}{.01} = \frac{.00137}{.01194} \approx .115$$

$$d \approx (.01)(.115) \approx .001$$

$$\theta \approx .41 + .001 = .411$$

TABLE A. Natural Logarithms

	.00	.01	.02	.03	.04	.05	.06	.07	.08	.09
1.0	0.0000	0.0100	0.0198	0.0296	0.0392	0.0488	0.0583	0.0677	0.0770	0.0862
1.1	0.0953	0.1044	0.1133	0.1222	0.1310	0.1398	0.1484	0.1570	0.1655	0.1740
1.2	0.1823	0.1906	0.1989	0.2070	0.2151	0.2231	0.2311	0.2390	0.2469	0.2546
1.3	0.2624	0.2700	0.2776	0.2852	0.2927	0.3001	0.3075	0.3148	0.3221	0.3293
1.4	0.3365	0.3436	0.3507	0.3577	0.3646	0.3716	0.3784	0.3853	0.3920	0.3988
1.5	0.4055	0.4121	0.4187	0.4253	0.4318	0.4383	0.4447	0.4511	0.4574	0.4637
1.6	0.4700	0.4762	0.4824	0.4886	0.4947	0.5008	0.5068	0.5128	0.5188	0.5247
1.7	0.5306	0.5365	0.5423	0.5481	0.5539	0.5596	0.5653	0.5710	0.5766	0.5822
1.8	0.5878	0.5933	0.5988	0.6043	0.6098	0.6152	0.6206	0.6259	0.6313	0.6366
1.9	0.6419	0.6471	0.6523	0.6575	0.6627	0.6678	0.6729	0.6780	0.6831	0.6881
2.0	0.6931	0.6981	0.7031	0.7080	0.7130	0.7178	0.7227	0.7275	0.7324	0.7372
2.1	0.7419	0.7467	0.7514	0.7561	0.7608	0.7655	0.7701	0.7747	0.7793	0.7839
2.2	0.7885	0.7930	0.7975	0.8020	0.8065	0.8109	0.8154	0.8198	0.8242	0.8286
2.3	0.8329	0.8372	0.8416	0.8459	0.8502	0.8544	0.8587	0.8629	0.8671	0.8713
2.4	0.8755	0.8796	0.8838	0.8879	0.8920	0.8961	0.9002	0.9042	0.9083	0.9123
2.5	0.9163	0.9203	0.9243	0.9282	0.9322	0.9361	0.9400	0.9439	0.9478	0.9517
2.6	0.9555	0.9594	0.9632	0.9670	0.9708	0.9746	0.9783	0.9821	0.9858	0.9895
2.7	0.9933	0.9969	1.0006	1.0043	1.0080	1.0116	1.0152	1.0188	1.0225	1.0260
2.8	1.0296	1.0332	1.0367	1.0403	1.0438	1.0473	1.0508	1.0543	1.0578	1.0613
2.9	1.0647	1.0682	1.0716	1.0750	1.0784	1.0818	1.0852	1.0886	1.0919	1.0953
3.0	1.0986	1.1019	1.1053	1.1086	1.1119	1.1151	1.1184	1.1217	1.1249	1.1282
3.1	1.1314	1.1346	1.1378	1.1410	1.1442	1.1474	1.1506	1.1537	1.1569	1.1600
3.2	1.1632	1.1663	1.1694	1.1725	1.1756	1.1787	1.1817	1.1848	1.1878	1.1909
3.3	1.1939	1.1970	1.2000	1.2030	1.2060	1.2090	1.2119	1.2149	1.2179	1.2208
3.4	1.2238	1.2267	1.2296	1.2326	1.2355	1.2384	1.2413	1.2442	1.2470	1.2499
3.5	1.2528	1.2556	1.2585	1.2613	1.2641	1.2669	1.2698	1.2726	1.2754	1.2782
3.6	1.2809	1.2837	1.2865	1.2892	1.2920	1.2947	1.2975	1.3002	1.3029	1.3056
3.7	1.3083	1.3110	1.3137	1.3164	1.3191	1.3218	1.3244	1.3271	1.3297	1.3324
3.8	1.3350	1.3376	1.3403	1.3429	1.3455	1.3481	1.3507	1.3533	1.3558	1.3584
3.9	1.3610	1.3635	1.3661	1.3686	1.3712	1.3737	1.3762	1.3788	1.3813	1.3838
4.0	1.3863	1.3888	1.3913	1.3938	1.3962	1.3987	1.4012	1.4036	1.4061	1.4085
4.1	1.4110	1.4134	1.4159	1.4183	1.4207	1.4231	1.4255	1.4279	1.4303	1.4327
4.2	1.4351	1.4375	1.4398	1.4422	1.4446	1.4469	1.4493	1.4516	1.4540	1.4563
4.3	1.4586	1.4609	1.4633	1.4656	1.4679	1.4702	1.4725	1.4748	1.4770	1.4793
4.4	1.4816	1.4839	1.4861	1.4884	1.4907	1.4929	1.4952	1.4974	1.4996	1.5019
4.5	1.5041	1.5063	1.5085	1.5107	1.5129	1.5151	1.5173	1.5195	1.5217	1.5239
4.6	1.5261	1.5282	1.5304	1.5326	1.5347	1.5369	1.5390	1.5412	1.5433	1.5454
4.7	1.5476	1.5497	1.5518	1.5539	1.5560	1.5581	1.5602	1.5623	1.5644	1.5665
4.8	1.5686	1.5707	1.5728	1.5748	1.5769	1.5790	1.5810	1.5831	1.5851	1.5872
4.9	1.5892	1.5913	1.5933	1.5953	1.5974	1.5994	1.6014	1.6034	1.6054	1.6074
5.0	1.6094	1.6114	1.6134	1.6154	1.6174	1.6194	1.6214	1.6233	1.6253	1.6273
5.1	1.6292	1.6312	1.6332	1.6351	1.6371	1.6390	1.6409	1.6429	1.6448	1.6467
5.2	1.6487	1.6506	1.6525	1.6544	1.6563	1.6582	1.6601	1.6620	1.6639	1.6658
5.3	1.6677	1.6696	1.6715	1.6734	1.6753	1.6771	1.6790	1.6808	1.6827	1.6845
5.4	1.6864	1.6882	1.6901	1.6919	1.6938	1.6956	1.6974	1.6993	1.7011	1.7029

$$\ln(N \cdot 10^m) = \ln N + m \ln 10, \qquad \ln 10 = 2.3026$$

TABLE A. Natural Logarithms

	.00	.01	.02	.03	.04	.05	.06	.07	.08	.09
5.5	1.7047	1.7066	1.7084	1.7102	1.7120	1.7138	1.7156	1.7174	1.7192	1.7210
5.6	1.7228	1.7246	1.7263	1.7281	1.7299	1.7317	1.7334	1.7352	1.7370	1.7387
5.7	1.7405	1.7422	1.7440	1.7457	1.7475	1.7492	1.7509	1.7527	1.7544	1.7561
5.8	1.7579	1.7596	1.7613	1.7630	1.7647	1.7664	1.7682	1.7699	1.7716	1.7733
5.9	1.7750	1.7766	1.7783	1.7800	1.7817	1.7834	1.7851	1.7867	1.7884	1.7901
6.0	1.7918	1.7934	1.7951	1.7967	1.7984	1.8001	1.8017	1.8034	1.8050	1.8066
6.1	1.8083	1.8099	1.8116	1.8132	1.8148	1.8165	1.8181	1.8197	1.8213	1.8229
6.2	1.8245	1.8262	1.8278	1.8294	1.8310	1.8326	1.8342	1.8358	1.8374	1.8390
6.3	1.8406	1.8421	1.8437	1.8453	1.8469	1.8485	1.8500	1.8516	1.8532	1.8547
6.4	1.8563	1.8579	1.8594	1.8610	1.8625	1.8641	1.8656	1.8672	1.8687	1.8703
6.5	1.8718	1.8733	1.8749	1.8764	1.8779	1.8795	1.8810	1.8825	1.8840	1.8856
6.6	1.8871	1.8886	1.8901	1.8916	1.8931	1.8946	1.8961	1.8976	1.8991	1.9006
6.7	1.9021	1.9036	1.9051	1.9066	1.9081	1.9095	1.9110	1.9125	1.9140	1.9155
6.8	1.9169	1.9184	1.9199	1.9213	1.9228	1.9242	1.9257	1.9272	1.9286	1.9301
6.9	1.9315	1.9330	1.9344	1.9359	1.9373	1.9387	1.9402	1.9416	1.9430	1.9445
7.0	1.9459	1.9473	1.9488	1.9502	1.9516	1.9530	1.9544	1.9559	1.9573	1.9587
7.1	1.9601	1.9615	1.9629	1.9643	1.9657	1.9671	1.9685	1.9699	1.9713	1.9727
7.2	1.9741	1.9755	1.9769	1.9782	1.9796	1.9810	1.9824	1.9838	1.9851	1.9865
7.3	1.9879	1.9892	1.9906	1.9920	1.9933	1.9947	1.9961	1.9974	1.9988	2.0001
7.4	2.0015	2.0028	2.0042	2.0055	2.0069	2.0082	2.0096	2.0109	2.0122	2.0136
7.5	2.0149	2.0162	2.0176	2.0189	2.0202	2.0215	2.0229	2.0242	2.0255	2.0268
7.6	2.0282	2.0295	2.0308	2.0321	2.0334	2.0347	2.0360	2.0373	2.0386	2.0399
7.7	2.0412	2.0425	2.0438	2.0451	2.0464	2.0477	2.0490	2.0503	2.0516	2.0528
7.8	2.0541	2.0554	2.0567	2.0580	2.0592	2.0605	2.0618	2.0631	2.0643	2.0656
7.9	2.0669	2.0681	2.0694	2.0707	2.0719	2.0732	2.0744	2.0757	2.0769	2.0782
8.0	2.0794	2.0807	2.0819	2.0832	2.0844	2.0857	2.0869	2.0882	2.0894	2.0906
8.1	2.0919	2.0931	2.0943	2.0956	2.0968	2.0980	2.0992	2.1005	2.1017	2.1029
8.2	2.1041	2.1054	2.1066	2.1078	2.1090	2.1102	2.1114	2.1126	2.1138	2.1150
8.3	2.1163	2.1175	2.1187	2.1199	2.1211	2.1223	2.1235	2.1247	2.1258	2.1270
8.4	2.1282	2.1294	2.1306	2.1318	2.1330	2.1342	2.1353	2.1365	2.1377	2.1389
8.5	2.1401	2.1412	2.1424	2.1436	2.1448	2.1459	2.1471	2.1483	2.1494	2.1506
8.6	2.1518	2.1529	2.1541	2.1552	2.1564	2.1576	2.1587	2.1599	2.1610	2.1622
8.7	2.1633	2.1645	2.1656	2.1668	2.1679	2.1691	2.1702	2.1713	2.1725	2.1736
8.8	2.1748	2.1759	2.1770	2.1782	2.1793	2.1804	2.1815	2.1827	2.1838	2.1849
8.9	2.1861	2.1872	2.1883	2.1894	2.1905	2.1917	2.1928	2.1939	2.1950	2.1961
9.0	2.1972	2.1983	2.1994	2.2006	2.2017	2.2028	2.2039	2.2050	2.2061	2.2072
9.1	2.2083	2.2094	2.2105	2.2116	2.2127	2.2138	2.2148	2.2159	2.2170	2.2181
9.2	2.2192	2.2203	2.2214	2.2225	2.2235	2.2246	2.2257	2.2268	2.2279	2.2289
9.3	2.2300	2.2311	2.2322	2.2332	2.2343	2.2354	2.2364	2.2375	2.2386	2.2396
9.4	2.2407	2.2418	2.2428	2.2439	2.2450	2.2460	2.2471	2.2481	2.2492	2.2502
9.5	2.2513	2.2523	2.2534	2.2544	2.2555	2.2565	2.2576	2.2586	2.2597	2.2607
9.6	2.2618	2.2628	2.2638	2.2649	2.2659	2.2670	2.2680	2.2690	2.2701	2.2711
9.7	2.2721	2.2732	2.2742	2.2752	2.2762	2.2773	2.2783	2.2793	2.2803	2.2814
9.8	2.2824	2.2834	2.2844	2.2854	2.2865	2.2875	2.2885	2.2895	2.2905	2.2915
9.9	2.2925	2.2935	2.2946	2.2956	2.2966	2.2976	2.2986	2.2996	2.3006	2.3016

TABLE B. Common Logarithms

n	0	1	2	3	4	5	6	7	8	9
1.0	.0000	.0043	.0086	.0128	.0170	.0212	.0253	.0294	.0334	.0374
1.1	.0414	.0453	.0492	.0531	.0569	.0607	.0645	.0682	.0719	.0755
1.2	.0792	.0828	.0864	.0899	.0934	.0969	.1004	.1038	.1072	.1106
1.3	.1139	.1173	.1206	.1239	.1271	.1303	.1335	.1367	.1399	.1430
1.4	.1461	.1492	.1523	.1553	.1584	.1614	.1644	.1673	.1703	.1732
1.5	.1761	.1790	.1818	.1847	.1875	.1903	.1931	.1959	.1987	.2014
1.6	.2041	.2068	.2095	.2122	.2148	.2175	.2201	.2227	.2253	.2279
1.7	.2304	.2330	.2355	.2380	.2405	.2430	.2455	.2480	.2504	.2529
1.8	.2553	.2577	.2601	.2625	.2648	.2672	.2695	.2718	.2742	.2765
1.9	.2788	.2810	.2833	.2856	.2878	.2900	.2923	.2945	.2967	.2989
2.0	.3010	.3032	.3054	.3075	.3096	.3118	.3139	.3160	.3181	.3201
2.1	.3222	.3243	.3263	.3284	.3304	.3324	.3345	.3365	.3385	.3404
2.2	.3424	.3444	.3464	.3483	.3502	.3522	.3541	.3560	.3579	.3598
2.3	.3617	.3636	.3655	.3674	.3692	.3711	.3729	.3747	.3766	.3784
2.4	.3802	.3820	.3838	.3856	.3874	.3892	.3909	.3927	.3945	.3962
2.5	.3979	.3997	.4014	.4031	.4048	.4065	.4082	.4099	.4116	.4133
2.6	.4150	.4166	.4183	.4200	.4216	.4232	.4249	.4265	.4281	.4298
2.7	.4314	.4330	.4346	.4362	.4378	.4393	.4409	.4425	.4440	.4456
2.8	.4472	.4487	.4502	.4518	.4533	.4548	.4564	.4579	.4594	.4609
2.9	.4624	.4639	.4654	.4669	.4683	.4698	.4713	.4728	.4742	.4757
3.0	.4771	.4786	.4800	.4814	.4829	.4843	.4857	.4871	.4886	.4900
3.1	.4914	.4928	.4942	.4955	.4969	.4983	.4997	.5011	.5024	.5038
3.2	.5051	.5065	.5079	.5092	.5105	.5119	.5132	.5145	.5159	.5172
3.3	.5185	.5198	.5211	.5224	.5237	.5250	.5263	.5276	.5289	.5302
3.4	.5315	.5328	.5340	.5353	.5366	.5378	.5391	.5403	.5416	.5428
3.5	.5441	.5453	.5465	.5478	.5490	.5502	.5514	.5527	.5539	.5551
3.6	.5563	.5575	.5587	.5599	.5611	.5623	.5635	.5647	.5658	.5670
3.7	.5682	.5694	.5705	.5717	.5729	.5740	.5752	.5763	.5775	.5786
3.8	.5798	.5809	.5821	.5832	.5843	.5855	.5866	.5877	.5888	.5899
3.9	.5911	.5922	.5933	.5944	.5955	.5966	.5977	.5988	.5999	.6010
4.0	.6021	.6031	.6042	.6053	.6064	.6075	.6085	.6096	.6107	.6117
4.1	.6128	.6138	.6149	.6160	.6170	.6180	.6191	.6201	.6212	.6222
4.2	.6232	.6243	.6253	.6263	.6274	.6284	.6294	.6304	.6314	.6325
4.3	.6335	.6345	.6355	.6365	.6375	.6385	.6395	.6405	.6415	.6425
4.4	.6435	.6444	.6454	.6464	.6474	.6484	.6493	.6503	.6513	.6522
4.5	.6532	.6542	.6551	.6561	.6571	.6580	.6590	.6599	.6609	.6618
4.6	.6628	.6637	.6646	.6656	.6665	.6675	.6684	.6693	.6702	.6712
4.7	.6721	.6730	.6739	.6749	.6758	.6767	.6776	.6785	.6794	.6803
4.8	.6812	.6821	.6830	.6839	.6848	.6857	.6866	.6875	.6884	.6893
4.9	.6902	.6911	.6920	.6928	.6937	.6946	.6955	.6964	.6972	.6981
5.0	.6990	.6998	.7007	.7016	.7024	.7033	.7042	.7050	.7059	.7067
5.1	.7076	.7084	.7093	.7101	.7110	.7118	.7126	.7135	.7143	.7152
5.2	.7160	.7168	.7177	.7185	.7193	.7202	.7210	.7218	.7226	.7235
5.3	.7243	.7251	.7259	.7267	.7275	.7284	.7292	.7300	.7308	.7316
5.4	.7324	.7332	.7340	.7348	.7356	.7364	.7372	.7380	.7388	.7396

TABLE B. Common Logarithms

n	0	1	2	3	4	5	6	7	8	9
5.5	.7404	.7412	.7419	.7427	.7435	.7443	.7451	.7459	.7466	.7474
5.6	.7482	.7490	.7497	.7505	.7513	.7520	.7528	.7536	.7543	.7551
5.7	.7559	.7566	.7574	.7582	.7589	.7597	.7604	.7612	.7619	.7627
5.8	.7634	.7642	.7649	.7657	.7664	.7672	.7679	.7686	.7694	.7701
5.9	.7709	.7716	.7723	.7731	.7738	.7745	.7752	.7760	.7767	.7774
6.0	.7782	.7789	.7796	.7803	.7810	.7818	.7825	.7832	.7839	.7846
6.1	.7853	.7860	.7868	.7875	.7882	.7889	.7896	.7903	.7910	.7917
6.2	.7924	.7931	.7938	.7945	.7952	.7959	.7966	.7973	.7980	.7987
6.3	.7993	.8000	.8007	.8014	.8021	.8028	.8035	.8041	.8048	.8055
6.4	.8062	.8069	.8075	.8082	.8089	.8096	.8102	.8109	.8116	.8122
6.5	.8129	.8136	.8142	.8149	.8156	.8162	.8169	.8176	.8182	.8189
6.6	.8195	.8202	.8209	.8215	.8222	.8228	.8235	.8241	.8248	.8254
6.7	.8261	.8267	.8274	.8280	.8287	.8293	.8299	.8306	.8312	.8319
6.8	.8325	.8331	.8338	.8344	.8351	.8357	.8363	.8370	.8376	.8382
6.9	.8388	.8395	.8401	.8407	.8414	.8420	.8426	.8432	.8439	.8445
7.0	.8451	.8457	.8463	.8470	.8476	.8482	.8488	.8494	.8500	.8506
7.1	.8513	.8519	.8525	.8531	.8537	.8543	.8549	.8555	.8561	.8567
7.2	.8573	.8579	.8585	.8591	.8597	.8603	.8609	.8615	.8621	.8627
7.3	.8633	.8639	.8645	.8651	.8657	.8663	.8669	.8675	.8681	.8686
7.4	.8692	.8698	.8704	.8710	.8716	.8722	.8727	.8733	.8739	.8745
7.5	.8751	.8756	.8762	.8768	.8774	.8779	.8785	.8791	.8797	.8802
7.6	.8808	.8814	.8820	.8825	.8831	.8837	.8842	.8848	.8854	.8859
7.7	.8865	.8871	.8876	.8882	.8887	.8893	.8899	.8904	.8910	.8915
7.8	.8921	.8927	.8932	.8938	.8943	.8949	.8954	.8960	.8965	.8971
7.9	.8976	.8982	.8987	.8993	.8998	.9004	.9009	.9015	.9020	.9025
8.0	.9031	.9036	.9042	.9047	.9053	.9058	.9063	.9069	.9074	.9079
8.1	.9085	.9090	.9096	.9101	.9106	.9112	.9117	.9122	.9128	.9133
8.2	.9138	.9143	.9149	.9154	.9159	.9165	.9170	.9175	.9180	.9186
8.3	.9191	.9196	.9201	.9206	.9212	.9217	.9222	.9227	.9232	.9238
8.4	.9243	.9248	.9253	.9258	.9263	.9269	.9274	.9279	.9284	.9289
8.5	.9294	.9299	.9304	.9309	.9315	.9320	.9325	.9330	.9335	.9340
8.6	.9345	.9350	.9355	.9360	.9365	.9370	.9375	.9380	.9385	.9390
8.7	.9395	.9400	.9405	.9410	.9415	.9420	.9425	.9430	.9435	.9440
8.8	.9445	.9450	.9455	.9460	.9465	.9469	.9474	.9479	.9484	.9489
8.9	.9494	.9499	.9504	.9509	.9513	.9518	.9523	.9528	.9533	.9538
9.0	.9542	.9547	.9552	.9557	.9562	.9566	.9571	.9576	.9581	.9586
9.1	.9590	.9595	.9600	.9605	.9609	.9614	.9619	.9624	.9628	.9633
9.2	.9638	.9643	.9647	.9652	.9657	.9661	.9666	.9671	.9675	.9680
9.3	.9685	.9689	.9694	.9699	.9703	.9708	.9713	.9717	.9722	.9727
9.4	.9731	.9736	.9741	.9745	.9750	.9754	.9759	.9763	.9768	.9773
9.5	.9777	.9782	.9786	.9791	.9795	.9800	.9805	.9809	.9814	.9818
9.6	.9823	.9827	.9832	.9836	.9841	.9845	.9850	.9854	.9859	.9863
9.7	.9868	.9872	.9877	.9881	.9886	.9890	.9894	.9899	.9903	.9908
9.8	.9912	.9917	.9921	.9926	.9930	.9934	.9939	.9943	.9948	.9952
9.9	.9956	.9961	.9965	.9969	.9974	.9978	.9983	.9987	.9991	.9996

TABLE C. Trigonometric Functions (degrees)

Deg.	Sin	Tan	Cot	Cos		Deg.	Sin	Tan	Cot	Cos	
0.0	0.00000	0.00000	∞	1.0000	**90.0**	**6.0**	0.10453	0.10510	9.514	0.9945	**84.0**
.1	.00175	.00175	573.0	1.0000	89.9	.1	.10626	.10687	9.357	.9943	83.9
.2	.00349	.00349	286.5	1.0000	.8	.2	.10800	.10863	9.205	.9942	.8
.3	.00524	.00524	191.0	1.0000	.7	.3	.10973	.11040	9.058	.9940	.7
.4	.00698	.00698	143.24	1.0000	.6	.4	.11147	.11217	8.915	.9938	.6
.5	.00873	.00873	114.59	1.0000	.5	.5	.11320	.11394	8.777	.9936	.5
.6	.01047	.01047	95.49	0.9999	.4	.6	.11494	.11570	8.643	.9934	.4
.7	.01222	.01222	81.85	.9999	.3	.7	.11667	.11747	8.513	.9932	.3
.8	.01396	.01396	71.62	.9999	.2	.8	.11840	.11924	8.386	.9930	.8
.9	.01571	.01571	63.66	.9999	89.1	.9	.12014	.12101	8.264	.9928	83.1
1.0	0.01745	0.01746	57.29	0.9998	**89.0**	**7.0**	0.12187	0.12278	8.144	0.9925	**83.0**
.1	.01920	.01920	52.08	.9998	88.9	.1	.12360	.12456	8.028	.9923	82.9
.2	.02094	.02095	47.74	.9998	.8	.2	.12533	.12633	7.916	.9921	.8
.3	.02269	.02269	44.07	.9997	.7	.3	.12706	.12810	7.806	.9919	.7
.4	.02443	.02444	40.92	.9997	.6	.4	.12880	.12988	7.700	.9917	.6
.5	.02618	.02619	38.19	.9997	.5	.5	.13053	.13165	7.596	.9914	.5
.6	.02792	.02793	35.80	.9996	.4	.6	.13226	.13343	7.495	.9912	.4
.7	.02967	.02968	33.69	.9996	.3	.7	.13399	.13521	7.396	.9910	.3
.8	.03141	.03143	31.82	.9995	.2	.8	.13572	.13698	7.300	.9907	.2
.9	.03316	.03317	30.14	.9995	88.1	.9	.13744	.13876	7.207	.9905	82.1
2.0	0.03490	0.03492	28.64	0.9994	**88.0**	**8.0**	0.13917	0.14054	7.115	0.9903	**82.0**
.1	.03664	.03667	27.27	.9993	87.9	.1	.14090	.14232	7.026	.9900	81.9
.2	.03839	.03842	26.03	.9993	.8	.2	.14263	.14410	6.940	.9898	.8
.3	.04013	.04016	24.90	.9992	.7	.3	.14436	.14588	6.855	.9895	.7
.4	.04188	.04191	23.86	.9991	.6	.4	.14608	.14767	6.772	.9893	.6
.5	.04362	.04366	22.90	.9990	.5	.5	.14781	.14945	6.691	.9890	.5
.6	.04536	.04541	22.02	.9990	.4	.6	.14954	.15124	6.612	.9888	.4
.7	.04711	.04716	21.20	.9989	.3	.7	.15126	.15302	6.535	.9885	.3
.8	.04885	.04891	20.45	.9988	.2	.8	.15299	.15481	6.460	.9882	.2
.9	.05059	.05066	19.74	.9987	87.1	.9	.15471	.15660	6.386	.9880	81.1
3.0	0.05234	0.05241	19.081	0.9986	**87.0**	**9.0**	0.15643	0.15838	6.314	0.9877	**81.0**
.1	.05408	.05416	18.464	.9985	86.9	.1	.15816	.16017	6.243	.9874	80.9
.2	.05582	.05591	17.886	.9984	.8	.2	.15988	.16196	6.174	.9871	.8
.3	.05756	.05766	17.343	.9983	.7	.3	.16160	.16376	6.107	.9869	.7
.4	.05931	.05941	16.832	.9982	.6	.4	.16333	.16555	6.041	.9866	.6
.5	.06105	.06116	16.350	.9981	.5	.5	.16505	.16734	5.976	.9863	.5
.6	.06279	.06291	15.895	.9980	.4	.6	.16677	.16914	5.912	.9860	.4
.7	.06453	.06467	15.464	.9979	.3	.7	.16849	.17093	5.850	.9857	.3
.8	.06627	.06642	15.056	.9978	.2	.8	.17021	.17273	5.789	.9854	.2
.9	.06802	.06817	14.669	.9977	86.1	.9	.17193	.17453	5.730	.9851	80.1
4.0	0.06976	0.06993	14.301	0.9976	**86.0**	**10.0**	0.1736	0.1763	5.671	0.9848	**80.0**
.1	.07150	.07168	13.951	.9974	85.9	.1	.1754	.1781	5.614	.9845	79.9
.2	.07324	.07344	13.617	.9973	.8	.2	.1771	.1799	5.558	.9842	.8
.3	.07498	.07519	13.300	.9972	.7	.3	.1788	.1817	5.503	.9839	.7
.4	.07672	.07695	12.996	.9971	.6	.4	.1805	.1835	5.449	.9836	.6
.5	.07846	.07870	12.706	.9969	.5	.5	.1822	.1853	5.396	.9833	.5
.6	.08020	.08046	12.429	.9968	.4	.6	.1840	.1871	5.343	.9829	.4
.7	.08194	.08221	12.163	.9966	.3	.7	.1857	.1890	5.292	.9826	.3
.8	.08368	.08397	11.909	.9965	.2	.8	.1874	.1908	5.242	.9823	.2
.9	.08542	.08573	11.664	.9963	85.1	.9	.1891	.1926	5.193	.9820	79.1
5.0	0.08716	0.08749	11.430	0.9962	**85.0**	**11.0**	0.1908	0.1944	5.145	0.9816	**79.0**
.1	.08889	.08925	11.205	.9960	84.9	.1	.1925	.1962	5.079	.9813	78.9
.2	.09063	.09101	10.988	.9959	.8	.2	.1942	.1980	5.050	.9810	.8
.3	.09237	.09277	10.780	.9957	.7	.3	.1959	.1998	5.005	.9806	.7
.4	.09411	.09453	10.579	.9956	.6	.4	.1977	.2016	4.959	.9803	.6
.5	.09585	.09629	10.385	.9954	.5	.5	.1994	.2035	4.915	.9799	.5
.6	.09758	.09805	10.199	.9952	.4	.6	.2011	.2053	4.872	.9796	.4
.7	.09932	.09981	10.019	.9951	.3	.7	.2028	.2071	4.829	.9792	.3
.8	.10106	.10158	9.845	.9949	.2	.8	.2045	.2089	4.787	.9789	.2
.9	.10279	.10334	9.677	.9947	84.1	.9	.2062	.2107	4.745	.9785	78.1
6.0	0.10453	0.10510	9.514	0.9945	**84.0**	**12.0**	0.2079	0.2126	4.705	0.9781	**78.0**
	Cos	Cot	Tan	Sin	Deg.		Cos	Cot	Tan	Sin	Deg.

TABLE C. Trigonometric Functions (degrees)

Deg.	Sin	Tan	Cot	Cos		Deg.	Sin	Tan	Cot	Cos	
12.0	0.2079	0.2126	4.705	0.9781	**78.0**	**18.0**	0.3090	0.3249	3.078	0.9511	**72.0**
.1	.2096	.2144	4.665	.9778	77.9	.1	.3107	.3269	3.060	.9505	71.9
.2	.2113	.2162	4.625	.9774	.8	.2	.3123	.3288	3.042	.9500	.8
.3	.2130	.2180	4.586	.9770	.7	.3	.3140	.3307	3.024	.9494	.7
.4	.2147	.2199	4.548	.9767	.6	.4	.3156	.3327	3.006	.9489	.6
.5	.2164	.2217	4.511	.9763	.5	.5	.3173	.3346	2.989	.9483	.5
.6	.2181	.2235	4.474	.9759	.4	.6	.3190	.3365	2.971	.9478	.4
.7	.2198	.2254	4.437	.9755	.3	.7	.3206	.3385	2.954	.9472	.3
.8	.2215	.2272	4.402	.9751	.2	.8	.3223	.3404	2.937	.9466	.2
.9	.2233	.2290	4.366	.9748	77.1	.9	.3239	.3424	2.921	.9461	71.1
13.0	0.2250	0.2309	4.331	0.9744	**77.0**	**19.0**	0.3256	0.3443	2.904	0.9455	**71.0**
.1	.2267	.2327	4.297	.9740	76.9	.1	.3272	.3463	2.888	.9449	70.9
.2	.2284	.2345	4.264	.9736	.8	.2	.3289	.3482	2.872	.9444	.8
.3	.2300	.2364	4.230	.9732	.7	.3	.3305	.3502	2.856	.9438	.7
.4	.2317	.2382	4.198	.9728	.6	.4	.3322	.3522	2.840	.9432	.6
.5	.2334	.2401	4.165	.9724	.5	.5	.3338	.3541	2.824	.9426	.5
.6	.2351	.2419	4.134	.9720	.4	.6	.3355	.3561	2.808	.9421	.4
.7	.2368	.2438	4.102	.9715	.3	.7	.3371	.3581	2.793	.9415	.3
.8	.2385	.2456	4.071	.9711	.2	.8	.3387	.3600	2.778	.9409	.2
.9	.2402	.2475	4.041	.9707	76.1	.9	.3404	.3620	2.762	.9403	70.1
14.0	0.2419	0.2493	4.011	0.9703	**76.0**	**20.0**	0.3420	0.3640	2.747	0.9397	**70.0**
.1	.2436	.2512	3.981	.9699	75.9	.1	.3437	.3659	2.733	.9391	69.9
.2	.2453	.2530	3.952	.9694	.8	.2	.3453	.3679	2.718	.9385	.8
.3	.2470	.2549	3.923	.9690	.7	.3	.3469	.3699	2.703	.9379	.7
.4	.2487	.2568	3.895	.9686	.6	.4	.3486	.3719	2.689	.9373	.6
.5	.2504	.2586	3.867	.9681	.5	.5	.3502	.3739	2.675	.9367	.5
.6	.2521	.2605	3.839	.9677	.4	.6	.3518	.3759	2.660	.9361	.4
.7	.2538	.2623	3.812	.9673	.3	.7	.3535	.3779	2.646	.9354	.3
.8	.2554	.2642	3.785	.9668	.2	.8	.3551	.3799	2.633	.9348	.2
.9	.2571	.2661	3.758	.9664	75.1	.9	.3567	.3819	2.619	.9342	69.1
15.0	0.2588	0.2679	3.732	0.9659	**75.0**	**21.0**	0.3584	0.3839	2.605	0.9336	**69.0**
.1	.2605	.2698	3.706	.9655	74.9	.1	.3600	.3859	2.592	.9330	68.9
.2	.2622	.2717	3.681	.9650	.8	.2	.3616	.3879	2.578	.9323	.8
.3	.2639	.2736	3.655	.9646	.7	.3	.3633	.3899	2.565	.9317	.7
.4	.2656	.2754	3.630	.9641	.6	.4	.3649	.3919	2.552	.9311	.6
.5	.2672	.2773	3.606	.9636	.5	.5	.3665	.3939	2.539	.9304	.5
.6	.2689	.2792	3.582	.9632	.4	.6	.3681	.3959	2.526	.9298	.4
.7	.2706	.2811	3.558	.9627	.3	.7	.3697	.3979	2.513	.9291	.3
.8	.2723	.2830	3.534	.9622	.2	.8	.3714	.4000	2.500	.9285	.2
.9	.2740	.2849	3.511	.9617	74.1	.9	.3730	.4020	2.488	.9278	68.1
16.0	0.2756	0.2867	3.487	0.9613	**74.0**	**22.0**	0.3746	0.4040	2.475	0.9272	**68.0**
.1	.2773	.2886	3.465	.9608	73.9	.1	.3762	.4061	2.463	.9265	67.9
.2	.2790	.2905	3.442	.9603	.8	.2	.3778	.4081	2.450	.9259	.8
.3	.2807	.2924	3.420	.9598	.7	.3	.3795	.4101	2.438	.9252	.7
.4	.2823	.2943	3.398	.9593	.6	.4	.3811	.4122	2.426	.9245	.6
.5	.2840	.2962	3.376	.9588	.5	.5	.3827	.4142	2.414	.9239	.5
.6	.2857	.2981	3.354	.9583	.4	.6	.3843	.4163	2.402	.9232	.4
.7	.2874	.3000	3.333	.9578	.3	.7	.3859	.4183	2.391	.9225	.3
.8	.2890	.3019	3.312	.9573	.2	.8	.3875	.4204	2.379	.9219	.2
.9	.2907	.3038	3.291	.9568	73.1	.9	.3891	.4224	2.367	.9212	67.1
17.0	0.2924	0.3057	3.271	0.9563	**73.0**	**23.0**	0.3907	0.4245	2.356	0.9205	**67.0**
.1	.2940	.3076	3.251	.9558	72.9	.1	.3923	.4265	2.344	.9198	66.9
.2	.2957	.3096	3.230	.9553	.8	.2	.3939	.4286	2.333	.9191	.8
.3	.2974	.3115	3.211	.9548	.7	.3	.3955	.4307	2.322	.9184	.7
.4	.2990	.3134	3.191	.9542	.6	.4	.3971	.4327	2.311	.9178	.6
.5	.3007	.3153	3.172	.9537	.5	.5	.3987	.4348	2.300	.9171	.5
.6	.3024	.3172	3.152	.9532	.4	.6	.4003	.4369	2.289	.9164	.4
.7	.3040	.3191	3.133	.9527	.3	.7	.4019	.4390	2.278	.9157	.3
.8	.3057	.3211	3.115	.9521	.2	.8	.4035	.4411	2.267	.9150	.2
.9	.3074	.3230	3.096	.9516	72.1	.9	.4051	.4431	2.257	.9143	66.1
18.0	0.3090	0.3249	3.078	0.9511	**72.0**	**24.0**	0.4067	0.4452	2.246	0.9135	**66.0**
	Cos	Cot	Tan	Sin	Deg.		Cos	Cot	Tan	Sin	Deg.

TABLE C. Trigonometric Functions (degrees)

Deg.	Sin	Tan	Cot	Cos		Deg.	Sin	Tan	Cot	Cos	
24.0	0.4067	0.4452	2.246	0.9135	**66.0**	**30.0**	0.5000	0.5774	1.7321	0.8660	**60.0**
.1	.4083	.4473	2.236	.9128	65.9	.1	.5015	.5797	1.7251	.8652	59.9
.2	.4099	.4494	2.225	.9121	.8	.2	.5030	.5820	1.7182	.8643	.8
.3	.4115	.4515	2.215	.9114	.7	.3	.5045	.5844	1.7113	.8634	.7
.4	.4131	.4536	2.204	.9107	.6	.4	.5060	.5867	1.7045	.8625	.6
.5	.4147	.4557	2.194	.9100	.5	.5	.5075	.5890	1.6977	.8616	.5
.6	.4163	.4578	2.184	.9092	.4	.6	.5090	.5914	1.6909	.8607	.4
.7	.4179	.4599	2.174	.9085	.3	.7	.5105	.5938	1.6842	.8599	.3
.8	.4195	.4621	2.164	.9078	.2	.8	.5120	.5961	1.6775	.8590	.2
.9	.4210	.4642	2.154	.9070	65.1	.9	.5135	.5985	1.6709	.8581	59.1
25.0	0.4226	0.4663	2.145	0.9063	**65.0**	**31.0**	0.5150	0.6009	1.6643	0.8572	**59.0**
.1	.4242	.4684	2.135	.9056	64.9	.1	.5165	.6032	1.6577	.8563	58.9
.2	.4258	.4706	2.125	.9048	.8	.2	.5180	.6056	1.6512	.8554	.8
.3	.4274	.4727	2.116	.9041	.7	.3	.5195	.6080	1.6447	.8545	.7
.4	.4289	.4748	2.106	.9033	.6	.4	.5210	.6104	1.6383	.8536	.6
.5	.4305	.4770	2.097	.9026	.5	.5	.5225	.6128	1.6319	.8526	.5
.6	.4321	.4791	2.087	.9018	.4	.6	.5240	.6152	1.6255	.8517	.4
.7	.4337	.4813	2.078	.9011	.3	.7	.5255	.6176	1.6191	.8508	.3
.8	.4352	.4834	2.069	.9003	.2	.8	.5270	.6200	1.6128	.8499	.2
.9	.4368	.4856	2.059	.8996	64.1	.9	.5284	.6224	1.6066	.8490	58.1
26.0	0.4384	0.4887	2.050	0.8988	**64.0**	**32.0**	0.5299	0.6249	1.6003	0.8480	**58.0**
.1	.4399	.4899	2.041	.8980	63.9	.1	.5314	.6273	1.5941	.8471	57.9
.2	.4415	.4921	2.032	.8973	.8	.2	.5329	.6297	1.5880	.8462	.8
.3	.4431	.4942	2.023	.8965	.7	.3	.5344	.6322	1.5818	.8453	.7
.4	.4446	.4964	2.014	.8957	.6	.4	.5358	.6346	1.5757	.8443	.6
.5	.4462	.4986	2.006	.8949	.5	.5	.5373	.6371	1.5697	.8434	.5
.6	.4478	.5008	1.997	.8942	.4	.6	.5388	.6395	1.5637	.8425	.4
.7	.4493	.5029	1.988	.8934	.3	.7	.5402	.6420	1.5577	.8415	.3
.8	.4509	.5051	1.980	.8926	.2	.8	.5417	.6445	1.5517	.8406	.2
.9	.4524	.5073	1.971	.8918	63.1	.9	.5432	.6469	1.5458	.8396	57.1
27.0	0.4540	0.5095	1.963	0.8910	**63.0**	**33.0**	0.5446	0.6494	1.5399	0.8387	**57.0**
.1	.4555	.5117	1.954	.8902	62.9	.1	.5461	.6519	1.5340	.8377	56.9
.2	.4571	.5139	1.946	.8894	.8	.2	.5476	.6544	1.5282	.8368	.8
.3	.4586	.5161	1.937	.8886	.7	.3	.5490	.6569	1.5224	.8358	.7
.4	.4602	.5184	1.929	.8878	.6	.4	.5505	.6594	1.5166	.8348	.6
.5	.4617	.5206	1.921	.8870	.5	.5	.5519	.6619	1.5108	.8339	.5
.6	.4633	.5228	1.913	.8862	.4	.6	.5534	.6644	1.5051	.8329	.4
.7	.4648	.5250	1.905	.8854	.3	.7	.5548	.6669	1.4994	.8320	.3
.8	.4664	.5272	1.897	.8846	.2	.8	.5563	.6694	1.4938	.8310	.2
.9	.4679	.5295	1.889	.8838	62.1	.9	.5577	.6720	1.4882	.8300	56.1
28.0	0.4695	0.5317	1.881	0.8829	**62.0**	**34.0**	0.5592	0.6745	1.4826	0.8290	**56.0**
.1	.4710	.5340	1.873	.8821	61.9	.1	.5606	.6771	1.4770	.8281	55.9
.2	.4726	.5362	1.865	.8813	.8	.2	.5621	.6796	1.4715	.8271	.8
.3	.4741	.5384	1.857	.8805	.7	.3	.5635	.6822	1.4659	.8261	.7
.4	.4756	.5407	1.849	.8796	.6	.4	.5650	.6847	1.4605	.8251	.6
.5	.4772	.5430	1.842	.8788	.5	.5	.5664	.6873	1.4550	.8241	.5
.6	.4787	.5452	1.834	.8780	.4	.6	.5678	.6899	1.4496	.8231	.4
.7	.4802	.5475	1.827	.8771	.3	.7	.5693	.6924	1.4442	.8221	.3
.8	.4818	.5498	1.819	.8763	.2	.8	.5707	.6950	1.4388	.8211	.2
.9	.4833	.5520	1.811	.8755	61.1	.9	.5721	.6976	1.4335	.8202	55.1
29.0	0.4848	0.5543	1.804	0.8746	**61.0**	**35.0**	0.5736	0.7002	1.4281	0.8192	**55.0**
.1	.4863	.5566	1.797	.8738	60.9	.1	.5750	.7028	1.4229	.8181	54.9
.2	.4879	.5589	1.789	.8729	.8	.2	.5764	.7054	1.4176	.8171	.8
.3	.4894	.5612	1.782	.8721	.7	.3	.5779	.7080	1.4124	.8161	.7
.4	.4909	.5635	1.775	.8712	.6	.4	.5793	.7107	1.4071	.8151	.6
.5	.4924	.5658	1.767	.8704	.5	.5	.5807	.7133	1.4019	.8141	.5
.6	.4939	.5681	1.760	.8695	.4	.6	.5821	.7159	1.3968	.8131	.4
.7	.4955	.5704	1.753	.8686	.3	.7	.5835	.7186	1.3916	.8121	.3
.8	.4970	.5727	1.746	.8678	.2	.8	.5850	.7212	1.3865	.8111	.2
.9	.4985	.5750	1.739	.8669	60.1	.9	.5864	.7239	1.3814	.8100	54.1
30.0	0.5000	0.5774	1.732	0.8660	**60.0**	**36.0**	0.5878	0.7265	1.3764	0.8090	**54.0**
	Cos	Cot	Tan	Sin	Deg.		Cos	Cot	Tan	Sin	Deg.

TABLE C. Trigonometric Functions (degrees)

Deg.	Sin	Tan	Cot	Cos		Deg.	Sin	Tan	Cot	Cos	
36.0	0.5878	0.7265	1.3764	0.8090	**54.0**	**40.5**	0.6494	0.8541	1.1708	0.7604	**49.5**
.1	.5892	.7292	1.3713	.8080	53.9	.6	.6508	.8571	1.1667	.7593	.4
.2	.5906	.7319	1.3663	.8070	.8	.7	.6521	.8601	1.1626	.7581	.3
.3	.5920	.7346	1.3613	.8059	.7	.8	.6534	.8632	1.1585	.7570	.2
.4	.5934	.7373	1.3564	.8049	.6	.9	.6547	.8662	1.1544	.7559	49.1
.5	.5948	.7400	1.3514	.8039	.5	**41.0**	0.6561	0.8693	1.1504	0.7547	**49.0**
.6	.5962	.7427	1.3465	.8028	.4	.1	.6574	.8724	1.1463	.7536	48.9
.7	.5976	.7454	1.3416	.8018	.3	.2	.6587	.8754	1.1423	.7524	.8
.8	.5990	.7481	1.3367	.8007	.2	.3	.6600	.8785	1.1383	.7513	.7
.9	.6004	.7508	1.3319	.7997	53.1	.4	.6613	.8816	1.1343	.7501	.6
37.0	0.6018	0.7536	1.3270	0.7986	**53.0**	.5	.6626	.8847	1.1303	.7490	.5
.1	.6032	.7563	1.3222	.7976	52.9	.6	.6639	.8878	1.1263	.7478	.4
.2	.6046	.7590	1.3175	.7965	.8	.7	.6652	.8910	1.1224	.7466	.3
.3	.6060	.7618	1.3127	.7955	.7	.8	.6665	.8941	1.1184	.7455	.2
.4	.6074	.7646	1.3079	.7944	.6	.9	.6678	.8972	1.1145	.7443	48.1
.5	.6088	.7673	1.3032	.7934	.5	**42.0**	0.6691	0.9004	1.1106	0.7431	**48.0**
.6	.6101	.7701	1.2985	.7923	.4	.1	.6704	.9036	1.1067	.7420	47.9
.7	.6115	.7729	1.2938	.7912	.3	.2	.6717	.9067	1.1028	.7408	.8
.8	.6129	.7757	1.2892	.7902	.2	.3	.6730	.9099	1.0990	.7396	.7
.9	.6143	.7785	1.2846	.7891	52.1	.4	.6743	.9131	1.0951	.7385	.6
38.0	0.6157	0.7813	1.2799	0.7880	**52.0**	.5	.6756	.9163	1.0913	.7373	.5
.1	.6170	.7841	1.2753	.7869	51.9	.6	.6769	.9195	1.0875	.7361	.4
.2	.6184	.7869	1.2708	7859	.8	.7	.6782	.9228	1.0837	.7349	.3
.3	.6198	.7898	1.2662	.7848	.7	.8	.6794	.9260	1.0799	.7337	.2
.4	.6211	.7926	1.2617	.7837	.6	.9	.6807	.9293	1.0761	.7325	47.1
.5	.6225	.7954	1.2572	.7826	.5	**43.0**	0.6820	0.9325	1.0724	0.7314	**47.0**
.6	.6239	.7983	1.2527	.7815	.4	.1	.6833	.9358	1.0686	.7302	46.9
.7	.6252	.8012	1.2482	.7804	.3	.2	.6845	.9391	1.0649	.7290	.8
.8	.6266	.8040	1.2437	.7793	.2	.3	.6858	.9424	1.0612	.7278	.7
.9	.6280	.8069	1.2393	.7782	51.1	.4	.6871	.9457	1.0575	.7266	.6
39.0	0.6293	0.8098	1.2349	0.7771	**51.0**	.5	.6884	.9490	1.0538	.7254	.5
.1	.6307	.8127	1.2305	.7760	50.9	.6	.6896	.9523	1.0501	.7242	.4
.2	.6320	.8156	1.2261	.7749	.8	.7	.6909	.9556	1.0464	.7230	.3
.3	.6334	.8185	1.2218	.7738	.7	.8	.6921	.9590	1.0428	.7218	.2
.4	.6347	.8214	1.2174	.7727	.6	.9	.6934	.9623	1.0392	.7206	46.1
.5	.6361	.8243	1.2131	.7716	.5	**44.0**	0.6947	0.9657	1.0355	0.7193	**46.0**
.6	.6374	.8273	1.2088	.7705	.4	.1	.6959	.9691	1.0319	.7181	45.9
.7	.6388	.8302	1.2045	.7694	.3	.2	.6972	.9725	1.0283	.7169	.8
.8	.6401	.8332	1.2002	.7683	.2	.3	.6984	.9759	1.0247	.7157	.7
.9	.6414	.8361	1.1960	.7672	50.1	.4	.6997	.9793	1.0212	.7145	.6
40.0	0.6428	0.8391	1.1918	0.7660	**50.0**	.5	.7009	.9827	1.0176	.7133	.5
.1	.6441	.8421	1.1875	.7649	49.9	.6	.7022	.9861	1.0141	.7120	.4
.2	.6455	.8451	1.1833	.7638	.8	.7	.7034	.9896	1.0105	.7108	.3
.3	.6468	.8481	1.1792	.7627	.7	.8	.7046	.9930	1.0070	.7096	.2
.4	.6481	.8511	1.1750	.7615	.6	.9	.7059	.9965	1.0035	.7083	45.1
40.5	0.6494	0.8541	1.1708	0.7604	**49.5**	**45.0**	0.7071	1.0000	1.0000	0.7071	**45.0**
	Cos	Cot	Tan	Sin	Deg.		Cos	Cot	Tan	Sin	Deg.

TABLE D. Trigonometric Functions (radians)

Rad.	Sin	Tan	Cot	Cos	Rad.	Sin	Tan	Cot	Cos
.00	.00000	.00000	∞	1.00000	**.50**	.47943	.54630	1.8305	.87758
.01	.01000	.01000	99.997	0.99995	.51	.48818	.55936	1.7878	.87274
.02	.02000	.02000	49.993	.99980	.52	.49688	.57256	1.7465	.86782
.03	.03000	.03001	33.323	.99955	.53	.50553	.58592	1.7067	.86281
.04	.03999	.04002	24.987	.99920	.54	.51414	.59943	1.6683	.85771
.05	.04998	.05004	19.983	.99875	.55	.52269	.61311	1.6310	.85252
.06	.05996	.06007	16.647	.99820	.56	.53119	.62695	1.5950	.84726
.07	.06994	.07011	14.262	.99755	.57	.53963	.64097	1.5601	.84190
.08	.07991	.08017	12.473	.99680	.58	.54802	.65517	1.5263	.83646
.09	.08988	.09024	11.081	.99595	.59	.55636	.66956	1.4935	.83094
.10	.09983	.10033	9.9666	.99500	**.60**	.56464	.68414	1.4617	.82534
.11	.10978	.11045	9.0542	.99396	.61	.57287	.69892	1.4308	.81965
.12	.11971	.12058	8.2933	.99281	.62	.58104	.71391	1.4007	.81388
.13	.12963	.13074	7.6489	.99156	.63	.58914	.72911	1.3715	.80803
.14	.13954	.14092	7.0961	.99022	.64	.59720	.74454	1.3431	.80210
.15	.14944	.15114	6.6166	.98877	.65	.60519	.76020	1.3154	.79608
.16	.15932	.16138	6.1966	.98723	.66	.61312	.77610	1.2885	.78999
.17	.16918	.17166	5.8256	.98558	.67	.62099	.79225	1.2622	.78382
.18	.17903	.18197	5.4954	.98384	.68	.62879	.80866	1.2366	.77757
.19	.18886	.19232	5.1997	.98200	.69	.63654	.82534	1.2116	.77125
.20	.19867	.20271	4.9332	.98007	**.70**	.64422	.84229	1.1872	.76484
.21	.20846	.21314	4.6917	.97803	.71	.65183	.85953	1.1634	.75836
.22	.21823	.22362	4.4719	.97590	.72	.65938	.87707	1.1402	.75181
.23	.22798	.23414	4.2709	.97367	.73	.66687	.89492	1.1174	.74517
.24	.23770	.24472	4.0864	.97134	.74	.67429	.91309	1.0952	.73847
.25	.24740	.25534	3.9163	.96891	.75	.68164	.93160	1.0734	.73169
.26	.25708	.26602	3.7591	.96639	.76	.68892	.95045	1.0521	.72484
.27	.26673	.27676	3.6133	.96377	.77	.69614	.96967	1.0313	.71791
.28	.27636	.28755	3.4776	.96106	.78	.70328	.98926	1.0109	.71091
.29	.28595	.29841	3.3511	.95824	.79	.71035	1.0092	.99084	.70385
.30	.29552	.30934	3.2327	.95534	**.80**	.71736	1.0296	.97121	.69671
.31	.30506	.32033	3.1218	.95233	.81	.72429	1.0505	.95197	.68950
.32	.31457	.33139	3.0176	.94924	.82	.73115	1.0717	.93309	.68222
.33	.32404	.34252	2.9195	.94604	.83	.73793	1.0934	.91455	.67488
.34	.33349	.35374	2.8270	.94275	.84	.74464	1.1156	.89635	.66746
.35	.34290	.36503	2.7395	.93937	.85	.75128	1.1383	.87848	.65998
.36	.35227	.37640	2.6567	.93590	.86	.75784	1.1616	.86091	.65244
.37	.36162	.38786	2.5782	.93233	.87	.76433	1.1853	.84365	.64483
.38	.37092	.39941	2.5037	.92866	.88	.77074	1.2097	.82668	.63715
.39	.38019	.41105	2.4328	.92491	.89	.77707	1.2346	.80998	.62941
.40	.38942	.42279	2.3652	.92106	**.90**	.78333	1.2602	.79355	.62161
.41	.39861	.43463	2.3008	.91712	.91	.78950	1.2864	.77738	.61375
.42	.40776	.44657	2.2393	.91309	.92	.79560	1.3133	.76146	.60582
.43	.41687	.45862	2.1804	.90897	.93	.80162	1.3409	.74578	.59783
.44	.42594	.47078	2.1241	.90475	.94	.80756	1.3692	.73034	.58979
.45	.43497	.48306	2.0702	.90045	.95	.81342	1.3984	.71511	.58168
.46	.44395	.49545	2.0184	.89605	.96	.81919	1.4284	.70010	.57352
.47	.45289	.50797	1.9686	.89157	.97	.82489	1.4592	.68531	.56530
.48	.46178	.52061	1.9208	.88699	.98	.83050	1.4910	.67071	.55702
.49	.47063	.53339	1.8748	.88233	.99	.83603	1.5237	.65631	.54869
.50	.47943	.54630	1.8305	.87758	**1.00**	.84147	1.5574	.64209	.54030
Rad.	Sin	Tan	Cot	Cos	Rad.	Sin	Tan	Cot	Cos

TABLE D. Trigonometric Functions (radians)

Rad.	Sin.	Tan	Cot	Cos	Rad.	Sin	Tan	Cot	Cos
1.00	.84147	1.5574	.64209	.54030	**1.50**	.99749	14.101	.07091	.07074
1.01	.84683	1.5922	.62806	.53186	1.51	.99815	16.428	.06087	.06076
1.02	.85211	1.6281	.61420	.52337	1.52	.99871	19.670	.05084	.05077
1.03	.85730	1.6652	.60051	.51482	1.53	.99917	24.498	.04082	.04079
1.04	.86240	1.7036	.58699	.50622	1.54	.99953	32.461	.03081	.03079
1.05	.86742	1.7433	.57362	.49757	1.55	.99978	48.078	.02080	.02079
1.06	.87236	1.7844	.56040	.48987	1.56	.99994	92.621	.01080	.01080
1.07	.87720	1.8270	.54734	.48012	1.57	1.00000	1255.8	.00080	.00080
1.08	.88196	1.8712	.53441	.47133	1.58	.99996	−108.65	−.00920	−.00920
1.09	.88663	1.9171	.52162	.46249	1.59	.99982	−52.067	−.01921	−.01920
1.10	.89121	1.9648	.50897	.45360	**1.60**	.99957	−34.233	−.02921	−.02920
1.11	.89570	2.0143	.49644	.44466	1.61	.99923	−25.495	−.03922	−.03919
1.12	.90010	2.0660	.48404	.43568	1.62	.99879	−20.307	−.04924	−.04918
1.13	.90441	2.1198	.47175	.42666	1.63	.99825	−16.871	−.05927	−.05917
1.14	.90863	2.1759	.45959	.41759	1.64	.99761	−14.427	−.06931	−.06915
1.15	.91276	2.2345	.44753	.40849	1.65	.99687	−12.599	−.07937	−.07912
1.16	.91680	2.2958	.43558	.39934	1.66	.99602	−11.181	−.08944	−.08909
1.17	.92075	2.3600	.42373	.39015	1.67	.99508	−10.047	−.09953	−.09904
1.18	.92461	2.4273	.41199	.38092	1.68	.99404	−9.1208	−.10964	−.10899
1.19	.92837	2.4979	.40034	.37166	1.69	.99290	−8.3492	−.11977	−.11892
1.20	.93204	2.5722	.38878	.36236	**1.70**	.99166	−7.6966	−.12993	−.12884
1.21	.93562	2.6503	.37731	.35302	1.71	.99033	−7.1373	−.14011	−.13875
1.22	.93910	2.7328	.36593	.34365	1.72	.98889	−6.6524	−.15032	−.14865
1.23	.94249	2.8198	.35463	.33424	1.73	.98735	−6.2281	−.16056	−.15853
1.24	.94578	2.9119	.34341	.32480	1.74	.98572	−5.8535	−.17084	−.16840
1.25	.94898	3.0096	.33227	.31532	1.75	.98399	−5.5204	−.18115	−.17825
1.26	.95209	3.1133	.32121	.30582	1.76	.98215	−5.2221	−.19149	−.18808
1.27	.95510	3.2236	.31021	.29628	1.77	.98022	−4.9534	−.20188	−.19789
1.28	.95802	3.3413	.29928	.28672	1.78	.97820	−4.7101	−.21231	−.20768
1.29	.96084	3.4672	.28842	.27712	1.79	.97607	−4.4887	−.22278	−.21745
1.30	.96356	3.6021	.27762	.26750	**1.80**	.97385	−4.2863	−.23330	−.22720
1.31	.96618	3.7471	.26687	.25785	1.81	.97153	−4.1005	−.24387	−.23693
1.32	.96872	3.9033	.25619	.24818	1.82	.96911	−3.9294	−.25449	−.24663
1.33	.97115	4.0723	.24556	.23848	1.83	.96659	−3.7712	−.26517	−.25631
1.34	.97348	4.2556	.23498	.22875	1.84	.96398	−3.6245	−.27590	−.26596
1.35	.97572	4.4552	.22446	.21901	1.85	.96128	−3.4881	−.28669	−.27559
1.36	.97786	4.6734	.21398	.20924	1.86	.95847	−3.3608	−.29755	−.28519
1.37	.97991	4.9131	.20354	.19945	1.87	.95557	−2.2419	−.30846	−.29476
1.38	.98185	5.1774	.19315	.18964	1.88	.95258	−3.1304	−.31945	−.30430
1.39	.98370	5.4707	.18279	.17981	1.89	.94949	−3.0257	−33.051	−.31381
1.40	.98545	5.7979	.17248	.16997	**1.90**	.94630	−2.9271	−.34164	−.32329
1.41	.98710	6.1654	.16220	.16010	1.91	.94302	−2.8341	−.35284	−.33274
1.42	.98865	6.5811	.15195	.15023	1.92	.93965	−2.7463	−.36413	−.34215
1.43	.99010	7.0555	.14173	.14033	1.93	.93618	−2.6632	−.37549	−.35153
1.44	.99146	7.6018	.13155	.13042	1.94	.93262	−2.5843	−.38695	−.36087
1.45	.99271	8.2381	.12139	.12050	1.95	.92896	−2.5095	−.39849	−.37018
1.46	.99387	8.9886	.11125	.11057	1.96	.92521	−2.4383	−.41012	−.37945
1.47	.99492	9.8874	.10114	.10063	1.97	.92137	−2.3705	−.42185	−.38868
1.48	.99588	10.983	.09105	.09067	1.98	.91744	−2.3058	−.43368	−.39788
1.49	.99674	12.350	.08097	.08071	1.99	.91341	−2.2441	−.44562	−.40703
1.50	.99749	14.101	.07091	.07074	**2.00**	.90930	−2.1850	−.45766	−.41615
Rad.	Sin	Tan	Cot	Cos	Rad.	Sin	Tan	Cot	Cos

Answers To Odd-Numbered Problems

PROBLEM SET 1-1 (Page 7)

1. > **3.** > **5.** > **7.** > **9.** < **11.** = **13.** $-3\sqrt{2}/2, -2, -\pi/2, -\sqrt{2}, \frac{3}{4}, \frac{43}{24}$ **15.** $\frac{8}{9}$

17. $-\frac{3}{4}$ **19.** $(2 + 6\sqrt{5})/3$ **21.** $(-8 + \sqrt{2})/4$ **23.** $\frac{23}{36}$ **25.** $\frac{53}{54}$ **27.** $\frac{9}{2}$ **29.** $\frac{5}{4}$ **31.** $\frac{3}{8}$ **33.** $\frac{17}{7}$

35. $72 - 4\sqrt{2}$ **37.** $\frac{7}{9}$ **39.** $\frac{1}{24}$ **41.** $\frac{22}{189}$ **43.** $\frac{4}{11}$ **45.** -5.41 **47.** $.55$ **49.** 3.88 **51.** 8.39

53. (a) $\frac{5}{9}$ (b) $\frac{31}{18}$ (c) $-\frac{5}{11}$ (d) $\frac{10}{33}$ (e) 1 (f) $\frac{-23}{18}$ (g) $\frac{1}{19}$ (h) 77

55. Additive inverse; zero is neutral element for addition; distributive property; associative property of addition; additive inverse; zero is neutral element for addition.

57. $(-a) \cdot b = (-a) \cdot b + 0$ **59.** a, c, e, g, h

$$= (-a) \cdot b + \{ab + [-(ab)]\}$$
$$= [(-a) \cdot b + ab] + [-(ab)]$$
$$= (-a + a)b + [-(ab)]$$
$$= 0 \cdot b + [-(ab)]$$
$$= 0 + [-(ab)]$$
$$= -(ab)$$

61. (a) $-4 < x < 4$ (b) $-2 < x < 6$ (c) $-\frac{5}{3} < x < 3$ **63.** $\frac{8}{37}$

65. Let r be a rational number and i an irrational number. Suppose that $s = r + i$ is rational. Then $i = s - r$. On the left, we have an irrational number, on the right, a rational number. This is a contradiction.

67. Follow a procedure similar to that in Problem 65. **69.** b, d, f

PROBLEM SET 1-2 (Page 17)

1. $3^3 = 27$ **3.** $2^2 = 4$ **5.** $2^3 = 8$ **7.** $1/5^2 = 1/25$ **9.** $-1/5^2 = -1/25$ **11.** $1/(-2)^5 = -1/32$

13. $1/(-\frac{2}{3})^3 = -27/8$ **15.** $\frac{27}{4}$ **17.** $\frac{81}{16}$ **19.** $\frac{3}{64}$ **21.** $\frac{1}{72}$ **23.** $81x^4$ **25.** x^6y^{12} **27.** $16x^8y^4/w^{12}$

29. $27y^6/x^3z^6$ **31.** $25x^8$ **33.** $1/16y^6$ **35.** $a/(5x^2b^2)$ **37.** $2z^3/x^6y^2$ **39.** $-x^3z^2/2y^7$ **41.** a^4b^3
43. $d^{40}/32b^{15}$ **45.** $a^3/(a+1)$ **47.** 0   **49.** 1.79   **51.** 432.20   **53.** .34   **55.** 3.41×10^8
57. 5.13×10^{-8} **59.** 1.245×10^{-10} **61.** 9.0×10^7 **63.** 8.43×10^{-17} **65.** 6.99×10^{-14} **67.** $9y^6/4x^4$
69. $\frac{5}{63}$ **71.** $a^8b^8/81$ **73.** $2x^3y^9$ **75.** $1/xy$ **77.** (a) 2^{-15} (b) 2   **79.** 158,000 miles
81. (a) $15\cancel{c}$, $31\cancel{c}$, $63\cancel{c}$ (b) $2^n - 1$ cents (c) About February 7.

PROBLEM SET 1-3 (Page 27)

1. Polynomial of degree 2.   **3.** Polynomial of degree 2.   **5.** Not a polynomial.   **7.** $-2x + 1$
9. $5y^2 - 9y + 9$ **11.** $33x^2 - 55x + 19$ **13.** $x^2 - x - 90$ **15.** $6y^2 + 7y - 3$ **17.** $2z^3 - 7z^2 + 11z - 4$
19. $v^2 + 10v + 25$ **21.** $4x^2 - 12x + 9$ **23.** $25t^2 - 9$ **25.** $h^3 + 9h^2 + 27h + 27$ **27.** $64y^3 - 1$
29. $4x^{10} - 9$ **31.** $2x^2 - xy - 3y^2$ **33.** $4a^2 + 12ab + 9b^2$ **35.** $9x^2 - 2y^2$ **37.** $y^3 + 8z^3$
39. $3x^2h + 3xh^2 + h^3 + 2h$ **41.** $a^2 + 2ab + b^2 - c^2$ **43.** $3x^2 + 24u^2w^2$ **45.** $x(x + 5)$
47. $(x + 6)(x - 1)$ **49.** $y^3(y - 6)$ **51.** $(y + 6)(y - 2)$ **53.** $(y + 4)^2$ **55.** $(2x - 3y)^2$
57. $(2a + 3)(2a - 3)$ **59.** $(2z + 1)(2z - 3)$ **61.** $(3a - 2x)(2a + x)$ **63.** $x^3(3 + y)(3 - y)$
65. $(2x + 3y)(4x^2 - 6xy + 9y^2)$ **67.** $a(x + b)^2$ **69.** $(x^2 - 3)(x^2 + 2)$ **71.** $(y + 2)(y - 2)(y^2 + 4)$
73. $(x + 2)(x - 2)(x^2 - 2x + 4)(x^2 + 2x + 4)$ **75.** $(x + 4y + 1)(x + 4y - 2)$ **77.** $(x + 1)(x^2 + 3)$
79. $(a + 5)(2x - 3)$ **81.** $(x + 3)(x + 3 + a)$ **83.** $1/(x - 6)$ **85.** $(y^2 - y + 1)/y$ **87.** $z(x + 2y)/(x + y)$
89. $(9x + 2)/(x - 2)(x + 2)$ **91.** $(-2x + 8)/(x - 2)^2(x + 2)$ **93.** $(4 - x)/(2x - 1)$
95. $(6y^2 + 9y + 2)/(3y - 1)(3y + 1)$ **97.** $5x/(2x - 1)(x + 1)$ **99.** $1/(x - 3)(x - 2)$ **101.** $5(x + 1)/x(2x - 1)$
103. $x/(x - 2)$ **105.** $-4/x(x + h)$ **107.** $4/(x + 4)(x + h + 4)$ **109.** $3xy + 2y^2$ **111.** $-4x$
113. $8x^3 + 6x$ **115.** $16x^4 + 81c^4$ **117.** $(2 - 3m)(2 + 3m)$ **119.** $4x(x - 2)$ **121.** $(3x + 2)(2x - 5)$
123. $3x(x - 3)(x^2 + 3x + 9)$ **125.** $(2b - 1)(2b + 1)(b - 4)(b + 4)$ **127.** $(x - 3y)(x + 5y)^2(x + 5y - 2)(x + 5y + 2)$
129. $(-11x - 12)/(x - 4)(3x - 4)$ **131.** $(x + 2)/(x - 2)(x - 4)$ **133.** $-4/(a + b)$
135. $(4x^2 + 4xh + 1)/x(x + h)$ **137.** $2\sqrt{5}$ **139.** (a) 94,000 (b) 7 (c) $\frac{15}{39}$

PROBLEM SET 1-4 (Page 37)

1. Identity.   **3.** Conditional equation.   **5.** $\frac{9}{2}$ **7.** $2/\sqrt{3}$ **9.** 7.57   **11.** $-\frac{1}{4}$ **13.** $\frac{22}{5}$ **15.** 3
17. No solution (2 is extraneous).   **19.** $P = A/(1 + rt)$ **21.** $r = (nE - IR)/nI$ **23.** $h = (A - 2\pi r^2)/2\pi r$
25. $R_1 = RR_2(R_2 - R)$ **27.** $-1; 2$ **29.** $\pm\frac{5}{3}$ **31.** $-2; \frac{1}{3}$ **33.** $-1; 7$ **35.** $-\frac{15}{2}; \frac{5}{2}$ **37.** $-9; 1$
39. $-\frac{1}{2}; \frac{3}{2}$ **41.** $-6; -2$ **43.** $(-5 \pm \sqrt{13})/2$ **45.** $(3 \pm \sqrt{42})/3$ **47.** $1 \pm \sqrt{6}$ **49.** $-.2714; 1.8422$
51. $x = -13; y = 13$ **53.** $x = 4; y = 3$ **55.** $x = 6; y = -8$ **57.** $s = 1; t = 4$ **59.** $x = 3; y = -1$
61. $x = 9; y = -2$ **63.** $x = \frac{1}{2}; y = \frac{1}{3}$ **65.** At 7:00 P.M.
67. At approximately 4:37 P.M.; approximately 277 miles.
69. $\frac{26}{5}$ **71.** $\frac{7}{5}$ **73.** $-\frac{1}{4}$ **75.** $-\frac{1}{4}, \frac{5}{4}$ **77.** 1   **79.** $-1 \pm \sqrt{5}$ **81.** $\pm 1, \pm\sqrt{6}$
83. $(1 - 2a)/(2a^2 - 1)$ **85.** $x = -1, y = 2$ **87.** $\frac{71}{4}$ feet   **89.** 32   **91.** $1511.36   **93.** About 1 : 49
95. $(1 + \sqrt{5})/2$

PROBLEM SET 1-5 (Page 47)

1. Conditional   **3.** Unconditional.   **5.** Conditional.   **7.** Conditional.   **9.** Conditional.   **11.** Unconditional.

13. $\{x : x < -6\}$ **15.** $\{x : x > -24\}$

17. $\{x : x < \frac{30}{7}\}$ **19.** $\{x : -5 \le x \le 2\}$

21. $\{x : x < -3 \text{ or } x > \frac{1}{2}\}$

23. $\{x : x \le 1 \text{ or } x \ge 4\}$

25. $\{x : \frac{1}{2} < x < 3\}$

27. $\{x : -\frac{5}{2} < x < -\frac{1}{2}\}$

29. $\{x : -1 \le x \le 0\}$

31. $\{x : -4 \le x \le 0 \text{ or } x \ge 3\}$

33. $\{x : x < 5 \text{ and } x \ne 2\}$

35. $\{x : -2 < x \le 5\}$

37. $\{x : -2 < x < 0 \text{ or } x > 5\}$

39. $\{x : -2 < x < 2 \text{ or } x > 3\}$

41. $|x - 3| < 3$ **43.** $|x - 3| \le 4$ **45.** $|x - 6.5| < 4.5$

47. $\{x : -\sqrt{7} < x < \sqrt{7}\}$

49. $\{x : x \le 2 - \sqrt{2} \text{ or } x \ge 2 + \sqrt{2}\}$

51. $\{x : x < -5.71 \text{ or } x > -.61\}$

53. 4 **55.** 100 **57.** $\{x : x < \frac{25}{2}\}$ **59.** $\{x : -3 < x < \frac{1}{2}\}$ **61.** $\{x : x < -1 \text{ or } -1 < x < 1 \text{ or } 4 < x < 8\}$

63. $\{x : -2 < x < 2\}$ **65.** $\{x : x \le \frac{1}{4} \text{ or } x \ge \frac{5}{4}\}$ **67.** $\{x : -1 < x < \frac{5}{2}\}$ **69.** $\{x : x > -\frac{1}{2}\}$

71. $\{x : x < -2 \text{ or } 0 < x < 2 \text{ or } x > 4\}$

73. (a) $\{k : k \le 4\}$ (b) $\{k : k \le -6 \text{ or } k \ge 6\}$ (c) $\{k : k \le 0 \text{ or } k \ge 4\}$ (d) $\{k : k = 0\}$

75. $25{,}000 < S < 53{,}000$ **77.** (a) 144 feet (b) $2 - \sqrt{3} < t < 2 + \sqrt{3}$ (c) $t = 5$ seconds

79. $c^n = c^2 c^{n-2} = (a^2 + b^2)c^{n-2} = a^2 c^{n-2} + b^2 c^{n-2} > a^2 a^{n-2} + b^2 b^{n-2} = a^n + b^n$

PROBLEM SET 1-6 (Page 56)

1.
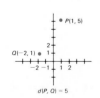
$d(P, Q) = 5$

3.
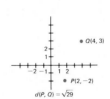
$d(P, Q) = \sqrt{29}$

5.
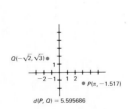
$d(P, Q) = 5.595686$

7.
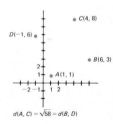
$d(A, C) = \sqrt{58} = d(B, D)$

9. $B(8, -1); D(2, 7)$ **11.** $(x - 1)^2 + (y + 2)^2 = 9$ **13.** $(-5, 3); 4$ **15.** $(4, -\frac{5}{2}); \sqrt{93}/2$

17. $(2, -4); \sqrt{22}$ **19.** $(-\pi/2, 1); \sqrt{\pi^2 + 4}/2$ **21.** $8; (x - 5)^2 + (y - 1)^2 = 16$

23.
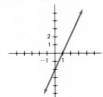

25.
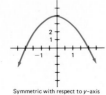
Symmetric with respect to y–axis

27.
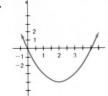

29.
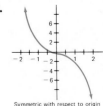
Symmetric with respect to origin

31.

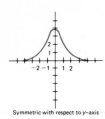

Symmetric with respect to y–axis

33.

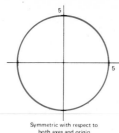

Symmetric with respect to both axes and origin

35.

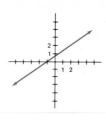

37.

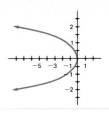

39.

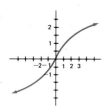

41. 7 or -1 **43.** $(x - 2)^2 + (y + 3)^2 = 34$ **45.** (a) $(-1, 5)$ (b) $(-10, 11)$

47. (a) Point: $(3, -2)$.   (b) Circle: radius 2, center $(\frac{1}{2}, -\frac{3}{2})$.   (c) Empty set.   (d) Circle: radius $\sqrt{5}$, center $(0, \sqrt{3})$.

49.

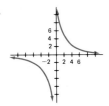

51.

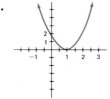

53.

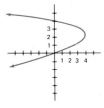

55. \$1.47 cheaper by truck.   **57.** 10   **59.** $60 + 2\pi$

CHAPTER 1. REVIEW PROBLEM SET (Page 60)

1. $\frac{27}{16}$, 1.7, $\frac{170}{99}$, $\sqrt{3}$, $1.\overline{7}$ **2.** (a) $\frac{1}{9}$ (b) $2\frac{7}{10}$ (c) $\sqrt{5} - 2$ **3.** 29.65   **4.** a, b, d

5. (a) $1/(\frac{5}{6})^2 = 36/25$ (b)$(\frac{6}{7})^2 = 36/49$ (c) $y^5/(2x^3)$ (d) $27a^8/(4b^{12})$

6. 1.6×10^{10} **7.** (a) $9x^2 - 6x + 5$ (b) $3y - 130$ (c) $10s^2 + s - 3$ (d) $9t^4 - 4$ (e) $x^2 - 4y^2 + 5x + 10y$

(f) $c^3 - 8d^3$ (g) $9x^2z^2 - 30xzh^2 + 25h^4$ (h) $x^3 + 6x^2m + 12xm^2 + 8m^3$

8. (a) $x^2(2x^2 - x + 11)$ (b) $(z - 3)(z + 9)$ (c) $(x^2 + 5)(x + 1)(x - 1)$ (d) $y^3(xy - 3)(x^2y^2 + 3xy + 9)$

(e) $(x + 2)(x - 1)(x^2 + x + 5)$

9. (a) 6   (b) $(-2x^2 + 2x - 3)/x(x - 1)$ (c) $-8/(2x + 2h + 5)(2x + 5)$ (d) $1/3x(x + 1)$

10. (a) $\frac{49}{48}$ (b) 7   (c) $-\frac{7}{3}$; 1   (d) -4; -5 (e) $(-1 \pm \sqrt{17})/4$ (f) -1; 4   **11.** (a) $z = 3(m - 4)/(m - 6)$

(b) $m = 6(z - 2)/(z - 3)$ **12.** $x = -\frac{3}{2}, y = 3$

13. (a) $x \le 6$ (b) $-\frac{4}{3} < x < 2$ (c) $-3 < x < \frac{1}{2}$ **14.** $(x - 3)^2 + (y - 4)^2 = 25$ **15.** Center: $(3, -7)$; radius: 2.

16. (a)

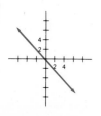

(b)

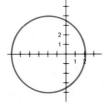

(c)

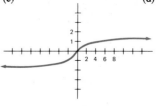

(d)

PROBLEM SET 2-1 (Page 68)

1. (a) 0 (b) −4 (c) −15/4 (d) −3.99 (e) −2 (f) $a^2 - 4$ (g) $(1 - 4x^2)/x^2$ (h) $x^2 + 2x - 3$ **3.** (a) $\frac{1}{4}$
(b) $-\frac{1}{2}$ (c) 2 (d) −8 (e) Undefined. (f) 100 (g) $x/(1 - 4x)$ (h) $1/(x^2 - 4)$ (i) $1/(h - 2)$ (j) $-1/(h + 2)$
5. All real numbers. **7.** $\{x : x \neq \pm 2\}$ **9.** $\{x : x \neq -2 \text{ and } x \neq 3\}$ **11.** All real numbers. **13.** $\{x : x \geq 2\}$
15. $\{x : x \geq 0 \text{ and } x \neq 25\}$ **17.** (a) 9 (b) 5 (c) 3; the positive integers.
19. $f(\#) = $ ⎡(⎤ # ⎡+⎤ 2 ⎡)⎤ x^2 ⎡=⎤; $f(2.9) = 24.01$ **21.** $f(\#) = 3$ ⎡×⎤ ⎡(⎤ # ⎡+⎤ 2 ⎡)⎤ x^2 ⎡−⎤ 4 ⎡=⎤;
$f(2.9) = 68.03$

23. $f(\#) = $ ⎡(⎤ 3 ⎡×⎤ # ⎡+⎤ 2 ⎡÷⎤ # ⎡$\sqrt{x}$⎤ ⎡)⎤ ⎡y^x⎤ 3 ⎡=⎤; $f(2.9) = 962.80311$
25. $f(\#) = $ ⎡(⎤ # ⎡y^x⎤ 5 ⎡−⎤ 4 ⎡)⎤ ⎡$\sqrt{x}$⎤ ⎡÷⎤ ⎡(⎤ 2 ⎡+⎤ # ⎡1/x⎤ ⎡)⎤ ⎡=⎤;$f(2.9) = 6.0479407$ **27.** $x + \sqrt{5}$
29. $2x^2 + 3x$ **31.** $\sqrt{3(x - 2)^2 + 9}$ **33.** 20 **35.** 5 **37.** 8 **39.** Undefined. **41.** $18xy - 10x$
43. $3 - 5x$ **45.** $y = 4x$ **47.** $y = 1/x$ **49.** $I = 324s/d^2$ **51.** (a) $R = 2v^2/45$ (b) About 28,444 feet.
53. (a) 3 (b) 3 (c) $-\frac{15}{4}$ (d) $2 - \sqrt{2}$ (e) $4 - 5\sqrt{2}$ (f) −199.9999 (g) $(1 - 2x^3)/x^2$ (h) $(a^4 - 2)/a^2$
(i) $(a^3 + 3a^2b + 3b^2 + b^3 - 2)/(a + b)$
55. (a) $\{t : t \neq 0, -3\}$ (b) $\{t : t \leq -2 \text{ or } t \geq 2\}$ (c) $\{t : t \geq 0, t \neq \frac{1}{2}\}$ (d) $\{(s, t) : -3 \leq s \leq 3, t \neq \pm 1\}$
57. Domain: $\{x : 0 < x < \frac{1}{\pi}\}$; range: $\{y : 0 < y < \frac{1}{\pi}\}$.
59. (a) $F(x) = \sqrt{3}x^2/36$ (b) $F(x) = 3\sqrt{3}x^2/2$ (c) $F(x) = 3\pi x^3/64$ (d) $F(x) = (1300 + 240x)/x$
(e) $F(x) = \begin{cases} 180 & \text{if } 0 \leq x \leq 100 \\ 180 + .22(x - 100) & \text{if } x > 100 \end{cases}$
61. $S(x, y, z) = 5000xy^2/3z$, $333\frac{1}{3}$ pounds **63.** $f(t) = 1 + t$

PROBLEM SET 2-2 (Page 77)

1. $\frac{5}{2}$ **3.** $-\frac{2}{7}$ **5.** $-\frac{5}{3}$ **7.** 0.1920 **9.** $4x - y - 5 = 0$ **11.** $2x + y - 2 = 0$ **13.** $2x + y - 4 = 0$
15. $5x - 2y - 4 = 0$ **17.** $x + 0 \cdot y - 1 = 0$ **19.** $5x - 4; 21$ **21.** $(-\frac{2}{5})x + \frac{16}{5}; \frac{6}{5}$ **23.** $(\frac{3}{2})x + \frac{7}{2}; 11$
25. 3; 5 **27.** $\frac{2}{3}; -\frac{4}{3}$ **29.** $-\frac{2}{3}; 2$ **31.** −4; 2
33. (a) $y + 3 = 2(x - 3)$ (b) $y + 3 = -\frac{1}{2}(x - 3)$ (c) $y + 3 = -\frac{2}{3}(x - 3)$ (d) $y + 3 = \frac{3}{2}(x - 3)$
(e) $y + 3 = -\frac{3}{4}(x - 3)$ (f) $x = 3$ (g) $y = -3$
35. $y + 4 = 2x$ **37.** $(-1, 2); y - 2 = \frac{3}{2}(x + 1)$
39. $(3, 1); y - 1 = -\frac{4}{3}(x - 3)$ **41.** $\frac{7}{5}$ **43.** $\frac{18}{13}$ **45.** $\frac{6}{5}$
47. (a) Parallel. (b) Perpendicular. (c) Neither. (d) Perpendicular.
49. $3x - 2y - 7 = 0$ **51.** $x + 2y - 9 = 0$
53. Note that $(a, 0)$ and $(0, b)$ satisfy the equation $x/a + y/b = 1$ and use the fact that two points determine a line.
55. $2x + y - 8 = 0$ **57.** $f(x) = -3x + 2, r = \frac{2}{3}$ **59.** $P(x) = 4x - 8500; -\$500$ (loss) **61.** $\frac{5}{13}$
63. Draw a picture. We may assume the triangle has vertices $(0, 0)$, $(a, 0)$, and (b, c). The midpoints are $(b/2, c/2)$ and
$(a + b)/2, c/2)$. The line joining the midpoints has slope 0, as does the base.
65. $(10 \pm 4\sqrt{2}, 0)$ **67.** $x = (1 + \sqrt{7})/2, y = (1 - \sqrt{7})/2$

PROBLEM SET 2-3 (Page 85)

1.

3.

5.

7.

9.

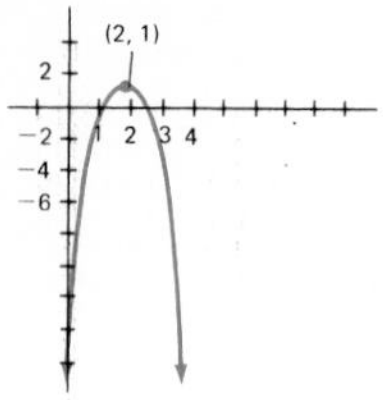

11. $y = 2x^2 - 4x + 9$ **13.** $y = -\frac{1}{2}x^2 + 5x - 10$ **15.**

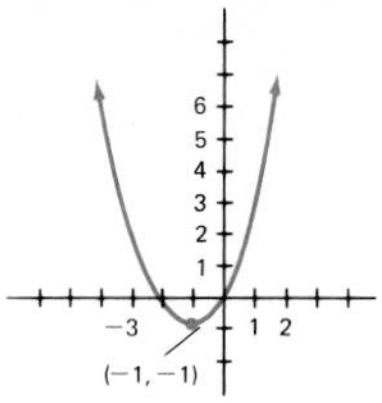

17.

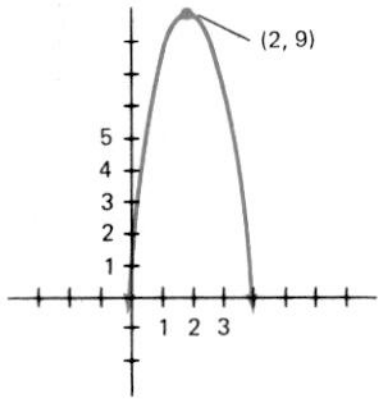

19.

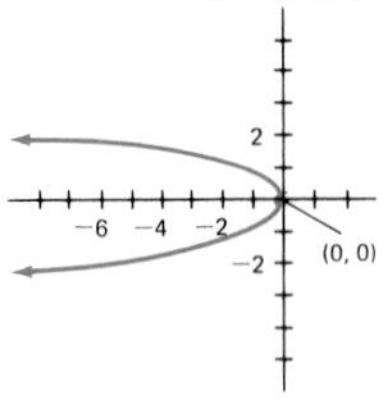

21.

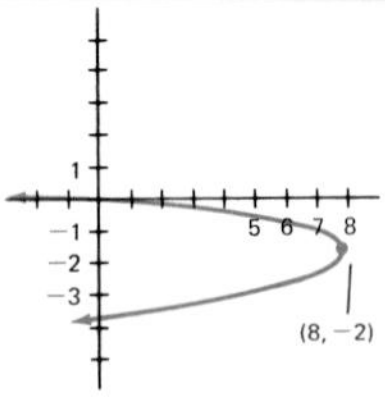

23.

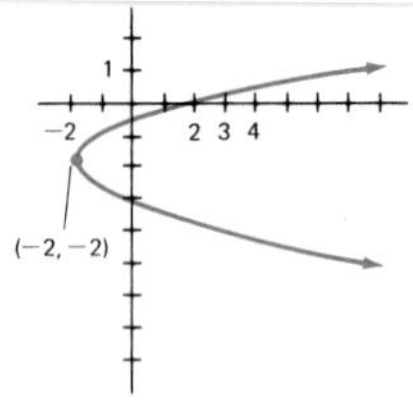

25. $(-3, 4); (0, 1)$ **27.** $(-1, 3); (2, -3)$ **29.** $(-.64, 2.25); (5.04, 10.75)$ **31.** $y = \frac{1}{12}x^2$
33. $y = -\frac{1}{8}(x + 2)^2$ **35.** $y = 2x^2 - 1$
37.

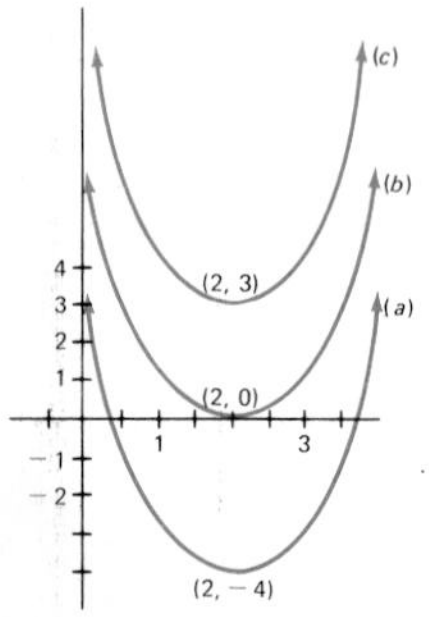

39. (a)

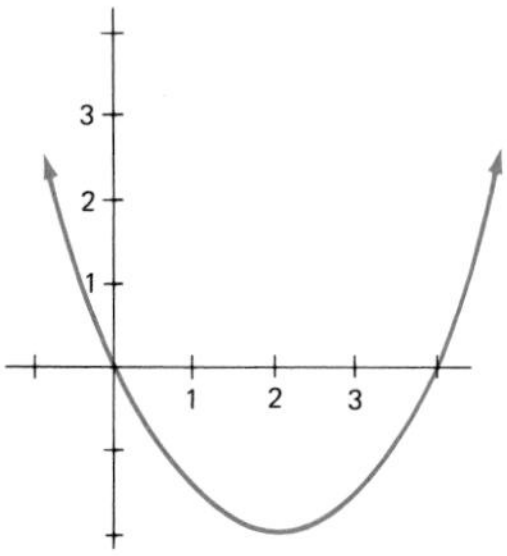

(b)

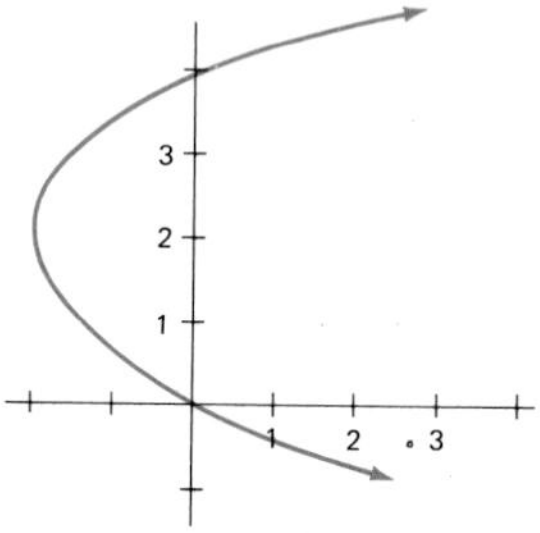

41. (a) $(-2, 14)$ and $(1, 5)$ (b) $(3, 3)$ (c) None. (d) $(-1, 10)$ and $(2, 7)$
43. $a = \frac{5}{2}; (5, -\frac{45}{2})$ **45.** $\frac{45}{4}$ **47.** $P = (300 - 100x)(x - 2)$; \$2.50 **49.** $(b - a)^3/8$ **51.** $L = 2p$

PROBLEM SET 2-4 (Page 93)

1.

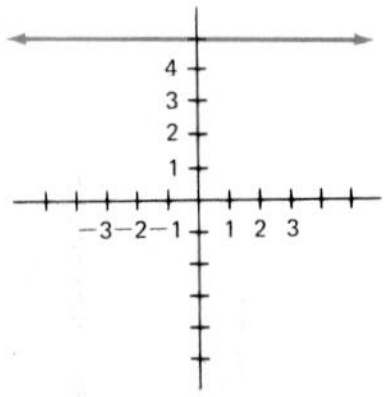

3.

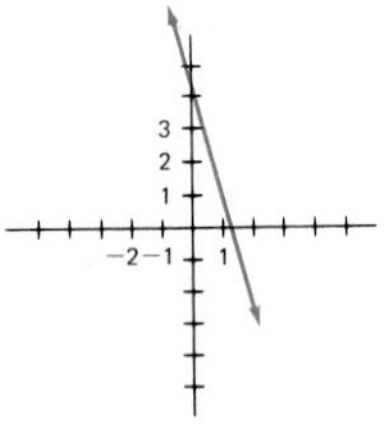

5.

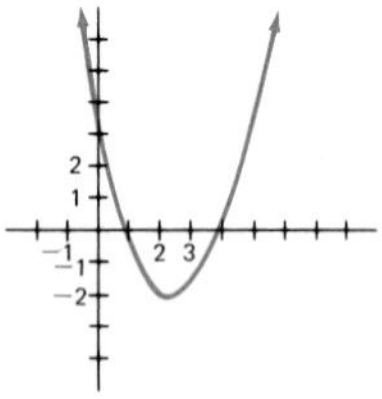

7.

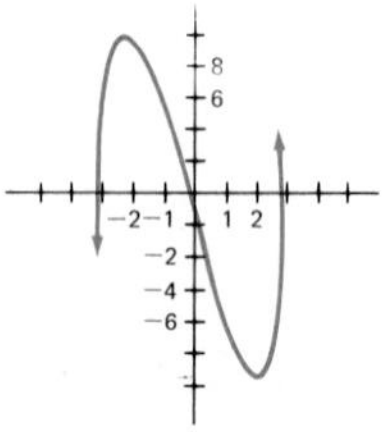

9.

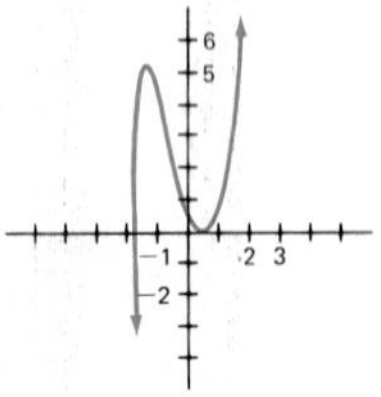

11.

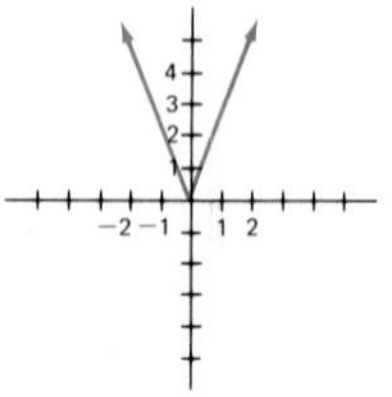

13.

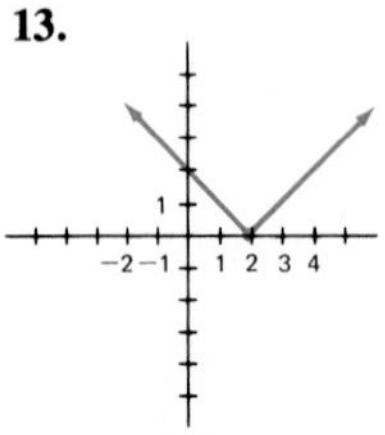

15.

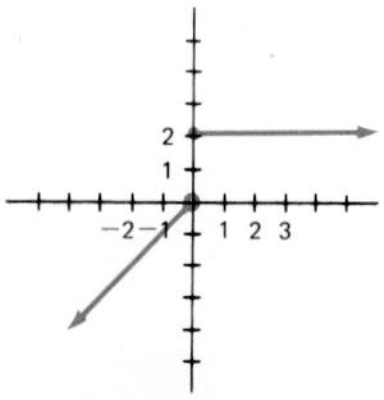

17.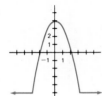

19. Even.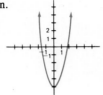

21. Neither even nor odd.

23. Odd.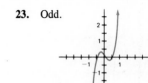

25. Even.

27.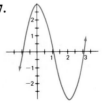

29.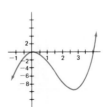

31.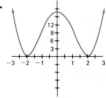

33.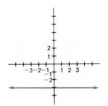

35.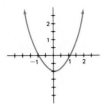

37.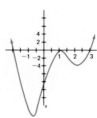

39.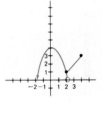

41.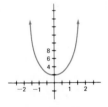

43.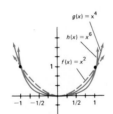

45. 55; 10.1923; 1033.9648

47.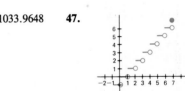

49. $C(x) = 15 + 10[x]$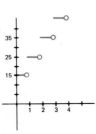

51.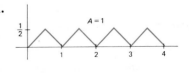

PROBLEM SET 2-5 (Page 101)

1.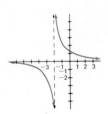

3.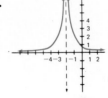

5.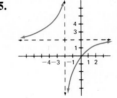

7.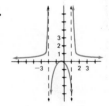

9.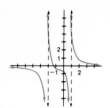

11.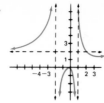

13.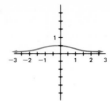

15.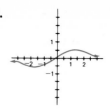

17.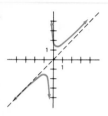

19.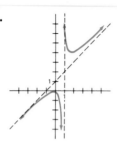

21.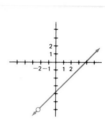

23.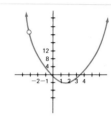

25.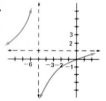

27.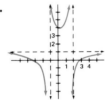

29.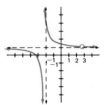

31.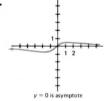

$y = 0$ is asymptote $y = 1$ is asymptote $y = x$ is asymptote

33.

n	Vertical	Horizontal	Oblique
1	$x = 1$	None	None
2	$x = -1, x = 1$	None	None
3	$x = 1$	None	$y = x$
4	$x = -1, x = 1$	$y = 1$	None
5	$x = 1$	$y = 0$	None
6	$x = -1, x = 1$	$y = 0$	None

35.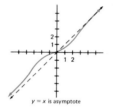

37. $f(x) = 2x + 3 + \dfrac{2}{x - 3} = \dfrac{2x^2 - 3x - 7}{x - 3}$

524   ANSWERS TO ODD-NUMBERED PROBLEMS

PROBLEM SET 2-6 (Page 109)

1. (a) 3 (b) Undefined. (c) −1 (d) −10 (e) 2 (f) $\frac{1}{2}$ (g) 10 (h) $\frac{2}{5}$ (i) 3

3. $(f + g)(x) = x^2 + x - 2$, all real numbers; $(f - g)(x) = x^2 - x + 2$, all real numbers; $(f \cdot g)(x) = x^3 - 2x^2$, all real numbers; $(f/g)(x) = x^2/(x - 2)$, $\{x : x \neq 2\}$.

5. $(f + g)(x) = x^2 + \sqrt{x}$, $\{x : x \geq 0\}$; $(f - g)(x) = x^2 - \sqrt{x}$, $\{x : x \geq 0\}$; $(f \cdot g)(x) = x^2\sqrt{x}$, $\{x : x \geq 0\}$; $(f/g)(x) = x^2/\sqrt{x}$, $\{x : x > 0\}$

7. $(f + g)(x) = (x^2 - x - 3)/(x - 2)(x - 3)$, $\{x : x \neq 2 \text{ and } x \neq 3\}$; $(f - g)(x) = (-x^2 + 3x - 3)/(x - 2)(x - 3)$, $\{x : x \neq 2 \text{ and } x \neq 3\}$; $(f \cdot g)(x) = x/(x - 2)(x - 3)$, $\{x : x \neq 2 \text{ and } x \neq 3\}$; $(f/g)(x) = (x - 3)/x(x - 2)$, $\{x : x \neq 0, x \neq 2, x \neq 3\}$.

9. $(g \circ f)(x) = x^2 - 2$, all real numbers; $(f \circ g)(x) = (x - 2)^2$, all real numbers.

11. $(g \circ f)(x) = (3x + 1)/x$, $\{x : x \neq 0\}$; $(f \circ g)(x) = 1/(x + 3)$, $\{x : x \neq -3\}$

13. $(g \circ f)(x) = x - 4$, $\{x : x \geq 2\}$; $(f \circ g)(x) = \sqrt{x^2 - 4}$, $\{x : |x| \geq 2\}$

15. $(g \circ f)(x) = x$, all real numbers; $(f \circ g)(x) = x$, all real numbers.   17. $g(x) = x^3; f(x) = x + 4$

19. $g(x) = x + 2; f(x) = \sqrt{x}$ 21. $g(x) = 1/x^3; f(x) = 2x + 5$ 23. $g(x) = |x|; f(x) = x^3 - 4$

25.

27.

29.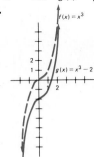

31.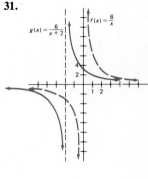

33. 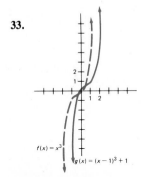   35. 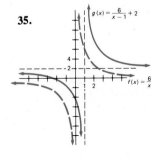   37. $2x + h; 7$ 39. $4x + 2h - 5; 9$ 41. 2; 2

43. $-2/(x + h + 1)(x + 1); -.1$ 45. $3x^2 + 3xh + h^2; 37$

47. (a) $x^3 + 2x + 3$ (b) $x^3 - 2x - 3$ (c) $2x^4 + 3x^3$ (d) $(2x + 3)/x^3$ (e) $2x^3 + 3$ (f) $(2x + 3)^3$
(g) $4x + 9$ (h) x^{27}

49. $1/|x^2 - 4|$; $\{x : x \neq -2 \text{ and } x \neq 2\}$

51. (a) 19 (b) 15.25 (c) 12.61 (d) 12.0001, the quotient approaches 12 and the slope of the tangent line to the graph of f at $x = 2$ is 12.

53. (a) .7053 (b) 1.6105   55. (a) $x; |x|$ (b) $x; x$ (c) $x^6; x^6$ (d) $1/x^6; 1/x^6$ 57. $\frac{1}{2}$

59.

61. (a) Odd. (b) Odd. (c) Even.
(d) Odd. (e) Neither. (f) Odd.
(g) Even. (h) Even. (i) Even.

63. (a)

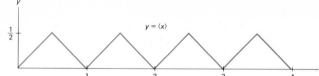

$y = \langle x \rangle$

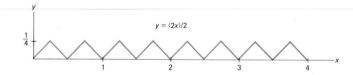

$y = \langle 2x \rangle / 2$

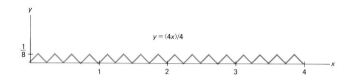

$y = \langle 4x \rangle / 4$

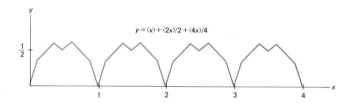

$y = \langle x \rangle + \langle 2x \rangle / 2 + \langle 4x \rangle / 4$

(b) $1; \frac{1}{2}; \frac{1}{4}; \frac{7}{4}$

PROBLEM SET 2-7 (Page 118)

1. (a) i; ii; iii; iv ; vii; viii (b) i; ii; viii; (c) i; ii; viii **3.** (a) 1 (b) $-\frac{1}{3}$ (c) $\frac{16}{3}$ **5.** $f^{-1}(x) = (\frac{1}{5})x$

7. $f^{-1}(x) = (\frac{1}{2})(x + 7)$ **9.** $f^{-1}(x) = (x - 2)^2$ **11.** $f^{-1}(x) = 3x/(x - 1)$ **13.** $f^{-1}(x) = 2 + \sqrt[3]{(x - 2)}$

15. 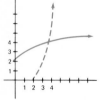 **17.** (a) 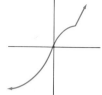 (b) 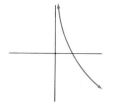 (c)

19. $(f \circ g)(x) = 3\left(\dfrac{2x}{3 - x}\right)\bigg/\left(\dfrac{2x}{3 - x} + 2\right) = \dfrac{6x}{3 - x}\bigg/\dfrac{6}{3 - x} = x;$

$(g \circ f)(x) = 2\left(\dfrac{3x}{x + 2}\right)\bigg/\left(3 - \dfrac{3x}{x + 2}\right) = \dfrac{6x}{x + 2}\bigg/\dfrac{6}{x + 2} = x$

21. $\{x : x \geq 1\}; f^{-1}(x) = 1 + \sqrt{x}$ **23.** $\{x : x \geq -1\}; f^{-1}(x) = -1 + \sqrt{x + 4}$

25. $\{x : x \geq -3\}; f^{-1}(x) = -3 + \sqrt{x + 2}$ **27.** $\{x : x \geq -2\}; f^{-1}(x) = x - 2$

29. $\{x : x \geq 1\}; f^{-1}(x) = 1 + \sqrt{2x/(1 + x)}$

31.

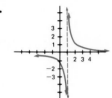

33. $f^{-1}(x) = (x + 1)/x$

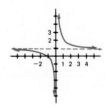

(a) $\frac{1}{2}$ (b) 3 (c) -1 (d) 0 (e) $\frac{4}{3}$ (f) $\frac{1}{2}$

35.

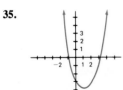

$\{x : x \geq 1\}; f^{-1}(x) = 1 + \sqrt{x + 4}$ **37.** $f^{-1}(x) = 1 + \sqrt{\dfrac{3}{2 - x}}$

39. The graph must be symmetric about the line $y = x$. This means the xy-equation determining f is unchanged if x and y are interchanged.

41. (a) $f^{-1}(x) = (b - dx)/(cx - a)$ (b) $(f \circ f^{-1})(x)$ should be x.

But $f(f^{-1}(x)) = \dfrac{a(b - dx)/(cx - a) + b}{c(b - dx)/(cx - a) + d} = \dfrac{(bc - ad)x}{bc - ad} = x$, provided $bc - ad \neq 0$.

(c) $a = -d$

CHAPTER 2. REVIEW PROBLEM SET (Page 122)

1. (a) 15 (b) 4 (c) Undefined. (d) 0 (e) $-\frac{3}{4}$ (f) $\frac{2}{15}$ **2.** $\{x : x \geq -1 \text{ and } x \neq 1\}$ **3.** $y = \frac{1}{8}x^3$ **4.** 12

5. (a) $x - y + 3 = 0$ (b) $x + 0 \cdot y - 4 = 0$ (c) $0 \cdot x + y - 2 = 0$ (d) $3x - 2y + 7 = 0$ (e) $5x + 2y + 1 = 0$

6. $(-1, -4), (1, -2)$

7. (a)

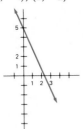

(b)

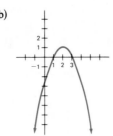

(c)

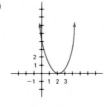

(d)

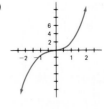

(e)

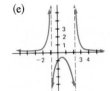

(f)

8.

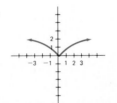

9.

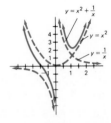

10. (a) $x^3 + 6x^2 + 12x + 9$ (b) $x^3 + 3$ (c) $x^9 + 3x^6 + 3x^3 + 2$ (d) $x + 4$ (e) $\sqrt[3]{x - 1}$ (f) $x - 2$ (g) $x + h + 2$

(h) 1 (i) $27x^3 + 1$ **11.** It is moved 2 units to the right and up 3 units.

12. (a) Even. (b) Odd, one-to-one. (c) Even. (d) One-to-one.

13. 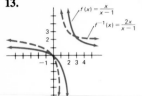 **14.** $\{x : x \geq -2\}$

PROBLEM SET 3-1 (Page 129)

1. 3 **3.** 2 **5.** 7 **7.** $\frac{3}{2}$ **9.** 25 **11.** 9 **13.** 2 **15.** $\frac{1}{100}$ **17.** $\sqrt{2}/2$ **19.** $\sqrt{5}$

21. $3xy\sqrt[3]{2xy^2}$ **23.** $(x + 2)y\sqrt[4]{y^3}$ **25.** $x\sqrt{1 + y^2}$ **27.** $x\sqrt[3]{x^3 - 9y}$ **29.** $(xz^2/y^2)\sqrt[3]{x}$

31. $2(\sqrt{x} - 3)/(x - 9)$ **33.** $2\sqrt{x + 3}/(x + 3)$ **35.** $\sqrt[4]{2x}/2x$ **37.** $(2y/x)\sqrt[3]{x^2}$ **39.** $\sqrt{2}$ **41.** 26

43. $\frac{9}{2}$ **45.** $-\frac{32}{15}$ **47.** 0 **49.** 4 **51.** $2[(\sqrt{x} - \sqrt{x + h})/\sqrt{x}\sqrt{x + h}]$ **53.** $(x + 7)/\sqrt{x + 6}$

55. $x/(2\sqrt[3]{x} + 2)$ **57.** $-9/(x^2\sqrt{x^2 + 9})$ **59.** (a) $2ab^2$ (b) $9b\sqrt[3]{b}$ (c) $3\sqrt{3}$ (d) $5a^2b^3\sqrt{10}$ (e) $-2y^2/x$

(f) $y^2/4x^2$ (g) $(2a^2 + 3a)\sqrt{2a}$ (h) $4\sqrt[4]{2} - 5\sqrt{2} + 2\sqrt[6]{2}$ (i) $a\sqrt[4]{1 + b^4}$ (j) $\sqrt[3]{49b^2}/(7bc)$ (k) $2(\sqrt{a} + b)/(a - b^2)$

(l) $a + \frac{1}{a} = (a^2 + 1)/a$ **61.** (a) 13 (b) 0; 2 (c) 6 (d) No solution. (e) 9 (f) -8; 64

63. They are reflections of each other in the line $y = x$. **65.** $\overline{AC} = \frac{15}{4}$

67. (a) Both sides are positive. $[(\sqrt{6} + \sqrt{2})/2)]^2 = (6 + 2\sqrt{12} + 2)/4 = 2 + \sqrt{3} = (\sqrt{2 + \sqrt{3}})^2$.

(b) Both sides are positive. $(\sqrt{2 + \sqrt{3}} + \sqrt{2 - \sqrt{3}})^2 = 2 + \sqrt{3} + 2\sqrt{1} + 2 - \sqrt{3} = 6$.

(c) $(\sqrt{3} - \sqrt{2})^3 = 3\sqrt{3} - 3 \cdot 3\sqrt{2} + 3\sqrt{3} \cdot 2 - 2\sqrt{2} = 9\sqrt{3} - 11\sqrt{2} = (\sqrt[3]{9\sqrt{3} - 11\sqrt{2}})^3$.

PROBLEM SET 3-2 (Page 132)

1. $7^{1/3}$ **3.** $7^{2/3}$ **5.** $7^{-1/3}$ **7.** $7^{-2/3}$ **9.** $7^{4/3}$ **11.** $x^{2/3}$ **13.** $x^{5/2}$ **15.** $(x + y)^{3/2}$

17. $(x^2 + y^2)^{1/2}$ **19.** $\sqrt[3]{16}$ **21.** $1/\sqrt{8^3} = \sqrt{2}/32$ **23.** $\sqrt[4]{x^4 + y^4}$ **25.** $y\sqrt[3]{x^4y}$ **27.** $\sqrt{\sqrt{x} + \sqrt{y}}$

29. 5 **31.** 4 **33.** $\frac{1}{27}$ **35.** .04 **37.** .000125 **39.** $\frac{1}{5}$ **41.** $\frac{1}{16}$ **43.** $\frac{1}{16}$ **45.** $-6a^2$ **47.** $8/x^4$

49. x^4 **51.** $4y^2/x^4$ **53.** y^9/x^{30} **55.** $(2y^3 - 1)/y$ **57.** $x + y + 2\sqrt{xy}$ **59.** $(7x + 2)/3(x + 2)^{1/5}$

61. $(1 - x^2)/(x^2 + 1)^{2/3}$ **63.** $\sqrt[6]{32}$ **65.** $\sqrt[12]{8x^2}$ **67.** $\sqrt[3]{x}$ **69.** 2.53151 **71.** 4.6364

73. 1.70777 **75.** .0050463

77. 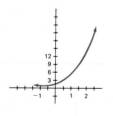 **79.** 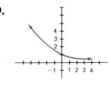 **81.**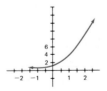

83. (a) $b^{3/5}$ (b) $x^{1/2}$ (c) $(a + b)^{2/3}$

85. (a) 72 (b) $3 \cdot 2^{1/6}$ (c) $a^{17/12}$ (d) $a^{5/6}$ (e) $a^3 + 2 + \frac{1}{a^3}$ (f) $\frac{a}{1 + a^2}$ (g) $a^{1/2}b^{1/12}$ (h) $a^8/b^{14/3}$ (i) 8

(j) $4a^{3/2}b^{9/4}$ (k) $3^{3/2}$ (l) $a - b$

87. (a) $-\frac{1}{2}$ (b) -1; 2 (c) All reals. (d) -4; 3 (e) 1; 8 (f) 2

89.

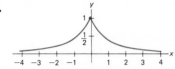

91. The graph of $y = f(x) = a^x$ has the x-axis as a horizontal asymptote. Also f is not a constant function. The graph of a nonconstant polynomial does not have a horizontal asymptote.

PROBLEM SET 3-3 (Page 142)

1. (a) Decays. (b) Grows. (c) Grows. (d) Decays. **3.** (a) 4.66095714 (b) 17.00006441
(c) 4801.02063 (d) 9750.87832 **5.** 1480 **7.** (a) 5.384 billion (b) 6.562 billion (c) 23.772 billion
9. (a) \$185.09 (b) \$247.60 **11.** (a) \$76,035.83 (b) \$325,678.40 **13.** $p(1 + r/100)^n$ **15.** \$7401.22
17. \$7102.09 **19.** \$7305.57 **21.** $\$1000(1 + .08/12)^{120} = \$2219.64$ **23.** 8100; 5400; 3600; 2400; 1600
25. 800 **27.** (a) 8680; 22,497 (b) About 44 years. **29.** (a) \$146.93 (b) \$148.59 (c) \$148.98 (d) \$149.18
31. (a) About 9 years. (b) About 11 years. **33.** (a) $k \approx .0005917$ (b) 10.76 milligrams **35.** 2270 years
37. \$320,057,300

PROBLEM SET 3-4 (Page 151)

1. $\log_4 64 = 3$ **3.** $\log_{27} 3 = \frac{1}{3}$ **5.** $\log_4 1 = 0$ **7.** $\log_{125}(1/25) = -\frac{2}{3}$ **9.** $\log_{10} a = \sqrt{3}$
11. $\log_{10} \sqrt{3} = a$ **13.** $5^4 = 625$ **15.** $4^{3/2} = 8$ **17.** $10^{-2} = .01$ **19.** $c^1 = c$ **21.** $c^y = Q$ **23.** 2
25. -1 **27.** $1/3$ **29.** -4 **31.** 0 **33.** $\frac{4}{3}$ **35.** 2 **37.** $\frac{1}{27}$ **39.** -2.9 **41.** 49 **43.** .778
45. 1.204 **47.** $-.602$ **49.** 1.380 **51.** $-.051$ **53.** .699 **55.** 1.5314789 **57.** $.08990511 - 2$
59. 3.9878003 **61.** $\log_{10}[(x + 1)^3(4x + 7)]$ **63.** $\log_2[8x(x + 2)^3/(x + 8)^2]$ **65.** $\log_6(\sqrt{x}\sqrt[3]{x^3 + 3})$
67. 47 **69.** $-\frac{11}{4}$ **71.** $\frac{16}{7}$ **73.** 5; 2 is extraneous. **75.** 7 **77.** 4.0111687 **79.** 2.0446727
81. (a) 2 (b) $\frac{1}{2}$ (c) 125 (d) 32 (e) 10 (f) 4 **83.** $\frac{8}{9}$
85. (a) 13 (b) -5 (c) No solution. (d) 20 (e) 3 (f) 16 **87.** (a) $y = x/(x - 1)$ (b) $y = \frac{1}{2}(a^x + a^{-x})$
89. $x = 10^c$, $\log_2 x = c \log_2 10$, $\log_2 x = (\log_{10} x)(\log_2 10)$
91. By Problem 90 $\log_a a = \log_b a \cdot \log_a b$, or $1 = \log_b a \cdot \log_a b$. Therefore, $\log_a b = \dfrac{1}{\log_b a}$.

93.

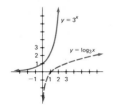

95.

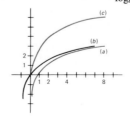

PROBLEM SET 3-5 (Page 159)

1. 1 **3.** 0 **5.** $\frac{1}{2}$ **7.** -3 **9.** 3.5 **11.** $-.2$ **13.** -7.5 **15.** 4.787 **17.** 6.537 **19.** .182
21. 9.1 **23.** .9 **25.** 90 **27.** 1.4609379 **29.** -2.0635682 **31.** -1.8411881 **33.** 50833303
35. 11.818915 **37.** 61.146135 **39.** 8.3311375 **41.** .8824969 **43.** 915.98 **45.** About 2.73.
47. About -0.737. **49.** About 6.84. **51.** Approximately 6.12 years. **53.** Approximately 4.71 years.
55. (a) 5^{10} (b) 9^{10} (c) 10^{20} (d) 10^{1000} **57.**

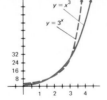

59. $y = ba^x$; $a \approx 1.5$, $b \approx 64$

61. $y = bx^a$; $a \approx 4$, $b \approx 12$　**63.** (a) 4.2　(b) 4　(c) $\frac{1}{2}$　**65.** (a) 7　(b) 12.25　(c) $-\frac{1}{2}$　(d) 0　(e) 125　(f) $\frac{1}{3}$
67. (a) $-.349$　(b) $-.823$　(c) $.633, -3.633$　(d) 2.166　(e) 4.560; .219　(f) $e^e \approx 15.154$　(g) $e \approx 2.718$
(h) $e \approx 2.718$　(i) $\pm\sqrt{e + 5} \approx \pm2.778$　**69.** $(\ln 2)/3 \approx .231$ years　**71.** $(\ln 2)/240 \approx .00289$

73.

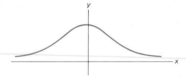

75. $e^{\pi/e - 1} > 1 + \pi/e - 1$, $e^{\pi/e}/e > \pi/e$, $e^{\pi/e} > \pi$, $e^{\pi} > \pi^e$
77. (a) $100(1 + .01)^{120} \approx \$330.04$　(b) $100(1 + .12/365)^{3650} \approx \$331.95$　(c) $100(1 + .12/(365)(24))^{(3650)(24)} \approx \$332.01$
(d) $100e^{(.12)(10)} \approx \$332.01$

PROBLEM SET 3-6　(Page 167)

1. 4　**3.** -2　**5.** $\frac{11}{2}$　**7.** -15　**9.** 10,000　**11.** .01　**13.** $10^{3/2}$　**15.** $10^{-3/4}$　**17.** .6355
19. 2.1987　**21.** $.5172 - 2$　**23.** 5.7505　**25.** 8.9652　**27.** 32.8　**29.** .0101　**31.** 3.98×10^8
33. 166　**35.** .838　**37.** .7191　**39.** 3.8593　**41.** $.0913 - 3$　**43.** 7.075　**45.** 8184　**47.** .03985
49. (a) $\frac{5}{4}$　(b) $-\frac{4}{3}$　(c) $\frac{5}{6}$　(d) -3　**51.** (a) 2.6926　(b) $.6726 - 2$　(c) 856.4　(d) .001861
53. (a) .0035703　(b) .0000845　**55.** $.2932 - 3$　**57.** .2123　**59.** 16　**61.** About 972.5 miles.

PROBLEM SET 3-7　(Page 172)

1. 128　**3.** .0959　**5.** .0208　**7.** 7.12×10^7　**9.** 3.50　**11.** .983　**13.** 6.05　**15.** 4762
17. 6.143　**19.** 3.530×10^{-6}　**21.** 8.90　**23.** 18.2　**25.** 5.19　**27.** $-.5984$　**29.** 2.24
31. (a) .3495　(b) 100.7　(c) .8274　**33.** (a) $\frac{2}{9}$　(b) 2　(c) 1　(d) 3　**35.** About 4.395 hours from now.
37. (a) About 6.17 billion.　(b) About the year 2018.

CHAPTER 3. REVIEW PROBLEM SET　(Page 175)

1. (a) $-2y^2\sqrt[3]{z}/z^5$　(b) $2xy^2\sqrt[4]{2x}$　(c) $2\sqrt[6]{5}$　(d) $2(\sqrt{x} + \sqrt{y})/(x - y)$　(e) $5\sqrt{2 + x^2}$　(f) $6\sqrt{2}$　**2.** (a) 12　(b) 4
3. (a) $125a^3$　(b) $1/a^{1/2}$　(c) $1/5^{7/4}$　(d) $3y^{13/6}/x^6$　(e) $x - 2x^{1/2}y^{1/2} + y$　(f) $2^{7/6}$

4.

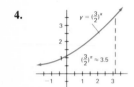

5. $(\frac{1}{2})^{81} \approx 4.14 \times 10^{-25}$　**6.** 16 million　**7.** \$220.80　**8.** (a) 3　(b) $\frac{1}{8}$　(c) 7
(d) 1　(e) $\frac{3}{2}$　(f) 5　(g) 10　(h) 1.14　**9.** $\log_4[(3x + 1)^2(x - 1)/\sqrt{x}]$
10. (a) $\frac{3}{2}$　(b) $\frac{4}{3}$　(c) $\frac{1}{2}$　(d) -2.773　**11.** (a) 1.680　(b) 9.3　(c) 3.517　(d) .9
12. 1.807　**13.** 13.9 years　**14.** .1204　**15.**　　　　　　　　　　　　　　**16.** 3999

PROBLEM SET 4-1 (Page 181)

1. .6600 **3.** .6534 **5.** 3.133 **7.** 12.5° **9.** 66.6° **11.** 69.3° **13.** 16.97 ≈ 17 **15.** 41.34 ≈ 41
17. 66.60 ≈ 67 **19.** $\beta = 48°$; $a = 23.42 ≈ 23$; $b = 26.01 ≈ 26$ **21.** $\alpha = 33.8°$; $a = 50.8$; $b = 75.9$
23. $\beta = 50.6°$; $b = 146$; $c = 189$ **25.** $c = 15$; $\alpha = 36.9°$; $\beta = 53.1°$ **27.** $b = 30$; $\alpha = 53.1°$; $\beta = 36.9°$
29. $\alpha = 26.7°$; $\beta = 63.3°$; $b = 29.0$ **31.** $\alpha = 32.9°$; $\beta = 57.1°$; $c = 17.5$ **33.** 14.6° **35.** 7.0°
37. 31.2 feet **39.** (a) .25862 (b) 6.6568 (c) 17.445 **41.** 725 feet **43.** 65.369 **45.** 364 feet
47. 448 meters **49.** 41.77° **51.** $P = 24$; $A = 24\sqrt{3}$ **53.** $\sqrt{3}$

PROBLEM SET 4-2 (Page 188)

1. $2\pi/3$ **3.** $4\pi/3$ **5.** $7\pi/6$ **7.** $7\pi/4$ **9.** 3π **11.** $-7\pi/3$ **13.** $8\pi/9$ **15.** $\frac{1}{9}$ **17.** 240°
19. $-120°$ **21.** 540° **23.** 259.0° **25.** 18.2° **27.** (a) 2 (b) 3.14 **29.** (a) 3 centimeters (b) 5.5 inches
31. II **33.** III **35.** II **37.** IV **39.** $16\pi/3 ≈ 16.76$ feet per second.
41. $320\pi ≈ 1005.3$ inches per minute. **43.** (a) -8π (b) 5π (c) $\frac{1}{3}$
45. (a) 25.5 centimeters (b) .0327 centimeters (c) 37.83 centimeters
47. $9600\pi ≈ 30,159$ centimeters **49.** $330\pi ≈ 1037$ miles per hour. **51.** 8.6×10^5 miles
53. $\frac{33\pi}{20} ≈ 5.184$ hours **55.** $264\pi ≈ 829$ miles **57.** $7\pi ≈ 21.99$ square inches. **59.** 130

PROBLEM SET 4-3 (Page 195)

1. $(\sqrt{3}/2, \frac{1}{2})$ **3.** $(-\sqrt{2}/2, \sqrt{2}/2)$ **5.** $(-\sqrt{2}/2, -\sqrt{2}/2)$ **7.** $(-\sqrt{3}/2, \frac{1}{2})$ **9.** $-\sqrt{2}/2$
11. $\sqrt{2}/2$ **13.** $-\sqrt{2}/2$ **15.** $-\frac{1}{2}$ **17.** 1 **19.** 0 **21.** $-\sqrt{3}/2$ **23.** $\frac{1}{2}$ **25.** $-\sqrt{2}/2$
27. $\frac{1}{2}$ **29.** $-\frac{1}{2}$ **31.** $-\sqrt{3}/2$ **33.** $-.95557$; $-.29476$ **35.** (a) $(1/\sqrt{5}, 2/\sqrt{5})$ (b) $2/\sqrt{5}, 1/\sqrt{5}$
37. (a) $\sin(\pi + t) = -y = -\sin t$ (b) $\cos(\pi + t) = -x = -\cos t$
39. (a) Negative. (b) Negative. (c) Positive. (d) Positive. (e) Positive. (f) Negative.
41. (a) $\pm\sqrt{3}/2$ (b) $7\pi/6$; $11\pi/6$
43. (a) $\pi/4$; $5\pi/4$ (b) $\pi/6 < t < \pi/3$; $2\pi/3 < t < 5\pi/6$ (c) $0 \le t \le \pi/3$; $2\pi/3 \le t \le 4\pi/3$;
$5\pi/3 \le t < 2\pi$ (d) $0 \le t < \pi/4$; $3\pi/4 < t < 5\pi/4$; $7\pi/4 < t < 2\pi$ **45.** (a) $\frac{3}{5}$ (b) $-\frac{7}{25}$
47. (a) $\frac{3}{5}$ (b) $\frac{4}{5}$ (c) $\frac{4}{5}$ (d) $\frac{4}{5}$ (e) $\frac{3}{5}$ (f) $-\frac{4}{5}$
49. (a) Period 1 (b) Period $\frac{1}{3}$ (c) Not periodic (d) Period 1 **51.** 0

PROBLEM SET 4-4 (Page 201)

1. (a) $-\frac{4}{3}$ (b) $-\frac{3}{4}$ (c) $-\frac{5}{3}$ (d) $\frac{5}{4}$ **3.** $-\sqrt{5}/2$; $\frac{3}{5}\sqrt{5}$ **5.** $\sqrt{3}/3$ **7.** $2\sqrt{3}/3$ **9.** 1
11. $2\sqrt{3}/3$ **13.** $-\sqrt{3}/2$ **15.** $\sqrt{3}$ **17.** 0 **19.** $-\sqrt{3}/3$ **21.** -2
23. (a) $\pi/2$; $3\pi/2$; $5\pi/2$; $7\pi/2$ (b) $\pi/2$; $3\pi/2$; $5\pi/2$; $7\pi/2$ (c) 0; π; 2π; 3π; 4π (d) 0; π; 2π; 3π; 4π
25. $-12/13$; $-12/5$; $13/5$ **27.** $-2\sqrt{5}/5$; 2; $-\sqrt{5}$ **29.** $\frac{3}{5}$; $\frac{5}{4}$ **31.** $-\frac{12}{13}$; $-\frac{12}{5}$ **33.** $(\frac{5}{13}, -\frac{12}{13})$
35. 111.8° **37.** (a) $-2\sqrt{3}/3$ (b) $\sqrt{3}$ (c) $\sqrt{2}$ (d) -1 (e) -2 (f) 1
39. (a) $\frac{24}{25}$ (b) $\frac{-7}{25}$ (c) $\frac{-24}{7}$ (d) $\frac{25}{24}$ (e) $\frac{-24}{7}$ (f) $\frac{-25}{7}$ **41.** (a) $3\pi/4$; $7\pi/4$ (b) $\pi/4$; $7\pi/4$ (c) $\pi/2$; $3\pi/2$
43. (a) 1 (b) $\sin\theta - 1/\cos\theta$ (c) $1 + 2\sin\theta\cos\theta$ (d) $1/\sin\theta$ (e) $\cos\theta + \sin\theta$ (f) $-(\sin^2\theta + 1)/(\cos^2\theta)$
45. (a) $\tan(t + \pi) = \sin(t + \pi)/\cos(t + \pi) = (-\sin t)/(-\cos t) = \tan t$
(b) $\cot(t + \pi) = \cos(t + \pi)/\sin(t + \pi) = (-\cos t)/(-\sin t) = \cot t$
(c) $\sec(t + \pi) = 1/\cos(t + \pi) = 1/(-\cos t) = -1/(\cos t) = -\sec t$
(d) $\csc(t + \pi) = 1/\sin(t + \pi) = 1/(-\sin t) = -1/(\sin t) = -\csc t$
47. $\frac{119}{169}$ **49.** -10 **51.** (a) -1.3764 (b) .3153 **53.** 428.98 centimeters

PROBLEM SET 4-5 (Page 206)

1. .98185 **3.** .7337 **5.** .93309 **7.** .9291 **9.** 1.30 **11.** .40 **13.** 1.10 **15.** 1.06 **17.** 1.12
19. .50 **21.** $3\pi/8$ **23.** $\pi/3$ **25.** .24 **27.** .24 **29.** $\pi/2$ **31.** .15023 **33.** 5.4707
35. .84147 **37.** -1.2885 **39.** $-.82534$ **41.** 1.25; 1.89 **43.** 1.65; 4.63 **45.** 1.37; 4.51
47. 1.84; 4.98 **49.** 40.4° **51.** 11.3° **53.** 80.2° **55.** .4051 **57.** $-.1962$ **59.** .4051 **61.** .15126
63. .9657 **65.** 21.3°; 158.7° **67.** 26.3°; 206.3° **69.** 155.3°; 204.7°
71. (a) .79608 (b) $-.79560$ (c) -1.5574 (d) $-.7513$ (e) -1.2349 (f) $-.9877$
73. (a) .9999997 (b) .744399 (c) 1.2338651
75. (a) .679996; 2.461597 (b) 1.222007; 5.061178 (c) 1.878966; 5.020558
77. (a) ϕ (b) $90° - \phi$ (c) $90° - \phi$ **79.** $-.514496$ **81.** $\phi \approx 126.9°$

PROBLEM SET 4-6 (Page 214)

1.

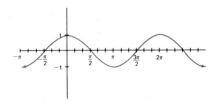

3.

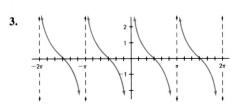

5. $\sec(t + 2\pi) = 1/\cos(t + 2\pi) = 1/\cos t = \sec t$

7. Domain: $\{t : t \neq \pi/2 + k\pi,\ k \text{ any integer}\}$; range: $\{y : |y| \geq 1\}$. **9.** $\pi; 2\pi$ **11.** $\cot(-t) = -\cot t$

13. 3; 2π

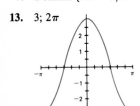

15. 1; 2π

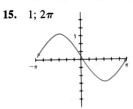

17. 1; $\pi/2$

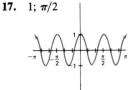

19. 2; 4π

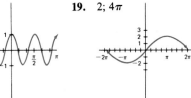

21. 2; $2\pi/3$

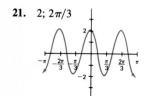

23.

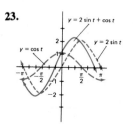

25.

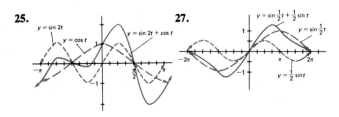

27.

29.

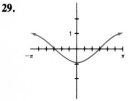

31.

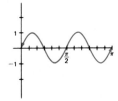

33. 1, 2π; 1, $\pi/2$

35. (a)

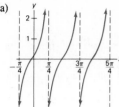

Period: $\pi/2$

(b)

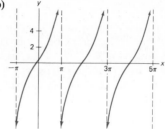

Period: 2π

37.

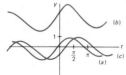

39.

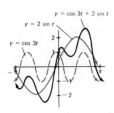

41.

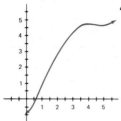

43. (a) $\frac{1}{60}$ seconds   (b) 60   (c) 30 amperes

45. (a) $1/\pi, 1/2\pi, 1/3\pi, 1/4\pi, \ldots$ (b) $1, -1, 1, -1, \ldots$ (c)

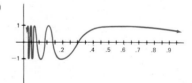

CHAPTER 4. REVIEW PROBLEM SET   (Page 218)

1. (a) $\beta = 42.9°$; $a = 27.0$; $b = 25.1$ (b) $\alpha = 46.7°$; $\beta = 43.3°$; $b = 393$ **2.** 5.01 feet   **3.** .576; 405°
4. 18,850 centimeters   **5.** (a) $-\frac{1}{2}$ (b) $\sqrt{3}/2$ (c) 1 (d) $\frac{1}{2}$ **6.** (a) .7771   (b) $-.6157$ (c) $-.5635$ (d) .5258
7. (a) $-\sin t$ (b) $\sin t$ (c) $-\sin t$ (d) $\sin t$ **8.** (a) $\{t : 0 \le t < \pi/2 \text{ or } 3\pi/2 < t \le 2\pi\}$
(b) $\{t : 0 \le t < \pi/4 \text{ or } 3\pi/4 < t < 5\pi/4 \text{ or } 7\pi/4 < t \le 2\pi\}$ **9.** (a) $\frac{5}{12}$ (b) $-\frac{13}{5}$ **10.** $-2/\sqrt{21}$

11.

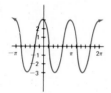

12.

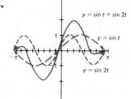

13. $\{y : -1 \le y \le 1\}$; $\{y : y \le -1 \text{ or } y \ge 1\}$

1. (a) $1 - \sin^2 t$ (b) $\sin t$ (c) $\sin^2 t$ (d) $(1 - \sin^2 t)/\sin^2 t$ **3.** (a) $1/\tan^2 t$ (b) $1 + \tan^2 t$ (c) $\tan t$ (d) 3

5. $\cos t \sec t = \cos t(1/\cos t) = 1$ **7.** $\tan x \cot x = \tan x(1/\tan x) = 1$ **9.** $\cos y \csc y = \cos y(1/\sin y) = \cot y$

11. $\cot \theta \sin \theta = (\cos \theta/\sin \theta) \sin \theta = \cos \theta$ **13.** $\tan u/\sin u = (\sin u/\cos u)(1/\sin u) = 1/\cos u$

15. $(1 + \sin z)(1 - \sin z) = 1 - \sin^2 z = \cos^2 z = 1/\sec^2 z$

17. $(1 - \sin^2 x)(1 + \tan^2 x) = \cos^2 x \sec^2 x = \cos^2 x(1/\cos^2 x) = 1$

19. $\sec t - \sin t \tan t = 1/(\cos t) - (\sin^2 t)/(\cos t) = (1 - \sin^2 t)/(\cos t) = (\cos^2 t)/(\cos t) = \cos t$

21. $(\sec^2 t - 1)/(\sec^2 t) = 1 - 1/(\sec^2 t) = 1 - \cos^2 t = \sin^2 t$

23. $\cos t(\tan t + \cot t) = \sin t + (\cos^2 t)/(\sin t) = (\sin^2 t + \cos^2 t)/(\sin t) = \csc t$

25. $\sin t = (1 - \cos^2 t)^{1/2}$; $\tan t = (1 - \cos^2 t)^{1/2}/\cos t$; $\cot t = \cos t/(1 - \cos^2 t)^{1/2}$; $\sec t = 1/\cos t$; $\csc t = 1/(1 - \cos^2 t)^{1/2}$

27. $\cos t = -3/5$; $\tan t = -4/3$; $\cot t = -3/4$; $\sec t = -5/3$; $\csc t = 5/4$

29. $\dfrac{\sec t - 1}{\tan t} \cdot \dfrac{\sec t + 1}{\sec t + 1} = \dfrac{\sec^2 t - 1}{\tan t(\sec t + 1)} = \dfrac{\tan^2 t}{\tan t(\sec t + 1)} = \dfrac{\tan t}{\sec t + 1}$

31. $\dfrac{\tan^2 x}{\sec x + 1} = \dfrac{\sec^2 x - 1}{\sec x + 1} = \dfrac{(\sec x - 1)(\sec x + 1)}{\sec x + 1} = \sec x - 1 = \dfrac{1 - \cos x}{\cos x}$

33. $\dfrac{\sin t + \cos t}{\tan^2 t - 1} \cdot \dfrac{\cos^2 t}{\cos^2 t} = \dfrac{(\sin t + \cos t) \cos^2 t}{\sin^2 t - \cos^2 t} = \dfrac{\cos^2 t}{\sin t - \cos t}$

35. (a) $2/\sin^2 x$ (b) $(2 + 2 \tan^2 x)/\tan^2 x$

37. $(1 + \tan^2 t)(\cos t + \sin t) = \sec^2 t \cos t + \sec^2 t \sin t = \sec t + \sec t \tan t = \sec t(1 + \tan t)$

39. $2 \sec^2 y - 1 = \dfrac{2}{\cos^2 y} - 1 = \dfrac{2 - \cos^2 y}{\cos^2 y} = \dfrac{1 + \sin^2 y}{\cos^2 y}$

41. $\dfrac{\sin z}{\sin z + \tan z} = \dfrac{\sin z}{\sin z + \sin z/\cos z} = \dfrac{1}{1 + 1/\cos z} = \dfrac{\cos z}{\cos z + 1}$

43. $(\csc t + \cot t)^2 = \left(\dfrac{1}{\sin t} + \dfrac{\cos t}{\sin t}\right)^2 = \dfrac{(1 + \cos t)^2}{1 - \cos^2 t} = \dfrac{1 + \cos t}{1 - \cos t}$

45. $\dfrac{1 + \tan x}{1 - \tan x} = \dfrac{1 + \sin x/\cos x}{1 - \sin x/\cos x} = \dfrac{\cos x + \sin x}{\cos x - \sin x}$

47. $(\sec t + \tan t)(\csc t - 1) = \left(\dfrac{1 + \sin t}{\cos t}\right)\left(\dfrac{1 - \sin t}{\sin t}\right) = \dfrac{\cos^2 t}{\cos t \sin t} = \cot t$

49. $\dfrac{\cos^3 t + \sin^3 t}{\cos t + \sin t} = \cos^2 t - \cos t \sin t + \sin^2 t = 1 - \sin t \cos t$

51. $\left(\dfrac{1 - \cos \theta}{\sin \theta}\right)^2 = \dfrac{(1 - \cos \theta)^2}{1 - \cos^2 \theta} = \dfrac{1 - \cos \theta}{1 + \cos \theta}$

53. $(\csc t - \cot t)^4(\csc t + \cot t)^4 = (\csc^2 t - \cot^2 t)^4 = 1^4 = 1$

55. $\sin^6 u + \cos^6 u = (1 - \cos^2 u)^3 + \cos^6 u = 1 - 3 \cos^2 u + 3 \cos^4 u$
$= 1 - 3 \cos^2 u(1 - \cos^2 u) = 1 - 3 \cos^2 u \sin^2 u$

57. $\cot 3x = \dfrac{1}{\tan 3x} = \dfrac{1 - 3 \tan^2 x}{3 \tan x - \tan^3 x}\left(\dfrac{\cot^3 x}{\cot^3 x}\right) = \dfrac{\cot^3 x - 3 \cot x}{3 \cot^2 x - 1} = \dfrac{3 \cot x - \cot^3 x}{1 - 3 \cot^2 x}$

1. (a) $(\sqrt{2} + 1)/2 \approx 1.21$ (b) $(\sqrt{2}\sqrt{3} + \sqrt{2})/4 \approx .97$

3. (a) $(\sqrt{2} - \sqrt{3})/2 \approx -.16$ (b) $(\sqrt{2}\sqrt{3} + \sqrt{2})/4 \approx .97$

5. $\sin(t + \pi) = \sin t \cos \pi + \cos t \sin \pi = -\sin t$

7. $\sin(t + 3\pi/2) = \sin t \cos(3\pi/2) + \cos t \sin(3\pi/2) = -\cos t$

9. $\sin(t - \pi/2) = \sin t \cos(\pi/2) - \cos t \sin(\pi/2) = -\cos t$

11. $\cos(t + \pi/3) = \cos t \cos(\pi/3) - \sin t \sin(\pi/3) = (1/2) \cos t - (\sqrt{3}/2) \sin t$

13. $\cos 2$ **15.** $\sin \pi$ **17.** $\cos 60°$ **19.** $\sin \alpha$ **21.** $\frac{56}{65}$; $-\frac{33}{65}$; in quadrant II.

23. $-(1 + 3\sqrt{3})/(2\sqrt{10}) \approx -.9797$; $(-3 + \sqrt{3})/(2\sqrt{10}) \approx -.2005$; quadrant III.

25. $\tan(s - t) = \tan(s + (-t)) = (\tan s + \tan(-t))/(1 - \tan s \tan(-t)) = (\tan s - \tan t)/(1 + \tan s \tan t)$

27. $\tan(t + \pi/4) = (\tan t + \tan \pi/4)/(1 - \tan t \tan \pi/4) = (1 + \tan t)/(1 - \tan t)$

29. (a) $-(\cos t + \sqrt{3} \sin t)/2$ (b) $(\sqrt{3} \cos t + \sin t)/2$

31. (a) $\sqrt{5}/3$ (b) $-2\sqrt{2}/3$ (c) $(4\sqrt{2} - \sqrt{5})/9$ (d) $(2\sqrt{10} - 2)/9$ (e) $(-\frac{2}{3})(\sqrt{2} + \sqrt{5})$ (f) $4\sqrt{2}/9$

33. (a) $\sqrt{3}/2$ (b) $-\sqrt{3}/2$ (c) $\sin 1 \approx .84147$

35. (a) $\sin(x + y) \sin(x - y) = (\sin x \cos y + \cos x \sin y)(\sin x \cos y - \cos x \sin y)$
$$= \sin^2 x \cos^2 y - \cos^2 x \sin^2 y = \sin^2 x(1 - \sin^2 y) - \cos^2 x \sin^2 y$$
$$= \sin^2 x - \sin^2 y(\sin^2 x + \cos^2 x) = \sin^2 x - \sin^2 y$$

(b) $\dfrac{\sin(x + y)}{\cos(x - y)} = \dfrac{\sin x \cos y + \cos x \sin y}{\cos x \cos y + \sin x \sin y} = \dfrac{\dfrac{\sin x \cos y}{\cos x \cos y} + \dfrac{\cos x \sin y}{\cos x \cos y}}{\dfrac{\cos x \cos y}{\cos x \cos y} + \dfrac{\sin x \sin y}{\cos x \cos y}} = \dfrac{\tan x + \tan y}{1 + \tan x \tan y}$

(c) $\dfrac{\cos 5t}{\sin t} - \dfrac{\sin 5t}{\cos t} = \dfrac{\cos 5t \cos t - \sin 5t \sin t}{\sin t \cos t} = \dfrac{\cos 6t}{\sin t \cos t}$

37.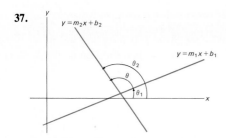

Since $\theta = \theta_2 - \theta_1$,
$$\tan \theta = \frac{\tan \theta_2 - \tan \theta_1}{1 + \tan \theta_2 \tan \theta_1} = \frac{m_2 - m_1}{1 + m_1 m_2}$$

39. (a) $\frac{1}{2}[\cos(s + t) + \cos(s - t)] = \frac{1}{2}[\cos s \cos t - \sin s \sin t + \cos s \cos t + \sin s \sin t] = \cos s \cos t$

(b) $-\frac{1}{2}[\cos s \cos t - \sin s \sin t - \cos s \cos t - \sin s \sin t] = \sin s \sin t$

(c) $\frac{1}{2}[\sin s \cos t + \cos s \sin t + \sin s \cos t - \cos s \sin t] = \sin s \cos t$

(d) $\frac{1}{2}[\sin s \cos t + \cos s \sin t - \sin s \cos t + \cos s \sin t] = \cos s \sin t$

41. (a) $(1 - \sqrt{3})/4$ (b) $-\sqrt{2}/2$ (c) $\dfrac{1 + \sqrt{2} + \sqrt{3} + \sqrt{6}}{2}$

43. $\tan(\alpha + \beta) = (\tan \alpha + \tan \beta)/(1 - \tan \alpha \tan \beta) = (\frac{1}{3} + \frac{1}{2})/(1 - \frac{1}{3} \cdot \frac{1}{2}) = 1 = \tan \gamma$. Thus $\alpha + \beta = \gamma$.

PROBLEM SET 5-3   (Page 237)

1. (a) $\frac{24}{25}$ (b) $\frac{7}{25}$ (c) $3\sqrt{10}/10$ (d) $\sqrt{10}/10$ **3.** $\sin 10t$ **5.** $\cos 3t$ **7.** $\cos(y/2)$

9. $\cos 1.2t$ **11.** $-\cos(\pi/4)$ **13.** $\cos^2(x/2)$ **15.** $\sin^2 2\theta$

17. (a) $\sin(\pi/8) = \sqrt{(1 - \cos \pi/4)/2} \approx .3827$ (b) $\cos 112.5° = -\sqrt{(1 + \cos 225°)/2} \approx -.3827$

19. $\tan 2t = \tan(t + t) = (\tan t + \tan t)/(1 - \tan^2 t) = 2 \tan t/(1 - \tan^2 t)$

21. $\tan \dfrac{t}{2} = \dfrac{\sin t/2}{\cos t/2} = \dfrac{\pm\sqrt{(1 - \cos t)/2}}{\pm\sqrt{(1 + \cos t)/2}} = \pm\sqrt{\dfrac{1 - \cos t}{1 + \cos t}}$

23. $\cos 3t = \cos(2t + t) = \cos 2t \cos t - \sin 2t \sin t = (2 \cos^2 t - 1) \cos t - 2 \sin^2 t \cos t$
$$= (2 \cos^2 t - 1) \cos t - 2(1 - \cos^2 t) \cos t = 4 \cos^3 t - 3 \cos t$$

25. $\csc 2t + \cot 2t = (1 + \cos 2t)/(\sin 2t) = (2 \cos^2 t)/(2 \sin t \cos t) = \cot t$

27. $\sin \theta/(1 - \cos \theta) = 2 \sin(\theta/2) \cos(\theta/2)/2 \sin^2(\theta/2) = \cot(\theta/2)$

29. $2 \tan \alpha/(1 + \tan^2 \alpha) = 2 \tan \alpha/\sec^2 \alpha = 2(\sin \alpha/\cos \alpha)\cos^2 \alpha = 2 \sin \alpha \cos \alpha = \sin 2\alpha$

31. $\sin 4\theta = 2 \sin 2\theta \cos 2\theta = 2(2 \sin \theta \cos \theta)(2 \cos^2 \theta - 1) = 4 \sin \theta(2 \cos^3 \theta - \cos \theta)$

33. (a) $\sin x$ (b) $\cos 6t$ (c) $-\cos(y/2)$ (d) $-\sin^2 2t$ (e) $\tan^2 2t$ (f) $\tan 3y$

35. (a) $120/169$ (b) $-2\sqrt{13}/13$ (c) $-\frac{3}{2}$ **37.** $\cos^4 z - \sin^4 z = (\cos^2 z + \sin^2 z)(\cos^2 z - \sin^2 z) = 1 \cdot \cos 2z$

39. $1 + (1 - \cos 8t)/(1 + \cos 8t) = 1 + \tan^2 4t = \sec^2 4t$

41. $\tan \frac{\theta}{2} - \sin \theta = (\sin \theta)/(1 + \cos \theta) - \sin \theta = (\sin \theta - \sin \theta - \sin \theta \cos \theta)/(1 + \cos \theta) = (-\sin \theta \cos \theta)/(1 + \cos \theta) = -(\sin \theta)/(\sec \theta + 1)$

43. $(3 \cos t - \sin t)(\cos t + 3 \sin t) = 3 \cos^2 t - 3 \sin^2 t + 8 \sin t \cos t = 3 \cos 2t + 4 \sin 2t$

45. $2(\cos 3x \cos x + \sin 3x \sin 3x)^2 = 2 \cos^2 2x = 1 + \cos 4x$

47. $\tan 3t = \tan(2t + t) = \dfrac{\tan 2t + \tan t}{1 - \tan 2t \tan t} = \dfrac{\dfrac{2 \tan t}{1 - \tan^2 t} + \tan t}{1 - \dfrac{2 \tan^2 t}{1 - \tan^2 t}} = \dfrac{3 \tan t - \tan^3 t}{1 - 3 \tan^2 t}$

49. $\sin^4 u + \cos^4 u = (\sin^2 u + \cos^2 u)^2 - 2 \sin^2 u \cos^2 u = 1 - \frac{1}{2} \sin^2 2u = 1 - \frac{1}{2} \cdot (1 - \cos 4u)/2 = \frac{3}{4} + \frac{1}{4} \cos 4u$

51. $\cos^2 x + \cos^2 2x + \cos^2 3x = \dfrac{1 + \cos 2x}{2} + \cos^2 2x + \dfrac{1 + \cos 6x}{2} = 1 + \frac{1}{2}(\cos 2x + \cos 6x) + \cos^2 2x =$

$1 + \cos 4x \cos 2x + \cos^2 2x = 1 + \cos 2x(\cos 4x + \cos 2x) = 1 + \cos 2x(2 \cos 3x \cos x) = 1 + 2 \cos x \cos 2x \cos 3x$

53. Since $\alpha + \beta + \gamma = 180°$, $2\gamma = 360° - 2\alpha - 2\beta$. Thus $\sin 2\alpha + \sin 2\beta + \sin 2\gamma = \sin 2\alpha + \sin 2\beta - \sin(2\alpha + 2\beta) = \sin 2\alpha + \sin 2\beta - \sin 2\alpha \cos 2\beta - \cos 2\alpha \sin 2\beta = \sin 2\alpha(1 - \cos 2\beta) + \sin 2\beta(1 - \cos 2\alpha) = 2 \sin \alpha \cos \alpha(2 \sin^2 \beta) + 2 \sin \beta \cos \beta(2 \sin^2 \alpha) = 4 \sin \alpha \sin \beta(\cos \alpha \sin \beta + \sin \alpha \cos \beta) = 4 \sin \alpha \sin \beta \sin(\alpha + \beta) = 4 \sin \alpha \sin \beta \sin \gamma$.

55. $(\frac{7}{9}, 4\sqrt{2}/9)$

PROBLEM SET 5-4 (Page 245)

1. $\pi/3$ **3.** $\pi/4$ **5.** 0 **7.** $\pi/3$ **9.** $2\pi/3$ **11.** $\pi/4$ **13.** $.2200$ **15.** $-.2200$ **17.** $.2037$
19. (a) $.7938$ (b) 1.9545 **21.** $.3486; 2.7930$ **23.** $1.2803; 4.4219$ **25.** $\frac{2}{3}$ **27.** 10 **29.** $\pi/3$
31. $\pi/4$ **33.** $\frac{3}{5}$ **35.** $2/\sqrt{5}$ **37.** $\frac{1}{3}$ **39.** $2\pi/3$ **41.** $.9666$ **43.** $.4508$ **45.** 2.2913 **47.** $\frac{24}{23}$
49. $\frac{7}{25}$ **51.** $\frac{56}{65}$ **53.** $(6 + \sqrt{35})/12 \approx .993$ **55.** $\tan(\sin^{-1} x) = \sin(\sin^{-1} x)/\cos(\sin^{-1} x) = x/\sqrt{1 - x^2}$
57. $\tan(2 \tan^{-1} x) = 2 \tan(\tan^{-1} x)/[1 - \tan^2(\tan^{-1} x)] = 2x/(1 - x^2)$ **59.** $\cos(2 \sec^{-1} x) = \cos[2 \cos^{-1}(1/x)] = 2/x^2 - 1$
61. (a) $-\pi/3$ (b) $-\pi/3$ (c) $2\pi/3$ **63.** (a) 43 (b) $\frac{12}{13}$ (c) $7\sqrt{2}/10$ (d) $(4 - 6\sqrt{2})/15$
65. (a) $\pm\sqrt{7}/4$ (b) $\pm.9$ (c) $\frac{11}{6}$ (d) $1; 2$ **67.** (a) $\pi/2$ (b) $\pi/4$ (c) $-\pi/4$ (d) $-\pi/2$ (e) π (f) $\pi/4$
69. (a) $\sin^{-1}(x/5)$ (b) $\tan^{-1}(x/3)$ (c) $\sin^{-1}(3/x)$ (d) $\tan^{-1}(3/x) - \tan^{-1}(1/x)$
71. (a) $.6435011$ (b) $-.3046927$ (c) $.6435011$ (d) 2.6905658 **73.** Show that the tangent of both sides is $120/119$.
75. (a) $\theta = \tan^{-1}(6/b) - \tan^{-1}(2/b)$ (b) $22.83°$ (c) $2\sqrt{3}$

PROBLEM SET 5-5 (Page 253)

1. $\{0, \pi\}$ **3.** $\{3\pi/2\}$ **5.** No solution.   **7.** $\{5\pi/6, 7\pi/6\}$ **9.** $\{\pi/4, 3\pi/4, 5\pi/4, 7\pi/4\}$
11. $\{\pi/4, 2\pi/3, 3\pi/4, 4\pi/3\}$ **13.** $\{0, \pi, 3\pi/2\}$ **15.** $\{0, \pi/3, \pi, 4\pi/3\}$ **17.** $\{\pi/3, \pi, 5\pi/3\}$
19. $\{.3649, 1.2059, 3.5065, 4.3475\}$ **21.** $\{0, \pi/2\}$ **23.** $\{\pi/6, \pi/2\}$ **25.** $\{0\}$
27. $\{\pi/6 + 2k\pi, 5\pi/6 + 2k\pi: k \text{ is an integer}\}$ **29.** $\{k\pi: k \text{ is an integer}\}$ **31.** $\{\pi/6 + k\pi, 5\pi/6 + k\pi: k \text{ is an integer}\}$
33. $\{0, \pi/2, \pi, 3\pi/2\}$ **35.** $\{\pi/8, 5\pi/8, 9\pi/8, 13\pi/8\}$ **37.** $\{3\pi/8, 7\pi/8, 11\pi/8, 15\pi/8\}$
39. $\{0, \pi/6, 5\pi/6, \pi\}$ **41.** $\{.9553, 2.1863, 4.0969, 5.3279\}$ **43.** $\{\pi/4, 5\pi/4\}$ **45.** $\{0, \pi/6, 5\pi/6, \pi, 7\pi/6, 11\pi/6\}$
47. $\{.3076, 2.8340\}$ **49.** $\{2\pi/3, 4\pi/3\}$ **51.** $\{2\pi/3, 5\pi/6, 5\pi/3, 11\pi/6\}$ **53.** $\{3\pi/2, 5.6397\}$
55. $\{\pi/4, 3\pi/4, 5\pi/4, 7\pi/4\}$ **57.** (a) 15 inches (b) $\tan \theta = \frac{2}{3}$ (c) $33.7°$ **59.** (a) $26.6°$ (b) $10.3°$
61. $\{k\pi/3, 2\pi/3 + 2k\pi, 4\pi/3 + 2k\pi: k \text{ is an integer}\}$ **63.** $\{\pi/6, \pi/3, 2\pi/3, 5\pi/6\}$

CHAPTER 5. REVIEW PROBLEM SET (Page 256)

1. (a) $\cot \theta \cos \theta = \cos^2 \theta/\sin \theta = (1 - \sin^2 \theta)/\sin \theta = 1/\sin \theta - \sin \theta = \csc \theta - \sin \theta$
(b) $(\cos x \tan^2 x)(\sec x - 1)/(\sec x + 1)(\sec x - 1) = \cos x \tan^2 x(\sec x - 1)/\tan^2 x = 1 - \cos x$
2. (a) $-\sin^3 x/(1 - \sin^2 x)$ (b) $1 - \sin x$ 3. (a) $\cos 45° = \sqrt{2}/2$ (b) $\sin 90° = 1$ (c) $\cos 45° = \sqrt{2}/2$
4. (a) $24/25 = .96$ (b) $3/\sqrt{10} \approx .95$ 5. (a) $\sin 2t \cos t - \cos 2t \sin t = \sin(2t - t) = \sin t$
(b) $\sec 2t + \tan 2t = (1 + \sin 2t)/\cos 2t = (\cos t + \sin t)^2/(\cos^2 t - \sin^2 t) = (\cos t + \sin t)/(\cos t - \sin t)$
(c) $\cos(\alpha + \beta)/\cos \alpha \cos \beta = \cos \alpha \cos \beta/\cos \alpha \cos \beta - \sin \alpha \sin \beta/\cos \alpha \cos \beta = 1 - \tan \alpha \tan \beta = \tan \alpha(\cot \alpha - \tan \beta)$
6. (a) $\{5\pi/6, 7\pi/6\}$ (b) $\{0, \pi, 7\pi/6, 11\pi/6\}$ (c) $\{0\}$ (d) $\{\pi/2, \pi\}$ (e) $\{\pi/6, 5\pi/6, 3\pi/2\}$
7. $-\pi/2 \le t \le \pi/2; 0 \le t \le \pi, -\pi/2 < t < \pi/2$ 8. (a) $-\pi/3$ (b) $5\pi/6$ (c) $-\pi/3$ (d) 6 (e) π (f) $\sqrt{5}/3$
(g) $-.02$ (h) $120/169$ 9. See the graph in the text on page 243.   10. -1.57
11. $\tan(\arctan \frac{1}{2} + \arctan \frac{1}{3}) = (\frac{1}{2} + \frac{1}{3})/(1 - \frac{1}{6}) = 1$. Since $\arctan \frac{1}{2}$ and $\arctan \frac{1}{3}$ are between 0 and $\pi/2$, their sum cannot be in quadrant III and so must equal $\pi/4$.

PROBLEM SET 6-1 (Page 263)

1. $\gamma = 55.5°; b \approx 20.9; c \approx 17.4$ 3. $\beta = 56°; a = c \approx 53$ 5. $\beta \approx 42°; \gamma \approx 23°; c \approx 20$
7. $\beta \approx 18°; \gamma \approx 132°; c \approx 12$ 9. Two triangles; $\beta_1 \approx 53°, \gamma_1 \approx 97°, c_1 \approx 9.9; \beta_2 \approx 127°, \gamma_2 \approx 23°, c_2 \approx 3.9$
11. 93.7 meters   13. 44.7°   15. 192.8   17. 265.3   19. 78.4°   21. 694.6 square feet
23. 1769 feet   25. 40   27. $6r^2 \sin \phi(\cos \phi + \sqrt{3} \sin \phi)$

PROBLEM SET 6-2 (Page 268)

1. $a \approx 12.5; \beta \approx 76°; \gamma \approx 44°$ 3. $c \approx 15.6; \alpha \approx 26°; \beta \approx 34°$ 5. $\alpha \approx 44.4°; \beta \approx 57.1°; \gamma \approx 78.5°$
7. $\alpha \approx 30.6°; \beta \approx 52.9°; \gamma \approx 96.5°$ 9. 98.8 meters   11. 24 miles   13. 106°
15. $s = 6, A = \sqrt{6 \cdot 3 \cdot 2 \cdot 1} = 6$ 17. 18.63   19. 41.68°   21. 42.60 miles
23. $\cos^{-1}(\frac{3}{4}) \approx 41.41°; \frac{1}{2}(\sqrt{3} + 3\sqrt{7})$ 27. $\sqrt{15}$

PROBLEM SET 6-3 (Page 273)

1.   3. 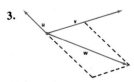   5. $w = \frac{1}{2}(u + v)$ 7. $\|w\| = 1$

9. $10\sqrt{2} + \sqrt{2} \approx 18.48$ 11. 243.7 kilometers; S43.9°W   13. .026 hours   15. 479 miles per hour; N2.68°E
17. 15.9; S7.5°W   19. 163.7, 118.9   21.   23. 0

25. (a) $\vec{AD} - \vec{AB}$ (b) $\frac{1}{2}(\vec{AB} + \vec{AD})$ (c) $\vec{AB} - \frac{1}{2}\vec{AD}$ (d) $\vec{AD} - \frac{1}{2}\vec{AB}$ 27. $\sqrt{7}/2 \approx 1.32$ miles per hour
29. N10.32°E   31. N1.019°W, 654.88 miles per hour   33. 651.3 pounds   35. 65.38°, 146.77 pounds

PROBLEM SET 6-4 (Page 280)

1. $4\mathbf{i} - 24\mathbf{j}$; -33; $-33/65$　**3.** $3\mathbf{i} + \mathbf{j}$; 10; $2/\sqrt{5}$　**5.** $101.385°$　**7.** $5\mathbf{i} + 2\mathbf{j}$; $4\mathbf{i} - 3\mathbf{j}$; 14
9. $-4\mathbf{i} - 5\mathbf{j}$; $-6\mathbf{i} + 5\mathbf{j}$; -1　**11.** $-5\mathbf{i} + 5\sqrt{3}\mathbf{j}$　**13.** $\frac{4}{3}$　**15.** $\frac{3}{5}\mathbf{i} - \frac{4}{5}\mathbf{j}$　**17.** $-4\mathbf{i} + 10\mathbf{j}$
19. $\mathbf{u} \cdot \mathbf{u} = a^2 + b^2 = \|\mathbf{u}\|^2$　**21.** 7　**23.** $(118/169)(5\mathbf{i} + 12\mathbf{j})$　**25.** $-56/5$　**27.** 100
29. $325\sqrt{2}$ dyne-centimeters　**31.** $\sqrt{34}$　**33.** $\pm(\frac{4}{5}\mathbf{i} + \frac{3}{5}\mathbf{j})$　**35.** $\frac{9}{25}\mathbf{i} - \frac{12}{25}\mathbf{j}$; $84.1°$
37. $|\mathbf{u} \cdot \mathbf{v}| = \|\mathbf{u}\|\,\|\mathbf{v}\|\,|\cos\theta| \le \|\mathbf{u}\|\,\|\mathbf{v}\|$ with equality when $\theta = 0°$ or $180°$.　**39.** $\|\mathbf{u}\| = \|\mathbf{v}\|$　**41.** $(-1 + \sqrt{5})/4$

PROBLEM SET 6-5 (Page 287)

1. (a)　　　　(b)　　　　(c)　　　　(d)

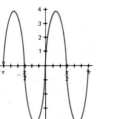　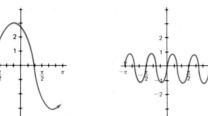　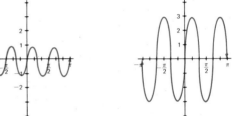

3. (a) π; 4; 0　　(b) 2π; 3, $-\pi/8$　　(c) $\pi/2$; 1; $-\pi/32$　　(d) $2\pi/3$; 3; $\pi/6$

5. 4π; 2; $\pi/8$　**7.**　　**9.**　　**11.**

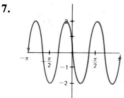　　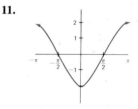

13. $(5\cos 4t,\ 5\sin 4t)$　**15.** $5\cos 4t$, $-8 + 5\sin 4t$
17. (a) $2\pi/5$; 1; 0　(b) 4π; $\frac{3}{2}$; 0　(c) $\pi/2$; 2; $\pi/4$　(d) $2\pi/3$; 4; $-\pi/4$　**19.** 4 feet; after 1.5 seconds.
21. $\sin t + \sqrt{25 - \cos^2 t}$
23. 156; 55　**25.** (a) 0　(b) 0　(c) 53.53　**27.**

$R = 1000 + 150\sin 2t$

$C = 200 + 50\sin (2t - .7)$

29. (a) $2\sqrt{2}\sin 2t + 2\sqrt{2}\cos 2t$　(b) $-\frac{3}{2}\sqrt{3}\sin 3t + \frac{3}{2}\cos 3t$
33. (a) $5\sin[2t + \tan^{-1}(4/3)]$　(b) $2\sqrt{3}(\sin 4t + 11\pi/6)$　**35.** $\sqrt{2}$; $-\sqrt{2}$

CHAPTER 6. REVIEW PROBLEM SET (Page 291)

1. (a) $\beta = 39.1°$; $a \approx 228$; $c \approx 139$ (b) $\alpha \approx 4.0°$; $\beta \approx 154.5°$; $\gamma \approx 21.5°$ (c) $c \approx 13.2$; $\alpha \approx 37.3°$; $\beta \approx 107.7°$
(d) $\gamma \approx 30.7°$; $\alpha \approx 100.7°$; $a \approx 75.5$ 2. $x \approx 37.1$; $A \approx 281$ 3. $6\mathbf{i} + (5\sqrt{3} - 3)\mathbf{j}$
4. (a) 5 (b) 13 (c) -33 (d) $120.5°$ (e) $-\frac{33}{13}$ (f) $(-\frac{33}{169})(5\mathbf{i} + 12\mathbf{j})$
5. (a) $120\mathbf{i} + 60\mathbf{j}$ (b) $(80\mathbf{i} + 60\mathbf{j}) \cdot (4\mathbf{i} + 2\mathbf{j})/\sqrt{5} \approx 196.8$ foot-pounds
6. (a) π; 1; 0 (b) $\pi/2$; 3; 0 (c) $2\pi/3$; 2; $\pi/6$ (d) 4π; 2; -2π
7. (a)   (b)   (c)   (d)

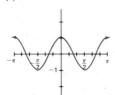 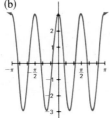 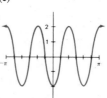 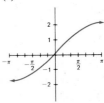

8. $(4 \cos(3\pi t/4 + \pi), 4 \sin(3\pi t/4))$ 9. (a) $2\pi/5$ seconds (b) At $x = -3$ feet. (c) When $t = \pi/5$ seconds.

PROBLEM SET 7-1 (Page 299)

1. $-2 + 8i$ 3. $-4 - i$ 5. $0 + 6i$ 7. $-4 + 7i$ 9. $11 + 4i$ 11. $14 + 22i$ 13. $16 + 30i$
15. $61 + 0i$ 17. $\frac{3}{2} + \frac{7}{2}i$ 19. $2 - 5i$ 21. $\frac{11}{2} + \frac{3}{2}i$ 23. $\frac{22}{25} - \frac{4}{25}i$ 25. $8 - 6i$ 27. $\frac{2}{5} + \frac{1}{5}i$
29. $\sqrt{3}/4 - \frac{1}{4}i$ 31. $\frac{1}{5} + \frac{8}{5}i$ 33. $(\sqrt{3} - \frac{1}{4}) + (-\sqrt{3}/4 - 1)i$ 35. $\frac{1}{2} \pm (\sqrt{3}/2)i$ 37. $\frac{3}{2}; \frac{1}{2}$
39. $\sqrt{3}/2 \pm \frac{5}{2}i$ 41. $-\frac{3}{5} \pm (\sqrt{11}/5)i$ 43. -1 45. $-i$ 47. $-729i$ 49. $-2 + 2i$
51. $i^4 = i^2 \cdot i^2 = (-1)(-1) = 1$. The four 4th roots of 1 are $1, -1, i$, and $-i$.
53. $(1 - i)^4 = (1 - i)^2(1 - i)^2 = (-2i)(-2i) = 4i^2 = -4$
55. (a) $-1 - i$ (b) $1 - 2i$ (c) $-8 + 6i$ (d) $-\frac{8}{29} + \frac{20}{29}i$ (e) $\frac{14}{25}$ (f) $2\sqrt{3} - i$
57. (a) $a = 2$, $b = -\frac{4}{5}$ (b) $a = 3$, $b = 2$ 59. (a) 1 (b) 0 (c) 3   61. (a) $4 \pm 3i$ (b) $4i, 5i$

PROBLEM SET 7-2 (Page 306)

1-11.   13. $\sqrt{13}$; $\sqrt{13}$; 5; 4; 1; 2   15. $0 - 4i$ 17. $-\sqrt{2} - \sqrt{2}i$ 19. $4(\cos \pi + i \sin \pi)$

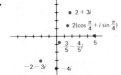

21. $5(\cos 270° + i \sin 270°)$ 23. $2\sqrt{2}(\cos 315° + i \sin 315°)$ 25. $4(\cos \pi/6 + i \sin \pi/6)$
27. $6.403(\cos .6747 + i \sin .6747)$ 29. $6(\cos 210° + i \sin 210°)$ 31. $\frac{3}{2}(\cos 125° + i \sin 125°)$
33. $\frac{2}{3}(\cos 70° + i \sin 70°)$ 35. $2(\cos 305° + i \sin 305°)$ 37. $16 + 0i$ 39. $-2 + 2\sqrt{3}i$
41. $16(\cos 60° + i \sin 60°)$ 43. $1(\cos 120° + i \sin 120°)$ 45. $16(\cos 270° + i \sin 270°)$
47.   (a) $|-5 + 12i| = 13$ (b) $|-4i| = 4$ (c) $|5(\cos 60° + i \sin 60°)| = 5$

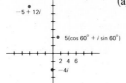

49. (a) 12(cos 0° + i sin 0°) (b) 2(cos 135° + i sin 135°) (c) 3(cos 270° + i sin 270°) (d) 4(cos 300° + i sin 300°)
(e) 8(cos 30° + i sin 30°) (f) 2(cos 315° + i sin 315°)
51. (a) 12(cos 160° + i sin 160°) (b) 3(cos 40° + i sin 40°) (c) cos 45° + i sin 45°
53. (a) 12 + 5i, −12 + 5i (b) ±(4√2 + 4√2i)
55. (a) r^3(cos 3θ + i sin 3θ) (b) r[cos(−θ) + i sin(−θ)] (c) r^2(cos 0 + i sin 0) (d) $\frac{1}{r}$[cos(−θ) + i sin(−θ)]
(e) r^{-2}[cos(−2θ) + i sin(−2θ)] (f) r[cos(θ + π) + i sin(θ + π)]
57. (a) The distance between **U** and **V** in the complex plane.
(b) The angle from the positive x-axis to the line joining **U** and **V**.
59. (a) − 2^8 (b) 0

PROBLEM SET 7-3 (Page 313)

1. 8[cos(3π/4) + i sin(3π/4)] **3.** 125(cos 66° + i sin 66°) **5.** 16(cos 0° + i sin 0°) **7.** 1 + 0i
9. −16√3 + 16i
11. 5(cos 15° + i sin 15°);
5(cos 135° + i sin 135°);
5(cos 255° + i sin 255°)

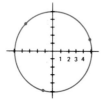

13. 2[cos(π/12) + i sin(π/12)]; 2[cos(5π/12) + i sin(5π/12)];
2[cos(9π/12) + i sin(9π/12)]; 2[cos(13π/12) + i sin(13π/12)];
2[cos(17π/12) + i sin(17π/12)]; 2[cos(21π/12) + i sin(21π/12)]

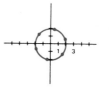

15. √2(cos 28° + i sin 28°); √2(cos 118° + i sin 118°)
√2(cos 208° + i sin 208°); √2(cos 298° + i sin 298°).

17. ±2; ±2i

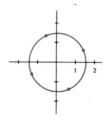

19. ±(√2 + √2i) **21.** ±(√2 + √6i)

23. ±1; ±i

25. cos(k · 36°) + i sin(k · 36°), k = 0, 1, . . . , 9

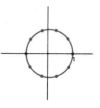

27. (a) 81(cos 80° + i sin 80°) (b) 90.09(cos 7.7 + i sin 7.7) (c) 8(cos 240° + i sin 240°)
(d) 16[cos(5π/3) + i sin(5π/3)]

29. $2(\cos 51° + i \sin 51°), 2(\cos 123° + i \sin 123°), 2(\cos 195° + i \sin 195°), 2(\cos 267° + i \sin 267°),$ $2(\cos 339° + i \sin 339°)$
31. Sum is 0; product is -1.   **33.** $\sqrt[5]{2}(\cos 27° + i \sin 27°) \approx 1.0235 + .5215i$
35. *Method 1.* Use the formula $\cos(k\pi/3) + i \sin(k\pi/3)$, $k = 0, 1, 2, 3, 4, 5$.   *Method 2.* Write $x^6 - 1 =$ $(x - 1)(x^2 + x + 1)(x + 1)(x^2 - x + 1) = 0$ and solve. Both methods give the answers $\pm 1, (-1 \pm \sqrt{3}i)/2, (1 \pm \sqrt{3}i)/2$.
37. $\pm i, \frac{1}{2}\sqrt{2} \pm \frac{1}{2}\sqrt{2}i, -\frac{1}{2}\sqrt{2} \pm \frac{1}{2}\sqrt{2}i$ **39.** (a) $\frac{1}{2}\sqrt{6} + \frac{1}{2}\sqrt{2}i$ (b) $\frac{1}{2}\sqrt{2} + \frac{1}{2}\sqrt{6}i$ **41.** -2^n

CHAPTER 7. REVIEW PROBLEM SET (Page 316)

1. (a) $-6 + 7i$ (b) $17 - 8i$ (c) $-10 - 5i$ (d) $\frac{3}{13} - \frac{11}{13}i$ (e) $2 + 12i$ (f) $\frac{5}{34} - \frac{31}{34}i$
2. (a) $\pm 2i$ (b) $-1 \pm \sqrt{3}i$ (c) $\frac{1}{3} \pm (\sqrt{11}/3)i$ (d) $(-1 \pm \sqrt{5})i$
3.

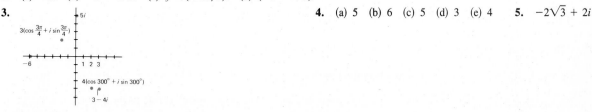

    4. (a) 5 (b) 6 (c) 5 (d) 3 (e) 4   **5.** $-2\sqrt{3} + 2i$

6. (a) $3[\cos(\pi/2) + i \sin(\pi/2)]$ (b) $6(\cos \pi + i \sin \pi)$ (c) $\sqrt{2}[\cos(5\pi/4) + i \sin(5\pi/4)]$
(d) $4[\cos(11\pi/6) + i \sin(11\pi/6)]$
7. (a) $32(\cos 145° + i \sin 145°)$ (b) $2(\cos 65° + i \sin 65°)$ (c) $512(\cos 315° + i \sin 315°)$
(d) $4096(\cos 330° + i \sin 330°)$
8. $2(\cos 20° + i \sin 20°); 2(\cos 80° + i \sin 80°); 2(\cos 140° + i \sin 140°); 2(\cos 200° + i \sin 200°); 2(\cos 260° + i \sin 260°);$ $2(\cos 320° + i \sin 320°)$
9. $\cos 0° + i \sin 0°; \cos 72° + i \sin 72°; \cos 144° + i \sin 144°; \cos 216° + i \sin 216°; \cos 288° + i \sin 288°$

PROBLEM SET 8-1 (Page 323)

1. $x + 1; 0$ **3.** $3x - 1; 10$ **5.** $2x^2 + x + 3; 0$ **7.** $x^2 + 4; 0$ **9.** $x + 2 + 5/x^2$
11. $x - 1 + (-x + 3)/(x^2 + x - 2)$ **13.** $2 + (3 - 4x)/(x^2 + 1)$ **15.** $2x^2 + x + 2; -2$
17. $3x^2 + 8x + 10; 0$ **19.** $x^3 + 3x^2 + 7x + 21; 62$ **21.** $x^2 + x - 4; 6$ **23.** $2x^3 + 4x + 5; \frac{3}{2}$
25. $x^2 - 3ix - 4 - 6i; 4$ **27.** $x^3 + 2ix^2 - 4x - 8i; -1$
29. $2x - 1 + (-x^2 - 2x - 1)/(x^3 + 1) = 2x - 1 + (-x - 1)/(x^2 - x + 1)$
31. (a) $2x + 3; -11x + 9$ (b) $2x + 16; -14$ (c) $x^2 - 8x + 16; x^2 + x + 1$ (d) $x^4 + 6x^2 + 2x + 9; 4x - 1$
33. (a) $x^4 + 3x^3 + 6x^2 + 12x + 8$ (b) $x^4 - 2x^3 + 4x^2 - 8x + 16$ (c) $x^3 - 2x^2 + 4x + 4$ (d) $x^2 + 1$
35. (a) -40 (b) 13 (c) 2   **37.** $a = 1, b = -9, c = 11$

PROBLEM SET 8-2 (Page 329)

1. -2 **3.** $-\frac{9}{4}$ **5.** -6 **7.** 14   **9.** $1, -2$, and 3, each of multiplicity 1.
11. $\frac{1}{2}$ (multiplicity 1); 2 (multiplicity 2); 0 (multiplicity 3).   **13.** $1 + 2i$ and $-\frac{2}{3}$, each of multiplicity 1.   **15.** $P(1) = 0$
17. $P(3) = 0$ **19.** $(5 \pm \sqrt{7}i)/4$ **21.** 3; 2 (multiplicity 2).   **23.** $(x - 2)(x - 3)$
25. $(x - 1)(x + 1)(x - 2)(x + 2)$ **27.** $(x - 6)(x - 2)(x + 5)$ **29.** $(x + 1)(x + 1 - \sqrt{13})(x + 1 + \sqrt{13})$
31. $x^3 + x^2 - 10x + 8$ **33.** $12x^2 + 4x - 5$ **35.** $x^3 - 2x^2 - 5x + 10$
37. $4x^5 + 20x^4 + 25x^3 - 10x^2 - 20x + 8$ **39.** $x^4 - 3x^2 - 4$ **41.** $x^4 - x^3 - 9x^2 + 79x - 130$

43. $-3, -2; (x - 1)^3(x + 3)(x + 2)$ **45.** $\pm i$ **47.** $(x - 3)(x + 3)(x - 2i)$ **49.** $(x + 1)^2(x - 1 - i)$
51. (a) 2   (b) -2 (c) 0
53. (a) ± 2 each of multiplicity 3   (b) 1 and 2 each of multiplicity 2   (c) $-1 \pm \sqrt{5}$ each of multiplicity 3;
-2 of multiplicity 4
55. (a) $(2x - 1)(3x + 1)(2x + 1)$ (b) $(2x - 1)(x - \sqrt{2})(x + \sqrt{2})$ (c) $(2x - 1)(x + i)(x - i)$
57.   **59.** 3, 2.35   **61.** $16x^4 - 32x^3 + 24x^2 - 8x + 1$

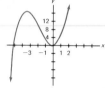

$y = 8$ at $x = -4, -\sqrt{2}, \sqrt{2}$

63. An nth degree polynomial has at most n zeros. Therefore, $P(x)$ must be the zero polynomial (without degree). From this it
follows that all coefficients are zero.
65. We are given that $c^6 - 5c^5 + 3c^4 + 7c^3 + 3c^2 - 5c + 1 = 0$. Note that $c \neq 0$ and so we may divide by c^6 to obtain

$$1 - 5\left(\frac{1}{c}\right) + 3\left(\frac{1}{c}\right)^2 + 7\left(\frac{1}{c}\right)^3 + 3\left(\frac{1}{c}\right)^4 - 5\left(\frac{1}{c}\right)^5 + \left(\frac{1}{c}\right)^6 = 0$$

which is the desired conclusion.
67. $a_n x^n + a_{n-1}x^{n-1} + \cdots + a_1 x + a_0 = a_n(x - c_1)(x - c_2) \cdots (x - c_n)$
$\qquad = a_n[x^n - (c_1 + c_2 + \cdots + c_n)x^{n-1} + (c_1 c_2 + c_1 c_3 + \cdots + c_{n-1}c_n)x^{n-2} + \cdots + (-1)^n c_1 c_2 \cdots c_n]$
$\qquad = a_n x^n - a_n(c_1 + c_2 + \cdots + c_n)x^{n-1} + a_n(c_1 c_2 + c_1 c_3 + \cdots + c_{n-1}c_n)x^{n-2} + \cdots + (-1)^n a_n c_1 c_2 \cdots c_n$
$\qquad$ Thus, $a_{n-1} = -a_n(c_1 + c_2 + \cdots + c_n)$, $a_{n-2} = a_n(c_1 c_2 + \cdots + c_{n-1}c_n)$, $a_0 = (-1)^n a_n c_1 c_2 \cdots c_n$

PROBLEM SET 8-3 (Page 338)

1. $2 - 3i$ **3.** $-4i$ **5.** $4 + \sqrt{6}$ **7.** $(2 + 3i)^8$ **9.** $2(1 - 2i)^3 - 3(1 - 2i)^2 + 5$ **11.** $5 + i$
13. $3 + 2i; 5 - 4i$ **15.** $-i; \frac{1}{2}$ **17.** $1 - 3i; -2; -1$ **19.** $x^2 - 4x + 29$ **21.** $x^3 + 3x^2 + 4x + 12$
23. $x^5 - 2x^4 + 18x^3 - 36x^2 + 81x - 162$ **25.** $-1; 1; 3$ **27.** $\frac{1}{2}; -1 \pm \sqrt{2}$ **29.** $\frac{1}{2}; (1 \pm \sqrt{5})/2$
31. $2 - i, (-1 \pm \sqrt{3}i)/2$ **33.** $-1, -2, 3 \pm \sqrt{13}$ **35.** $1, 1, 1, 1, -10$
37. This follows from the fact that nonreal solutions occur in pairs.
39. If $u = a + bi$, $\bar{u} = a - bi$, then $u + \bar{u} = 2a$, $u\bar{u} = a^2 + b^2$, both of which are real.
41. $(x - 1)(x + 1)(x^2 + 3x + 4)$ **43.** $(x - 1)(x + 1)(x^2 + 1)(x^2 - \sqrt{2}x + 1)(x^2 + \sqrt{2}x + 1)$
45. Let $r = \sqrt{5} - 2$ and $s = \sqrt{5} + 2$. Then $x = r^{1/3} - s^{1/3}$. When we cube x and simplify, we obtain
$x^3 = -4 - 3x(rs)^{1/3} = -4 - 3x$. Thus $x^3 + 3x + 4 = 0$, that is, $(x + 1)(x^2 - x + 4) = 0$. Since x is real, x must be -1.

PROBLEM SET 8-4 (Page 345)

1. $-3/(x + 1) + 2/(x - 1)$ **3.** $2/(x - 5) - 1/(x + 1)$ **5.** $-5/(x - 4) + 6/(x + 1)$
7. $2/(x - 1) + 3/(x + 2) - 1/(x - 3)$ **9.** $3/x - 1/x^2 + 2/(x - 3)$
11. $-2/(x + 2) - 3/(x + 2)^2 + 2/(x - 2) + 1/(x - 2)^2$ **13.** $3/(x + 2) - 1/(x^2 + 2)$
15. $2/(x - 2) + 5/(x^2 + 2x + 4)$ **17.** $1 + (-\frac{5}{4})/(x + 2) + \frac{5}{4}/(x - 2)$ **19.** $2 - 1/x + 5/(x - 2)$
21. $x/(x^2 + 1) - x/(x^2 + 1)^2$ **23.** $2/x + (-x + 1)/(x^2 + x + 1) + 4/(x^2 + x + 1)^2$ **25.** $3/(x - 4) - 2/(x + 5)$
27. $-4/x + 3/(x + 2) + 1/(x - 1)$ **29.** $4/(x + 2) - 3/(x + 2)^2 + 10/(x - 5)$
31. $-2/(x + 3) + 2x/(x^2 - 3x + 9)$ **33.** $3x - 2/x + 5/x^3 + 4/(x - 2)$

1. (a) $2x - 5$; $20x - 20$ (b) $x^2 - 3x + 1$; $-3x + 5$ (c) $x^4 - x^3 + x^2 - x + 1$; 2
2. (a) $x^2 - 4$; -1 (b) $2x^3 - 6x^2 + 3x - 5$; 12 (c) $x^2 + 5x + 1$; -1 **3.** 1; 97
4. (a) -1 (two); 1 (two); i (one); $-i$ (one). (b) 0 (one); $1 + \sqrt{3}i$ (one); $1 - \sqrt{3}i$ (one); $-\pi$ (three).
(c) $-\sqrt{2}i$ (four); $\sqrt{2}i$ (four); $-\sqrt{3}i$ (four); $\sqrt{3}i$ (four).   **5.** $2(x - 3)(x - \frac{1}{2})(x + 3)$ **6.** $(x + 4)(x - \sqrt{7})(x + \sqrt{7})$
7. $x^4 - 9x^3 + 18x^2 + 32x - 96$ **8.** $x^4 - 10x^3 + 31x^2 - 38x + 10$ **9.** Remaining zeros; $-2, 4$.
10. $6x^3 - 25x^2 + 3x + 4$ **11.** -6 **12.** $x^3 - 6x^2 + 9x + 50$; $(x + 2)(x^2 - 8x + 25)$
13. $(x^2 - 4x + 13)^2(x + 4)$ **14.** $2 \pm 5i$; $\pm\sqrt{5}$ **15.** $\frac{3}{2}$; $3 \pm 2\sqrt{2}$
16. A cubic equation has three solutions (counting multiplicities) and the nonreal solutions for an equation with real coefficients occur in conjugate pairs. The only possible rational roots are ± 1 and ± 7, but none of them work.
17. $1/x - 3/(x - 2) + 2/(x + 5)$ **18.** $(x + 2)/(x^2 + 5) + 2/(x - 3) - 4/(x - 3)^2$
19. $2x - x/(x^2 + 1) + 4/(x^2 + 1)^2$

PROBLEM SET 9-1   (Page 355)

1. $(2, -1)$ **3.** $(-2, 4)$ **5.** $(1, -2)$ **7.** $(0, 0, -2)$ **9.** $(1, 4, -1)$ **11.** $(2, 1, 4)$ **13.** $(0, 0, 0)$
15. $(5, 6, 0, -1)$ **17.** $(15z - 110, 4z - 32, z)$ **19.** $(2y - 3z - 2, y, z)$ **21.** $(-z, 2z, z)$ **23.** Inconsistent.
25. $(-z + \frac{2}{5}, z + \frac{16}{5}, z)$ **27.** $(2, 4)$; $(10, 0)$ **29.** $(5, -7)$; $(6, 0)$ **31.** $(-1, 2)$; $(1, 2)$ **33.** $(3, -2)$
35. $(0, -2)$ **37.** $(-6, -12, 24)$ **39.** $(2\sqrt{5}/5, 4\sqrt{5}/5)$ **41.** $a = \frac{3}{2}$, $b = 6$ **43.** 285
45. $x^2 + y^2 - 4x - 3y = 0$, $r = \frac{5}{2}$ **47.** $a = -6$, $b = 8$, $c = -3$ **49.** 15 inches by 8 inches

PROBLEM SET 9-2   (Page 364)

1. $\begin{bmatrix} 2 & -1 & 4 \\ 1 & -3 & -2 \end{bmatrix}$ **3.** $\begin{bmatrix} 1 & -2 & 1 & 3 \\ 2 & 1 & 0 & 5 \\ 1 & 1 & 3 & -4 \end{bmatrix}$ **5.** $\begin{bmatrix} 2 & -3 & -4 \\ 3 & 1 & -2 \end{bmatrix}$ **7.** $\begin{bmatrix} 1 & 0 & 0 & 5 \\ 1 & 2 & -1 & 4 \\ 3 & -1 & -5 & -13 \end{bmatrix}$
9. Unique solution.   **11.** No solution.   **13.** Unique solution.   **15.** Infinitely many solutions.
17. No solution.   **19.** $(1, 2)$ **21.** $(x, \frac{3}{2}x - \frac{1}{2})$ **23.** $(1, 4, -1)$ **25.** $(\frac{16}{3}z + \frac{32}{3}, -\frac{7}{3}z - \frac{2}{3}, z)$
27. $(3, 0, 0)$ **29.** $(4.36, 1.26, -.97)$ **31.** Unique solution.   **33.** Unique solution.
35. No solution.   **37.** $(0, 2, 1)$ **39.** $(\frac{21}{2}z - 48, -5z + 26, z)$ **41.** $a = -4$, $b = 8$, $c = 0$
43. $\alpha = 80°$, $\beta = 30°$, $\gamma = 110°$, $\delta = 50°$ **45.** A: 10 pounds, B: 40 pounds, C: 50 pounds.

PROBLEM SET 9-3   (Page 371)

1. $\begin{bmatrix} 8 & 4 \\ 1 & 10 \end{bmatrix}$; $\begin{bmatrix} -4 & -6 \\ 5 & 4 \end{bmatrix}$; $\begin{bmatrix} 6 & -3 \\ 9 & 21 \end{bmatrix}$ **3.** $\begin{bmatrix} 5 & 4 & 4 \\ 8 & 3 & -6 \end{bmatrix}$; $\begin{bmatrix} 1 & -8 & 6 \\ 0 & -3 & 0 \end{bmatrix}$; $\begin{bmatrix} 9 & -6 & 15 \\ 12 & 0 & -9 \end{bmatrix}$
5. $\begin{bmatrix} 14 & 7 \\ 4 & 36 \end{bmatrix}$; $\begin{bmatrix} 27 & 29 \\ 5 & 23 \end{bmatrix}$ **7.** $\begin{bmatrix} -3 & -4 & 2 \\ 8 & 22 & -13 \\ -2 & 0 & 9 \end{bmatrix}$; $\begin{bmatrix} 0 & 5 & -17 \\ 13 & 10 & 3 \\ 1 & -3 & 18 \end{bmatrix}$
9. **AB** not possible; **BA** $= \begin{bmatrix} 7 & 2 & -7 & 6 \\ 15 & 2 & -11 & 16 \end{bmatrix}$
11. **AB** $= \begin{bmatrix} 2 \\ 16 \\ -2 \end{bmatrix}$; **BA** not possible.   **13.** **AB** $=$ **BA** $= \begin{bmatrix} 0 & 0 \\ 0 & 0 \end{bmatrix}$ **15.** $\begin{bmatrix} -4 & 7 & 9 \\ -5 & -5 & 8 \end{bmatrix}$

17. $A(B + C) = AB + AC = \begin{bmatrix} -7 & -3 \\ 39 & 34 \end{bmatrix}$; the distributive property.   **19.** 93.5917

21. $\begin{bmatrix} 2 & 5 & -1 \\ -8 & 5 & -3 \\ 16 & -2 & -1 \end{bmatrix}$; $\begin{bmatrix} -16 & -6 & 8 \\ 12 & 0 & 22 \\ 11 & -16 & 19 \end{bmatrix}$; $\begin{bmatrix} 32 & -3 & 12 \\ 36 & 29 & 24 \\ 34 & 6 & 25 \end{bmatrix}$

23. $AB = -7$; $BA = \begin{bmatrix} 2 & 4 & 6 & 8 \\ 1 & 2 & 3 & 4 \\ -1 & -2 & -3 & -4 \\ -2 & -4 & -6 & -8 \end{bmatrix}$

25. A and B are square, of the same size, and $AB = BA$.

27. $A^2 = 0$, $B^3 = 0$. The nth power of a strictly upper triangular $n \times n$ matrix is the zero matrix.

29. Multiplication of B on the left by A multiplies the three rows by 3, -4, and 5, respectively. Similarly, multiplication on the right by A multiplies the columns of B by 3, -4, and 5, respectively.

31. (a) $\begin{bmatrix} 16 \\ 13 \\ 14 \end{bmatrix}$ $\rightarrow$ Art's wages on Monday.   (b) $\begin{bmatrix} 18 \\ 14 \\ 16 \end{bmatrix}$ Each man's corresponding wages on Tuesday.

(c) $\begin{bmatrix} 7 & 9 & 3 \\ 9 & 3 & 4 \\ 8 & 5 & 4 \end{bmatrix}$ The combined output for Monday and Tuesday.   (d) $\begin{bmatrix} 34 \\ 27 \\ 30 \end{bmatrix}$ Each man's combined wages for the two days.

33. (a) $U + V = \begin{bmatrix} u_1 + v_1 & u_2 + v_2 \\ -(u_2 + v_2) & u_1 + v_1 \end{bmatrix}$, $UV = \begin{bmatrix} u_1v_1 - u_2v_2 & u_1v_2 + u_2v_1 \\ -(u_1v_2 + u_2v_1) & u_1v_1 - u_2v_2 \end{bmatrix}$

(b) $I^2 = I$, $J^2 = -I$ (c) $U + V = (u_1 + v_1)I + (u_2 + v_2)J$, $UV = (u_1v_1 - u_2v_2)I + (u_1v_2 + u_2v_1)J$

(d) This system of matrices behaves just like the complex number system provided we identify I with 1 and J with i.

PROBLEM SET 9-4   (Page 379)

1. $\begin{bmatrix} -1 & -3 \\ 1 & 2 \end{bmatrix}$ **3.** $\begin{bmatrix} \frac{1}{6} & \frac{7}{6} \\ 0 & \frac{1}{2} \end{bmatrix}$ **5.** $\begin{bmatrix} 1 & 0 \\ 0 & 1 \end{bmatrix}$ **7.** $\begin{bmatrix} 1/a & 0 \\ 0 & 1/b \end{bmatrix}$ **9.** $\begin{bmatrix} -2 & \frac{3}{2} \\ 1 & -\frac{1}{2} \end{bmatrix}$

11. $\begin{bmatrix} -\frac{4}{7} & \frac{2}{7} & \frac{3}{7} \\ \frac{6}{7} & -\frac{3}{7} & -\frac{1}{7} \\ \frac{5}{7} & \frac{1}{7} & -\frac{2}{7} \end{bmatrix}$ **13.** $\begin{bmatrix} -\frac{1}{9} & \frac{1}{9} & \frac{8}{9} \\ \frac{10}{9} & -\frac{1}{9} & -\frac{26}{9} \\ \frac{1}{9} & -\frac{1}{9} & \frac{1}{9} \end{bmatrix}$ **15.** $\begin{bmatrix} 1 & -1 & 2 & -\frac{5}{4} \\ 0 & \frac{1}{2} & -\frac{3}{2} & \frac{7}{8} \\ 0 & 0 & 1 & -\frac{3}{4} \\ 0 & 0 & 0 & \frac{1}{4} \end{bmatrix}$ **17.** $(\frac{5}{7}, \frac{10}{7}, -\frac{1}{7})$

19. $(\frac{47}{9}, -\frac{128}{9}, \frac{7}{9})$ **23.** Inverse does not exist.

25. $\begin{bmatrix} -4 & 3 & -4 \\ \frac{1}{2} & -\frac{1}{2} & 1 \\ 6 & -4 & 6 \end{bmatrix}$ **27.** $\begin{bmatrix} \frac{1}{2} & 0 & 0 \\ 0 & \frac{1}{3} & 0 \\ 0 & 0 & -\frac{1}{4} \end{bmatrix}$

29. $(-4a + 3b - 4c, \frac{1}{2}a - \frac{1}{2}b + c, 6a - 4b + 6c)$

33. $\begin{bmatrix} 2 & -1 & 0 & 0 \\ -1 & 0 & 1 & 0 \\ 0 & 1 & 0 & -1 \\ 0 & 0 & -1 & 1 \end{bmatrix}$ **35.** $\begin{bmatrix} 1 & a & b \\ 0 & 1 & c \\ 0 & 0 & 1 \end{bmatrix}^{-1} = \begin{bmatrix} 1 & -a & ac - b \\ 0 & 1 & -c \\ 0 & 0 & 1 \end{bmatrix}$

PROBLEM SET 9-5   (Page 387)

1. -8 **3.** 22   **5.** -50 **7.** 0   **9.** (a) 12   (b) -12 (c) 36   (d) 12   **11.** -30 **13.** -16

15. 7   **17.** 42.3582   **19.** $(2, -3)$ **21.** $(2, -2, 1)$ **23.** (a) 0   (b) 0   (c) 0   (d) 6   (e) 1   (f) 2

25. (a) $k \neq \pm 1$ (b) $k = 1$ (c) $k = -1$ **27.** (a) -32 (b) $-\frac{1}{2}$ (c) -2 (d) -216

29.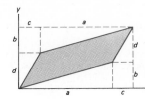

$$A = (a + c)(b + d) - 2bc - 2(\tfrac{1}{2}ab) - 2(\tfrac{1}{2}cd)$$
$$= ab + ad + bc + cd - 2bc - ab - cd$$
$$= ad - bc = \begin{vmatrix} a & b \\ c & d \end{vmatrix}$$

31. $P(a, b, c) = ab + ac + bc - a^2 - b^2 - c^2$

PROBLEM SET 9-6 (Page 393)

1. -20 **3.** -1 **5.** 39 **7.** 4 **9.** 57 **11.** $x = 2$ **13.** 6 **15.** -72 **17.** -960
19. $aehj$ **21.** 156.8659 **23.** (a) -60 (b) $ef(ad - bc)$ **25.** 0 **27.** $D_2 = D_3 = D_4 = 1$
29. $x = -1, y = 3, z = -3, w = 1$
31. (a) $|C_2| = |C_3| = |C_4| = 0; |C_n| = 0$ for $n \geq 2$ (b) $|C_2| = (a_2 - a_1)(b_2 - b_1); |C_n| = 0$ for $n \geq 3$

PROBLEM SET 9-7 (Page 401)

1. 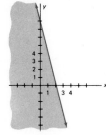   **3.** 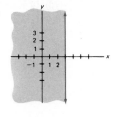   **5.**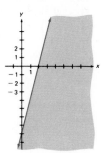

7. 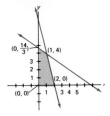   **9.**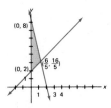

11. Maximum value: 6; minimum value: 0.   **13.** Maximum value: $-\frac{4}{3}$; minimum value: -8.
15. Minimum value of 14 at $(2, 2)$.   **17.** Minimum value of 4 at $(\frac{3}{2}, 1)$.

19. 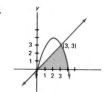   **21.** 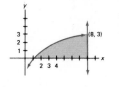   **23.**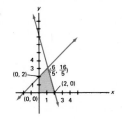

25. Maximum value of 4 at (2, 0); minimum value of -2 at (0, 2).

27. Maximum value of $\frac{11}{2}$ at $(\frac{9}{4}, \frac{13}{4})$; minimum value of 0 at (0, 0).   **29.** 266   **31.** 2 camper units and 6 house trailers.

33. 10 pounds of type A and 5 pounds of type B.

35.

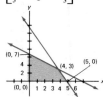

On A, $|y - 2x| + y + x = 2y - x$, which achieves its maximum value of 9 at (3, 6).
On B, $|y - 2x| + y + x = 3x$, which achieves its maximum value of 12 at (4, 7).

Therefore, the maximum value on the entire polygon is 12.

CHAPTER 9. REVIEW PROBLEM SET (Page 406)

1. (2, 6)   **2.** (5, 2, 3)   **3.** No solution.   **4.** No solution.   **5.** $(20w - 36, -9w + 22, -2w + 4, w)$

6. (a) $\begin{bmatrix} -1 & 8 & -2 \\ 8 & -1 & 3 \\ 5 & 13 & 0 \end{bmatrix}$ (b) $\begin{bmatrix} 7 & -11 & 9 \\ -1 & 7 & -11 \\ 0 & -16 & -5 \end{bmatrix}$ (c) $\begin{bmatrix} 0 & 12 & 3 \\ -8 & 6 & -9 \\ -3 & -3 & 0 \end{bmatrix}$ (d) $\begin{bmatrix} 7 & -5 & -5 \\ 3 & 9 & 0 \\ 32 & 14 & -10 \end{bmatrix}$

7. $\begin{bmatrix} \frac{1}{6} & \frac{1}{2} & -\frac{1}{3} \\ \frac{1}{6} & -\frac{1}{2} & \frac{2}{3} \\ \frac{2}{3} & 0 & -\frac{1}{3} \end{bmatrix}$ **8.** $(-1, 2, 3)$ **9.** 2   **10.** 0   **11.** -114 **12.** 0   **13.** -144 **14.** $(1, -2, 4)$

15.

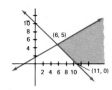

16.   **17.** 14   **18.** 11   **19.** 24   **20.** 30 suits; 65 dresses

PROBLEM SET 10-1 (Page 415)

1. (a) 27; 32 (b) 32; 41 (c) $-5; -7$ (d) 32; 64 (e) 36; 49   **3.** $\frac{6}{37}$ **5.** -240 **7.** 30

9. $-1 + (n - 1)6 = 6n - 7$; 299   **11.** $2 + (n - 1)(.3) = .3n + 1.7$; 17

13. $28 + (n - 1)(-4) = -4n + 32$; -172 **15.** $.5n + 4.5$; 30   **17.** 7599

19. (a) 10,100 (b) 10,000 (c) 6633   **21.** 382.5   **23.** 11; 2772   **25.** (a) 90 (b) $\frac{25}{6}$ (c) 15,350 (d) 9,801

27. (a) $\sum_{i=3}^{20} b_i$ (b) $\sum_{i=1}^{19} i^2$ (c) $\sum_{i=1}^{n} 1/i$ **29.** $\frac{11}{125}$ **31.** -4 **33.** (a) $-.75$ (b) -17.50 (c) 63.75

35. 902   **37.** 375   **39.** (a) 10,200 (b) $\frac{100}{101}$ (c) $\frac{1}{2} - \frac{1}{10302} \approx .4999029$ **41.** 44,850   **43.** 2440

45. 2475   **47.** 5400π inches

PROBLEM SET 10-2 (Page 423)

1. (a) 8; 16 (b) $\frac{1}{2}; \frac{1}{4}$ (c) .00003; .000003   **3.** (a) 2; 2^{n-2} (b) $\frac{1}{2}$; $8(\frac{1}{2})^{n-1} = 1/2^{n-4}$ (c) .1; .3$(.1)^{n-1} = 3 \times 10^{-n}$

5. (a) $2^{28} \approx 2.68 \times 10^8$ (b) $(\frac{1}{2})^{26} \approx 1.49 \times 10^{-8}$ (c) 3×10^{-30} **7.** (a) $\frac{31}{2}$ (b) $\frac{31}{2}$ (c) .33333

9. (a) $\frac{1}{2}(2^{30} - 1) \approx 5.3687 \times 10^8$ (b) $16(1 - 1/2^{30}) = 16 - (\frac{1}{2})^{26}$ (c) $.3\{[1 - (.1)^{30}]/.9\} = \frac{1}{3}(1 - (.1)^{30})$

11. $100(2)^{10} = 102,400$ **13.** $\$(2^{31} - 1)$, which is over $2 billion.   **15.** (a) $\frac{1}{2}$ (b) $\frac{4}{15}$ **17.** 30 feet   **19.** $\frac{1}{9}$

21. $\frac{25}{99}$ **23.** $611/495$ **25.** $625; 125; 25; 5; 1$ **27.** $a_n = 130(\frac{1}{2})^{n-1}$, $c_n = 103 + 2n$, $d_n = 100(1.05)^n$, $f_n = 3(-2)^{n-1}$

29. $\$215.89$, 12 years   **31.** $\$1564.55$ **33.** $3 billion   **35.** $3\sqrt{3}/14$ **37.** $(3^{32} - 1)/2 \approx 9.2651 \times 10^{14}$

39. 20   **41.** 125 miles

PROBLEM SET 10-3 (Page 430)

(*Note:* In the text, several proofs by mathematical induction are given in complete detail. To save space, we show only the key step here, namely, that P_{k+1} is true if P_k is true.)

1. $(1 + 2 + \cdots + k) + (k + 1 = k(k + 1)/2 + k + 1 = [k(k + 1) + 2(k + 1)]/2 = (k + 1)(k + 2)/2$

3. $(3 + 7 + \cdots + (4k - 1)) + (4k + 3) = k(2k + 1) + (4k + 3) = 2k^2 + 5k + 3 = (k + 1)(2k + 3)$

5. $(1 \cdot 2 + 2 \cdot 3 + \cdots + k(k + 1)) + (k + 1)(k + 2) = \frac{1}{3}k(k + 1)(k + 2) + (k + 1)(k + 2) = \frac{1}{3}(k + 1)(k + 2)(k + 3)$

7. $(2 + 2^2 + \cdots + 2^k) + 2^{k+1} = 2(2^k - 1) + 2^{k+1} = 2^{k+1} - 2 + 2^{k+1} = 2(2^{k+1} - 1)$

9. P_n is true for $n \geq 8$.   **11.** P_1 is true.   **13.** P_n is true whenever n is odd.

15. P_n is true for every positive integer n.   **17.** P_n is true whenever n is a positive integer power of 4.

19. $n = 4$. If $k + 5 < 2^k$, then $k + 6 < 2^k + 1 < 2^k + 2^k = 2^{k+1}$.

21. $n = 1$. Since $k + 1 < 10k$, $\log(k + 1) < 1 + \log k < 1 + k$.

23. $n = 1$. Multiply both sides of $(1 + x)^k \geq 1 + kx$ by $(1 + x)$: $(1 + x)^{k+1} \geq (1 + x)(1 + kx) = 1 + (k + 1)x + kx^2 > 1 + (k + 1)x$.

25. $x^{2k+2} - y^{2k+2} = x^{2k}(x^2 - y^2) + (x^{2k} - y^{2k})y^2$. Now $(x - y)$ is a factor of both $x^2 - y^2$ and $x^{2k} - y^{2k}$, the latter by assumption.

27. $(k + 1)^2 - (k + 1) = k^2 + 2k + 1 - k - 1 = (k^2 - k) + 2k$. Now 2 divides $k^2 - k$ by assumption and clearly divides $2k$.

29. (a) $(1 + 2 + 3 + \cdots + k) + (k + 1) = \frac{1}{2}k(k + 1) + k + 1 = (k + 1)(\frac{1}{2}k + 1) = \frac{1}{2}(k + 1)(k + 2)$

(c) $(1^3 + 2^3 + \cdots + k^3) + (k + 1)^3 = \frac{1}{4}k^2(k + 1)^2 + (k + 1)^3 = [(k + 1)^2/4][k^2 + 4(k + 1)] = \frac{1}{4}(k + 1)^2(k + 2)^2$

(d) $(1^4 + 2^4 + \cdots + k^4) + (k + 1)^4 = \frac{1}{30}k(k + 1)(6k^3 + 9k^2 + k - 1) + (k + 1)^4$

$= \frac{k + 1}{30}(6k^4 + 9k^3 + k^2 - k + 30k^3 + 90k^2 + 90k + 30)$

$= \frac{k + 1}{30}(6k^4 + 39k^3 + 91k^2 + 89k + 30) = \frac{1}{30}(k + 1)(k + 2)(6k^3 + 27k^2 + 37k + 15)$

$= \frac{1}{30}(k + 1)(k + 2)[6(k + 1)^3 + 9(k + 1)^2 + (k + 1) - 1]$

31. (a) 15,250 (b) 220 (c) 4355 (d) $2n(n + 1)^2$

33. $\left(1 - \frac{1}{4}\right)\left(1 - \frac{1}{9}\right) \cdots \left(1 - \frac{1}{k^2}\right)\left(1 - \frac{1}{(k + 1)^2}\right) = \frac{k + 1}{2k}\left[1 - \frac{1}{(k + 1)^2}\right] = \frac{k + 1}{2k}\frac{k^2 + 2k}{(k + 1)^2} = \frac{k + 2}{2(k + 1)}$

35. Let $S_k = \frac{1}{k + 1} + \frac{1}{k + 2} + \cdots + \frac{1}{2k}$ and assume $S_k > \frac{3}{5}$. Then $S_{k+1} = \frac{1}{k + 2} + \frac{1}{k + 3} + \cdots + \frac{1}{2k} + \frac{1}{2k + 1} + \frac{1}{2k + 2}$

$= S_k + \frac{1}{2k + 1} + \frac{1}{2k + 2} - \frac{1}{k + 1} > S_k + \frac{2}{2k + 2} - \frac{1}{k + 1} = S_k > \frac{3}{5}$.

37. The statement is true when $n = 3$ since it asserts that the angles of a triangle have a sum of 180°. Now any $(k + 1)$-sided convex polygon can be dissected into a k-sided polygon and a triangle. Its angles add up to $(k - 2)180° + 180° = (k - 1)180°$.

39. $f_1^2 + f_2^2 + \cdots + f_k^2 + f_{k+1}^2 = f_k f_{k+1} + f_{k+1}^2 = f_{k+1}(f_k + f_{k+1}) = f_{k+1}f_{k+2}$

41. Assume the equality holds for a_k and a_{k+1}. Then $a_{k+2} = (a_k + a_{k+1})/2 = \frac{2}{3}[(1 - (-\frac{1}{2})^k + 1 - (-\frac{1}{2})^{k+1})/2] = \frac{2}{3}[1 - \frac{1}{2}(-\frac{1}{2})^k - \frac{1}{2}(-\frac{1}{2})^{k+1}] = \frac{2}{3}[1 - (-\frac{1}{2})^{k+2}]$.

PROBLEM SET 10-4 (Page 438)

1. 21   **3.** 35   **5.** 1140   **7.** 75,287,520   **9.** 5040   **11.** 1.3077×10^{12} **13.** $x^3 + 3x^2y + 3xy^2 + y^3$

15. $x^3 - 6x^2y + 12xy^2 - 8y^3$ **17.** $c^8 - 12c^6d^3 + 54c^4d^6 - 108c^2d^9 + 81d^{12}$

19. $a^7 + 7a^6b + 21a^5b^2 + 35a^4b^3 + 35a^3b^4 + 21a^2b^5 + 7ab^6 + b^7$ **21.** $x^{20} + 20x^{19}y + 190x^{18}y^2$
23. $x^{20} + 20x^{14} + 190x^8$ **25.** $a^{50} - 50a^{49}b + 1225a^{48}b^2$ **27.** $-120y^{14}z^9$ **29.** $-1760a^3b^9$ **31.** 23.424
33. 552.375 **35.** 149 **37.** (a) $x^8 - 12x^5 + 54x^2 - 108/x + 81/x^4$ (b) $5x^4 + 10x^3h + 10x^2h^2 + 5xh^3 + h^4$
39. $-448x^5y^6$ **41.** (a) $-38 + 41i$ (b) $-7 + 24i$ **45.** 2^n
47. (a) $x^3 + y^3 + z^3 + 3xy^2 + 3xz^2 + 3x^2y + 3yz^2 + 3x^2z + 3y^2z + 6xyz$ (b) $420x^2y^4z$ **49.** 3^n
51. (a) 2^k (b) $2^{n+1} - 1$ (c) $2^{n+2} - n - 3$

CHAPTER 10. REVIEW PROBLEM SET (Page 442)

1. (a) and (c) are arithmetic; (b) and (e) are geometric.   **2.** (a) $a_n = a_{n-1} + 3$ (b) $b_n = 3b_{n-1}$ (c) $c_n = c_{n-1} - .5$
3. (a) $a_n = 2 + (n - 1)3 = 3n - 1$ (b) $b_n = 2 \cdot 3^{n-1}$ **4.** 6767 **5.** $3^{100} - 1$ **6.** 3 **7.** 89 **8.** 40
9. $250,500$ **10.** $\frac{5}{9}$ **11.** $100(1.02)^{48}$ **13.** P_n is true for n a multiple of 3.   **14.** (a) $40,320$ (b) 84 (c) 780
15. $x^{10} + 20x^9y + 180x^8y^2 + 960x^7y^3$ **16.** $-20a^3b^6$ **17.** 1.0408

PROBLEM SET 11-1 (Page 447)

1. Upward.   **3.** To the right.   **5.** To the left.   **7.** $p = 2$

9. $p = \frac{1}{8}$ **11.** $p = \frac{1}{2}$ **13.** $p = \frac{9}{16}$

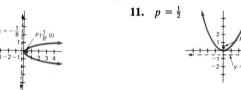

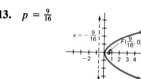

15. $(-\frac{1}{2}, 1); (\frac{1}{2}, 1)$ **17.** $3x^2 = 2y$ **19.** $x^2 = -12y$ **21.** $y^2 = 4x$ **23.** $(-\frac{1}{5}, 0); x = \frac{1}{5}$ **25.** $4p$
27. 2.5 feet   **29.** 225 meters   **31.** $y^2 = 4(p + r)x$ **33.** $8\sqrt{3}p$

PROBLEM SET 11-2 (Page 453)

1. Vertical; $8, 2\sqrt{7}$.   **3.** Horizontal; $12, 4\sqrt{5}$.   **5.** Horizontal; $2, \frac{4}{3}$.   **7.** Vertical; $2/k, 1/k$.

9. Horizontal ellipse; $a = 5, b = 3, c = 4$.   **11.** Vertical ellipse; $a = 2, b = 1$ $c = \sqrt{3}$.

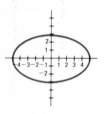

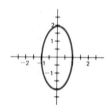

13. $x^2/16 + y^2/25 = 1$; $e = \frac{3}{5}$

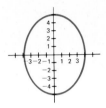

15. $x^2/49 + y^2/40 = 1$; $e = \frac{3}{7}$

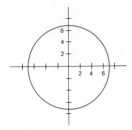

17. $x^2/49 + y^2/4$; $e = 3\sqrt{5}/7$

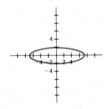

19. $x^2/81 + y^2/9 = 1$; $e = 2\sqrt{2}/3$

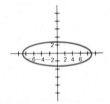

21. $a = 5$, $b = 2$, $c = \sqrt{21}$, $e = \sqrt{21}/5$ **23.** $x^2/128 + y^2/144 = 1$ **25.** $5\sqrt{3}$ feet

27. 20,000 miles, $4000\sqrt{21}$ miles   **29.** $\pi\sqrt{77}$ **31.** (a) 80π square feet (b) 176π square feet

33. (a) $\overline{PR} + \overline{RQ} = 2a$; $\overline{PR'} + \overline{R'Q} > 2a$ since R' is outside the ellipse.

(b) Let Q' be the mirror image of Q about the line l. Show that $\angle Q'RR' = \alpha$.

PROBLEM SET 11-3 (Page 459)

1. Horizontal; $a = 4$, $b = 6$, $c = 2\sqrt{13}$.   **3.** Vertical; $a = 3$, $b = 4$, $c = 5$.

5. Horizontal; $a = \frac{1}{2}$, $b = \frac{1}{4}$, $c = \sqrt{5}/4$.   **7.** Horizontal; $a = 2$, $b = 4$, $c = 2\sqrt{5}$.

9. Horizontal hyperbola; $a = 5$, $b = 3$, $c = \sqrt{34}$.   **11.** Vertical hyperbola; $a = 8$, $b = 6$, $c = 10$.

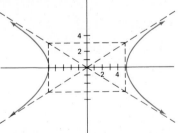

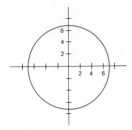

13. $-x^2/\frac{9}{4} + y^2/9 = 1$

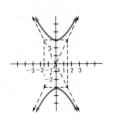

15. $x^2/1 - y^2/15 = 1$

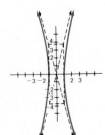

17. $x^2/9 - y^2/36 = 1$

19. $-x^2/28 + y^2/36 = 1$

21. $x^2/16 - y^2/128 = 1$

23. $\frac{32}{3}$ **25.** $\sqrt{2}$ **27.** $5\sqrt{3}$

29. $-x^2/3,630,000 + y^2/1,210,000 = 1$

PROBLEM SET 11-4 (Page 466)

1. $u^2 + 2v^2 = 2$; ellipse. **3.** $u^2 + v^2 = 1$; circle **5.** $u^2 = 4v$; parabola. **7.**

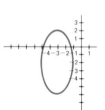

9.

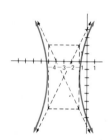

11.

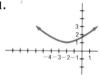

13.

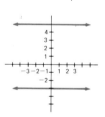

15. $(x + 2)^2 + (y - 1)^2 = 5$; circle: center $(-2,1)$, radius $\sqrt{5}$.

17. $(x - 2)^2/\frac{26}{4} + (y - 4)^2/26 = 1$; vertical ellipse; center $(2, 4)$. **19.** $4(x - 2)^2 + (y - 4)^2 = 0$; point: $(2, 4)$.

21. $(x - 2)^2 = -\frac{1}{4}(y - 24)$; vertical parabola: vertex $(2, 24)$.

23. $4(x - 2)^2 - 9(y - 1)^2 = 0$; two lines intersecting at $(2, 1)$. **25.**

27. $-(x - 1)^2/4 + (y + 3)^2/9 = 1$; 6 **29.** Focus; $(21/20, 1)$; directrix: $x = -29/20$.

31. (a) 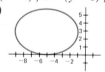 (b)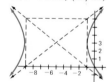

33. $a < 0$ (hyperbola); $a = 0$ (parabola); $a > 0$, $a \neq 1$ (ellipse); $a = 1$ (circle).

35. $-4(x - 4) = (y - 5)^2$ **37.** $x^2/8 + (y - 2)^2/4 = 1$ **39.** $-5(x - 3)^2/36 + 5(y + 4)^2/144 = 1$

41. (a) $y = x^2 - x$ (b) $x = \frac{1}{4}y^2 - y$ (c) $x^2 + y^2 - 5x - 5y = 0$ **43.** $x^2 + (y + \frac{7}{3})^2 = \frac{625}{9}$

PROBLEM SET 11-5 (Page 473)

1. $v = 0$ **3.** $4u^2 + v^2 = 16$ **5.** $u^2 + 2uv + v^2 - 8u + 8v = 0$ **7.** $u^2 + 3v^2 = 8$ **9.** $-2u^2 + 3v^2 = 6$

11. $3x^2 + 10xy + 3y^2 + 8 = 0$; $x = (\sqrt{2}/2)(u - v)$ and $y = (\sqrt{2}/2)(u + v)$; $-u^2 + v^2/4 = 1$

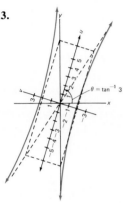

13. $4x^2 - 3xy = 18$; $x = (1/\sqrt{10})(u - 3v)$ and $y = (1/\sqrt{10})(3u + v)$; $-u^2/36 + v^2/4 = 1$

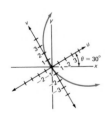

15. $x^2 - 2\sqrt{3}xy + 3y^2 - 12\sqrt{3}x - 12y = 0$; $x = \frac{1}{2}(\sqrt{3}u - v)$ and $y = \frac{1}{2}(u + \sqrt{3}v)$; $v^2 = 6u$

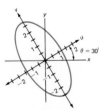

17. $13x^2 + 6\sqrt{3}xy + 7y^2 - 32 = 0$; $x = \frac{1}{2}(\sqrt{3}u - v)$ and $y = \frac{1}{2}(u + \sqrt{3}v)$; $u^2/2 + v^2/8 = 1$

19. $9x^2 - 24xy + 16y^2 - 60x + 80y + 75 = 0$; $x = \frac{1}{5}(4u - 3v)$ and $y = \frac{1}{5}(3u + 4v)$; $v^2 + 4v + 3 = 0$

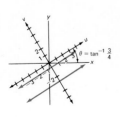

21. $u = (5 - 3\sqrt{3})/2$; $v = (-5\sqrt{3} - 3)/2$ **23.** $u = 4$; $v = -2$ **25.** $u = 5$; $v = 0$

27. $x^2 + 2\sqrt{3}xy + 3y^2 = 8(-\sqrt{3}x + y)$ **29.** $u^2/2 + v^2/10 = 1$; ellipse. **31.** $(u - 4)^2 + v^2 = 16$

33. $u^2/5 - v^2/21 = 1$; hyperbola. **35.** Equation transforms to $u = d$, a line $|d|$ units from the origin. **37.** 3

PROBLEM SET 11-6 (Page 480)

1–11.

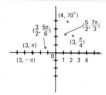

13. $(2\sqrt{2}, 2\sqrt{2})$ **15.** $(-3, 0)$ **17.** $(-5, -5\sqrt{3})$ **19.** $(\sqrt{2}, -\sqrt{2})$

21. $(4, 0)$ **23.** $(2, \pi)$ **25.** $(2\sqrt{2}, \pi/4)$ **27.** $(2\sqrt{2}, 3\pi/4)$ **29.** $(2, -\pi/3)$ **31.** $(2\sqrt{3}, 11\pi/6)$

33.

35.

37.

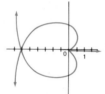

39.

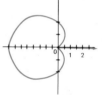

41.

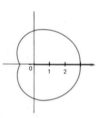

43. $r = 2$ **45.** $r = \tan \theta \sec \theta$ **47.** $y = 2x$ **49.** $(x^2 + y^2)^{3/2} = x^2 - y^2$

51.

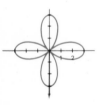

53.

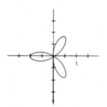

55.

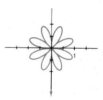

57. (a) $3y - 2x = 5$; a line. (b) $(x - 2)^2 + (y + 3)^2 = 13$; a circle.

59.

61.

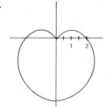

63.

65. (a) 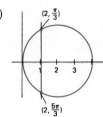 $(2, \pi/3), (2, 5\pi/3)$ (b) 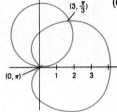 $(0, \pi), (3, \pi/3)$

67. Use the law of cosines; $4\sqrt{5}$ **69.** $\frac{1}{2}(\beta - \alpha)(b - a)(b + a)$

71.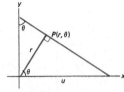

Since $\cos \theta = \dfrac{r}{u}$ and $\sin \theta = \dfrac{u}{4}$, it follows that

$r = u \cos \theta = 4 \sin \theta \cos \theta = 2 \sin 2\theta$.

Now compare with Example B.

PROBLEM SET 11-7 (Page 488)

1. $r \cos \theta = 4$ **3.** $r \cos \theta = -3$ **5.**

7.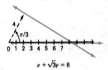

9.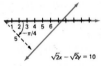

11. $r = 8 \cos \theta, \; x^2 + y^2 = 8x$

13. $r = 10 \cos (\theta - \pi/3); \; x^2 + y^2 = 5x + 5\sqrt{3}\,y$ **15.** Ellipse; $e = \frac{2}{3}; \; x = 6$.   **17.** Hyperbola; $e = 2; \; x = \frac{5}{4}$.

19. Parabola; $e = 1; \; x = -7$.   **21.** Ellipse; $e = \frac{2}{3}; \; x = -\frac{1}{2}$.   **23.** Parabola; $e = 1; \; y = 5$.

25. Ellipse; $e = \frac{1}{2}; \; y = -6$.   **27.** Hyperbola; $e = \frac{5}{4}; \; y = \frac{8}{3}$.

29. 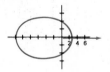   **31.** 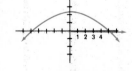   **33.**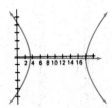

35. (a) $r = 3 \csc \theta$ (b) $r = 3$ (c) $r = -18 \cos \theta$ (d) $r = 6\sqrt{2} \cos (\theta - \pi/4)$

37. (a) $r = 2/(1 - \cos \theta)$ (b) $r = 4/(1 + \sin \theta)$ **39.** $(x + 3y)(x - 2y) = 0$ **41.** $d = 8, \; e = \frac{1}{2}; \frac{32}{3}, \; 16\sqrt{3}/3$

43. 8 **45.** (a)

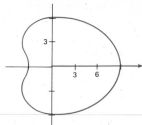

(b) $2a + a \cos \theta + 2a + a \cos(\theta + \pi) = 4a$

PROBLEM SET 11-8 (Page 498)

1. $x = 2t;\ y = -3t$ **3.** $x = 1 + 3t;\ y = 2 - 7t$ **5.** $-2;\ 1$ **7.** $2x + 3y = 17$ **9.** $2y = x^2 + 3x + 2$
11. $x^2 + y^2 = 4$ **13.** $x^2/4 + y^2/9 = 1$ **15.** $27y = x^3 - 3x^2 + 3x - 1$ **17.**

19.

21.

23.
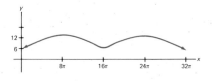

25. (a) $y = -x^2/768 + x/\sqrt{3}$ (b) 4 (c) $256\sqrt{3}$ (d) 64
27. (a) $3x + 2y = 17$ (b) $y = 4\cos(x/3)$ (c) $x^2/4 - y^2/9 = 1$ (d) $8x + (y + 1)^3 = 8$ (e) $x = (x + y)^6 + 2(x + y)^3$
29. Show that $x^2 + y^2 = 4$ in each case. Verify that the parameter interval gives the same quarter circle in each case.
31. $(x - 2)^2/9 + (y - 1)^2/16 = 1$
33. (a) $y = (\tan \alpha)x - 16x^2/(v_0{}^2 \cos^2 \alpha)$ (b) $(v_0 \sin \alpha)/16$ (c) $(v_0 \cos \alpha)(v_0 \sin \alpha)/16 = (v_0{}^2 \sin 2\alpha)/32$ (d) $\pi/4$
35. Eliminate t to get $-x + 9y = 40$. Endpoints are $(-4, 4)$ and $(5, 5)$. Parameter interval $\pi/2$ to π gives same segment traced in the reverse direction.

CHAPTER 11. REVIEW PROBLEM SET (Page 501)

1. (a) xii (b) iii (c) viii (d) ii (e) vii (f) vi (g) v (h) x (i) ix (j) xi (k) x
2. Focus $(-2, -\frac{11}{4})$; vertex $(-2, -3)$. **3.** $x^2 = \frac{4}{3}y$ **4.** $(x - 2)^2/9 + (y - 1)^2/25 = 1$
5. **6.** $-x^2/9 + y^2/9 = 1$ **7.** $(3 - \sqrt{34}, -1);\ (3 + \sqrt{34}, -1)$

$e = \sqrt{3}/2$

8. $u^2/8 + v^2/\frac{56}{3} = 1;\ 4\sqrt{2}$

9.

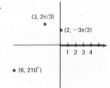

10. (a) $(-3/2, 3\sqrt{3}/2)$ (b) $(0, 2)$ (c) $(-3\sqrt{3}, -3)$

11. (a) $(5, 0)$ (b) $(4, 7\pi/4)$ (c) $(4, 5\pi/6)$

12. (a)

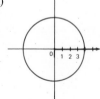

(b)

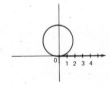

(c)

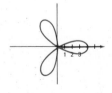

13. $r^2 = 4 \sec \theta \csc \theta; (x^2 + y^2)^{3/2} = 2xy$ **14.** (a) iii (b) vii (c) ii (d) v (e) iii (f) ii (g) iv (h) vi

15. (a) $x^2 + y^2 = 25$ (b) $r = 5$ (c) $x = 5 \cos \theta; y = 5 \sin \theta$ **16.**

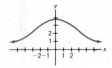

Index
of Teaser Problems

Index
of Names and Subjects

GEOMETRY

Triangles

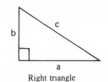

Right triangle

Pythagorean Theorem

$$a^2 + b^2 = c^2$$

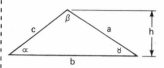

Any triangle

Angles $\quad \alpha + \beta + \gamma = 180°$

Area $\quad A = \frac{1}{2}bh$

Circles

Circumference $\quad C = 2\pi r$

Area $\quad\quad\quad\quad A = \pi r^2$

Cylinders

Surface area $\quad S = \pi r^2 + \pi r\sqrt{r^2 + h^2}$

Volume $\quad\quad V = \frac{1}{3}\pi r^2 h$

Spheres

Surface area $\quad S = 4\pi r^2$

Volume $\quad\quad V = \frac{4}{3}\pi r^3$

Conversions

1 inch = 2.54 centimeters

1 liter = 1000 cubic centimeters

1 kilogram = 2.20 pounds

1 kilometer = .62 miles

1 liter = 1.057 quarts

1 pound = 453.6 grams

π radians = 180 degrees

FORMULA CARD

to accompany

ALGEBRA AND TRIGONOMETRY, 3rd ed.
COLLEGE ALGEBRA, 3rd ed.
PRE CALCULUS MATHEMATICS, 2nd ed.
PLANE TRIGONOMETRY, 2nd ed.

Walter Fleming and Dale Varberg
PRENTICE HALL, Englewood Cliffs, N. J. 07632

ALGEBRA

Exponents

$a^m a^n = a^{m+n}$

$(a^m)^n = a^{mn}$

$\dfrac{a^m}{a^n} = a^{m-n}$

$(ab)^n = a^n b^n$

$\left(\dfrac{a}{b}\right)^n = \dfrac{a^n}{b^n}$

Radicals

$(\sqrt[n]{a})^n = a$

$\sqrt[n]{a^n} = a \quad$ if $\ a \geq 0$

$\sqrt[n]{ab} = \sqrt[n]{a}\,\sqrt[n]{b}$

$\sqrt[n]{\dfrac{a}{b}} = \dfrac{\sqrt[n]{a}}{\sqrt[n]{b}}$

Logarithms

$\log_a MN = \log_a M + \log_a N \qquad \log_a(M/N) = \log_a M - \log_a N$

$$\log_a(N^P) = P \log_a N$$

Quadratic Formula

Solutions to $ax^2 + bx + c = 0 \quad$ are $\quad x = \dfrac{-b \pm \sqrt{b^2 - 4ac}}{2a}$

Factoring Formulas

$$x^2 - y^2 = (x - y)(x + y)$$
$$x^2 + 2xy + y^2 = (x + y)^2$$
$$x^2 - 2xy + y^2 = (x - y)^2$$
$$x^3 - y^3 = (x - y)(x^2 + xy + y^2)$$
$$x^3 + y^3 = (x + y)(x^2 - xy + y^2)$$
$$x^3 + 3x^2y + 3xy^2 + y^3 = (x + y)^3$$

Binomial Formula

$$(x + y)^n = {}_nC_0 x^n y^0 + {}_nC_1 x^{n-1} y^1 + \cdots + {}_nC_{n-1} x^1 y^{n-1} + {}_nC_n x^0 y^n$$

$${}_nC_r = \frac{n!}{(n - r)!\, r!} = \frac{n(n - 1)\cdots(n - r + 1)}{r(r - 1)\cdots 3 \cdot 2 \cdot 1}$$

$${}_nC_0 = {}_nC_n = 1$$

TRIGONOMETRY

Basic Identities

$$\tan t = \frac{\sin t}{\cos t} \qquad \cot t = \frac{\cos t}{\sin t} \qquad \cot t = \frac{1}{\tan t}$$

$$\sec t = \frac{1}{\cos t} \qquad \csc t = \frac{1}{\sin t} \qquad \sin^2 t + \cos^2 t = 1$$

$$1 + \tan^2 t = \sec^2 t \qquad\qquad 1 + \cot^2 t = \csc^2 t$$

Confunction Identities

$$\sin\left(\frac{\pi}{2} - t\right) = \cos t \qquad \cos\left(\frac{\pi}{2} - t\right) = \sin t \qquad \tan\left(\frac{\pi}{2} - t\right) = \cot t$$

Odd-even Identities

$$\sin(-t) = -\sin t \qquad \cos(-t) = \cos t \qquad \tan(-t) = -\tan t$$

Addition Formulas

$$\sin(s + t) = \sin s \cos t + \cos s \sin t \qquad \sin(s - t) = \sin s \cos t - \cos s \sin t$$
$$\cos(s + t) = \cos s \cos t - \sin s \sin t \qquad \cos(s - t) = \cos s \cos t + \sin s \sin t$$
$$\tan(s + t) = \frac{\tan s + \tan t}{1 - \tan s \tan t} \qquad \tan(s - t) = \frac{\tan s - \tan t}{1 + \tan s \tan t}$$

Double Angle Formulas

$$\sin 2t = 2 \sin t \cos t$$

$$\tan 2t = \frac{2 \tan t}{1 - \tan^2 t}$$

$$\cos 2t = \cos^2 t - \sin^2 t = 1 - 2 \sin^2 t = 2 \cos^2 t - 1$$

Half Angle Formulas

$$\sin\frac{t}{2} = \pm\sqrt{\frac{1 - \cos t}{2}} \qquad \cos\frac{t}{2} = \pm\sqrt{\frac{1 + \cos t}{2}} \qquad \tan\frac{t}{2} = \frac{1 - \cos t}{\sin t}$$

Product Formulas

$$2 \sin s \cos t = \sin(s + t) + \sin(s - t)$$
$$2 \cos s \cos t = \cos(s + t) + \cos(s - t)$$
$$2 \cos s \sin t = \sin(s + t) - \sin(s - t)$$
$$2 \sin s \sin t = \cos(s - t) - \cos(s + t)$$

Factoring Formulas

$$\sin s + \sin t = 2 \cos\frac{s - t}{2} \sin\frac{s + t}{2}$$

$$\cos s + \cos t = 2 \cos\frac{s + t}{2} \cos\frac{s - t}{2}$$

$$\sin s - \sin t = 2 \cos\frac{s + t}{2} \sin\frac{s - t}{2}$$

$$\cos s - \cos t = -2 \sin\frac{s + t}{2} \sin\frac{s - t}{2}$$

Laws of Sines and Cosines

$$\frac{\sin \alpha}{a} = \frac{\sin \beta}{b} = \frac{\sin \gamma}{c}$$
$$a^2 = b^2 + c^2 - 2bc \cos \alpha$$

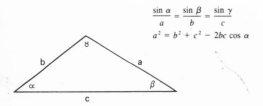

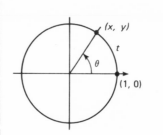

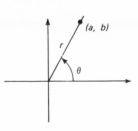

$$\sin t = \sin \theta = y = \frac{b}{r}$$

$$\cos t = \cos \theta = x = \frac{a}{r}$$

$$\tan t = \tan \theta = \frac{y}{x} = \frac{b}{a}$$

$$\cot t = \cot \theta = \frac{x}{y} = \frac{a}{b}$$

Graphs

$y = \cos t$

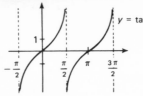

$y = \tan t$

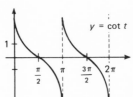

$y = \cot t$

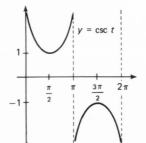

$y = \csc t$

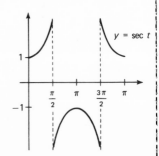

$y = \sec t$

GRAPHS OF TRIGONOMETRIC FUNCTIONS

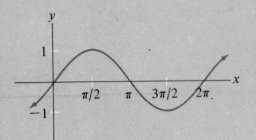

$y = \sin x$

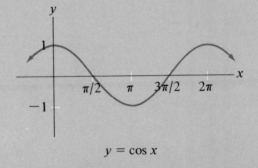

$y = \cos x$

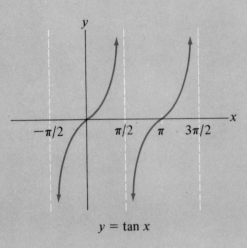

$y = \tan x$

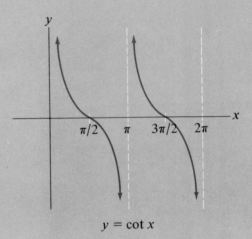

$y = \cot x$

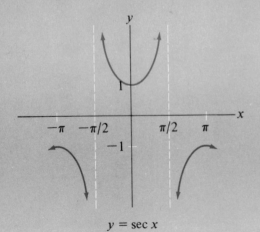

$y = \sec x$

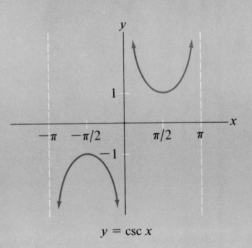

$y = \csc x$